Graduate Texts in Mathematics 26

Ernest G. Manes

Algebraic Theories

Springer-Verlag New York Heidelberg Berlin

Ernest G. Manes

University of Massachusetts
Department of Mathematics and Statistics
Graduate Research Center Tower
Amherst, Massachusetts 01002

AMS Subject Classifications
Primary: 02H10, 08A25, 18–02, 18C15
Secondary: 06A20, 18A40, 18B20, 18D30, 18H05, 22A99, 54D30, 54H20, 68–02, 93B25, 93E99, 94A30, 94A35

Library of Congress Cataloging in Publication Data

Manes, Ernest G. 1943–
Algebraic theories.
(Graduate texts in mathematics;v.26)
Bibliography: p.341
Includes index.
1. Algebra, Universal. I. Title. II. Series.
QA251.M365 512 75–11991

Softcover reprint of the hardcover 1st edition 1976

ISBN-13:978-1-4612-9862-5 e-ISBN-13:978-1-4612-9860-1
DOI: 10.1007/978-1-4612-9860-1

To Mainzy and Regina

"*Universal algebra has been looked on with some suspicion by many mathematicians as being comparatively useless as an engine of investigation.*"

Alfred North Whitehead
[Whitehead 1897, preface]

"*General classifications of abstract systems are usually characterized by a wealth of terminology and illustration, and a scarcity of consequential deduction.*"

Garrett Birkhoff
[Birkhoff 1935, page 438]

"*Since Hilbert and Dedekind, we have known very well that large parts of mathematics can develop logically and fruitfully from a small number of well-chosen axioms. That is to say, given the bases of a theory in an axiomatic form, we can develop the whole theory in a more comprehensible way than we could otherwise. This is what gave the general idea of the notion of mathematical structure. Let us say immediately that this notion has since been superseded by that of category and functor, which includes it under a more general and convenient form.*"

Jean Dieudonné
[Dieudonné 1970, page 138]

Preface

In the past decade, category theory has widened its scope and now interacts with many areas of mathematics. This book develops some of the interactions between universal algebra and category theory as well as some of the resulting applications.

We begin with an exposition of equationally defineable classes from the point of view of "algebraic theories," but without the use of category theory. This serves to motivate the general treatment of algebraic theories in a category, which is the central concern of the book. (No category theory is presumed; rather, an independent treatment is provided by the second chapter.) Applications abound throughout the text and exercises and in the final chapter in which we pursue problems originating in topological dynamics and in automata theory.

This book is a natural outgrowth of the ideas of a small group of mathematicians, many of whom were in residence at the Forschungsinstitut für Mathematik of the Eidgenössische Technische Hochschule in Zürich, Switzerland during the academic year 1966–67. It was in this stimulating atmosphere that the author wrote his doctoral dissertation. The "Zürich School," then, was Michael Barr, Jon Beck, John Gray, Bill Lawvere, Fred Linton, and Myles Tierney (who were there) and (at least) Harry Appelgate, Sammy Eilenberg, John Isbell, and Saunders Mac Lane (whose spiritual presence was tangible.)

I am grateful to the National Science Foundation who provided support, under grants GJ 35759 and DCR 72-03733 A01, while I wrote this book.

I wish to thank many of my colleagues, particularly Michael Arbib, Michael Barr, Jack Duskin, Hartmut Ehrig, Walter Felscher, John Isbell, Fred Linton, Saunders Mac Lane, Robert Paré, Michael Pfender, Walter Tholen, Donovan Van Osdol, and Oswald Wyler, whose criticisms and suggestions made it possible to improve many portions of this book; and Saunders Mac Lane, who provided encouragement on many occasions.

Preface

Table of Contents

Introduction

"Groups," "rings," and "lattices" are definable in the language of finitary operations and equations. "Compact Hausdorff spaces" are also equationally definable except that the requisite operations (of ultrafilter convergence) are quite infinitary. On the other hand, systems of structured sets such as "topological spaces" cannot be presented using only operations and equations. While "topological groups" is not equational when viewed as a system of sets with structure, when viewed as a system of "topological spaces with structure" the additional structure *is* equational; here we must say equational "over topological spaces."

The program of this book is to define for a "base category" $\mathscr{K}$—a system of mathematical discourse consisting of objects whose structure we "take for granted"—categories of $\mathscr{K}$-objects with "additional structure," to classify where the additional structure is "algebraic over $\mathscr{K}$," to prove general theorems about such algebraic situations, and to present examples and applications of the resulting theory in diverse areas of mathematics.

Consider the finitary equationally definable notion of a "semigroup," a set X equipped with a binary operation $x \cdot y$ which is associative:

$$(x \cdot y) \cdot z = x \cdot (y \cdot z).$$

For any set A, the two "derived operations" or *terms*

$$a_1 \cdot [a_2 \cdot ((a_3 \cdot a_4) \cdot (a_5 \cdot a_6))], \qquad (a_1 \cdot a_2) \cdot [a_3 \cdot (a_4 \cdot (a_5 \cdot a_6))]$$

(with $a_1, \ldots, a_6$ in A) are "equivalent" in the sense that one can be derived from the other with (two) applications of associativity. The quotient set of all equivalence classes of terms with "variables" in A may be identified with the set of all parenthesis-free strings $a_1 \cdots a_n$ with $n > 0$; call this set AT. A function $\beta: B \longrightarrow CT$ extends to the function

$$\begin{array}{ccc} BT & \xrightarrow{\beta^{\#}} & CT \\ b_1 \cdots b_n & \longrightarrow & \beta_{b_1} \cdots \beta_{b_n} \end{array}$$

whose syntactic interpretation is performing "substitution" of terms with variables in C for variables of terms in BT. Thus, for each A, B, C, there is the composition

$$(A \xrightarrow{\alpha} BT, B \xrightarrow{\beta} CT) \longrightarrow A \xrightarrow{\alpha \circ \beta} CT = A \xrightarrow{\alpha} BT \xrightarrow{\beta^{\#}} CT$$

There is also the map

$$A \xrightarrow{A\eta} AT, \qquad a \longmapsto a$$

which expresses "variables are terms." $\mathbf{T} = (T, \eta, \circ)$ is the "algebraic theory" corresponding to "semigroups."

In general, an *algebraic theory* (of sets) is any construction $\mathbf{T} = (T, \eta, \circ)$ of the above form such that $\circ$ is associative, η is a two-sided unit for $\circ$ and

$$(A \xrightarrow{f} B \xrightarrow{B\eta} BT) \circ (B \xrightarrow{\beta} CT) = A \xrightarrow{f} B \xrightarrow{\beta} CT$$

A **T**-*algebra* is then a pair (X, ξ) where $\xi : XT \longrightarrow X$ satisfies two axioms, and a **T**-*homomorphism* $f:(X, \xi) \longrightarrow (Y, \theta)$ is a function $f:X \longrightarrow Y$ which "preserves" the algebra structure; see section 1.4 for the details.

If **T** is the algebraic theory for semigroups then "semigroups" and "**T**-algebras" are isomorphic categories of sets with structure in the sense that for each set X the passage from semigroup structures $(X, \bullet)$ to **T**-algebra structures (X, ξ) defined by

$$(x_1 \cdots x_n)\xi = x_1 \bullet \cdots \bullet x_n$$

is bijective in such a way that $f:(X, \bullet) \longrightarrow (Y, *)$ is a semigroup homomorphism if and only if it is a **T**-homomorphism between the corresponding **T**-algebras.

The situation "over sets," then, is as follows. Every finitary equational class induces its algebraic theory **T** via a terms modulo equations construction generalizing that for semigroups, and the **T**-algebras recover the original class. The "finitary" theories—those which are induced by a finitary equational class—are easily identified abstractly. More generally, any algebraic theory of sets corresponds to a (possibly infinitary) equationally-definable class. While the passage from finitary to infinitary increases the syntactic complexity of terms, there is no increase in complexity from the "algebraic theories" point of view. It is also true that many algebraic theories arise as natural set-theoretic constructions before it is clear what their algebras should be. Also, algebraic theories are interesting algebraic objects in their own right and are subject to other interpretations than the one we have used to motivate them (see section 4.3).

An examination of the definition of the algebraic theory **T** and its algebras and their homomorphisms reveals that only superficial aspects of the theory of sets and functions between them are required. Precisely what is needed is that "sets and functions" forms a category (as defined in the section on preliminaries). Generalization to the "base category" is immediate.

The relationship between the four chapters of the book is depicted below:

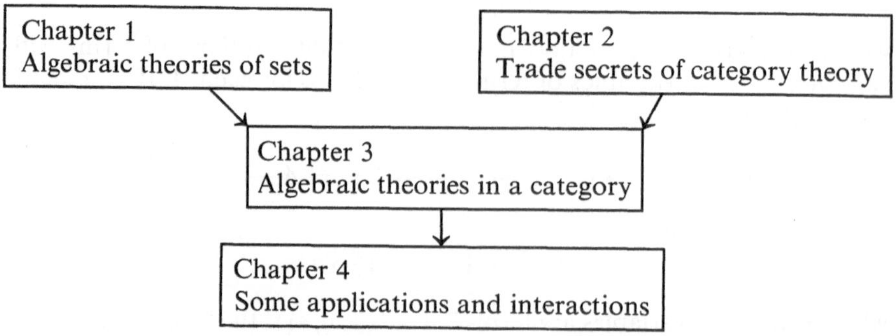

The first chapter is a selfcontained exposition (without the use of category theory) of the relationships between algebraic theories of sets and universal algebra, finitary and infinitary. The professional universal algebraist wishing to learn about algebraic theories will find this chapter very easy reading.

The second chapter may be read independently of the rest of the book, although some of the examples there relate to Chapter 1. We present enough category theory for our needs and at least as much as every pure mathematician should know! The section on "objects with structure" uses a less "puristic" approach than is currently fashionable in category theory; we hope that the reader will thereby be more able to generalize from previous knowledge of mathematical structures.

The third chapter, which develops the topics of central concern, draws heavily from the first two. The choice of applications in the fourth chapter has followed the author's personal tastes.

Why is the material of the third chapter useful? Well, to suggest an analogy, it is dramatic to announce that a concrete structure of interest (such as a plane cubic curve) is a group in a natural way. After all, many naturally-arising binary operations do not satisfy the group axioms; and, moreover, a lot is known about groups. In a similar vein, it is useful to to know that a category of objects with structure is algebraic because this is a special property with nice consequences and about which much is known.

Many exercises are provided, sometimes with extended hints. We have avoided the noisome practice of framing crucial lemmas used in the text as "starred" exercises of earlier sections. For lack of space we have, however, developed many important topics entirely in the exercises.

Reference a.b.c refers to item c of section b in Chapter a. Depending on context, d.e refers to section e of Chapter d or to item e of section d of the current chapter.

Preliminaries

The reader is expected to have some background in set-theoretic pure mathematics. We assume familiarity with the concept of *function* $f: X \longrightarrow Y$ between sets and a minimum of experience with algebra and topology, e.g. the definitions of "topological space," "continuous mapping of topological spaces," "group," and "homomorphism of groups."

A variety of notations are employed for the evaluation of a function f on its argument x. Usually we write xf instead of fx or $f(x)$ (although $d(x, y)$, for the distance between two points in a metric space, is chosen over $(x, y)d$). Another notation for xf is $\langle x, f \rangle$. This notation is especially convenient when x or f is a long expression. We also employ the "passage arrow" $\mapsto$ and write $x \longmapsto xf$ which is read "x is sent to xf". This notation is useful when defining functions.

The composition of functions

$$X \xrightarrow{f} Y \xrightarrow{g} Z$$

will be written fg or $f.g$. Thus $x(fg) = x(f.g) = (xf)g$. For any set X, the *identity function of* X is the function $\mathrm{id}_X: X \longrightarrow X$ defined by $x(\mathrm{id}_X) = x$. It is clear that for any $f: X \longrightarrow Y$ we have $\mathrm{id}_X.f = f = f.\mathrm{id}_Y$. This may be expressed by the *commutative diagram*. We say the diagram *commutes*

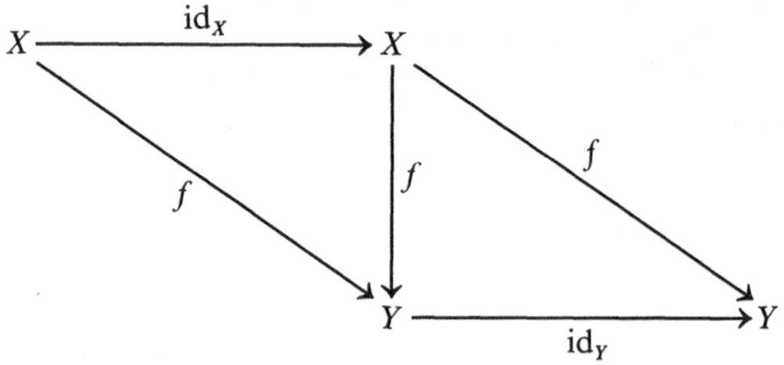

because all composition paths between the same sets in the diagram are the same function. Similarly, the familiar associative law of composition, $(fg)h = f(gh)$, is expressed with a commutative diagram. Because of the

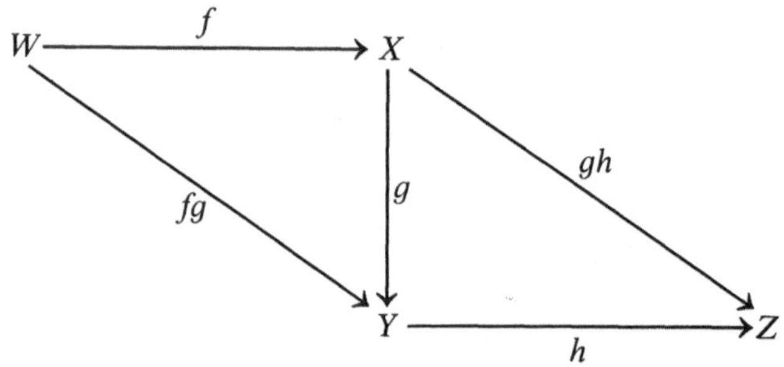

associative law, $fgh: W \longrightarrow Z$ is well defined; and it is this principle that allows commutative diagrams to display effectively the result of composing long chains of functions.

The theory of categories, functors between categories, and natural transformations between functors—to the extent that it is needed—is developed gradually beginning with Chapter 2. Since certain functors and natural transformations arise naturally in Chapter 1, these concepts are defined here. A *category* $\mathscr{K}$ is defined by the following data and axioms.

Datum 1. There is given a class $\mathrm{Obj}(\mathscr{K})$ *of* $\mathscr{K}$*-objects.*

Datum 2. For each ordered pair (A, B) *of* $\mathscr{K}$*-objects there is given a class* $\mathscr{K}(A, B)$ *of* $\mathscr{K}$*-morphisms from* A *to* B*. If* $f \in \mathscr{K}(A, B)$*,* A *is the domain of* f *and* B *is the codomain of* f *(see Axiom 3).*

Datum 3. For each $\mathscr{K}$*-object* A *there is given a distinguished* $\mathscr{K}$*-morphism* $\mathrm{id}_A \in \mathscr{K}(A, A)$ *called the identity of* A*.*

Datum 4. For each ordered triple (A, B, C) *of* $\mathscr{K}$*-objects there is given a composition law*

$$\mathscr{K}(A, B) \times \mathscr{K}(B, C) \longrightarrow \mathscr{K}(A, C)$$
$$(f, g) \longmapsto fg$$

Axiom 1. Composition is associative, that is given $f \in \mathscr{K}(A, B)$*,* $g \in \mathscr{K}(B, C)$ *and* $h \in \mathscr{K}(C, D)$ *then* $(fg)h = f(gh) \in \mathscr{K}(A, D)$*.*

Axiom 2. If $f \in \mathscr{K}(A, B)$ *then* $(\mathrm{id}_A)f = f = f(\mathrm{id}_B)$*.*

Axiom 3. If $(A, B) \neq (A', B')$ *then* $\mathscr{K}(A, B) \cap \mathscr{K}(A', B') = \varnothing$*.*

"Sets and functions" form a category which we will denote henceforth by **Set**. Thus a **Set**-object is an arbitrary set and **Set**(A, B) is the set of functions from A to B. Identities and composition are defined in the way already discussed. Axiom 3 asserts that for the purposes of category theory, a function is not properly defined unless the set it maps from and the set it maps to are included in the definition. Thus the polynomial x^2 thought of as mapping all the real numbers into itself is a different function from x^2 thought of as mapping all the nonzero real numbers into the set of all real numbers.

The reader should recognize at once that "topological spaces and continuous mappings" as well as "groups and group homomorphisms" are two further examples of categories.

If $\mathscr{K}$ is an arbitrary category we will write $f: A \longrightarrow B$ to denote $f \in \mathscr{K}(A, B)$. We will also use $f.g$ as an alternate notation to fg. Axioms 1, 2 can be expressed as commutative diagrams just as we did earlier for the category **Set**. Let $\mathscr{K}$ and $\mathscr{L}$ be two categories. A *functor, H, from $\mathscr{K}$ to $\mathscr{L}$* is defined by the following data and axioms:

Datum 1. For each $\mathscr{K}$*-object* A*, there is given an* $\mathscr{L}$*-object* AH*.*

Datum 2. For each $\mathscr{K}$*-morphism of form* $f: A \longrightarrow B$ *there is given an* $\mathscr{L}$*-morphism of form* $fH: AH \longrightarrow BH$*.*

Axiom 1. *H preserves identities; that is, for every $\mathscr{K}$-object A,* $(\mathrm{id}_A)H = \mathrm{id}_{AH}$.

Axiom 2. *H preserves composition, that is, given $f: A \longrightarrow B$ and $g: B \longrightarrow C$ in $\mathscr{K}$, $(f.g)H = fH.gH: AH \longrightarrow CH$ in $\mathscr{L}$.*

We use the notation $H: \mathscr{K} \longrightarrow \mathscr{L}$ if H is a functor from $\mathscr{K}$ to $\mathscr{L}$.

Suppose now that H, $H': \mathscr{K} \longrightarrow \mathscr{L}$ are two functors between the same two categories. A *natural transformation α from H to H'* is defined by the following datum and axiom:

Datum. *For each $\mathscr{K}$-object A there is given an $\mathscr{L}$-morphism $A\alpha: AH \longrightarrow AH'$.*

Axiom. *For each $\mathscr{K}$-morphism $f: A \longrightarrow B$ the following square of $\mathscr{L}$-morphisms is commutative:*

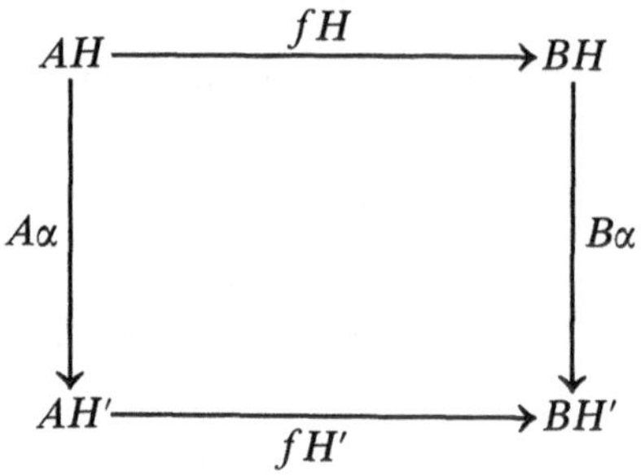

i.e., $A\alpha.fH' = fH.B\alpha$.

We use the notation $\alpha: H \longrightarrow H'$ when α is a natural transformation from H to H'.

Chapter 1

Algebraic Theories of Sets

This chapter is a selfcontained introduction to algebraic theories of sets. Category theory is not used in the development. The motivating example of equationally-definable classes is eventually seen to be coextensive with algebraic theories with rank. Compact Hausdorff spaces and complete atomic Boolean algebras arise as algebras over theories (without rank) whereas complete Boolean algebras do not.

1. Finitary Universal Algebra

In this section we define (finitary) *equationally-definable classes.* Further systematic study of finitary universal algebra is referred to the literature (see the notes at the end of this section) but some of the standard examples are developed in the exercises.

There are a number of ways to define the concept of a group. Here are three of them:

1.1 Definition. *A group is a set X equipped with a binary operation $m: X \times X \longrightarrow X$ (multiplication), a unary operation $i: X \longrightarrow X$ (inversion) and a distinguished element $e \in X$ (the unit) subject to the equations*

$$xymzm = xyzmm \qquad (m \text{ is associative})$$
$$xem = x = exm \qquad (e \text{ is a two-sided unit for } m)$$
$$xixm = e = xxim \qquad (xi \text{ is the multiplicative inverse of } x)$$

for all x, y, z in X.

(Notice the use, in 1.1, of parenthesis-free "Polish notation," e.g. $xymzm$ instead of $((x, y)m, z)m$. A formal proof that this notation works is given below in 1.11.)

1.2 Definition. *A group is a set X equipped with a binary operation $d: X \times X \longrightarrow X$ (division) subject to the single incredible equation*

$$xxxdydzdxxdxdzddd = y$$

for all x, y, z in X. It is proved in [Higman & Neumann, '52] that a bijective passage from 1.1 to 1.2 is obtained by $xyd = xyim$. The structure of "$xxxdydzdxxdxdzddd$" is examined in 1.13 below.

1.3 Definition. *A group is a set X equipped with a binary operation m such that m is associative and admits unit and inverses, i.e., such that there exists* *a unary operation i and a distinguished element e of X subject to the equations of 1.1.*

Very roughly speaking, group theory is an algebraic theory and 1.1, 1.2, 1.3 are presentations of that theory. (Actually, the empty set is a group according to 1.2 but not according to 1.1 and 1.3; to remedy this one should modify 1.2 by requiring a distinguished element e satisfying $xed = x$.) The first two are equational presentations in that they take the form of a set of operations subject to a set of equations, whereas the third is not an equational presentation because existential quantification is not equationally expressible. We devote this section to setting down, in precise terms, the definition of a finitary equational presentation (Ω, E) and the resulting equationally-definable class (or variety) of all (Ω, E)-algebras.

1.4 Definition. *An operator domain is a disjoint sequence of sets, $\Omega = (\Omega_n : n = 0, 1, 2 \ldots)$. Ω_n is the set of n-ary operator labels of Ω.*

We remark, as an aside, that an operator domain may be viewed as a directed graph whose nodes are natural numbers and whose edges terminate at 1. Thus a directed graph suitable for "groups" as in 1.1 is

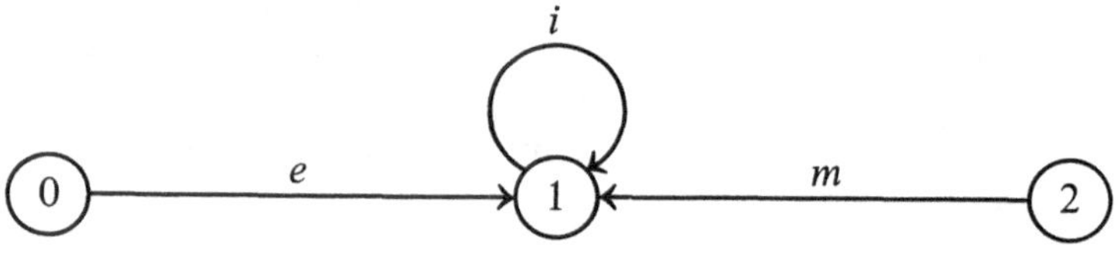

This point of view is a natural precursor to viewing an operator domain as a category, an approach which receives only brief treatment in this book (see 1.5.35, the notes to section 3, Exercises 2.1.25–27 and Exercise 3.2.7).

An Ω-algebra is a pair (X, δ) where X is a set and δ assigns to each ω in Ω_n an n-ary operation $\delta_\omega : X^n \longrightarrow X$. Given Ω-algebras (X, δ) and (Y, γ), an Ω-homomorphism from (X, δ) to (Y, γ) is a function $f : X \longrightarrow Y$ which commutes with the Ω-operations, that is, for all $\omega \in \Omega_n$ and n-tuples $(x_1, \ldots, x_n)$ of X, we have $(x_1, \ldots, x_n)\delta_\omega f = (x_1 f, \ldots, x_n f)\gamma_\omega$. Denoting the passage of $(x_1, \ldots, x_n)$ to $(x_1 f, \ldots, x_n f)$ by $f^n : X^n \longrightarrow Y^n$, this may be equivalently written as the commutative square:

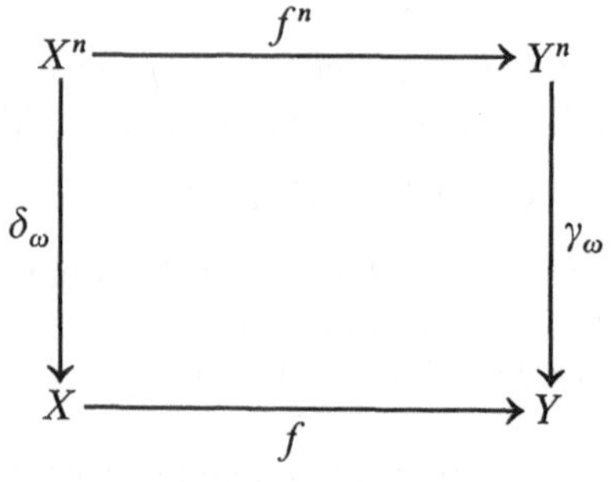

(1.5)

1.6 Example. Define $\Omega_0 = \{e\}$, $\Omega_1 = \{i\}$, $\Omega_2 = \{m\}$, $\Omega_n = \phi$ for all $n > 2$. Then every group (as in 1.1) is an Ω-algebra, but not conversely. The Ω-homomorphisms between groups are ordinary group homomorphisms.

An equational presentation, as is yet to be defined, should consist of a pair (Ω, E) where Ω is an operator domain and E is a set of Ω-equations. To properly formulate "Ω-equation" we must formalize the construction of expressions such as *xxmzm* and *xxim*.

1.7 Definition. *Let A be a set. A word in A is an n-tuple of elements of A with n an integer* >0; *n is the length of the word.* We will write $a_1a_2 \cdots a_n$ instead of $(a_1, \ldots, a_n)$ to convey the feeling of "word in the alphabet A." An expression such as a_1a_2m is a word in the appropriate "alphabet" A. In general, let Ω be an operator domain, set $|\Omega|$ to be the union of all Ω_n, and define an *Ω-word in A* to be a word in the disjoint union $A + |\Omega|$; (the *disjoint union* of the sets X, Y is the set $X + Y = (X \times \{0\}) \cup (Y \times \{1\})$). Notationally, we will use separate symbols for elements of A and elements of $|\Omega|$ and write Ω-words as words in $A \cup |\Omega|$. If Ω is as in 1.6, *abmcm*, *eam*, and *ei* are all Ω-words in A; unfortunately, so are nonsense words such as *mmamib*. An *Ω-term in A* is an Ω-word in A which can be derived by finitely many applications of 1.8 and 1.9 below:

(*1.8*) *a is an Ω-term in A for all $a \in A$.*

(*1.9*) *If $\omega \in \Omega_n$ and $p_1, \ldots, p_n$ are Ω-terms in A, then $p_1 \cdots p_n\omega$ is an Ω-term in A.*

The set of all Ω-terms in A will be denoted $A\Omega$.

Intuitively, an Ω-word in A is a term if and only if it has the appearance of a well-defined function in finitely-many variables of A. For example, if Ω is as in 1.6 and if A has at least three distinct elements a, b, c then the doubleton $\{abmcm, abcmm\}$ is the essence of the associative law; for if (X, δ) is any Ω-algebra and if (x_1, x_2, x_3) is any 3-tuple of elements of X then by virtue of the substitution "x_1 for a, x_2 for b, x_3 for c", *abmcm* induces the ternary operation $((x_1, x_2)\delta_m, x_3)\delta_m$ on X and *abcmm* similarly induces a ternary operation on X; (X, δ) satisfies the associative law if and only if these ternary operations are the same. This motivates

1.10 Definition. *Fix any convenient* (effectively enumerated, see, e.g., [Hermes '65, page 11]) *set* V *of abstract variables,* $\mathsf{V} = \{v_1, v_2, \ldots, v_n \ldots\}$. *For example,* V *might be the set of positive integers. An Ω-equation is a doubleton $\{e_1, e_2\}$ of Ω-terms in* V. *An equational presentation is a pair (Ω, E) where Ω is an operator domain and E is a set of Ω-equations.*

The equational presentation corresponding to 1.1 is Ω as in 1.6 and $E = \{\{v_1v_2mv_3m, v_1v_2v_3mm\}, \{v_1em, v_1\}, \{ev_1m, v_1\}, \{v_1iv_1m, e\}, \{v_1v_1im, e\}\}$. This overly formal notation is difficult to read and in most situations we use the more colloquial "$e_1 = e_2$," use parenthetical notation instead of Polish notation, and write $x, y, z \ldots$ for $v_1, v_2, v_3 \ldots$. Thus, E as above is written:

$$((x, y)m, z)m = (x, (y, z)m)m$$
$$(x, e)m = x = (e, x)m$$
$$(xi, x)m = e = (x, xi)m$$

We now set forth to formalize the means which allowed us to make actual operations out of terms in the style that we accomplished this for *abmcm* in the preceding paragraph.

1.11 Uncoupling Lemma. Let A be a set and let Ω be an operator domain. Then for each $p \in A\Omega$ of word length greater than 1 there exists a unique integer n greater than 0 and unique $\omega \in \Omega_n$ and n-tuple $(p_1, \ldots, p_n) \in A\Omega^n$ such that $p = p_1 \cdots p_n\omega$.

Proof. Since p is constructed from (1.8) and (1.9) and has more than one symbol, it is clear that there exists a representation $p = p_1 \cdots p_n\omega$ as in the statement and that n and ω are unique. We must prove that if $p = q_1 \cdots q_n\omega$ is another such representation, then $p_i = q_i$ for all i. It is helpful to define the integer-valued *valency map*, val ([Cohn '65, p. 118]), on the set of all Ω-words in A by $\mathrm{val}(\tilde{\omega}) = 1 - m$ (for all $\tilde{\omega} \in \Omega_m$), $\mathrm{val}(a) = 1$ (for all $a \in A$), $\mathrm{val}(b_1 \cdots b_m) = \mathrm{val}(b_1) + \cdots + \mathrm{val}(b_m)$. Since an Ω-formula q can be constructed from (1.8) and (1.9), $\mathrm{val}(q) = 1$ and $\mathrm{val}(s) > 0$ for any left segment s of q (where, if $q = b_1 \cdots b_m$, the *left segments of* q are the m Ω-words $b_1 \cdots b_k$ for $1 \leqslant k \leqslant m$). The crucial observation is:

(1.12) If s is a proper left segment of $p_i \cdots p_n$ and if $s \in A\Omega$, then s is a left segment of p_i. (For otherwise, there exists $i \leqslant k < n$ and a left segment t of p_{k+1} such that $s = p_i \cdots p_k t$; it follows that $1 = \mathrm{val}(s) = \mathrm{val}(p_i \cdots p_k t) = k - i + 1 + \mathrm{val}(t)$ and $i - k = \mathrm{val}(t) \geqslant 0$ (i.e., if t is empty then $k > i$), the desired contradiction).

Applying 1.12 to $s = q_1$, we see that q_1 is a left segment of p_1. Symmetrically p_1 is a left segment of q_1, so $p_1 = q_1$. Therefore, $p_2 \cdots p_n\omega = q_2 \cdots q_n\omega$ and we can apply 1.12 to prove $p_2 = q_2$. Similarly, $p_3 = q_3, \ldots, p_n = q_n$. □

The uncoupling process of 1.11 can be geometrically depicted by the "tree"

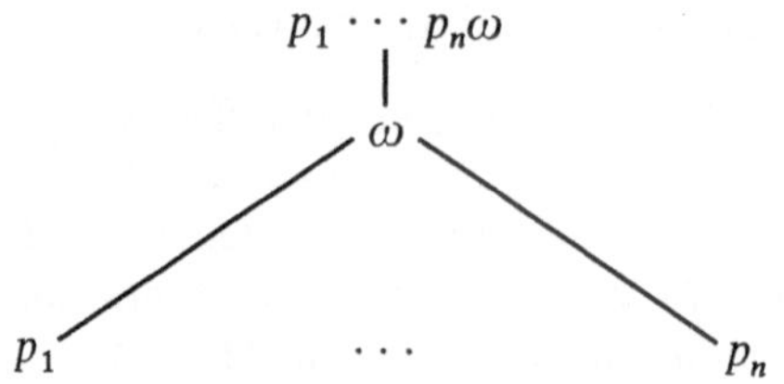

Each p_i has shorter length than the original term. Each p_i of length greater than 1 can be similarly decoupled until we obtain the complete *derivation tree* of the term in which all terminal branches are terms of length 1, that is variables or 0-ary operations.

1.13 Example. The derivation tree of *xxxdydzdxxdxdzddd* as in 1.2 is

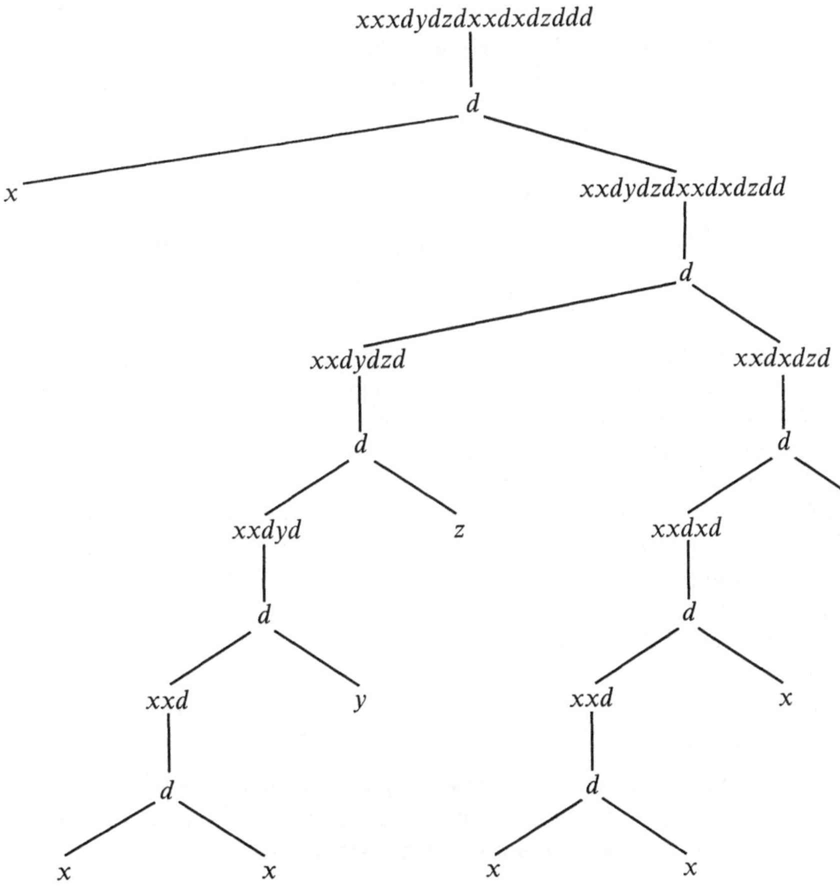

Since the derivation of Ω-terms is unique we have:

1.14 Principle of Finitary Algebraic General Recursion. Let Ω be an operator domain and let A be a set. To define a function ψ on $A\Omega$ it suffices to specify

(1.15) $a\psi$ for all $a \in A$.

(1.16) $(p_1 \cdots p_n\omega)\psi$ in terms of $p_i\psi$ and ω. □

1.17 Example (Substitution of Variables in Terms). Let $f: A \longrightarrow B$ be a function (substituting variables in B for variables in A). By algebraic general recursion we may define the function $f\Omega: A\Omega \longrightarrow B\Omega$ by

$$\langle a, f\Omega \rangle = af$$
$$\langle (p_1 \cdots p_n\omega), f\Omega \rangle = \langle p_1, f\Omega \rangle \cdots \langle p_n, f\Omega \rangle \omega$$

Thus, $\langle aaadbdcdaadadcddd, f\Omega\rangle = xxxdydzdxxdxdzddd$ if $af = x$, $bf = y$, $cf = z$. In the picture of 1.13, we plug in the appropriate terminal branches x, y, z and chase up the tree.

1.18 Example (The Total Description Map). Let (X, δ) be an Ω-algebra. The *total description map* $\delta^{@}: X\Omega \longrightarrow X$ is defined by algebraic general recursion:

$$x\delta^{@} = x$$
$$(p_1 \cdots p_n\omega)\delta^{@} = (p_1\delta^{@}, \ldots, p_n\delta^{@})\delta_\omega$$

Clearly, the total description map accomplishes what we wanted: it makes operations out of formulas, although we should note the role of 1.17 in interpreting variables as arguments. We are finally ready for:

1.19 Definition. Let Ω *be an operator domain, and let* (X, δ) *be an* Ω-*algebra. For each* V-*tuple* $r: \mathsf{V} \longrightarrow X$ *there is an interpretation map* $r^{\#}$ *defined by* $r^{\#}: \mathsf{V}\Omega \longrightarrow X = r\Omega.\delta^{@}$. Notice that $r^{\#}$ can be defined directly by algebraic general recursion: $vr^{\#} = vr$, $(p_1 \cdots p_n\omega)r^{\#} = (p_1r^{\#}, \cdots p_nr^{\#})\delta_\omega$. If $\{e_1, e_2\}$ is an Ω-equation, say that (X, δ) *satisfies* $\{e_1, e_2\}$ if $e_1r^{\#} = e_2r^{\#}$ for all $r: \mathsf{V} \longrightarrow X$. If (Ω, E) is an equational presentation, an (Ω, E)-*algebra* is an Ω-algebra which *satisfies* E, that is satisfies every equation in E. *The class of all* (Ω, E)-*algebras is said to be an equationally-definable class of algebras, or a variety of algebras.* For example, the equationally-definable class defined by the presentation in 1.10 is "groups" as in 1.1.

The above construction of interpretation maps is based on an important principle. Notice, first, that 1.9 defines an Ω-algebra structure on $A\Omega$ (and we will always regard $A\Omega$ as an algebra in this way). We can now state

1.20 Principle of Finitary Algebraic Simple Recursion. Let Ω be an operator domain, let (X, δ) be an Ω-algebra and let $f: A \longrightarrow X$ be a function. Then there exists a unique Ω-homomorphism $f^{\#}: A\Omega \longrightarrow (X, \delta)$ extending f.

Proof. By 1.14 there exists unique function $f^{\#}$ such that $af^{\#} = af$ and $(p_1 \cdots p_n\omega)f^{\#} = (p_1f^{\#}, \ldots, p_nf^{\#})\delta_\omega$. □

To help explain the terminology of 1.20, recall that a sequence $x: \mathbf{N} \longrightarrow X$ (where $\mathbf{N} = \{0, 1, 2, 3 \ldots\}$) is defined by *simple recursion* if there exists an endomorphism $\delta: X \longrightarrow X$ such that $x_{n+1} = x_n\delta$. The general recursion of 1.14 amounts to "mathematical induction" (see the notes at the end of this section). Observe that if $X = \{a, b\}$ and if x is defined by $x_0 = x_1 = a$, $x_n = b$ for $n > 1$, then x is not definable by simple recursion. This situation is an instance of 1.20 and 1.14, corresponding to the operator domain

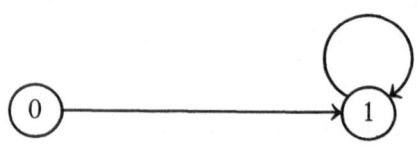

Notes for Section 1

The founder of universal algebra—the study of finitary equational classes—is Garrett Birkhoff [Birkhoff '35]. We refer the reader to the texts of [Cohn '65], [Grätzer '67], and [Pierce '68]. Of these, Grätzer's is the most complete, in our judgment, and has the largest bibliography. Pierce is recommended for infinitary universal algebra (not treated in the other two) which we will study in section 5, but in a different way. In the three texts cited above, the concept of "variety" is defined relatively late in the book:

Author	Number of pages	Page "variety" is first defined
Cohn	333	162
Grätzer	368	152
Pierce	143	124

Thus, section 1 provides a rapid introduction not available in the expository literature, to our knowledge, at this writing. (We hasten to add that, in the three books above, "variety" is viewed as but one of a number of central topics.)

Lemma 1.11 was proved for unary and binary operations by [Menger '30], for arbitrary finitary operations by [Schröter '43] and [Gerneth '48], and for infinitary operations by [Felscher '65].

Ω-terms are the usual terms of mathematical logic (cf. [Bell and Slomson '71, page 70]), $A\Omega$ is known as an "absolutely free Ω-algebra" in the literature of universal algebra.

In set theory, "recursion" and "induction" have taken on special meanings (see [Monk '69, Chapter 13]). In particular, it would appear to be inappropriate to call 1.14 "algebraic induction."

Exercises for Section 1

1. If you do not already know them, look up the definitions of "monoid," "ring," "lattice," and "real vector space." Give finitary equational presentations for these objects. Further hints can be found in [Cohn '65, pages 50–55].
2. In 1.7 we defined the set of words in A to be the union $A \cup A^2 \cup A^3 \cdots$. More properly, we should have insisted on the disjoint union $A + A^2 + A^3 \cdots$. To prove this is necessary, give an example of a set A such that A and A^2 have elements in common.
3. Give an example of an equational presentation such that every algebra has exactly one element.
4. (Jónsson and Tarski). Give an example of an equational presentation with one binary operation and two unary operations such that every algebra with at least two elements is infinite. [Hint: make the binary operation bijective.]

5. Let Ω have one unary operation and no other operations. Show that the Ω-terms in A may be identified with the set $A \times \mathbf{N}$.
6. Let S be the set of all Ω-homomorphisms from (X, δ) to (X', δ'). Then S is a subset of the topological space $(X')^X$, the Tychanoff cartesian power of copies of X' with the discrete topology. Prove that S is closed.
7. Let $g: A\Omega \longrightarrow A\Omega$ be a bijective Ω-homomorphism. Prove that g^{-1} is also an Ω-homomorphism. Prove that g maps A bijectively onto A and that $g = f\Omega$ for $f: A \longrightarrow A$, $af = ag$. [Hint: use 1.11.]
8. Ianov's program schemata (see [Rutledge '64 and the bibliography there]) provide a "dual" concept to Ω-terms. Fix an operator domain Ω with $\Omega_0 = \varnothing$. An *initialized Ω-flowchart scheme* is a finite directed graph, with a distinguished "initial" node, in which every node of outdegree $n > 0$ is labelled with an element of Ω_n; the nodes of outdegree 0 are called *exits*. A *partial function* from X to Y is a function from a subset of X to Y. An *Ω-coalgebra* is a pair (X, δ) where X is a set and δ assigns to $\omega \in \Omega_n$ a partial function $\delta_\omega: X \longrightarrow n \cdot X$ [where $n \cdot X = X + \cdots + X$ (n times)].
   (a) Regarding a flowchart scheme as an "abstract program" and an Ω-coalgebra as a "machine," show that "running the program" results in a partial function $X \longrightarrow s \cdot X$ (where s is the number of exits), a *semantic interpretation* of the scheme. [Hint: to compose partial functions, $x(fg)$ is defined if xf and $(xf)g$ are.]
   (b) Let $\Omega_1 = \{\alpha\}$, $\Omega_2 = \{\beta, \gamma\}$. Formalize how the flowchart scheme on the left can have the semantic interpretation shown on the right.

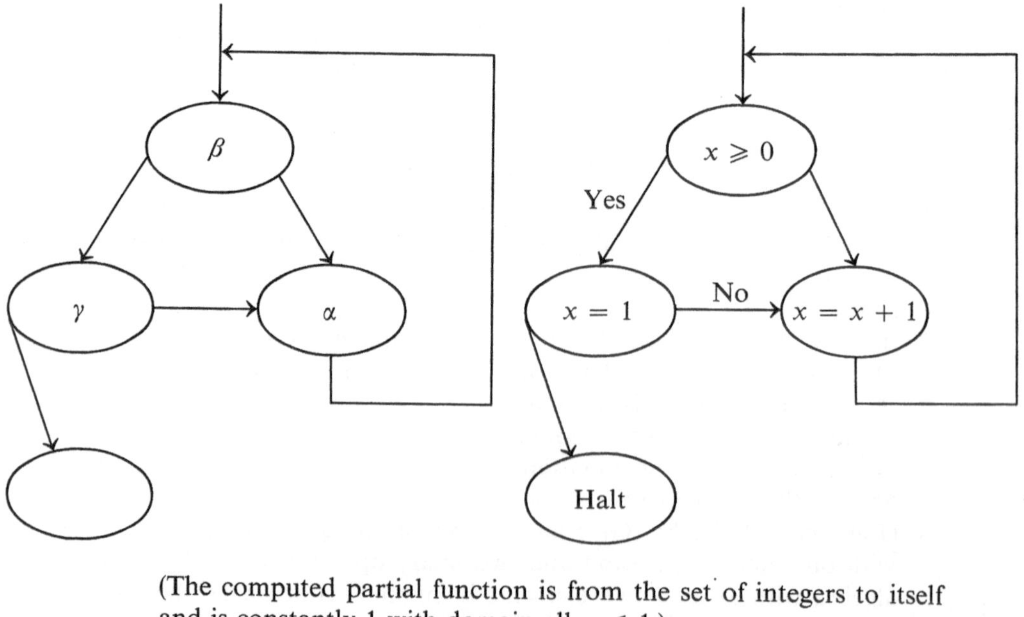

(The computed partial function is from the set of integers to itself and is constantly 1 with domain all $x \leqslant 1$.)

(c) Construct a scheme with four nodes which is equivalent to that in (b) in the sense that both have the same interpretation in all co-algebras. Why can't two distinct Ω-terms have the same interpretation in every Ω-algebra?

9. "Lattices" constitute an equationally-definable class (e.g., see exercise 3.2.10d). A lattice is *modular* if $(x \wedge b) \vee a = (x \vee a) \wedge b$ whenever $a \leqslant b$. Prove that modular lattices constitute an equationally-definable class.

2. The Clone of an Equational Presentation

We opened the book with the observation that two equational presentations, 1.1 and 1.2, were "equivalent." There are three ways to make this precise and, happily, they coincide (see Theorem 2.17 below). An equational presentation (Ω, E) provides us explicitly with sets of terms $A\Omega$ and equivalence relations E_A on $A\Omega$, where pE_Aq means "p and q have the same interpretation in all (Ω, E)-algebras"; for example if (Ω, E) corresponds to 1.1, *aibimi* and *bam* have the same meaning in all groups. The set $AT = A\Omega/E_A$ of equivalence classes turns out to be presentation independent and to possess all the algebraic invariants so long as we include the formal description of the ways in which formulas combine with each other. Making all of this precise is the goal of this section.

For the time being, fix an equational presentation (Ω, E).

2.1 Definition. *For each set A define an equivalence relation E_A on $A\Omega$ by $E_A = \{(p, q)$: for all (Ω, E)-algebras (X, δ) and all functions $f: A \longrightarrow X$, $pf^{\#} = qf^{\#}\}$. It is obvious that E_A is an equivalence relation. We denote the quotient set $A\Omega/E_A$ by AT (T for "theory"), and the canonical projection by $A\rho: A\Omega \longrightarrow AT$. We will also adopt the notation $[p] \in AT$ for the equivalence class $\langle p, A\rho\rangle$ of p.*

2.2 Proposition. For each set A, there exists a unique Ω-algebra structure on AT making $A\rho$ an Ω-homomorphism, that is $[p_1] \cdots [p_n]\omega = [p_1 \cdots p_n\omega]$ is well defined. Moreover, AT is an (Ω, E)-algebra.

Proof. The first statement is obvious from the definition of E_A and the fact (1.20) that each $f^{\#}$ is an Ω-homomorphism. Enroute to the second statement we make two observations:

(2.3) Whenever $f:(X_1, \delta_1) \longrightarrow (X_2, \delta_2)$ and $g:(X_2, \delta_2) \longrightarrow (X_3, \delta_3)$ are Ω-homomorphisms, so is $fg:(X_1, \delta_1) \longrightarrow (X_3, \delta_3)$. (This is obvious from Definition 1.5; notice that $(fg)^n = f^n g^n$.)

(2.4) For all $g: V \longrightarrow A\Omega$ and $\{e_1, e_2\}$ in E, $(e_1 g^{\#}, e_2 g^{\#}) \in E_A$. (Proof: For every (Ω, E)-algebra (X, δ) and function $f: A \longrightarrow X$, $g^{\#}f^{\#}: V\Omega \longrightarrow (X, \delta)$ is an Ω-homomorphism by 2.3 so that by 1.20 $g^{\#}f^{\#}$ has the form $h^{\#}$ where h is the restriction of $g^{\#}f^{\#}$ to V. Since (X, δ) satisfies $\{e_1, e_2\}$, $e_1 g^{\#}f^{\#} = e_2 g^{\#}f^{\#}$; since f is arbitrary, we are done with 2.4.)

To complete the proof, let $r: \mathsf{V} \longrightarrow AT$ be a function. By the axiom of choice (but see exercise 1) there exists a function $g: \mathsf{V} \longrightarrow A\Omega$ such that $g.A\rho = r$. Since $g^{\#}.A\rho$ is an Ω-homomorphism (by 2.3 and the first part of the proof), and the restriction of $g^{\#}.A\rho$ to V coincides with r, we have from 1.20 that $g^{\#}.A\rho = r^{\#}$. Since $g^{\#}$ maps equations into E_A (2.4) and $A\rho$ identifies elements of E_A, AT satisfies E as desired. □

AT is called the *free* (Ω, E)*-algebra generated by* A. The map $A\rho: A\Omega \longrightarrow AT$ presents AT by "generators and relations," and "free" means that there are just enough relations to satisfy E, but no more. To a category theorist, this intuitively correct formulation would be justified by the following result:

2.5 The Universal Property of AT. For each set A define the *insertion-of-the-variables map* $A\eta: A \longrightarrow AT$ by $\langle a, A\eta \rangle = [a]$. Then for every (Ω, E)-algebra (X, δ) and every function $f: A \longrightarrow X$ there exists a unique Ω-homomorphism $f^{\#\#}: AT \longrightarrow (X, \delta)$ extending f.

Proof. Consider the diagram below. The unique Ω-homomorphism

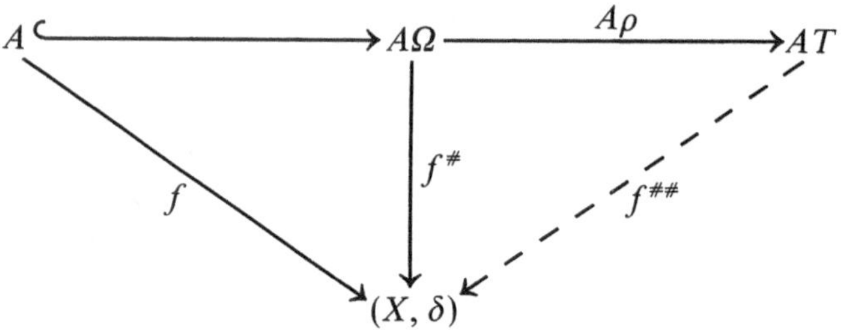

$f^{\#}: A\Omega \longrightarrow (X, \delta)$ of 1.20 respects E_A by the definition of E_A, and this induces a unique function $f^{\#\#}: AT \longrightarrow (X, \delta)$ which is a homomorphism because

(2.6) Given a surjective Ω-homomorphism $g:(X_1, \delta_1) \longrightarrow (X_2, \delta_2)$, an Ω-algebra (X_3, δ_3), and a function $h: X_2 \longrightarrow X_3$, if gh is an Ω-homomorphism, then so is h.

The proof of 2.6 and the uniqueness of $f^{\#\#}: AT \longrightarrow (X, \delta)$ may be safely left to the reader. □

When E is empty, 2.5 reduces to 1.20. It will gradually become clear that 1.20 is the pivotal theorem in transforming a set-theoretic symbol-manipulative analysis of algebra into a categorical one.

Let us turn now to the promised formalization of the way formulas combine with each other. To this end, let us think of $p_1 \cdots p_n\omega$ in $A\Omega$ not so much as an n-ary operation indexed by ω as an $n + 1$-ary operation in $p_1, \ldots, p_n, \omega$. More specifically, if $p \in B\Omega$ with variables (see exercise 10) $b_1, \ldots, b_n$ and if $q_1, \ldots, q_n \in C\Omega$, we may substitute q_i for b_i to get $(q_i)p \in C\Omega$, the *clone composition of* $q_1, \ldots, q_n$, p. This includes the case of $p_1 \cdots p_n\omega$ above if we set $B = \mathsf{V}$, $p = v_1 \cdots v_n\omega$, $C = A$, $q_i = p_i$. For another example, if Ω has binary $+$, unary i, and nullary e, $p = b_3ib_1+b_2+$, $q_1 = e$, $q_2 = c_1i$,

$q_3 = c_1c_2+$ then $(q_i)p = c_1c_2+ie+c_1i+$. We expect clone composition to be "associative." To make this come true, let us think of $(q_i)p$ as a binary operation $p \circ (q_i)$. For uniformity, we should replace p by a tuple (p_j) of p's and define $(p_j) \circ (q_i)$ as the tuple $p_j \circ (q_i)$. Here is the formal definition:

2.7 Definition. *The clone of* (Ω, E) *is the category* **Set**(Ω, E) *whose objects are sets* $A, B, C \ldots$ *and whose morphisms* $\alpha: A \longrightarrow B$ *are functions* $\alpha: A \longrightarrow BT$. *Composition is defined by*

$$(A \xrightarrow{\alpha} B) \circ (B \xrightarrow{\beta} C) = A \xrightarrow{\alpha} BT \xrightarrow{\beta^{\#}} CT \tag{2.8}$$

Identity morphisms are defined by

$$A\eta: A \longrightarrow AT \tag{2.9}$$

Throughout the book we will adopt the following notational conventions: morphisms in **Set**(Ω, E) are distinguished from functions by the use of single-headed arrows. The symbol for the **Set**(Ω, E)-identity of A will always be $A\eta$, whereas the symbol for the identity function of A will always be id_A; thus id_{AT} will never mean $AT\eta: AT \longrightarrow AT$, but one of $\mathrm{id}_{AT}: AT \longrightarrow AT$ or $\mathrm{id}_{AT}: AT \longrightarrow A$. The symbol $\circ$ will be used for clone composition as in 2.8, whereas ordinary composition of functions can be denoted with a period.

Let us verify that the formal definition meshes with what motivated it. First suppose that E is empty so that a map $\beta: B \longrightarrow C$ is a function $\beta: B \longrightarrow C\Omega$, that is a B-tuple $(q_b: b \in B)$ of terms in C. If $p \in B\Omega$ has set of variables (specifically, the terminal branches of the derivation tree of p which are not nullary operations) $\{b_1, \ldots, b_n\}$ then $p\beta^{\#}$ is clearly the unique term in $C\Omega$ whose derivation tree is built down from that of p by substituting the derivation tree of q_{b_i} for each occurrence of b_i; in short, $p\beta^{\#} = p \circ (q_b)$. Moreover, if we have a function $\alpha: A \longrightarrow B\Omega$, that is an entire tuple (p_a), then $\alpha \circ \beta = \alpha \cdot \beta^{\#}$ is the A-tuple $p_a \circ (q_b)$ as advertised. Expression (2.9) asserts that "variables are terms."

Even if E is arbitrary, clone composition is at the level of representatives, that is

$$([p_a]) \circ ([q_b]) = ([p_a \circ (q_b)]) \tag{2.10}$$

Proof. Given $\alpha: A \longrightarrow B\Omega$ and $\beta: B \longrightarrow C\Omega$ we must show that the two paths from A to CT shown below are equal:

$$\begin{array}{ccccc}
A & \xrightarrow{\alpha} & B\Omega & \xrightarrow{\beta^{\#}} & C\Omega \\
 & & \Big\downarrow{\scriptstyle B\rho} & & \Big\downarrow{\scriptstyle C\rho} \\
 & & BT & \xrightarrow[(\beta.C\rho)^{\#}]{} & CT
\end{array}$$

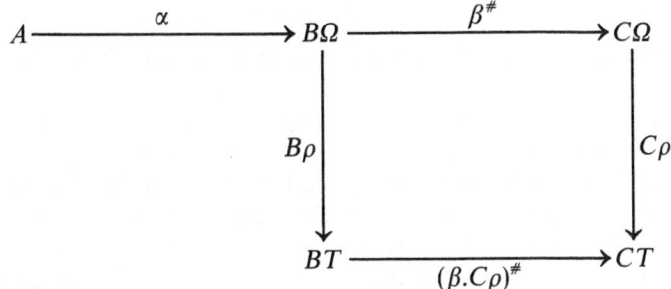

It suffices to prove that the square is commutative. This is true for variables $b \in B\Omega$ since $\langle b, B\rho.(\beta.C\rho)^{\#}\rangle = \langle b, B\eta.(\beta.C\rho)^{\#}\rangle = \langle b, \beta.C\rho\rangle = \langle b, \beta^{\#}.C\rho\rangle$. Since all four maps in the square are Ω-homomorphisms, we are done by 2.3 and 1.20. □

2.11 Proposition. **Set**(Ω, E), *as defined in 2.7, is indeed a category.*

Proof. Consider $\alpha: A \longrightarrow B$, $\beta: B \longrightarrow C$ and $\gamma: C \longrightarrow D$. It is immediately clear from the diagram that we have

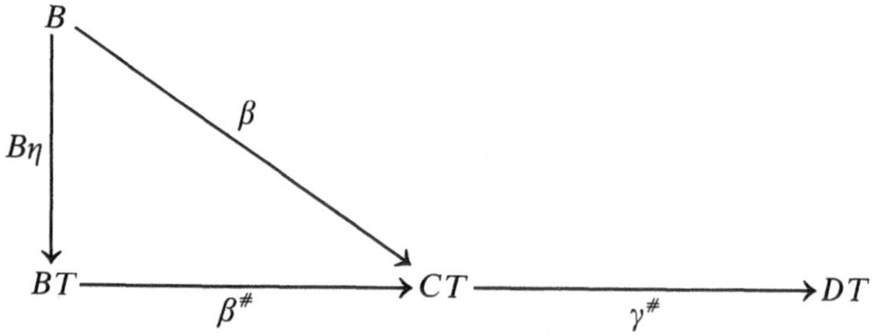

$$(\beta.\gamma^{\#})^{\#} = \beta^{\#}.\gamma^{\#}. \tag{2.12}$$

Thus $(\alpha \circ \beta) \circ \gamma = (\alpha.\beta^{\#}) \circ \gamma = \alpha.\beta^{\#}.\gamma^{\#} = \alpha.(\beta.\gamma^{\#})^{\#} = \alpha \circ (\beta \circ \gamma)$. That $A\eta.\alpha^{\#} = \alpha$ is clear. We now make explicit the trivially true but important principle:

$$\text{If } (X, \delta) \text{ is an } \Omega\text{-algebra, } \mathrm{id}_X:(X, \delta) \longrightarrow (X, \delta) \tag{2.13}$$

is an Ω-homomorphism.

From 2.13 and 2.5 it follows that

$$(A\eta)^{\#} = \mathrm{id}_{AT}, \qquad \text{for all sets } A. \tag{2.14}$$

In particular, $\alpha \circ B\eta = \alpha.(B\eta)^{\#} = \alpha$. □

Notice that 2.5 provides a bijection between morphisms $A \longrightarrow B$ and Ω-homomorphisms $AT \longrightarrow BT$; the requisite passages are

$$A \xrightarrow{\alpha} B \longmapsto AT \xrightarrow{\alpha^{\#}} BT$$

$$AT \xrightarrow{f} BT \longmapsto A \xrightarrow{A\eta.f} B$$

since $(A\eta)^{\#} = id_{AT}$ and $(\alpha \circ \beta)^{\#} = (\alpha.\beta^{\#})^{\#} = \alpha^{\#}.\beta^{\#}$. Thus **Set**$(\Omega, E)$ may be identified with the category of (Ω, E)-algebras of form AT.

2.15 Example. Let Ω have binary $+$, unary i and nullary e. Set $A = \{1\}$, $B = \{b, x\}$, $C = \{c, y\}$, $D = \{d, z\}$. Set $\alpha = xie+b+$, $\beta_b = cc+$, $\beta_x = y$, $\gamma_c = dz+i$, $\gamma_y = e$. Then $\alpha \circ \beta = yie+cc++$, $(\beta \circ \gamma)_b = dz+idz+i+$, $(\beta \circ \gamma)_x = e$, and $\alpha \circ (\beta \circ \gamma) = (\alpha \circ \beta) \circ \gamma = eie+dz+idz+i++$. Notice "$B\eta \circ \beta = \beta$" reduces to the tautology that if we substitute b for b and x for x then (β_b, β_x) is transformed into itself, whereas "$\alpha \circ A\eta = \alpha$" says that $xie+b+$ is left invariant by substituting x for x and b for b.

The reader should notice that our proof of 2.11 is a formal consequence of 2.5 and the fact that sets and functions form a category.

2.16 Definition. *The monoid of* (Ω, E), *denoted* $\mathsf{V}(\Omega, E)$, *is the endomorphism monoid of* V *in the category* $\mathbf{Set}(\Omega, E)$.

2.17 Theorem. *Let* (Ω, E) *and* (Ω', E') *be finitary equational presentations as defined in 1.10. Then the following three statements are equivalent and define the equivalence relation of structural equivalence on finitary equational presentations.*

(2.18) The category (Ω, E)-alg of (Ω, E)-algebras and Ω-homomorphisms (see 2.3 and 2.13) is *isomorphic as a category of sets with structure to* (Ω', E')-alg, that is, for each set X there exists a bijection ψ_X from the set of all (Ω, E) structures, δ, on X to the set of all (Ω', E') structures, δ', on X subject to the joint condition that for every function $f: X \longrightarrow Y$, (Ω, E)-structure δ on X and (Ω, E)-structure γ on Y then $f:(X, \delta) \longrightarrow (Y, \gamma)$ is an Ω-homomorphism if and only if $f:(X, \delta\psi_X) \longrightarrow (Y, \gamma\psi_Y)$ is an Ω'-homomorphism.

(2.19) The clones $\mathbf{Set}(\Omega, E)$, $\mathbf{Set}(\Omega', E')$ are *isomorphic*; that is, for each set X there exists a bijection $X\lambda: XT \longrightarrow XT'$ subject to *preservation of variables* and *preservation of composition*

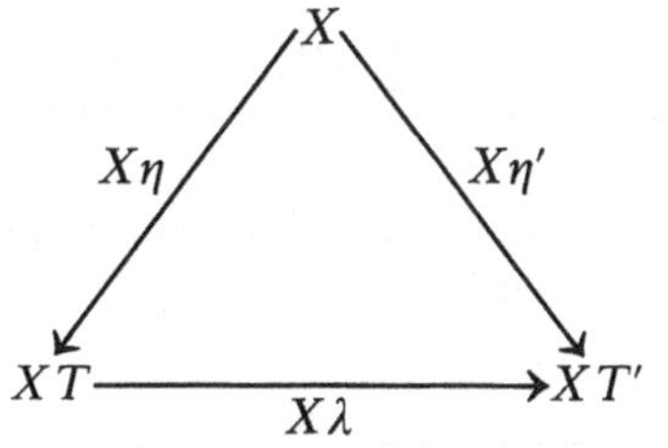

$$(\alpha \circ \beta).C\lambda = (\alpha.B\lambda) \circ' (\beta.C\lambda) \text{ for all}$$
$$\alpha: A \longrightarrow BT, \qquad \beta: B \longrightarrow CT.$$

(2.20) The monoids $\mathsf{V}(\Omega, E)$ and $\mathsf{V}(\Omega', E')$ are *isomorphic* not just as abstract monoids but in the stronger sense that there exists a bijective map $\Gamma: \mathsf{V}T \longrightarrow \mathsf{V}T'$ composing with which yields a monoid isomorphism:

$$\mathsf{V}\eta.\Gamma = \mathsf{V}\eta'$$
$$(\alpha \circ \beta).\Gamma = (\alpha.\Gamma) \circ' (\beta.\Gamma)$$

and subject to the requirement that Γ *preserves true constants*, in the sense that there exists a bijection Γ_0 such that

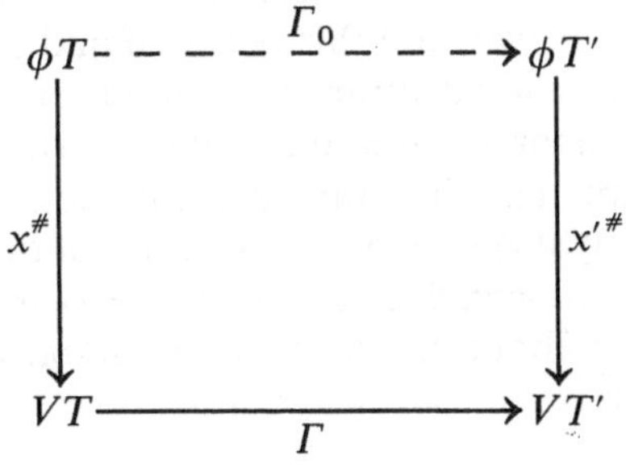

where x, x' are the unique morphisms $\phi \longrightarrow VT$, $\phi \longrightarrow VT'$. (The terminology "true constants" is explained in 5.13–.)

Proof. (2.18) *implies* (2.19). By 2.5, there exists a unique Ω-homomorphism $X\lambda : XT \longrightarrow XT'$ (where XT' is an (Ω, E)-algebra by virtue of $\psi_{XT'}^{-1}$) such that preservation of variables holds. To prove that composition is preserved, consider the diagram

$$\begin{array}{ccccc} A & \xrightarrow{\ \alpha\ } & BT & \xrightarrow{\ \beta^{\#}\ } & CT \\ & & \Big\downarrow{\scriptstyle B\lambda} & & \Big\downarrow{\scriptstyle C\lambda} \\ & & BT' & \xrightarrow[\ (\beta.C\lambda)^{\#}\ ]{} & CT' \end{array}$$

where the bottom morphism of the square can be regarded as Ω-homomorphism with the help of ψ^{-1}. The remaining details are similar to the proof of 2.10 and need no repeating.

(2.19) *implies* (2.20). Set $\Gamma = V\lambda$. That composing with Γ is a monoid isomorphism is clear, and it suffices to prove

$$\begin{array}{ccc} \phi T & \xrightarrow{\ \phi\lambda\ } & \phi T' \\ \Big\downarrow{\scriptstyle x^{\#}} & & \Big\downarrow{\scriptstyle x'^{\#}} \\ VT & \xrightarrow[\ \Gamma\ ]{} & VT' \end{array}$$

To this end, $x^{\#}.\Gamma = (\mathrm{id}_{\phi T} \circ x).\Gamma = (\mathrm{id}_{\phi T}.\phi\lambda) \circ' (x.\Gamma) = \phi\lambda \circ' x' = \phi\lambda.x'^{\#}$.

(2.20) *implies* (2.18). Let us begin by observing that we have symmetry: it is easy to check that Γ^{-1} enjoys the same properties as Γ. Next, let us make some comments about the empty set as an algebra. ϕ can be an Ω-algebra in at most one way and this occurs if and only if $\Omega_0 = \phi$; in this case, ϕ is an (Ω, E)-algebra for arbitrary E, the unique function $\phi \longrightarrow (X, \delta)$ is an Ω-homomorphism for any δ, and the same is true in the other direction $(X, \delta) \longrightarrow \phi$, except that the only such function is id_{ϕ}. Notice further that $\Omega_0 = \phi$ if and only if $\phi T = \phi$. Our sole use of the preservation of true constants consists in the conclusion that $\Omega_0 = \phi$ if and only if $\Omega'_0 = \phi$, and we may now forget about the empty set in establishing 2.18. (The reader may wish to consult exercise 7 at this point.)

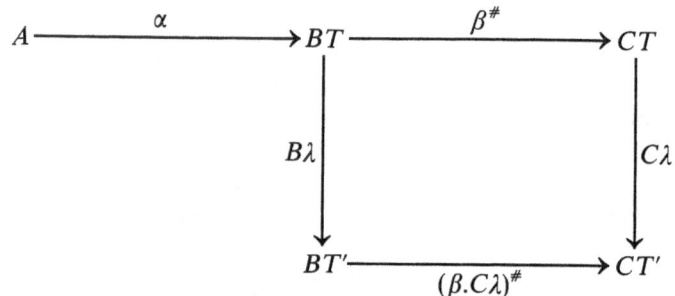

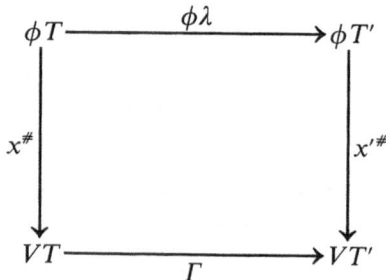

For each nonempty set X define the desired ψ_X of 2.18 by $\delta\psi_X = \delta'$ where for all ω' in Ω'_n,

$$(x_1, \ldots, x_n)\delta'_{\omega'} = \langle [v_1 \cdots v_n\omega'], \Gamma^{-1}.r^{\#}\rangle \tag{2.21}$$

where $r\colon \mathsf{V} \longrightarrow X$ is any function such that $v_i r = x_i$; in case $n = 0$ let r be arbitrary (but here we require that X be nonempty). For an example see exercise 9. We need:

2.22 Lemma. *$(x_1, \ldots, x_n)\delta'_{\omega'}$ as in 2.21 is independent of the choice of r.*

Proof. Observe that for any $\alpha\colon \mathsf{V} \longrightarrow \mathsf{V}T$ we have

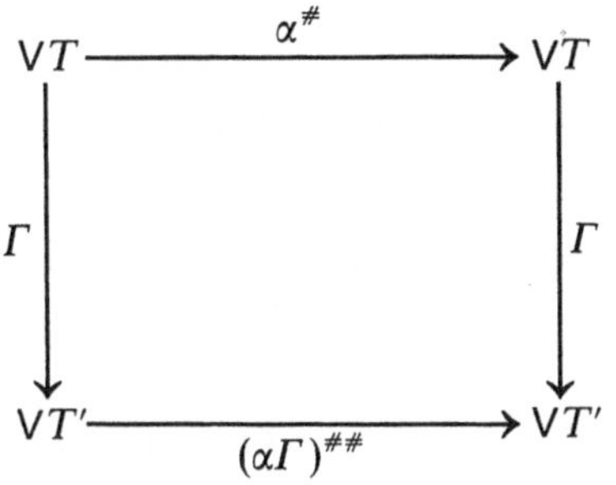

(2.23)

since $\alpha^{\#}.\Gamma = (\mathrm{id}_{\mathsf{V}T} \circ \alpha).\Gamma = (\mathrm{id}_{\mathsf{V}T}.\Gamma) \circ' (\alpha\Gamma) = (\alpha\Gamma)^{\#\#}$. Define $f\colon\mathsf{V} \longrightarrow \mathsf{V}$ by

$$v_i f = \begin{cases} v_i & \text{if } n \geqslant 1 \text{ and } 1 \leqslant i \leqslant n \\ v_1 & \text{otherwise} \end{cases}$$

and set $\alpha\colon\mathsf{V} \longrightarrow \mathsf{V}T$ to be $f.\mathsf{V}\eta$. Let X, δ, ω', r be as in 2.21. Observing that $\alpha^{\#}.r^{\#} = (f.r)^{\#}$ (use 2.5), we have

$$\begin{aligned}
&\langle [v_1 \cdots v_n\omega'], \Gamma^{-1}.r^{\#}\rangle \\
&\quad = \langle ([v_1](\alpha\Gamma)^{\#\#}, \ldots, [v_n](\alpha\Gamma)^{\#\#})\omega', \Gamma^{-1}.r^{\#}\rangle && (\Gamma \text{ preserves variables; notation as in 2.2)} \\
&\quad = \langle [v_1 \cdots v_n\omega'](\alpha\Gamma)^{\#\#}, \Gamma^{-1}.r^{\#}\rangle \\
&\quad = \langle [v_1 \cdots v_n\omega'], \Gamma^{-1}.\alpha^{\#}.r^{\#}\rangle && \text{(by 2.23)} \\
&\quad = \langle [v_1 \cdots v_n\omega'], \Gamma^{-1}.(f.r)^{\#}\rangle
\end{aligned}$$

which completes the proof of 2.22 since the final expression depends only on the restriction of r to $\{v_1, \ldots, v_n\}$.

Lemma 2.22 plays an essential role in showing that if $\delta' = \delta\psi_X$ as in 2.21 then (X, δ') satisfies E'. To do this it suffices to establish

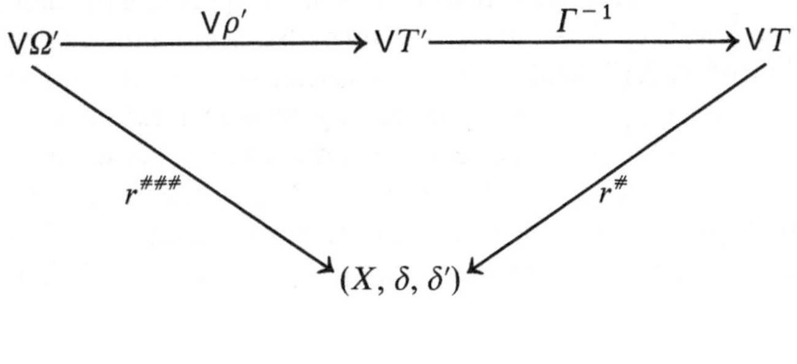

(2.24)

for arbitrary $r: \mathsf{V} \longrightarrow X$ (i.e., if $\{e_1, e_2\}$ is in E' then $e_1 r^{\#\#\#} = e_2 r^{\#\#\#}$ because $\langle e_1, \mathsf{V}\rho' \rangle = \langle e_2, \mathsf{V}\rho' \rangle$). We prove this by algebraic general recursion:

Basis step: Each path sends v to vr.

Recursive step: Let $p = p_1 \cdots p_n \omega'$ with both paths agreeing on p_i.

Let $\alpha, \beta: \mathsf{V} \longrightarrow \mathsf{V}T$ satisfy $v_1\alpha = [v_1 \cdots v_n\omega']$, $v_i\beta = [p_i]$. Then $\langle v_1, \alpha \circ' \beta \rangle = [p]$ by 2.10. Then

$$
\begin{aligned}
&\langle [p], \Gamma^{-1}.r^{\#} \rangle \\
&= \langle v_1, (\alpha\Gamma^{-1}) \circ (\beta\Gamma^{-1}) \rangle r^{\#} \\
&= \langle [v_1 \cdots v_n\omega']\Gamma^{-1}, (\beta.\Gamma^{-1}.r^{\#})^{\#} \rangle \qquad \text{(use 2.5)} \\
&= (\langle v_1, \beta.\Gamma^{-1}.r^{\#} \rangle, \ldots, \langle v_n, \beta.\Gamma^{-1}.r^{\#} \rangle)\delta'_{\omega'} \qquad \text{(by 2.22)} \\
&= (p_1 r^{\#\#\#}, \ldots, p_n r^{\#\#\#})\delta'_{\omega'} \\
&= p r^{\#\#\#}
\end{aligned}
$$

The remaining details of the proof that 2.18 holds are easy. If $f:(X, \delta) \longrightarrow (Y, \gamma)$ is an Ω-homomorphism between (Ω, E)-algebras, then $r^{\#}.f = (r.f)^{\#}$ by 2.5, so that

$$
\begin{aligned}
\langle (x_1, \ldots, x_n)\delta'_{\omega'}, f \rangle &= \langle [v_1 \cdots v_n\omega'], \Gamma^{-1}.r^{\#}.f \rangle \\
&= \langle [v_1 \cdots v_n\omega'], \Gamma^{-1}.(r.f)^{\#} \rangle \\
&= (x_1 f, \ldots, x_n f)\gamma'_{\omega'}
\end{aligned}
$$

so that $f:(X, \delta\psi_X) \longrightarrow (Y, \gamma\psi_Y)$ is an Ω'-homomorphism. ψ_X^{-1} is defined symmetrically. If $\gamma' = \delta'\psi_X^{-1}\psi_X$ then

$$
\begin{aligned}
(x_1, \ldots, x_n)\gamma'_{\omega'} &= \langle [v_1 \cdots v_n\omega'], \Gamma^{-1}.r^{\#} \rangle \\
&= \langle [v_1 \cdots v_n\omega'], r^{\#\#\#} \rangle \qquad \text{(by 2.24)} \\
&= (x_1, \ldots, x_n)\delta'_{\omega'}
\end{aligned}
$$

The proof of 2.17 is complete. □

Notes for Section 2

The term "clone"—an acronym for "closed set of operations"—is attributed to P. Hall in [Cohn '65, III.3]. The earliest published paper we know of which deals with the algebra of clone composition is [Menger '46]. See also [Felscher '68, '72 2.1], [Menger '59], [Schmidt '62], [Schweizer and Sklar '69], [Whitlock '64], and the bibliographies there.

The general concept of "isomorphism of categories of sets with structure" used in 2.18 will be formalized in section 3 of Chapter 2. This definition was given by [Mal'cev '71, p. 59] under the term "structural equivalence" which we have adopted in 2.17. Mal'cev indicated that such isomorphisms between categories of (Ω, E)-algebras are induced by transformations at the level of syntax ([Mal'cev '58]) and precise statements and proofs were provided by [Felscher, '68, '69, '72] and [Hoehnke, '66]. In this connection, $\Gamma: \mathsf{V}T \longrightarrow \mathsf{V}T'$ of 2.20 should be regarded as "interpreting T-terms as T'-terms" and 2.21 asserts that "the Ω-interpretation of ω' is provided by Γ^{-1}". (Cf. also the logicians' "interpretations between theories," e.g. [Enderton '72, section 2.7].) More general clone homomorphisms will appear in 3.2.8.

Theorem 2.17 is a "folklore theorem" which has not, to our knowledge, appeared in print before; however, see [Felscher '72, 2.1.8, 2.3.2].

Exercises for Section 2

1. The *axiom of choice* asserts that for every onto function $f:X \longrightarrow Y$ there exists a "choice function" $c:Y \longrightarrow X$ such that $c.f = \mathrm{id}_Y$. From the point of view of "naive" set theory one simply chooses for each y any c_y with $c_y f = y$ (since f is onto) and defines $yc = c_y$. From the point of view of axiomatic set theory (see [Monk '69, Chapter 3], [Lawvere '64], [Jech '73, Chapter 1]) this cannot be established from the other "standard" axioms unless Y is finite. Show that the axiom of choice in the proof of 2.2 can be avoided by restricting $r:V \longrightarrow AT$ to the finite set of variables occurring in e_1 and e_2.
2. Prove that the axiom of choice, as defined in exercise 1, is equivalent to the following: for every function $f:X \longrightarrow Y$ with X nonempty there exists $g:Y \longrightarrow X$ with $fgf = f$.
3. Let Ω have a single n-ary operation ω and let E possess the equation $\{(w_1, \ldots, w_n)\omega, w_1\}$ whenever $w_1, \ldots, w_n \in \mathsf{V}$ are such that not all w_i are distinct. Show that $AT = \{[a]:a \in A\}$ if A has less than n elements and that AT is infinite otherwise.
4. Let Ω be a finitary operator domain and let E, E' be two sets of Ω-equations. Prove that (Ω, E) and (Ω, E') are structurally equivalent if and only if $E_\mathsf{V} = E'_\mathsf{V}$
5. Exercise 4 may be generalized as follows. Let (Ω, E_1), (Ω_2, E_2) be two finitary equational presentations. Let Ω be the operator domain defined by $\Omega_n = (\Omega_1)_n + (\Omega_2)_n$. Prove that (Ω_1, E_1) and (Ω_2, E_2) are structurally equivalent if and only if $(E_1)_\mathsf{V} = (E_2)_\mathsf{V}$.
6. Attempt to prove 2.11 by generalizing the notations of 2.15.
7. Let T, T' correspond respectively to the equational presentations "one nullary operation e and no equations" and "one unary operation u together with the equation $\{v_1 u, v_2 u\}$." Prove that there exists a bijection $\Gamma:\mathsf{V}T \longrightarrow \mathsf{V}T'$ composition with which is a monoid isomorphism (i.e., the first two conditions of 2.20 hold). Observe that these equational presentations are not structurally equivalent by considering the empty algebra.
8. Write out an explicit description of $\mathsf{V}(\Omega, E)$ for the (Ω, E) of exercise 5 of section 1.
9. Let Ω have one binary operation m and no equations and let Ω' have one ternary operation t with single equation $(x, y, a)t = (x, y, b)t$. Define $\Gamma:\mathsf{V}\Omega \longrightarrow \mathsf{V}T'$ by the recursion

$$v\Gamma = v$$
$$pqm\Gamma = [p][q][v_1]t$$

Show that Γ is a structural equivalence in the sense of 2.20 and that δ', as in 2.21, is given by $(x, y, z)\delta'_t = (x, y)\delta_m$.

10. Given $p \in A\Omega$, the set var(p) of *variables occurring in p* is defined in the obvious way by algebraic recursion:

$$\begin{aligned} \mathrm{var}(a) &= \{a\} \\ \mathrm{var}(p_1 \cdots p_n\omega) &= \mathrm{var}(p_1) \cup \cdots \cup \mathrm{var}(p_n) \end{aligned}$$

Given an (Ω, E)-algebra (X, δ), prove that if $r, s: A \longrightarrow X$ agree on var(p) then $[p]r^{\#} = [p]s^{\#}$. Verify that this description of var(p) agrees with the one given in 2.10−.

11. Verify that 2.23 is equivalent to $(\alpha \circ \beta).\Gamma = (\alpha.\Gamma) \circ' (\beta.\Gamma)$.
12. Verify that "groups" as in 1.1 is structurally equivalent to "groups" as in 1.2 (as modified in 1.3+).
13. Without peeking at Chapter 2, formulate a definition of "isomorphism of categories" to formalize our assertion of 2.14+ that **Set**(Ω, E) may be "identified" with the category of (Ω, E)-algebras of form AT.
14. Verify that the proof of 2.22 is valid when $n = 0$.
15. Let Ω have a single nullary operation and let E be empty. Show that **Set**(Ω, E) may be identified with the category of sets and partial functions.

3. Algebraic Theories

Roughly speaking, the algebraic theory of an equational presentation (Ω, E) is its equivalence class in the various senses of 2.17. In this section we describe **Set**(Ω, E) as an "algebraic" object without reference to any (Ω, E). The definition is so elementary that, unlike the situation in 1.10, no intrinsic structure of sets is referred to; we need only to know that sets and functions form a category. To this end:

3.1 Definition. Fix an arbitrary category $\mathscr{K}$. $\mathscr{K}$ is the *base category*.

Not until Chapter 3 will we use the full generality of 3.1. Right now, the reader will do well to pretend that $\mathscr{K}$ is a familiar category such as sets, topological spaces, or groups.

3.2 Definition. An *algebraic theory* (*in clone form*) *in* $\mathscr{K}$ *is a triple* $\mathbf{T} = (T, \eta, \circ)$, where

T is an object function, assigning to each object A of $\mathscr{K}$ another object, AT, "of T-terms with variables in A."

η is an assignment to each object A of $\mathscr{K}$ an "insertion-of-the-variables" map $A\eta: A \longrightarrow AT$.

$\circ$ is an assignment to each ordered triple (A, B, C) of objects in $\mathscr{K}$ a "clone-composition" function

$$\mathscr{K}(A, BT) \times \mathscr{K}(B, CT) \xrightarrow{\quad\circ\quad} \mathscr{K}(A, CT)$$

Before stating the axioms on $\mathbf{T}$ we establish some notations. First of all, we use the same notational conventions as in 2.7. In addition, we recognize that each $\mathscr{K}$-morphism $f:A \longrightarrow B$ induces $f^{\Delta}:A \longrightarrow B$ defined by

$$f^{\Delta} = A \xrightarrow{f} B \xrightarrow{B\eta} BT$$

The axioms on **T**, then, are:

(3.3) $(\alpha \circ \beta) \circ \gamma = \alpha \circ (\beta \circ \gamma)$ for all $\alpha:A \longrightarrow B$, $\beta:B \longrightarrow C$ and $\gamma:C \longrightarrow D$

$$\alpha \circ B\eta = \alpha \qquad \text{for all} \qquad \alpha:A \longrightarrow B$$

$$f^{\Delta} \circ \alpha = f.\alpha \qquad \text{for all} \qquad f:A \longrightarrow B, \qquad \alpha:B \longrightarrow C$$

This defines a category, $\mathscr{K}_{\mathbf{T}}$, with the same objects as $\mathscr{K}$, composition $\circ$, and identity morphisms η (set $f = \mathrm{id}_A$ to prove that $A\eta \circ \alpha = \alpha$). $\mathscr{K}_{\mathbf{T}}$ is called the *Kleisli category of* **T**.

3.4 Example. Let (Ω, E) be an equational presentation and let T, η, $\circ$ be as in 2.1, 2.9, and 2.8. Then $(T, \eta, \circ)$ is an algebraic theory in **Set** and $\mathbf{Set_T} = \mathbf{Set}(\Omega, E)$. To prove this we must check 3.3; indeed, $f^{\Delta} \circ \alpha = f^{\Delta}.\alpha^{\#} = f.(B\eta.\alpha^{\#}) = f.\alpha$.

3.5 Example. Sets and Relations. A *relation from* a set A *to* a set B is a subset α of $A \times B$. We may write $a\alpha b$ for $(a, b) \in \alpha$. Since $a\alpha b$ if and only if $b \in a\alpha$, where $a\alpha = \{b \in B: a\alpha b\}$, there is a natural bijective correspondence between relations from A to B and functions from A to BT, where BT is the power set of B, that is, the set of all subsets of B. Given another relation from B to C, call it β, there is a well-known composition $a(\alpha \circ \beta)c$ if and only if there exists $b \in B$ with $a\alpha b$ and $b\beta c$; or, in the second notation, $\langle a, \alpha \circ \beta \rangle = \{c \in C: \text{there exists } b \in a\alpha \text{ with } c \in b\beta\}$. Define $A\eta: A \longrightarrow AT$ by $\langle a, A\eta \rangle = \{a\}$. It is easy to check that $(T, \eta, \circ)$ is an algebraic theory in **Set**.

3.6 Example. Matrices. Let R be a ring with unit. If B is a set, a *vector in* B is a B-tuple of "scalars" (that is, elements of R) $(\lambda_b : b \in B)$ such that all but finitely many $\lambda_b = 0$. Let BT denote the set of all vectors in B. An $A \times B$ *matrix* is a function $\alpha: A \longrightarrow BT$ (where we think of $a\alpha$ as a row vector and b as indexing columns). Given an $A \times B$ matrix α and a $B \times C$ matrix β define their composition $\alpha \circ \beta$ by the usual matrix multiplication formula $\langle a, \alpha \circ \beta \rangle_c = \sum_{b \in B} (a\alpha)_b (b\beta)_c$. Define $A\eta: A \longrightarrow AT$ by $\langle a, A\eta \rangle = \delta^a$, where δ^a is the Kronecker δ, $\delta^a_b = 0$ if $a \neq b$, $\delta^a_a = 1$. It is routine to check that $(T, \eta, \circ)$ is an algebraic theory in **Set**. In fact, **T** comes from a suitable (Ω, E) (see exercise 1).

3.7 Example. Let $\mathscr{K}$ be the category of topological spaces and continuous maps and let G be a topological group. For each space B let BT be the topological space $B \times G$. A map $\alpha: A \longrightarrow BT$ amounts to a pair $\begin{pmatrix} f \\ f' \end{pmatrix}$ of continuous maps, $f: A \longrightarrow B$, $f': A \longrightarrow G$. If $\beta = \begin{pmatrix} g \\ g' \end{pmatrix}: B \longrightarrow CT$, define $\alpha \circ \beta = \begin{pmatrix} fg \\ fg' * f' \end{pmatrix}$, where $*: G \times G \longrightarrow G$ is the (continuous) group-multiplication map. Define $A\eta: A \longrightarrow AT$ by $A = \begin{pmatrix} \mathrm{id}_A \\ e \end{pmatrix}$, where $e: A \longrightarrow G$ is constantly the group unit. It is not hard to prove that $(T, \eta, \circ)$ is an algebraic theory in $\mathscr{K}$.

In the first two sections of this book we extracted the algebraic theory of a bunch of algebras. In section 4, we will learn how algebras can be defined in terms of their theory. The point of Examples 3.5, 3.6, and 3.7 is that algebraic theories can arise naturally without first knowing what the algebras are. The reader might ponder just what kind of algebras are defined by these three examples; they are all relatively famous examples of "structures."

Our immediate goal is to reformulate algebraic theories in clone form as algebraic theories in monoid form, with considerable technical gains. We begin by fixing an arbitrary algebraic theory $\mathbf{T} = (T, \eta, \circ)$ (in clone form) in a category $\mathscr{K}$ and studying some formal consequences of Axiom 3.3.

3.8 Triple Product Law. Given $f: A \longrightarrow B$, $\beta: B \longrightarrow C$ and $\gamma: C \longrightarrow D$, $f.\beta \circ \gamma$ is well defined. For $(f.\beta) \circ \gamma = (f^{\Delta} \circ \beta) \circ \gamma = f^{\Delta} \circ (\beta \circ \gamma) = f.(\beta \circ \gamma)$. □

3.9 Compatibility of Compositions. Given $f: A \longrightarrow B$ and $g: B \longrightarrow C$, then $(f.g)^{\Delta} = f^{\Delta} \circ g^{\Delta}$. Proof: $(f.g)^{\Delta} = f.g.C\eta = f.g^{\Delta} = f^{\Delta} \circ g^{\Delta}$. □

3.10 Proposition. *If for $f: A \longrightarrow B$ we define $fT: AT \longrightarrow BT = \mathrm{id}_{AT} \circ f^{\Delta}$, T is a functor.*

Proof. $(\mathrm{id}_A)T = \mathrm{id}_{AT} \circ (\mathrm{id}_A)^{\Delta} = \mathrm{id}_{AT} \circ A\eta = \mathrm{id}_{AT}$. If also $g: B \longrightarrow C$, $(f.g)T = \mathrm{id}_{AT} \circ (f.g)^{\Delta} = \mathrm{id}_{AT} \circ f^{\Delta} \circ g^{\Delta} = fT \circ g^{\Delta} = fT.\mathrm{id}_{BT} \circ g^{\Delta} = fT.gT$. □

Let us explore this new construction in some previous examples. In the (Ω, E) case, $fT = \mathrm{id}_{AT} \circ f^{\Delta} = (f^{\Delta})^{\#}$. From the diagram of 2.5 and 1.20, $fT: AT \longrightarrow BT$ is determined by

$$\begin{aligned} [a]fT &= [af] \\ [p_1 \cdots p_n\omega]fT &= [p_1]fT \cdots [p_n]fT\omega \end{aligned}$$

When E is empty, fT is the $f\Omega$ of 1.17. In some sense, then, fT is "substitution of variables"; but this must be taken with a grain of salt since an equation such as $\{v_1 v_1 im, e\}$ in group theory makes it impossible to define the variables of an equivalence class of formulas.

Let us pin down fT in the context of Example 3.5. The change in point of view in passing from $\mathrm{id}_{AT}: AT \longrightarrow AT$ to $\mathrm{id}_{AT}: AT \longrightarrow A$ converts an uninteresting identity function into the $\in$-relation: $S \ \mathrm{id}_{AT}\ a$ if and only if $a \in S$. $f^{\Delta}: A \longrightarrow BT$ sends a to $\{af\}$. Therefore, $\langle S, \mathrm{id}_{AT} \circ f^{\Delta} \rangle = \{b \in B$: there exists $a \in S$ with $b \in \{af\}\} = \{af: a \in S\}$, and fT is the direct image map.

Since T is now a functor, it makes sense to ask if η is a natural transformation:

3.11 Proposition. *η is a natural transformation from the identity functor of $\mathscr{K}$ to T; that is, for all $f: A \longrightarrow B$ we have*

$$\begin{array}{ccc} A & \xrightarrow{A\eta} & AT \\ {\scriptstyle f}\downarrow & & \downarrow{\scriptstyle fT} \\ B & \xrightarrow[B\eta]{} & BT \end{array}$$

Proof. $A\eta.fT = A\eta \circ f^{\Delta}$ (see 3.12) $= f^{\Delta} = f.B\eta$. □

The functoriality of T also allows us to state the counterpart of "$f^{\Delta} \circ \alpha = f.\alpha$" on the other side:

3.12 Proposition. *For all $\alpha: A \longrightarrow B$, $g: B \longrightarrow C$ we have $\alpha \circ g^{\Delta} = \alpha.gT$.*

Proof. $\alpha \circ g^{\Delta} = \alpha.\mathrm{id}_{BT} \circ g^{\Delta} = \alpha.gT$. □

Let us return briefly to $A\Omega$. While 1.11 provided us with a unique derivation tree for each formula, there is more than one way in which one could choose to assemble the tree from its pieces. The associative law of clone composition may be regarded as the statement that different assembling procedures build the same tree. To be specific, let us consider the example of 2.15 where the formula $eie+dz+idz+i++$ in $D\Omega$ was broken up in two different ways:

$$\begin{array}{ll} \alpha \circ (\beta \circ \gamma) & (e)ie+(dz+idz+i+)+ \\ (\alpha \circ \beta) \circ \gamma & (e)ie+(dz+i)(dz+i)++ \end{array}$$

It is not necessary to introduce parentheses as formal symbols to make these distinctions. Let us agree to interpret the parentheses enclosing p in (p) as the desire to consider the formula $p \in A\Omega$ as a mere variable in $(A\Omega)\Omega$. Thus $(e)ie+(dz+idz+i+)+$ and $(e)ie+(dz+i)(dz+i)++$ become elements of $(D\Omega)\Omega$ of word lengths 6 and 8, respectively. Since $A\Omega$ is an Ω-algebra, it has a total description map (1.18) $A\mu:(A\Omega)\Omega \longrightarrow A\Omega$ which is the unique Ω-homomorphism that preserves variables (i.e., sends (p) to p). In short, $A\mu$ is the desired map which removes the parentheses. $(A\Omega)\Omega$ has enough structure to define clone composition and $A\mu$ can express the associative law!

Let us return now to the general $\mathbf{T}$. There is no problem in defining "formulas of formulas." It is obvious that we in fact get a functor TT by defining $ATT = (AT)T$, $fTT = (fT)T$. To see how to define μ, observe that in the Ω-case, $A\mu = (\mathrm{id}_{A\Omega})^{\#} = \mathrm{id}_{A\Omega\Omega}.(\mathrm{id}_{A\Omega})^{\#} = \mathrm{id}_{A\Omega\Omega} \circ \mathrm{id}_{A\Omega}$. This motivates the general definition of μ. We sum up the structure of μ in four axioms and go on to see that (T, η, μ) is coextensive with $(T, \eta, \circ)$.

3.13 Definition. *For each object A of $\mathscr{K}$ define a $\mathscr{K}$-morphism $A\mu: ATT \longrightarrow AT$ by $A\mu = (\mathrm{id}_{ATT}: ATT \longrightarrow AT) \circ (\mathrm{id}_{AT}: AT \longrightarrow A)$.*

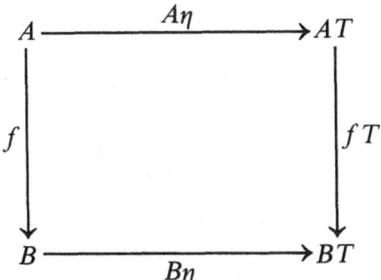

The reader may check that in 3.7, $A\mu : A \times G \times G \longrightarrow A \times G$ is the continuous map which sends (a, g, h) to $(a, g*h)$. In 3.5, $A\mu$ is the union map sending the collection $\mathscr{A}$ of subsets of A to its union $\{a \in A:$ there exists $S \in \mathscr{A}$ with $a \in S\}$.

3.14 Kleisli Composition Law. For each $\beta : B \longrightarrow C$ define $\beta^{\#} : BT \longrightarrow CT = (\beta T : BT \longrightarrow CTT).(C\mu : CTT \longrightarrow CT)$. Then for all $\alpha : A \longrightarrow B$ and $\beta : B \longrightarrow C$, $\alpha \circ \beta = \alpha.\beta^{\#}$.

Proof. $\alpha \circ \beta = \alpha \circ (\beta.\mathrm{id}_{CT}) = \alpha \circ \beta^{\Delta} \circ \mathrm{id}_{CT} = \alpha.\beta T \circ \mathrm{id}_{CT} = \alpha.(\beta T.\mathrm{id}_{CTT}) \circ \mathrm{id}_{CT} = \alpha.(\beta T^{\Delta} \circ \mathrm{id}_{CTT} \circ \mathrm{id}_{CT}) = \alpha.(\beta T^{\Delta} \circ C\mu) = \alpha.\beta T.C\mu$. □

We will see in the next section that the correct interpretation of 3.14 is the obvious one: algebraic theories really are like the motivating example **Set**(Ω, E). The next two statements establish the four axioms on μ we mentioned earlier.

3.15 Proposition. *$\mu : TT \longrightarrow T$ is a natural transformation, that is for all $f : A \longrightarrow B$ we have*

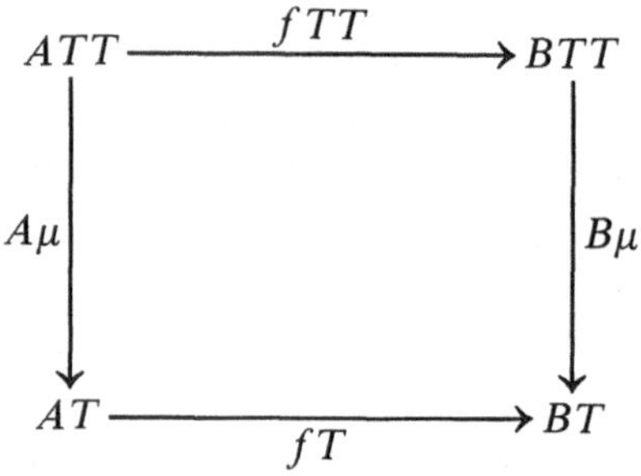

Proof. $A\mu.fT = A\mu \circ f = \mathrm{id}_{ATT} \circ \mathrm{id}_{AT} \circ f^{\Delta} = \mathrm{id}_{ATT} \circ fT = \mathrm{id}_{ATT}.fTT.B\mu$ (by 3.14). □

3.16 Proposition. *For every object $A \in \mathscr{K}$, the following diagrams commute:*

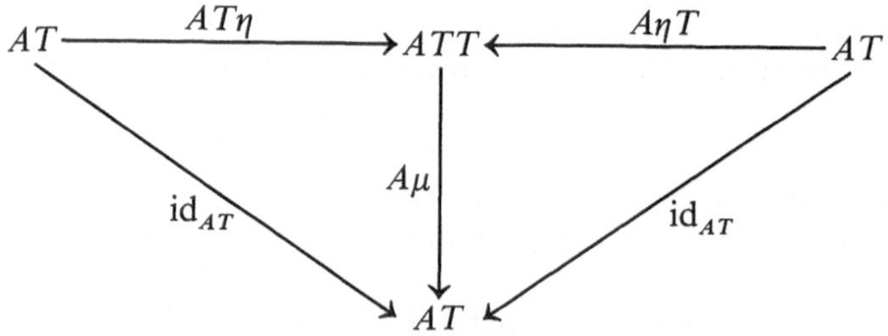

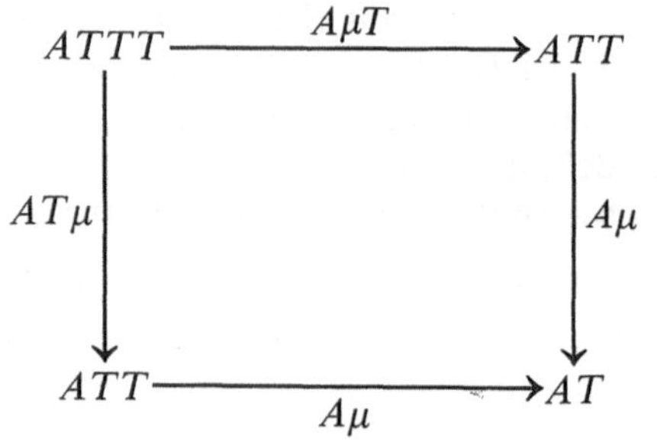

Proof. $AT\eta.A\mu = AT\eta.\mathrm{id}_{ATT} \circ \mathrm{id}_{AT} = AT\eta \circ \mathrm{id}_{AT} = \mathrm{id}_{AT}$. $A\eta T.A\mu = \mathrm{id}_{AT} \circ A\eta$(by 3.14) $= \mathrm{id}_{AT}$. $A\mu T.A\mu = \mathrm{id}_{ATTT} \circ A\mu = \mathrm{id}_{ATTT} \circ \mathrm{id}_{ATT} \circ \mathrm{id}_{AT} = AT\mu \circ \mathrm{id}_{AT} = AT\mu.(\mathrm{id}_{AT})T.A\mu = AT\mu.A\mu$. □

3.17 Definition. *An algebraic theory (in monoid form) in a category $\mathscr{K}$ is a triple (T, η, μ) where T is a functor from $\mathscr{K}$ to $\mathscr{K}$, η is a natural transformation from the identity functor of $\mathscr{K}$ to T, and μ is a natural transformation from TT to T, all such that the three diagrams of 3.16 commute for every object A of $\mathscr{K}$.*

The reason for the terminology "monoid form" lies in the analogy with the diagrams (see exercise 4) that define an ordinary monoid. Think of μ as a "binary multiplication" and η as a "unit"; the diagrams of 3.16 say that "μ is associative and η is a two-sided unit for μ." We agree with [Mac Lane '71, p. v] that "monoid" is one of the central concepts of category theory and we will see in Chapter 3 (3.2.6) how (T, η, μ) really is a monoid.

3.18 Theorem. *In any category $\mathscr{K}$, the passage from algebraic theories in clone form $(T. \eta, \circ)$ to algebraic theories in monoid form (T, η, μ) defined by 3.10, 3.11, 3.13, 3.15, and 3.16 is bijective.*

Proof. The inverse passage can only be achieved by the formula of 3.14. Let us prove this is well defined. We have $\alpha \circ B\eta = \alpha.B\eta T.B\mu = \alpha$. The associative law is proved as in 2.11 because we can recapture 2.12: $(\beta.\gamma^{\#})^{\#} = (\beta.\gamma^{\#})T.D\mu = (\beta.\gamma T.D\mu)T.D\mu = \beta T.\gamma TT.D\mu T.D\mu = \beta T.\gamma TT.DT\mu.D\mu = \beta T.C\mu.\gamma T.D\mu = \beta^{\#}.\gamma^{\#}$. The rest of 3.3: $f^{\Delta} \circ \alpha = (f.B\eta).\alpha T.C\mu = f.\alpha.CT\eta.C\mu = f.\alpha$. Passing from $(T, \eta, \circ)$ to (T, η, μ) to $(T, \eta, \circ')$, $\circ = \circ'$ by 3.14. Now let us pass from (T, η, μ) to $(T, \eta. \circ)$ to (T, η, μ'). To prove that $\mu = \mu'$ we use the only axiom about (T, η, μ) we have not used already, namely that T preserves identity maps: $A\mu' = \mathrm{id}_{ATT} \circ \mathrm{id}_{AT} = \mathrm{id}_{ATT}.(\mathrm{id}_{AT})T.A\mu = A\mu$. □

We close this section with two fascinating examples of algebraic theories in **Set**. In section 5 we will identify the algebras of these theories respectively as complete atomic Boolean algebras and compact Hausdorff spaces!

3.19 Example. The Double Power-Set Theory. For each set X let XT be the set of all collections of subsets of X. $XTTT$ is quite complicated. For future and present convenience we established some helpful notations:

$\mathscr{A} \in XT$	a collection of subsets of X
$A \in \mathscr{A}$	a subset of X
$x \in A$	an element of X
$\bar{\mathscr{B}} \in YTT$	a collection of subsets of YT
$\bar{B} \in \bar{\mathscr{B}}$	a subset of YT
$\mathscr{B} \in \bar{B}$	an element of YT
$\hat{\mathscr{C}} \in ZTTT$	a collection of subsets of ZTT
$\hat{C} \in \hat{\mathscr{C}}$	a subset of ZTT
$\bar{\mathscr{C}} \in \hat{C}$	an element of ZTT

$x \in A \in \mathscr{A} \in \bar{A} \in \bar{\mathscr{A}} \in \hat{A} \in \hat{\mathscr{A}} \in ATTT$

Given an X-tuple $(\mathscr{B}_x : x \in X)$ of elements of YT and a Y-tuple $(\mathscr{C}_y : y \in Y)$ of elements of ZT define $(\mathscr{B}_x) \circ (\mathscr{C}_y)$ by $((\mathscr{B}_x) \circ (\mathscr{C}_y))_{\hat{x}} = \{C \subset Z : \{y \in Y : C \in \mathscr{C}_y\}$

$\in \mathscr{B}_{\hat{x}}\}$. Define $\langle x, X\eta\rangle = \text{prin}(x) \in XT$, where prin($x$) is the *principal ultrafilter on* x (see 3.21) defined by $\text{prin}(x) = \{A \subset X : x \in A\}$. The proof that $(T, \eta, \circ)$ is an algebraic theory is left as an exercise. For $f: X \longrightarrow Y$, fT is defined by $\langle \mathscr{A}, fT\rangle = \{B \subset Y : Bf^{-1} \in \mathscr{A}\}$. $X\mu: XTT \longrightarrow XT$ is defined by $\langle \bar{\mathscr{A}}, X\mu\rangle = \{A \subset X : \{\mathscr{A} \in XT : A \in \mathscr{A}\} \in \bar{\mathscr{A}}\}$.

3.20 Definition. *Let* **T** *be an algebraic theory in* **Set**, *and let* T' *assign to each set* A *a subset* AT' *of* AT. T' *is a subtheory of* T *if for all* A, *the image of* $A\eta$ *is a subset of* AT' (*thereby defining the map* $A\eta': A \longrightarrow AT'$) *and if for all* $\alpha: A \longrightarrow BT'$ *and* $\beta: B \longrightarrow CT'$, *the image of*

$$(A \xrightarrow{\alpha} BT' \hookrightarrow BT) \circ (B \xrightarrow{\beta} CT' \hookrightarrow CT)$$

is a subset of CT' (*thereby defining* $\alpha \circ' \beta: A \longrightarrow CT'$). Clearly $(T', \eta', \circ')$ is a theory if T' is a subtheory.

3.21 Example. The Ultrafilter Theory. The reader has probably heard of ultrafilters before, but no matter if not. We will postpone a discussion of the elementary properties of ultrafilters until 5.24, taking for the moment one of the many well-known equivalent definitions. An *ultrafilter* $\mathscr{U}$ *on* a set X is a collection of subsets of X satisfying

(3.22) If $n > 0$ and $A_1, \ldots, A_n \in \mathscr{U}$ then $A_1 \cap \cdots \cap A_n$ is nonempty. This is called the *finite intersection property*.

(3.23) If A is any subset of X then either $A \in \mathscr{U}$ or $X - A \in \mathscr{U}$, where $X - A = \{x \in X : x \notin A\}$ is the complement of A in X.

The ultrafilter theory is sufficiently interesting to deserve the special symbol $\boldsymbol{\beta}$. (We will learn in 2.2.8 how "β" comes from "beta-compactification".) For each set X define $X\beta$ to be the set of all ultrafilters on X. If **T** is the theory of 3.19, $X\beta$ is a subset of XT. We show now that β is a subtheory of **T**. It is obvious that prin(x) is an ultrafilter. Now let $\mathscr{U} = ((\mathscr{B}_x) \circ (\mathscr{C}_y))_{\hat{x}} \in CT$ where each $\mathscr{B}_x$ is an ultrafilter on B and each $\mathscr{C}_y$ is an ultrafilter on Z. Let $C_1, \ldots, C_n \in \mathscr{U}$. Therefore, $B_i = \{y \in Y : C_i \in \mathscr{C}_y\} \in \mathscr{B}_{\hat{x}}$ for all $1 \leqslant i \leqslant n$. Since $\mathscr{B}_{\hat{x}}$ is an ultrafilter, there exists $y \in B_1 \cap \cdots \cap B_n$. As $\mathscr{C}_y$ is an ultrafilter and $C_1, \ldots, C_n \in \mathscr{C}_y$, $C_1 \cap \cdots \cap C_n \neq \phi$. Now let C be any subset of Z and suppose that $C \notin \mathscr{U}$. Therefore, $\{y \in Y : C \in \mathscr{C}_y\} \notin \mathscr{B}_{\hat{x}}$. Since $\mathscr{B}_{\hat{x}}$ and each $\mathscr{C}_y$ are ultrafilters, $\{y \in Y : Z - C \in \mathscr{C}_y\} = \{y \in Y : C \notin \mathscr{C}_y\} \in \mathscr{B}_{\hat{x}}$, and $Z - C \in \mathscr{U}$. The proof that $\boldsymbol{\beta}$ is an algebraic theory is complete. □

Notes for Section 3

Our Definition 3.2 of "algebraic theory in clone form" has its origins not so much in the "abstract clones" of P. Hall [Cohn '65, p. 132] or the work on clones cited in section 2 as in the fundamentally different approach of [Lawvere '63]. A version of Lawvere's definition is given in 5.35. A brief textbook treatment of universal algebra in Lawvere's formalism appears in [Pareigis '70] and in [Schubert '72, Chapter 18]. See also [Kock '68] and [Wraith '70].

The "algebraic theories in monoid form" of 3.17 were first defined in the appendix to [Godement '58] where they were called "standard constructions." The first paper relating these standard constructions with universal algebra was provided by [Eilenberg and Moore '65] who called them "triples." The term "monad" is favored by [Mac Lane '71, chapter VI]. At this writing, whether to call them triples or monads is regarded by some as controversial. Our use of "algebraic theory" counters the argument that this term is pre-empted by Lawvere on the grounds that Lawvere's theories and triples-monads are coextensive.

The passage from **T** (in monoid form) to $\mathscr{K}_{\mathbf{T}}$ was given by [Kleisli '65]. This suggested Definitions 3.2 and 3.18 (which are new). For an axiomatization of $(T, \eta, (-)^{\#})$ see Exercise 12.

Exercises for Section 3

1. Construct a finitary equational presentation whose associated algebraic theory is that of Example 3.6. (Hint: (Ω, E)-algebras are R-modules; AT is the weak direct sum of A copies of R.)
2. The construction of 3.7 can be done in **Set** for each abstract group G. Show that this algebraic theory comes from a finitary equational presentation. (Hint: Ω has only unary operations and they are the elements of G.)
3. Show that $\beta^{\#} = \mathrm{id}_{BT} \circ \beta$ for $\beta: B \longrightarrow C$. Using this as a definition, reprove the theory of 3.14. See Exercise 12 below.
4. A *monoid* (X, m, e) is a set X equipped with a binary associative operation $m: X \times X \longrightarrow X$ with a two-sided unit $e \in X$. Express the associative law as a commutative diagram. Similarly, recognizing that e is a function $\mathrm{e}: 1 \longrightarrow X$ from the one-element set 1, express the unit laws as a commutative diagram. (Hint: see 3.2.3.)
5. An algebraic theory **T** in **Set** is *affine* ([Wraith '70]) if $1T = 1$ (i.e., 1 is a one-element set and "$1T = 1$" means "$1T$ again has one element"). Let **T** be arbitrary, let $t_A: A \longrightarrow 1$ be the unique function and identify $1\eta: 1 \longrightarrow 1T$ with the corresponding element of $1T$. Prove that $AT_0 = \{p \in AT: \langle p, t_A T\rangle = 1\eta\}$ is a subtheory of **T** which is affine.
6. [Wraith '70] Let R be the field of real numbers and let **T** be the algebraic theory of real matrices as in 3.6. Show that the affine subtheory of **T** (as in exercise 5) is given by $AT_0 = \{(\lambda_a): \sum \lambda_a = 1\}$. Show that stochastic matrices (all $\lambda_a \geqslant 0$) is an affine subtheory of T_0.
7. Let **T** be the algebraic theory of 3.5. Show that "finite subsets" and "nonempty subsets" form subtheories of **T**.
8. Let **T** correspond to an equational presentation such that, in each equation, the same set of variables is used on both sides. Given $p \in AT$ define $\mathscr{A}_p = \{S \subset A: p$ is in the image of $(\mathrm{inc}_S)T\}$ where inc_S denotes the inclusion map $S \longrightarrow A$. Show that $AT_0 = \{\mathscr{A}_p: p \in AT\}$ is a subtheory of the double power-set theory of 3.19.
9. Let T: **Set** $\longrightarrow$ **Set** be an arbitrary functor. A *composition law on* T is an assignment to each pair (X, Y) of sets a function $c_{X,Y}: XT^Y \times YT \longrightarrow XT$ which is natural in X and dinatural [Mac Lane '71, Chapter IX, section 4] in Y; that is, given $f: X \longrightarrow X'$ and $g: Y \longrightarrow Y'$ we have the commutative squares (where fT^Y and XT^g mean, respectively, compose

$$\begin{array}{ccc} XT^Y \times YT & \xrightarrow{\;c_{X,Y}\;} & XT \\ {\scriptstyle fT^Y \times \mathrm{id}}\downarrow & & \downarrow{\scriptstyle fT} \\ X'T^Y \times YT & \xrightarrow[\;c_{X',Y}\;]{} & X'T \end{array} \qquad \begin{array}{ccc} XT^{Y'} \times YT & \xrightarrow{\;XT^g \times \mathrm{id}\;} & XT^Y \times YT \\ {\scriptstyle \mathrm{id} \times gT}\downarrow & & \downarrow{\scriptstyle c_{X,Y}} \\ XT^{Y'} \times Y'T & \xrightarrow[\;c_{X,Y'}\;]{} & XT \end{array}$$

with fT, g on the appropriate side). Prove that there is a bijective correspondence between composition laws on T and natural transformations $\mu: TT \longrightarrow T$. (Hint: to define $X\mu$, inject XTT into $XT^{XT} \times XTT$ by using id_{XT} as the first coordinate and follow with $c_{X,XT}$; conversely, given $a: Y \longrightarrow XT$ and $b \in YT$ define $(a, b)c_{X,Y} = \langle b, aT.X\mu \rangle$.)

10. Show that any intersection of subtheories is a subtheory.
11. Show that $n\eta: n \longrightarrow n\beta$ is a bijection if n is finite.
12. The motivation for Definition 3.2 stressed $(-)^{\#}$ as more basic than $\circ$, and an alternate definition is easily given. An *algebraic theory in extension form* is $\mathbf{T} = (T, \eta, (-)^{\#})$ with T and η as in 3.2, and with $(-)^{\#}$ assigning to each $\alpha: A \longrightarrow BT$ an "extension" $\alpha^{\#}: AT \longrightarrow BT$ subject to the three axioms "$A\eta.\alpha^{\#} = \alpha$," 2.14, and 2.12. Show that extension form and clone form are coextensive via the passages 2.8 and $\alpha^{\#} = \mathrm{id}_{AT} \circ \alpha$. We have preferred clone form because the associative law for $\circ$ is more natural in appearance than 2.12.

4. The Algebras of a Theory

The example of (Ω, E)-algebras raises the question if an arbitrary algebraic theory $\mathbf{T}$ in a category $\mathscr{K}$ has algebras. Theorem 2.17 teaches us that if $\mathbf{T}$ is coextensive with its algebras, the way to describe them is as a category "of $\mathscr{K}$-objects with structure"; specifically, for each object A of $\mathscr{K}$ we should provide a set $\{\xi\}$ of $\mathbf{T}$-algebra structures ξ on A and, more important, we should define when a $\mathscr{K}$-morphism $f: A \longrightarrow B$ is a $\mathbf{T}$-homomorphism from the $\mathbf{T}$-algebra (A, ξ) to the $\mathbf{T}$-algebra (B, θ). We begin by reexamining (Ω, E)-algebras and discover how to describe them in the language of $\mathbf{T}, \eta, \circ, \mu$ which gives rise to the concept of a $\mathbf{T}$-algebra. The main result of this section is that the algebras of a finitary equational presentation are the same thing as the algebras of a finitary algebraic theory. Enroute, we introduce product algebras, subalgebras, and quotient algebras and prove the Birkhoff variety theorem.

4.1 Proposition. *Let (Ω, E) be a finitary equational presentation (1.10) and let $\mathbf{T} = (T, \eta, \circ, \mu)$ be the algebraic theory (3.2, 3.17, 3.18) corresponding to (Ω, E) as in (2.1, 2.9, 2.8, 3.13). Then*

(4.2) For each Ω-algebra (X, δ), (X, δ) satisfies E if and only if there exists a function $\xi: XT \longrightarrow X$, called the *structure map of* (X, δ), such that

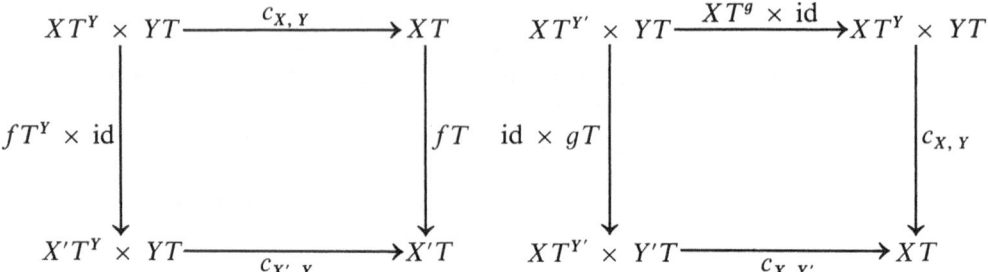

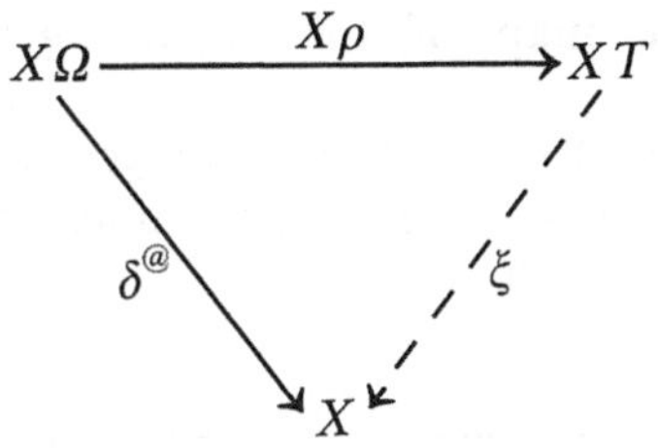

where $\delta^@$ is defined in 1.18 and $X\rho$ is defined in 2.1.

(4.3) If (X, δ) is an (Ω, E)-algebra then its structure map $\xi : XT \longrightarrow X$ is the unique Ω-homomorphism such that

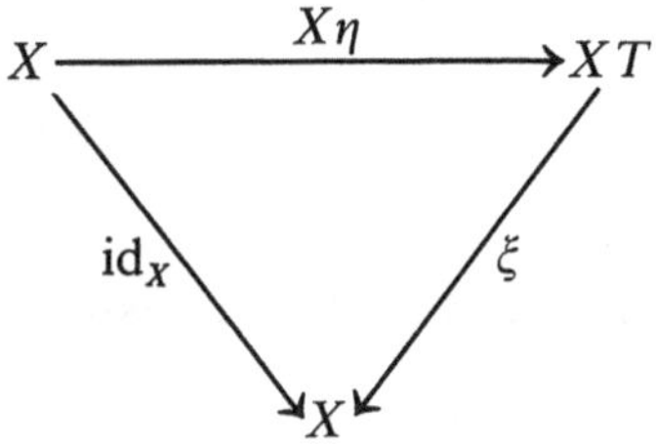

that is, $\xi = (\mathrm{id}_X)^\#$ (as in 2.5).

(4.4) For fixed $\omega \in \Omega_n$, define, for each set X, a function $X\hat{\omega}: X^n \longrightarrow XT$ by $\langle (x_1, \ldots, x_n), X\hat{\omega} \rangle = [x_1 \cdots x_n\omega]$. Then $\hat{\omega}$ is a natural transformation, that is for every function $f: X \longrightarrow Y$ we have

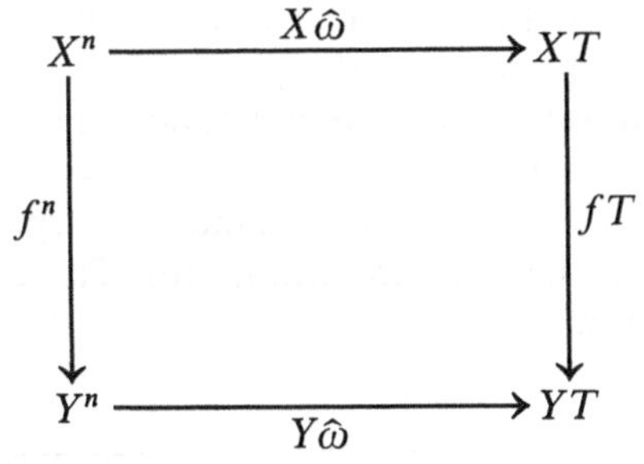

where f^n is defined in 1.4 and fT is defined by 3.10. The Ω-structure, δ, on X is recaptured by

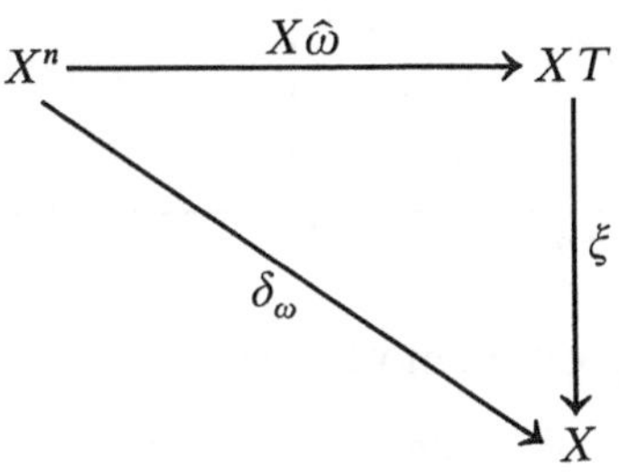

(4.5)

for all $\omega \in \Omega_n$.

(4.6) The structure map of the (Ω, E)-algebra XT (2.2) is $X\mu$: $XTT \longrightarrow XT$.

(4.7) If (X, δ) and (Y, γ) are (Ω, E)-algebras with structure maps $\xi : XT \longrightarrow X$ and $\theta : YT \longrightarrow Y$ then a function $f : X \longrightarrow Y$ is an Ω-homomorphism if and only if

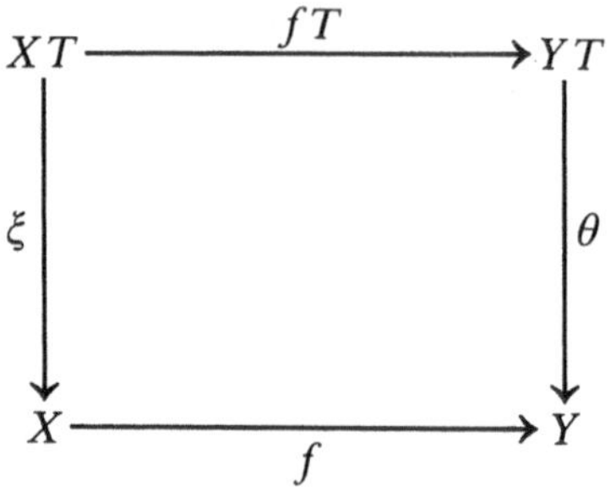

Proof. Let (X, δ) be an Ω-algebra and assume that ξ exists as in 4.2. For each $r : V \longrightarrow X$, $r^{\#} = r\Omega.\delta^{@}$ (1.19), so that if $\{e_1, e_2\} \in E$ we have $e_1 r^{\#} = \langle e_1, r\Omega.X\rho.\xi\rangle = \langle e_2, r\Omega.X\rho.\xi\rangle$ (2.4 with $g = V\eta.r\Omega$) $= e_2 r^{\#}$, and (X, δ) satisfies E. Conversely, if (X, δ) is an (Ω, E)-algebra then, using 2.5, define $\xi = (\mathrm{id}_X)^{\#}$ and observe that the diagram of 4.2 commutes because all are Ω-homomorphisms and the diagram clearly commutes on the variables $x \in X$. The naturality condition on $\hat{\omega}$ is clear from the remarks on fT in 3.10+. Expression 4.5 is checked by $\langle (x_1, \ldots, x_n), X\hat{\omega}.\xi\rangle = \langle [x_1 \cdots x_n\omega], \xi\rangle = \langle x_1 \cdots x_n\omega, \delta^{@}\rangle = (x_1, \ldots, x_n)\delta_\omega$, the last being the Definition (1.18) of $\delta^{@}$. Since $X\mu = \mathrm{id}_{XTT} \circ \mathrm{id}_{XT} = \mathrm{id}_{XTT}.(\mathrm{id}_{XT})^{\#} = (\mathrm{id}_{XT})^{\#}$, $X\mu$ is the structure map of XT. Let us turn to 4.7. For any function f, the diagram commutes restricted to variables, that is we have $X\eta.\xi.f = f = f.Y\eta.\theta = X\eta.fT.\theta$. For any function f, ξ, fT, and θ are always Ω-homomorphisms. Thus if f is also an Ω-homomorphism, the diagram commutes. The converse is clear from 4.4, 4.5, and the diagram

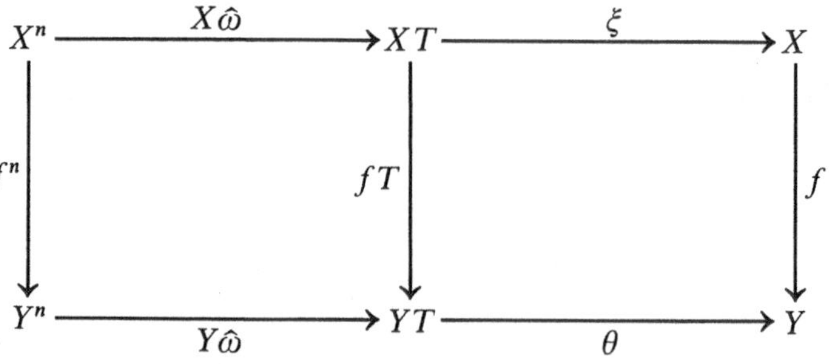

□

The preceding proposition motivates:

4.8 Definition. *Let $\mathscr{K}$ be an arbitrary category and let* $\mathbf{T}$ *be an algebraic theory in $\mathscr{K}$. A* $\mathbf{T}$*-algebra is a pair (X, ξ) where X is an object of $\mathscr{K}$ and*

$\xi : XT \longrightarrow X$ is a morphism in $\mathscr{K}$, called the structure map of (X, ξ), which satisfies 4.9 and 4.10 below.

$$\begin{array}{ccc} X & \xrightarrow{X\eta} & XT \\ & \searrow^{\mathrm{id}_X} & \downarrow \xi \\ & & X \end{array} \tag{4.9}$$

$$\begin{array}{ccc} XTT & \xrightarrow{\xi T} & XT \\ X\mu \downarrow & & \downarrow \xi \\ XT & \xrightarrow[\xi]{} & X \end{array} \tag{4.10}$$

(See Exercise 11 for an alternate axiomitization.)
*If (X, ξ) and (Y, θ) are **T**-algebras, a **T**-homomorphism from (X, ξ) to (Y, θ) is a $\mathscr{K}$-morphism $f : X \longrightarrow Y$* such that

$$\begin{array}{ccc} XT & \xrightarrow{fT} & YT \\ \xi \downarrow & & \downarrow \theta \\ X & \xrightarrow[f]{} & Y \end{array} \tag{4.11}$$

Because T is a functor, $\mathrm{id}_X : (X, \xi) \longrightarrow (X, \xi)$ is a **T**-homomorphism, and $f.g : (X, \xi) \longrightarrow (X'', \xi'')$ is a **T**-homomorphism so long as $f : (X, \xi) \longrightarrow (X', \xi')$ and $g : (X', \xi') \longrightarrow (X'', \xi'')$ are. This gives us a category $\mathscr{K}^{\mathbf{T}}$ of **T**-algebras and **T**-homomorphisms and a "forgetful $\mathscr{K}$-object" functor $U^{\mathbf{T}} : \mathscr{K}^{\mathbf{T}} \longrightarrow \mathscr{K}$.

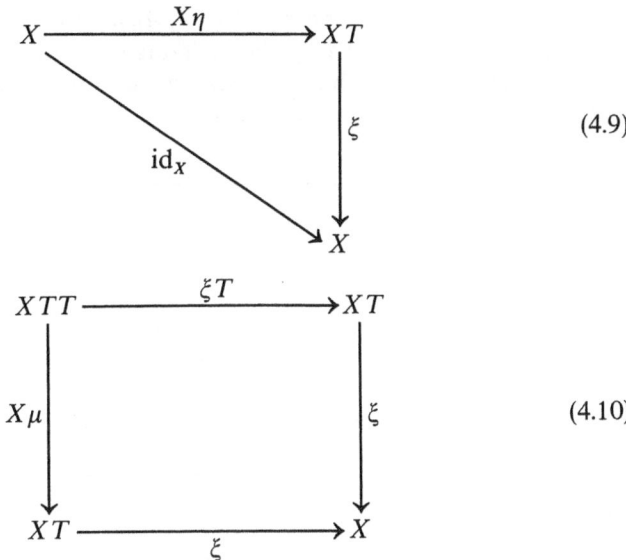

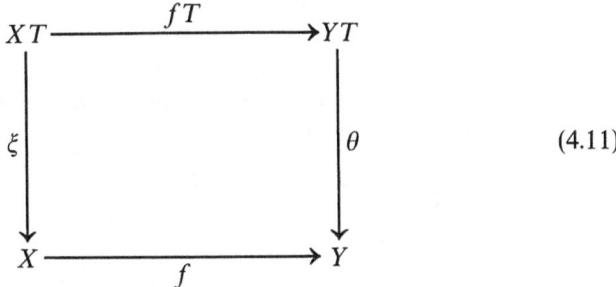

While all definitions in 4.8 were motivated by the considerations of 4.1, it is surprising that we do not have to say more. Let us examine the heuristics somewhat further. Expressions 4.9 and 4.10 represent the idea that "$\xi = (\mathrm{id}_X)^{\#}$." The role of 4.9 here is clear, and 4.10 is a special case of 4.11: "ξ is a **T**-homomorphism from $(XT, X\mu)$ to (X, ξ)." It is reasonable to want $X\mu$ to be the algebra structure of XT in view of 4.6, and it is consistent with our philosophy to assert so since $(XT, X\mu)$ is a **T**-algebra (two of the diagrams of 3.16). Even more striking is the fact that we can characterize algebraic simple recursion by a universal mapping property:

4.12 The Universal Property of $(AT, A\mu)$. Let $\mathbf{T}$ be an algebraic theory in a category $\mathscr{K}$, let (X, ξ) be a $\mathbf{T}$-algebra and let $f: A \longrightarrow X$ be a $\mathscr{K}$-morphism. Then there exists a unique $\mathbf{T}$-homomorphism $f^{\#}$: $(AT, A\mu) \longrightarrow (X, \xi)$ such that

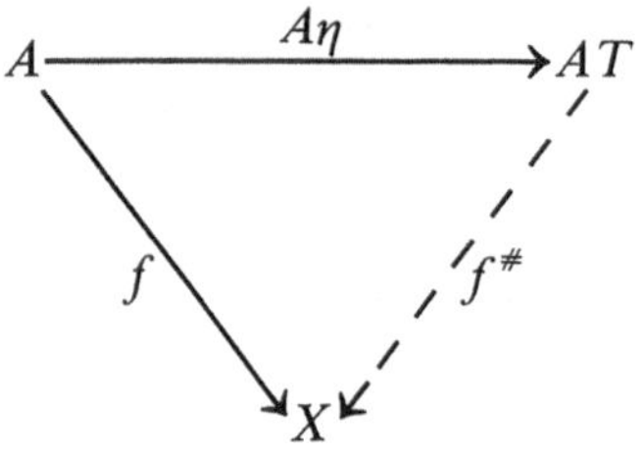

(cf. 2.5). Moreover, the formula for $f^{\#}$ is given by

$$f^{\#} = AT \xrightarrow{fT} XT \xrightarrow{\xi} X \tag{4.13}$$

(cf. $r^{\#} = r\Omega.\delta^{@}$ in 1.19).

Proof. We have already remarked that $(AT, A\mu)$ is a $\mathbf{T}$-algebra. fT: $(AT, A\mu) \longrightarrow (XT, X\mu)$ is a $\mathbf{T}$-homomorphism precisely because μ is natural (3.15). $\xi:(XT, X\mu) \longrightarrow (X, \xi)$ is a $\mathbf{T}$-homomorphism by (4.10). Therefore, $f^{\#}$ as in 4.13 is a $\mathbf{T}$-homomorphism. Because η is natural (3.11), we have $A\eta.f^{\#} = A\eta.fT.\xi = f.X\eta.\xi = f$ (using 4.9). It remains to show uniqueness, and this is where (for the first time) the law "$A\eta T.A\mu = \mathrm{id}_{AT}$" of 3.16 gets used. Suppose $g: AT \longrightarrow X$ satisfies $A\eta.g = f$ and $A\mu.g = gT.\xi$. Since T is a functor we have $A\eta T.gT = fT$. That $g = f^{\#}$ is now clear from the diagram

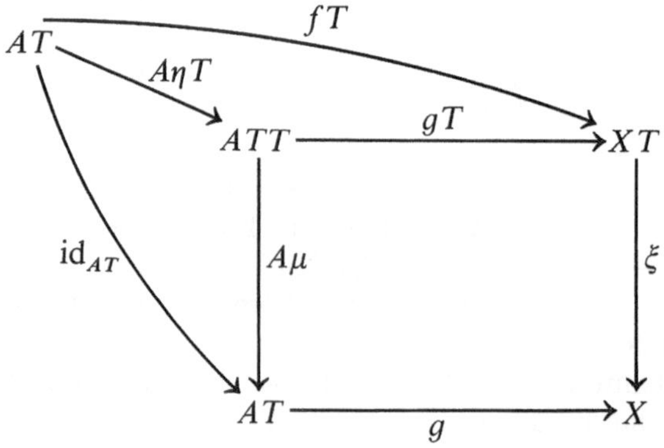

The proof is complete. □

We have now completely justified the motivation for 4.9 and 4.10: if (X, ξ) is a $\mathbf{T}$-algebra then $\xi = (\mathrm{id}_X)^{\#}$. If the reader has been keeping score, she will have noticed that the axiom "$(\mathrm{id}_A)T = \mathrm{id}_{AT}$" is the only axiom about

$\mathbf{T} = (T, \eta, \mu)$ that has not been interpreted. As it turns out, this axiom is crucial in proving that $\mathscr{K}^{\mathbf{T}}$ determines $\mathbf{T}$ (cf. 2.17). While we will not prove this result until Chapter 3 (3.2.11) the idea is basically as follows: Suppose (AT', μ') is a $\mathbf{T}$-algebra and $\eta': A \longrightarrow AT'$ is a $\mathscr{K}$-morphism such that $(AT', \mu'; \eta')$ also enjoys the same universal property as $(AT, A\mu; A\eta)$ as in 4.12. Consider the diagram:

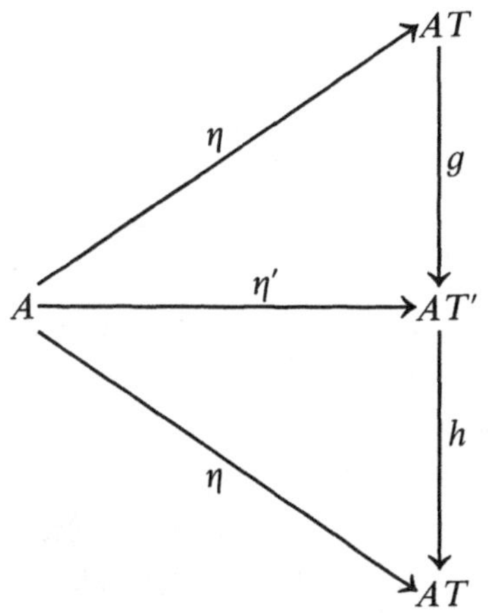

(4.14)

where g and h are the unique $\mathbf{T}$-homomorphisms making their respective triangles commute. Thus $g.h$ is the unique $\mathbf{T}$-homomorphism leaving η invariant. id_{AT} is a $\mathscr{K}$-morphism leaving η invariant. Because T preserves identities, id_{AT} is a $\mathbf{T}$-homomorphism, so must coincide with $g.h$. Symmetrically, $h.g = \mathrm{id}_{AT'}$. In most categories (such as $\mathscr{K} = \mathbf{Set}$) g and h are considered mutually inverse isomorphisms, that is $(AT, A\mu, A\eta)$ and (AT', μ', η') are as "abstractly equal" as any two such things could be. We will comment further on "isomorphisms in a category" in chapter 2 (2.1.4). Right now, we put the finishing touch on 4.1:

4.15 Theorem. *Let (Ω, E) be a finitary equational presentation and let* $\mathbf{T}$ *be its algebraic theory as in 4.1. Then* $\mathbf{Set}^{\mathbf{T}}$ *and the category of (Ω, E)-algebras are isomorphic categories of sets with structure, that is for each set X the passage from an (Ω, E)-structure δ to its structure map ξ as in 4.2 is bijective onto the* $\mathbf{T}$*-algebra structures on X, and for each function $f: X \longrightarrow Y$, (Ω, E)-structures δ and γ and corresponding structure maps ξ and θ, $f:(X, \delta) \longrightarrow (Y, \gamma)$ is an Ω-homomorphism if and only if $f:(X, \xi) \longrightarrow (Y, \theta)$ is a* $\mathbf{T}$*-homomorphism.*

Proof. Most of the work was done in 4.1. We have only to prove that if (X, ξ) is a $\mathbf{T}$-algebra then there exists an (Ω, E)-algebra (X, δ) whose structure map is ξ. Define $g: X\Omega \longrightarrow X$ by $g = X\rho.\xi$. On variables, $xg = \langle x, X\eta.\xi \rangle$ (see the definition of η in 2.5) $= x$ (by 4.9). Now consider the formula $p_1 \cdots p_n\omega$ in $X\Omega$. As in the discussion of 3.12+ the elements $[p_i]$ in XT may be thought of as variables in $XT\Omega$ giving rise to

$([p_1])\cdots([p_n])\omega$ in $XT\Omega$. Since $X\mu{:}XTT \longrightarrow XT = (\mathrm{id}_{XT})^{\#}$ we have $\langle[([p_1])\cdots([p_n])\omega], X\mu\rangle = [p_1\cdots p_n\omega]$. From 4.10 we have a commutative diagram

$$\begin{array}{ccccc}
XT\Omega & \xrightarrow{\;XT\rho\;} & XTT & \xrightarrow{\;X\mu\;} & XT \\
\downarrow{\scriptstyle \xi\Omega} & & \downarrow{\scriptstyle \xi T} & & \downarrow{\scriptstyle \xi} \\
X\Omega & \xrightarrow[\;X\rho\;]{} & XT & \xrightarrow[\;\xi\;]{} & X
\end{array}$$

(where the first square can be proved by the reasoning of 2.10, and is actually a special case of the diagram of 2.10). Putting it together:

$$\begin{aligned}
(p_1\cdots p_n\omega)g &= \langle[p_1\cdots p_n\omega], \xi\rangle \\
&= \langle([p_1])\cdots([p_n])\omega, XT\rho.X\mu.\xi\rangle \\
&= \langle([p_1])\cdots([p_n])\omega, \xi\Omega.X\rho.\xi\rangle \\
&= \langle[[p_1]\xi\cdots[p_n]\xi\omega], \xi\rangle \qquad \text{(by 1.17)} \\
&= ([p_1]\xi, \ldots, [p_n]\xi)\delta_\omega
\end{aligned}$$

where the Ω-algebra (X, δ) is defined by 4.5. Comparing these facts about g with 1.18, we have $g = \delta^{@}$. It is then immediately clear from the equation $X\rho.\xi = \delta^{@}$ and from 4.2 both that (X, δ) satisfies E and that its structure map is ξ. □

4.16 Example. Topological Transformation Groups. Let $\mathscr{K}$ be the category of topological spaces and continuous maps and let **T** be the algebraic theory corresponding to the topological group G as in 3.7; μ for this theory is described in 3.13+ A **T**-algebra is a topological space X together with a continuous map $\xi{:}X \times G \longrightarrow X$ subject to

$$\begin{aligned}
xe &= x && \text{(for all } x \in X) \\
x(g*h) &= (xg)h && \text{(for all } x \in X,\ g, h \in G)
\end{aligned}$$

where we have written xg for $(x, g)\xi$. This is a well-known mathematical structure known as a *topological transformation group with phase group G* [Gottschalk and Hedlund '55, Definition 1.01]. The **T**-homomorphisms are continuous maps f which are *equivariant*, i.e. $(xg)f = (xf)g$. This example is one of the rare instances where the **T**-description coincides with the traditional one.

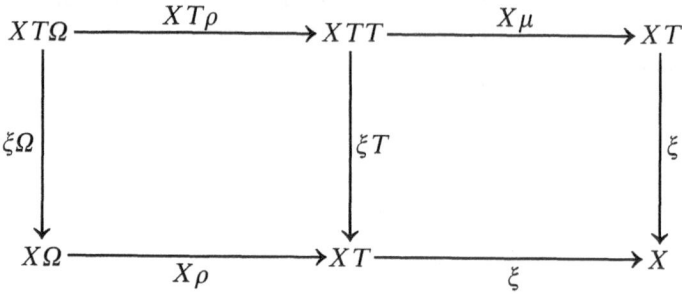

4.17 Example. Semigroups. A semigroup is a set X together with a binary operation $*{:}X \times X \longrightarrow X$ which is associative: $(x*y)*z = x*(y*z)$. For example, every monoid is, in part, a semigroup but if X has at least two elements and $x*y = x$ then $(X, *)$ is a semigroup which cannot be made into a monoid. Clearly, semigroups are the same thing as (Ω, E)-algebras with one binary operation symbol $*$ and one equation $\{v_1v_2{*}v_3{*}, v_1v_2v_3{*}{*}\}$. Let us describe the algebraic theory. For each set X, define XT to be the set of all words in X, as in 1.7. XT is a semigroup under the binary operation of *concatenation*:

$$(x_1 \cdots x_n, y_1 \cdots y_m) \longmapsto x_1 \cdots x_n y_1 \cdots y_m.$$

That concatenation is associative is obvious. Define $X\eta$ by $\langle x, X\eta \rangle = x$. If $f: X \longrightarrow (Y, *)$ is a function to the underlying set of a semigroup, $(x_1 \cdots x_n)f^{\#} = x_1 f * \cdots * x_n f$ is clearly the unique Ω-homomorphism extending f. By the uniqueness argument of 4.14 we have described the algebraic theory of (Ω, E); for example $\circ$ is defined by 2.8. It is becoming clearer what we meant by the "pivotal" role of the universal property in our remarks of 2.5+. The structure map ξ of the semigroup $(X, *)$ maps $x_1 \cdots x_n$ to $x_1 * \cdots * x_n$. Expression 4.5 amounts to the recovery: $x*y = (xy)\xi$. Conversely, let us start with a **T**-algebra (X, ξ), define $x*y = (xy)\xi$ and see how the associative law gets proved. Note first of all that $X\mu = (\mathrm{id}_{XT})^{\#}$ converts words of words to words by deleting parentheses; for example, the word $(x_1x_2)(y)(z_1z_2z_3)$ of length 3 in XTT is mapped to the word $x_1x_2yz_1z_2z_3$ of length 6 in XT. The essence of the associative law is that the word xyz can be broken up both into $(xy)(z)$ and into $(x)(yz)$. Thus, $(xyz)\xi = \langle (xy)(z), X\mu.\xi \rangle = \langle (xy)(z), \xi T.\xi \rangle = \langle (xy)\xi(z)\xi, \xi \rangle = (x*y)*z$. Similarly, using $(x)(yz)$, $(xyz)\xi = x*(y*z)$.

The reader must be curious as to the meaning of **T**-algebras for arbitrary **T** in the category of sets; in fact, for **T** as in 3.5, 3.19, or 3.21. By the time we have finished section 5, it will be clear that **T**-algebras are always (Ω, E)-algebras so long as Ω is not restricted to finitary operations. The technical convenience of the finitary restriction has been great. The uncoupling Lemma 1.11 and its many successive consequences are much more cumbersome with infinitary formulas and the reader would have perhaps been much confused if we had attempted this. Let us devote the rest of this section to isolating the "finitary" algebraic theories and proving that they are coextensive with finitary universal algebra.

4.18 Definition. *Let* **T** *be an algebraic theory in* **Set**. **T** *is finitary if for every set X and every element $\bar{x} \in XT$ there exists an integer $n \geqslant 0$, a function $r: \mathsf{V}_n \longrightarrow X$ (where V_n denotes the set of the first n variables (1.10), $\{v_1, \ldots, v_n\}$), and an element $p \in \mathsf{V}_nT$ such that $\langle p, rT \rangle = \bar{x}$.* Our reference to V_n provides the interpretation "**T** is finitary if formulas in XT have only finitely-many variables." It is an easy exercise, however, to show that any set with n elements can replace V_n.

4.19 Proposition. Let (Ω, E) be a finitary equational presentation and let **T** be the corresponding algebraic theory as in 4.1. Then **T** is finitary.

Proof. Let $[q] \in XT$. Let $\{x_1, \ldots, x_n\}$ be the finite set of variables occuring in q. Define the obvious bijection $s: \mathsf{V}_n \longrightarrow \{x_1, \ldots, x_n\}$, $v_is = x_i$, and define $p = \langle q, (s^{-1})\Omega \rangle$. Define $r: \mathsf{V}_n \longrightarrow X$ by $v_ir = x_i$. The proof is easily completed (and is also valid when $n = 0$). □

4.20 Example. The algebraic theory of 3.5 is not finitary. Given any $r: \mathsf{V}_n \longrightarrow X$, rT is the direct image map, as was pointed out in 3.10+. Since every subset of V_n is finite, the image of rT consists only of finite subsets of X and this will not exhaust XT if X is infinite. □

While we have postponed the definition of "isomorphism of theories" for Chapter 3, the uniqueness argument of 4.14 practically gives the definition away and certainly makes it clear that if two algebraic theories are so different that one is finitary and the other is not, then they cannot have the same categories of algebras. In particular, if **T** is not finitary (as in 4.20) its algebras cannot be presented using finitary operations and equations because of 4.19. Our terminology suggests that finitary theories do not have this problem; before proving this we establish a well-known theorem of Garrett Birkhoff.

4.21 Definition. *Let Ω be an operator domain as in 1.4. Given a family $((X_i, \delta_i): i \in I)$ of Ω-algebras, the cartesian product set $X = \prod X_i$ (see 2.1.5) admits a unique Ω-algebra structure δ such that for all $i \in I$ the projection pr_i: $(X, \delta) \longrightarrow (X_i, \delta_i)$ is an Ω-homomorphism, namely $((x_i^1), \ldots, (x_i^n))\delta_\omega = ((x_i^1, \ldots, x_i^n)(\delta_i)_\omega : i \in I)$. (X, δ) is called the cartesian product algebra of the algebras (X_i, δ_i).* For example, if $(X, *)$ and $(Y, \$)$ are semigroups, the binary operation on the product $X \times Y$ is $((x^1, y^1), (x^2, y^2)) \longmapsto (x^1 * x^2, y^1 \$ y^2)$. In case I is empty, the cartesian product algebra is the one-element set 1 provided with its unique Ω-algebra structure. Let (X, δ) be an Ω-algebra. *A subset A of X is a subalgebra of (X, δ) if A is closed under the operations of X, that is for each $\omega \in \Omega_n$, $\delta_\omega : X^n \longrightarrow X$ maps A^n into A. If A is a subalgebra of (X, δ) there exists a unique algebra structure δ_0 on A such that the inclusion map $(A, \delta_0) \hookrightarrow (X, \delta)$ is an Ω-homomorphism; A qua algebra will still be called a subalgebra.* For example, if (X, δ) is a group (with respect to either of 1.1 or 1.2) then its subalgebras are more usually called subgroups. *An Ω-algebra (Y, γ) is a quotient algebra of (X, δ) if there exists a surjective Ω-homomorphism of (X, δ) onto (Y, γ).* For example, the two element group is a quotient group of the group of integers. A bijective Ω-homomorphism $f: (X, \delta) \longrightarrow (Y, \gamma)$ is an *isomorphism* (note: f^{-1} is also one) and such (X, δ) and (Y, γ) are said to be *isomorphic*. Isomorphic algebras are "abstractly the same" (see 2.1.4 and exercise 10). By 2.17, since "product," "subalgebra," "quotient algebra," and "isomorphic" are described in the language of homomorphisms, these concepts do not depend on the presentation (Ω, E).

4.22 Birkhoff Variety Theorem. *Let Ω be an operator domain and let $\mathscr{A}$ be a class of Ω-algebras. Then a necessary and sufficient condition that $\mathscr{A}$ is the class of (Ω, E)-algebras for some set E of equations is that $\mathscr{A}$ is closed under products subalgebras and quotients (that is, the product algebra of any family of algebras in $\mathscr{A}$ is again in $\mathscr{A}$ and whenever (X, δ) is in $\mathscr{A}$ so are all its subalgebras and quotient algebras). In either case, $\mathscr{A}$ is said to be a variety of Ω-algebras.*

Proof. To prove that the conditions are necessary, examine the diagrams:

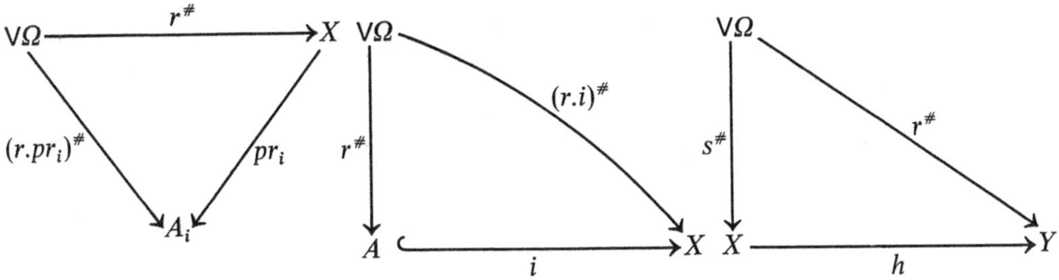

and use the following principles: To prove that two elements of the product X are equal, it is necessary and sufficient to prove this followed by each projection (perfectly true even when the index set is empty). To prove that two elements of A are equal it is necessary and sufficient to prove that they are identified by the inclusion map i. Because h is onto we may invoke the axiom of choice (section 2, exercise 1) to choose $s{:}\mathsf{V} \longrightarrow X$ such that $s.h = r$. Let us turn to the proof of sufficiency. Let E be the set of all equations $\{e_1, e_2\} \subset \mathsf{V}\Omega$ such that for every algebra (X, δ) in $\mathscr{A}$ and every function $r{:}\mathsf{V} \longrightarrow X$ it is the case that $r^\#$ identifies e_1 and e_2. Trivially, all algebras in $\mathscr{A}$ satisfy E. Let (X, δ) be an arbitrary (Ω, E)-algebra. We will show that (X, δ) is a quotient of a subalgebra of a product of elements of $\mathscr{A}$. If $\mathbf{T}$ is the algebraic theory of (Ω, E) as in 4.1, the structure map of (X, δ) is an Ω-homomorphism onto, so it suffices to show that XT is isomorphic to a subalgebra of a product of elements of $\mathscr{A}$. To do this we resort to a standard argument and show that XT admits enough homomorphisms to elements of $\mathscr{A}$ to separate points, that is:

(4.23) If $[p] \neq [q] \in XT$ then there exists an algebra (A, γ) in $\mathscr{A}$ and a map $r{:}X \longrightarrow A$ such that $[p]r^\# \neq [q]r^\#$.

To prove 4.23, let $X_n = \{x_1, \ldots, x_n\}$ be the finite set of all variables in X occurring in either p or q and let $a{:}\mathsf{V}_n \longrightarrow X_n$ be the obvious bijection, $v_i a = x_i$. Then there exist formulas $p', q' \in \mathsf{V}_n\Omega$ with $\langle p', a\Omega \rangle = p$ and $\langle q', a\Omega \rangle = q$ (cf. the proof of 4.19). Reasoning as in 2.10, we have the commutative diagram

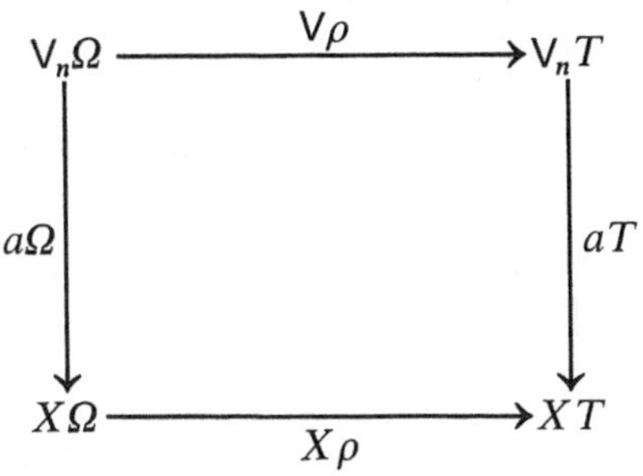

so that, in particular, $[p'] \neq [q']$. By the definition of E there exists (A, γ) in $\mathscr{A}$ and $s{:}\mathsf{V}_n \longrightarrow A$ such that $s^\#{:}\mathsf{V}_n\Omega \longrightarrow (A, \gamma)$ distinguishes p' and q'. Since A cannot be empty, $a^{-1}.s$ extends (in many ways perhaps) to a function r

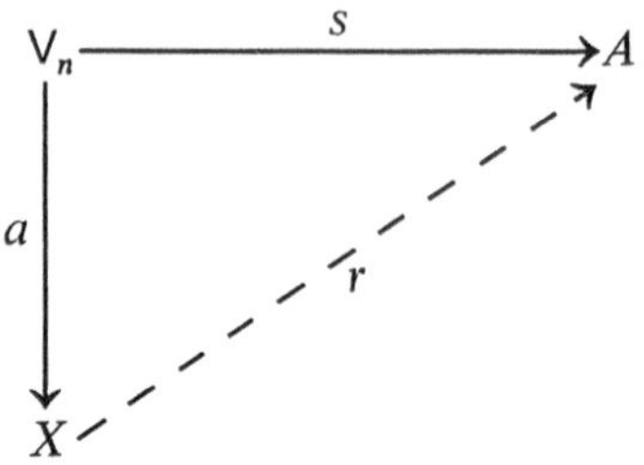

Since $[p]r^{\#} = \langle p', s^{\#} \rangle$ the proof of 4.23 is complete. The rest of the proof is based on very general principles. For each pair (t, u) of distinct elements of XT choose $A_{t,u}$ in $\mathscr{A}$ and a homomorphism $r_{t,u}: XT \longrightarrow A_{t,u}$ which maps t and u to different values. Let A be the product algebra of $(A_{t,u}: t \neq u \in XT)$ and define a single homomorphism $r: XT \longrightarrow A$ by $([p]r)_{t,u} = [p]r_{t,u}$. By construction, r is injective. The proof is complete by the following standard fact which we leave to the reader for verification.

(4.24) If $f:(X, \delta) \longrightarrow (Y, \gamma)$ is an injective Ω-homomorphism then its image $Xf \subset Y$ is a subalgebra of (Y, γ) and the map $g: X \longrightarrow Xf$, $xg = xf$ is an isomorphism. (For a hint, see 4.32.) □

We have already seen that the passage from a finitary equational presentation to its theory is a well-defined injection from equivalence classes of presentations as in 2.18 into isomorphism classes of finitary theories ("isomorphism" being informally defined by 4.14). We conclude this section with a proof that this passage is bijective so that "finitary universal algebra is the study of finitary algebraic theories in **Set**."

4.25 Theorem. *Let* **T** *be a finitary algebraic theory in* **Set**. *Then there exists an equational presentation* (Ω, E) *such that* **T**-*algebras and* (Ω, E)-*algebras are isomorphic as categories of sets with structure (as defined in 4.15).*

Proof. Define $\Omega_n = \{n\} \times \mathsf{V}_n T$ where V_n is as defined in 4.18. The first coordinates assure that $(\Omega_n : n = 0, 1, \ldots)$ is a disjoint sequence of sets as required by 1.4; for convenience we will drop the "n" from the notation, however. Fix $\omega \in \Omega_n$. Define a map $X\hat{\omega}: X^n \longrightarrow XT$ for each set X by $\langle (x_1, \ldots, x_n), X\hat{\omega} \rangle = \langle \omega, rT: \mathsf{V}_n T \longrightarrow XT \rangle$ where $r: \mathsf{V}_n \longrightarrow X$ is defined by $v_i r = x_i$. Then

(4.26) $\hat{\omega}$ is a natural transformation; that is for every function $f: X \longrightarrow Y$ we have

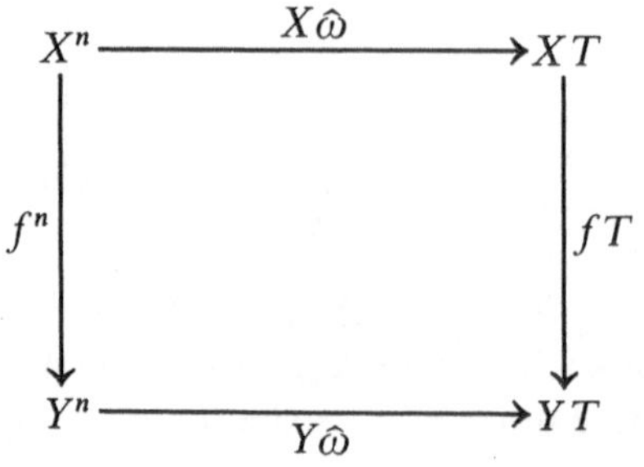

The proof of 4.26 is an immediate consequence of the fact that $(r.f)T = rT.fT$. As might have been expected, we assign an Ω-algebra structure δ to the **T**-algebra (X, ξ) by setting $\delta_\omega = X\hat{\omega}.\xi$. It is immediately clear from 4.26 and a glance at the diagram

$$\begin{array}{ccccc} X^n & \xrightarrow{X\hat{\omega}} & XT & \xrightarrow{\xi} & X \\ \downarrow{\scriptstyle f^n} & & \downarrow{\scriptstyle fT} & & \downarrow{\scriptstyle f} \\ Y^n & \xrightarrow[Y\hat{\omega}]{} & YT & \xrightarrow[\theta]{} & Y \end{array}$$

that a **T**-homomorphism is an Ω-homomorphism. To prove, conversely, that an Ω-homomorphism is a **T**-homomorphism it is sufficient to prove that every element $\bar{x} \in XT$ is in the image of $X\hat{\omega}$ for some ω, that is, there must exist some integer n, some element $\omega \in \mathsf{V}_n T$, and some function $r: \mathsf{V}_n \longrightarrow X$ such that $\langle \omega, rT \rangle = \bar{x}$; but this is precisely the definition "**T** is finitary." In particular, consideration of $f = \mathrm{id}_X$ and the fact that T is a functor allows us to see that if $\xi \neq \xi'$ then $\delta \neq \delta'$. Let $\mathscr{A}$ be the class of all Ω-algebras which arise from **T**-algebras as above. It is clear that, to finish the proof, it is sufficient to find a set E of equations such that $\mathscr{A} =$ all (Ω, E)-algebras. By 4.22 we need only show that $\mathscr{A}$ is closed under products, subalgebras, and quotients. We will give particular attention to the verification since it gives us our first encounter with "universal algebra in the language of **T**-algebras."

(4.27) Let (X_i, ξ_i) be a family of **T**-algebras, and let $X = \prod X_i$ be the cartesian product set. Then there exists a unique $\xi: XT \longrightarrow X$ such that (X, ξ) is a **T**-algebra and each projection $pr_i:(X, \xi) \longrightarrow (X_i, \xi_i)$ is a **T**-homomorphism. Not surprisingly, (X, ξ) is called the *cartesian product* **T**-algebra.

To prove 4.27, observe that there exists a unique function $\xi: XT \longrightarrow X$ such that

$$\begin{array}{ccc} XT & \xrightarrow{(pr_i)T} & X_iT \\ \downarrow{\scriptstyle \xi} & & \downarrow{\scriptstyle \xi_i} \\ X & \xrightarrow[pr_i]{} & X_i \end{array}$$

is commutative for all i, namely $(\langle \bar{x}, \xi \rangle)_i = \langle \bar{x}, (pr_i)T.\xi_i \rangle$. As usual, this is consistent with the case that the set of indices i is empty. We must show that

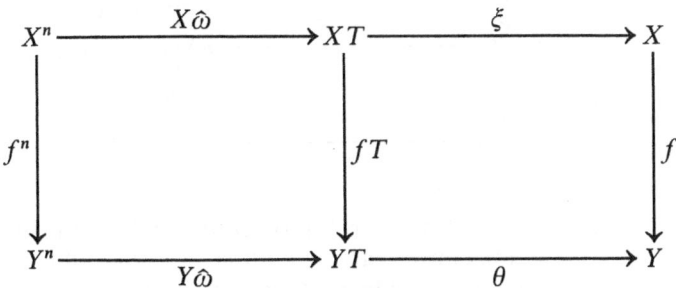

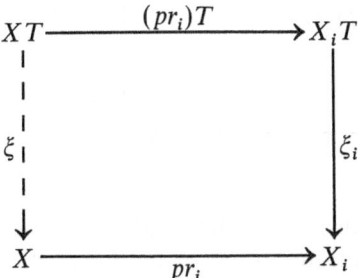

(X, ξ) is a **T**-algebra. Consider

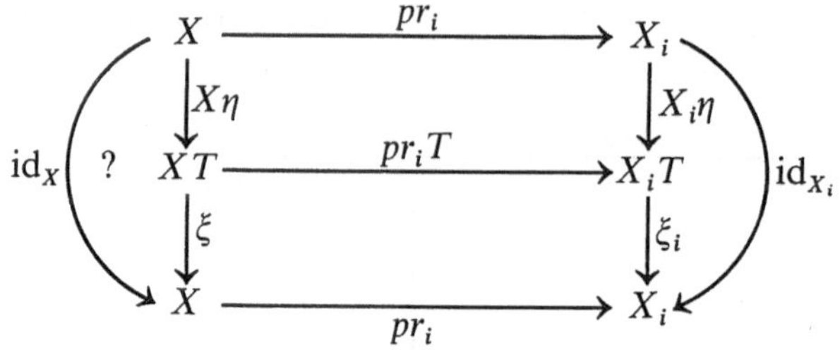

Everything commutes except perhaps (?). But now apply the principle that to prove that two functions into X are equal it is necessary and sufficient to prove that they are equal followed by each projection; which is exactly what we know. The other algebra law is proved by the same reasoning:

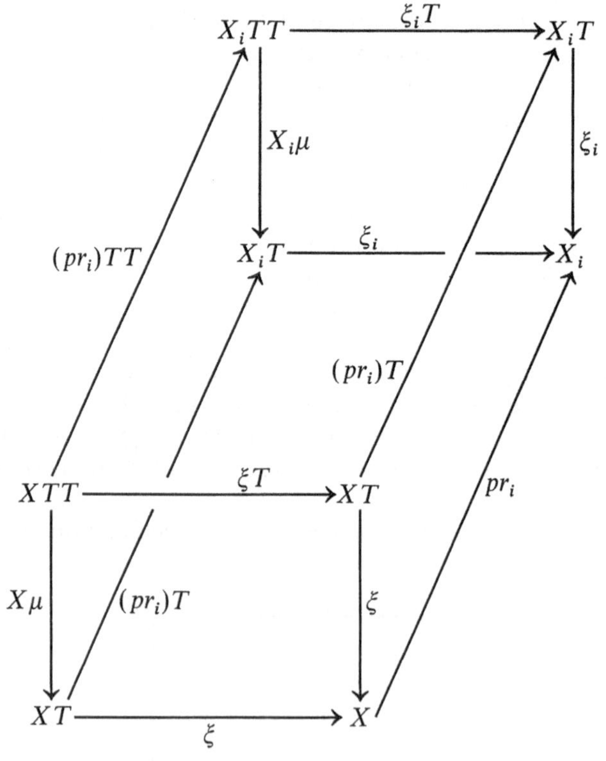

This finishes 4.27 and also establishes that $\mathscr{A}$ is closed under products, since both products are the unique structure making projections homomorphisms.

(4.28) Let (X, ξ) be a **T**-algebra and let A be a subset of X with inclusion map $i: A \hookrightarrow X$. Say that A is a **T**-*subalgebra of* (X, ξ) if there exists a factorization:

$$\begin{array}{ccc} AT & \xrightarrow{\quad iT \quad} & XT \\ {\scriptstyle \xi_0}\Big\downarrow & & \Big\downarrow{\scriptstyle \xi} \\ A & \xrightarrow[\quad i \quad]{} & X \end{array}$$

In common with the family (pr_i) of 4.27, $i:A \hookrightarrow X$ has the virtue that to prove two elements of A are the same it is necessary and sufficient to prove this followed by i. The same reasoning as in 4.27 guarantees, then, that (A, ξ_0) is a $\mathbf{T}$-algebra, so that A is a $\mathbf{T}$-subalgebra of (X, ξ) if and only if there exists a $\mathbf{T}$-algebra structure on A making i a $\mathbf{T}$-homomorphism. To prove that $\mathscr{A}$ is closed under subalgebras it is still necessary to show that an Ω-subalgebra of (X, ξ) is a $\mathbf{T}$-subalgebra. Let A be an Ω-subalgebra. For $\bar{a} \in AT$ there exists $r:\mathsf{V}_n \longrightarrow A$ and $\omega \in \mathsf{V}_nT$ with $\langle \omega, rT\rangle = \bar{a}$. For $1 \leqslant j \leqslant n$ set $a_j = v_jr$. By hypothesis, $\delta_\omega: X^n \longrightarrow X$ maps A^n into A. Therefore, $\langle \bar{a}, iT.\xi\rangle = \langle \omega, rT.iT.\xi\rangle = \langle \omega, (r.i)T.\xi\rangle = (a_1, \ldots, a_n)\delta_\omega \in A$. As $\bar{a} \in AT$ is arbitrary, the proof that $\mathscr{A}$ is closed under subalgebras is complete.

(4.29) If H: **Set** $\longrightarrow$ **Set** is any functor and if $f:X \longrightarrow Y$ is surjective than $fH:XH \longrightarrow YH$ is also surjective. This is a consequence of the axiom of choice. Let $d: Y \longrightarrow X$ be a choice function such that $d.f = \mathrm{id}_Y$. As H is functorial, $dH.fH = \mathrm{id}_{YH}$. It follows immediately that fH is surjective, since if $\bar{y} \in YH$ then $\langle \bar{x}, fH\rangle = \bar{y}$ if $x = \langle \bar{y}, dH\rangle$.

(4.30) Let (X, ξ) be a $\mathbf{T}$-algebra. A surjection $f:X \longrightarrow Y$ is a $\mathbf{T}$-*quotient algebra of* (X, ξ) if there exists a factorization θ

$$\begin{array}{ccc} XT & \xrightarrow{\quad fT \quad} & YT \\ {\scriptstyle \xi}\Big\downarrow & & \Big\downarrow{\scriptstyle \theta} \\ X & \xrightarrow[\quad f \quad]{} & Y \end{array}$$

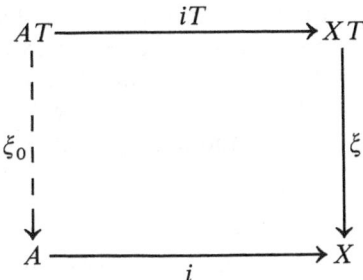

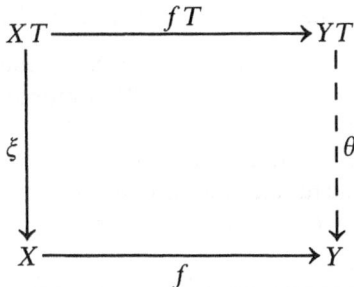

This definition is reasonable precisely because both fT and fTT are surjective and surjective maps $g:A \longrightarrow B$ have the property that to prove h, $h':B \rightrightarrows C$ are equal it suffices to check that $g.h = g.h'$. Thus, for example, if θ exists it is unique. Moreover, the algebra laws are quite clear from essentially the same two diagrams used in 4.27 (substitute Y for X_i, f for pr_i, and θ for ξ_i). Therefore f is a $\mathbf{T}$-quotient algebra of (X, ξ) if and only if there is a necessarily unique $\mathbf{T}$-algebra structure θ on Y making f a $\mathbf{T}$-homomorphism. To show that $\mathscr{A}$ is closed under quotient algebras it suffices to show that if f is an Ω-quotient algebra via the Ω-structure γ, then f is a **T**-quotient algebra. To this end we invoke the axiom of choice to get a choice function $d: Y \longrightarrow X$ with $d.f = \mathrm{id}_Y$. Define $\theta: YT \longrightarrow Y = (dT: YT \longrightarrow XT).(\xi: XT \longrightarrow X).(f: X \longrightarrow Y)$. We will show $fT.\theta = \xi.f$. Let $x \in XT$. There exists $r: \mathsf{V}_n \longrightarrow X$ and $\omega \in \mathsf{V}_nT$ with $\langle \omega, rT \rangle = \bar{x}$. Then

$$\begin{aligned}
\langle \bar{x}, \xi.f \rangle &= \\
\langle \omega, rT.\xi.f \rangle &= \\
(v_1r, \ldots, v_nr)\delta_\omega f &= \qquad \text{(definition of } \delta_\omega) \\
(v_1rf, \ldots, v_nrf)\gamma_\omega &= \\
(v_1rf.df, \ldots, v_nrf.df)\gamma_\omega &= \\
(v_1rfd, \ldots, v_nrfd)\delta_\omega f &= \\
\langle \omega, (rfd)T.\xi.f \rangle &= \\
\langle \bar{x}, fT.dT.\xi.f \rangle &
\end{aligned}$$

The proof of 4.25 is complete. □

For use in the next section, we prove a further result about **T**-subalgebras. One expects that each subset A of a **T**-algebra (X, ξ) generates a **T**-subalgebra $\langle A \rangle$ of (X, ξ) by "closing up A under the **T**-operations." For example, if **T** is as in 4.17, $\langle A \rangle = \{a_1 \cdots a_n : a_1, \ldots, a_n \in A\}$. In general, one expects to consider those "terms" in XT which "have variables in A," that is, are in the image of $iT: AT \longrightarrow XT$ for $i: A \longrightarrow X$ the inclusion map, and then define $\langle A \rangle$ to be the image in X of $iT.\xi: AT \longrightarrow X$. The following theorem shows that this works.

4.31 Theorem on Generated Subalgebras. *Let* **T** *be an arbitrary algebraic theory of sets, let* (X, ξ) *be a* **T***-algebra and for each subset* A *of* X *with inclusion map* $i: A \longrightarrow X$ *define* $\langle A \rangle \subset X$ *to be the image of* $iT.\xi$. *Then* $A \subset \langle A \rangle$, *if* $A \subset B$ *then* $\langle A \rangle \subset \langle B \rangle$, *and* $\langle\langle A \rangle\rangle = \langle A \rangle$. $\langle A \rangle$ *is a subalgebra of* (X, ξ) *and is contained in any other subalgebra of* (X, ξ) *which contains* A.

Proof. Let us first record another expected general fact:

(4.32) If $f: (X, \xi) \longrightarrow (Y, \theta)$ is a **T**-homomorphism, then the image of f is a subalgebra of (Y, θ).

To see why, let I be the image of f with inclusion map $i: I \longrightarrow Y$ and let $p: X \longrightarrow I$ be the unique function with $f = p.i$, that is $xp = xf$. As p is onto, there exists $d: I \longrightarrow X$ with $d.p = \mathrm{id}_I$ by the axiom of choice. Define γ:

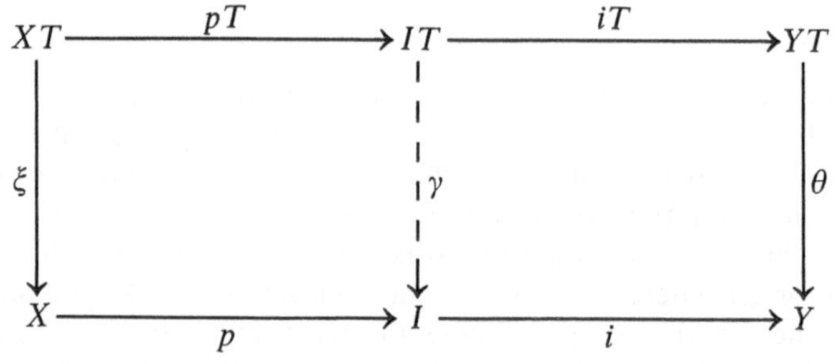

$IT \longrightarrow I$ by $\gamma = dT.\xi.p$. Then $pT.\gamma.i = pT.dT.\xi.p.i = pT.dT.pT.iT.\theta = pT.iT.\theta$. As pT is surjective (by 4.29), $\gamma.i = iT.\theta$ as desired.

Now we return to the proof of 4.31 proper. The diagram

$$\begin{array}{ccc} AT & \xrightarrow{\quad iT \quad} & XT \\ {\scriptstyle A\eta}\Big\uparrow & & \Big\downarrow{\scriptstyle \xi} \\ A & \xrightarrow[\quad i \quad]{} & X \end{array}$$

is commutative because η is natural and $X\eta.\xi = \mathrm{id}_X$. This proves that $A \subset \langle A \rangle$. It is obvious that $\langle A \rangle \subset \langle B \rangle$ whenever $A \subset B$. Since $iT.\xi$: $(AT, A\mu) \longrightarrow (X, \xi)$ is a **T**-homomorphism, $\langle A \rangle$ is a subalgebra of (X, ξ) by 4.32. From the definition of "subalgebra" it is obvious that if B is a subalgebra, $B = \langle B \rangle$. Therefore $\langle\langle A \rangle\rangle = \langle A \rangle$ and whenever A is contained in the subalgebra B, $\langle A \rangle \subset \langle B \rangle = B$. □

Notes for Section 4

Algebras of an algebraic theory in monoid form were defined by [Eilenberg and Moore '65, (2.6)]. While they recognized that groups arise as $\mathbf{Set}^{\mathbf{T}}$, their main example is $AT = A \otimes \Lambda$ (where $\mathscr{K}$ is the category of modules over the commutative ring R and Λ is an R-algebra) whose algebras are the Λ-modules.

It was Jon Beck who first perceived that "triples" describe universal algebra in the category of sets. The atmosphere at that time is best conveyed by quoting two paragraphs from Beck's thesis [Beck '67, pages 72–73]. The quote is verbatim (except that our reference numbers have been used), and immediately follows a discussion of groups in the style of 4.17.

> The example of groups is typical. It is known that all *algebraic categories* in the sense of [Lawvere '63] are tripleable over sets, with respect to their usual underlying set functors. [Linton '66] has shown that over sets this is almost the whole story: admitting infinitary operations one gets *equational categories* of algebras, and over the base category of sets tripleableness is equivalent to equationality.
>
> Over other base categories, tripleableness does not seem to have any such standard interpretations. It is the proposal of this paper that tripleableness be regarded as a new type of mathematical structure, such as algebraic, equational, topological, ordered, ...

In the above, "tripleable over sets" means "of the form $\mathbf{Set}^{\mathbf{T}}$." The first published proof that triples capture equational classes is the "isomorphism theorem" of [Linton '69, pages 36–50]. This book, with the theorems culminating in 5.40 and 5.45, offers the first expository proof of these results. See also [Felscher '72, 4.1].

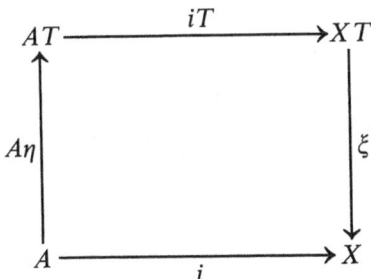

The Birkhoff variety theorem was proved in [Birkhoff '35, theorems 6, 9]. The ideas in 4.21–4.31 are interesting in their own right; for a different proof of 4.25 see 5.40 below. A much more general proof of 4.32 will be given in 3.4.17.

Exercises for Section 4

1. Starting from the point of view that the structure of an (Ω, E)-algebra (X, δ) can be described by a function $\xi: XT \longrightarrow X$ (e.g. $x*y = (xy)\xi$ in 4.17), expand the heuristics of 4.11 + into a motivation for the definition of "algebraic theory in monoid form."
2. In the proof of 4.15 we did not explicitly show that, for a fixed set X, the passage from δ to ξ is injective. Show that this follows from 4.7.
3. A *semilattice* is a partially ordered set in which every pair of elements has a supremum. Let $\mathbf{T}$ be the algebraic theory of nonempty finite subsets (cf. exercises 7, 10 of section 3). Show that $\mathbf{Set}^{\mathbf{T}}$ may be identified with the category of semilattices and functions which preserve binary suprema. [Hint: the structure map is "supremum."]
4. Prove that the double power-set theory of 3.19 is not finitary.
5. Why is "groups" not a variety in "monoids"?
6. Show that a subsemigroup of a group need not be a subgroup even if it is a group. (Hint: the units are different!)

   For the following three exercises (implicit in [Birkhoff '35, page 141]) fix an algebraic theory $\mathbf{T}$ in $\mathbf{Set}$. A *variety in* $\mathbf{Set}^{\mathbf{T}}$ is a collection of $\mathbf{T}$-algebras closed under the formation of products, subalgebras, and quotients.
7. For any collection $\mathscr{X}$ of $\mathbf{T}$-algebras, show that the class $\mathrm{Var}(\mathscr{X})$ of all quotients of subalgebras of products of elements of $\mathscr{X}$ is a variety and is the smallest variety containing $\mathscr{X}$.
8. Given $(X, \xi; A)$ where (X, ξ) is a $\mathbf{T}$-algebra and A is a subset of X such that $\langle A \rangle = X$, define $\mathrm{Var}[X, \xi; A]$ to be the class of all $\mathbf{T}$-algebras (Y, θ) such that every function $f: A \longrightarrow Y$ extends to a unique $\mathbf{T}$-homomorphism $f^{\#}: (X, \xi) \longrightarrow (Y, \theta)$. Show that $\mathrm{Var}[X, \xi; A]$ is a variety.
9. Let $\mathscr{X}$ be any collection of $\mathbf{T}$-algebras and let A be a fixed set. Define $\mathscr{F}$ to be the set of all $(X, \xi; f)$ such that $(X, \xi) \in \mathscr{X}$ and $f: A \longrightarrow X$. Consider the product $\mathbf{T}$-algebra P and function δ defined by

$$\begin{array}{ccc} A & \xrightarrow{\quad\delta\quad} & P = \prod((X, \xi):(X, \xi; f) \in \mathscr{F}) \\ & {}_{f}\searrow \qquad \swarrow_{pr_{(X, \xi; f)}} & \\ & X & \end{array}$$

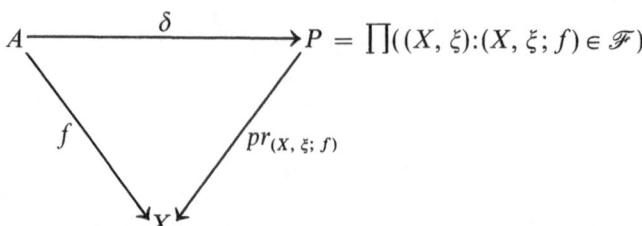

   Define AT' to be the subalgebra of P generated by the image of δ. Show that T' extends to an algebraic theory and that $\mathbf{Set}^{\mathbf{T}'}$ may be identified with $\mathrm{Var}(\mathscr{X})$. Conclude that if $\mathscr{X}$ consists of finitely many finite algebras then, if A is finite then AT' is finite. (Hint: show that AT' has a universal property in Var($\mathscr{X}$) and proceed as in 2.7.) This generalizes [Birkhoff '35, Theorem 11]. $\mathscr{X}$ must be a "small set" in order that P be definable; see the "primer on set theory" at the end of this chapter.

10. This exercise should help the reader to appreciate why isomorphic algebras are "abstractly the same."
    (a) There are four monoid structures on the two-element set $\{x, y\}$ whose multiplication tables are shown below:

	xx	xy	yx	yy
1	x	y	y	x
2	x	y	y	y
3	y	x	x	y
4	x	x	x	y

    Show that 1 and 3 are isomorphic and that 2 and 4 are isomorphic, but that no other two are isomorphic.
    (b) Prove that "isomorphism" is an equivalence relation.
    (c) (Cf. 2.3.1 below.) If (Y, γ) is an (Ω, E)-algebra and if $f:X \longrightarrow Y$ is a bijection, prove that there exists unique δ such that $f:(X, \delta) \longrightarrow (Y, \gamma)$ is an Ω-homomorphism and then that (X, δ) is an (Ω, E)-algebra and that f is an isomorphism.
    (d) Isomorphic structures "enjoy the same properties." Verify this for groups with respect to the following properties: "possesses three normal subgroups"; "has no elements of finite order"; "admits a surjective homomorphism from the group of integers."
11. Let $\mathbf{T}$ be an algebraic theory in $\mathscr{K}$. Show that the following axioms on $\xi:XT \longrightarrow X$ (suitable for theories presented in extension form as in exercise 3.12) are equivalent to 4.9 and 4.10.
    Axiom 1. $X\eta.\xi = \mathrm{id}_X$ (same as 4.9).
    Axiom 2. For all $\alpha, \beta:A \longrightarrow XT$, if $\alpha.\xi = \beta.\xi$ then $\alpha^{\#}.\xi = \beta^{\#}.\xi:AT \longrightarrow X$.
    This version of the algebra axioms is sometimes more useful than the original one in discovering what the $\mathbf{T}$-algebras are because T need not be iterated and because, when $\mathbf{T}$ is finitary, A can be assumed finite (see exercise 5.21.)
12. In any category $\mathscr{K}$, let **id** denote the theory $AT = A$, $A\eta = \mathrm{id}_A$, $f \circ g = f.g$. Show that $U^{\mathbf{id}}:\mathscr{K}^{\mathbf{id}} \longrightarrow \mathscr{K}$ may be identified with the identity functor of $\mathscr{K}$.
13. Using the diagram

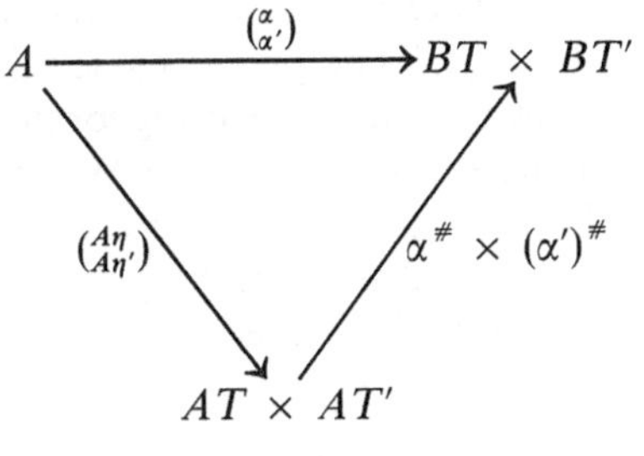

as a hint, define the product of two algebraic theories in **Set** and prove that it is again a theory. If $\mathbf{T} = \mathbf{id} \times \mathbf{id}$ (see exercise 12) show that $\mathbf{Set}^{\mathbf{T}}$ is the equationally-definable class corresponding to one binary operation and the equations $aa = a$, $(ab)(cd) = ad$.

14. [Jónsson and Tarski '61]. Let **T** be a theory in **Set** admitting a finite algebra with at least two elements. Let A be a finite set.
   (a) If **T** corresponds to "groups," show that there exists finite $B \subset AT$ such that $\langle B \rangle = AT$ and $A \cap B = \varnothing$.
   (b) If $B \subset AT$ and $\langle B \rangle = AT$, prove that $\text{card}(A) \leqslant \text{card}(B)$. [Hint: let (X, ξ) be a finite algebra with at least two elements; the map $X^A \longrightarrow X^B$ sending f to the restriction of $f^\#$ is injective.]

5. Infinitary Theories

In this section we restrict our attention, once again, to algebraic theories in the base category **Set** of sets and functions. We define the syntactic rank (the number of variables needed to write formulas) and the semantic rank (the number of variables needed by the operations on actual algebras) for an algebraic theory, and prove they are equal. Examples such as complete semilattices, complete atomic Boolean algebras, and compact Hausdorff spaces demonstrate that interesting mathematical structures arise as the algebras of infinitary theories. Bounded theories are coextensive with equationally definable classes of algebras (with perhaps infinitary operations). In general, theories are coextensive with "tractable large" equational presentations. We prove the theorem of [Gaifmann '64] and [Hales '64] that complete Boolean algebras do not constitute a tractable equational class.

Some useful facts about set theory which relate to this section are presented in a "primer" at the end.

Let us fix an algebraic theory $\mathbf{T} = (T, \eta, \circ, \mu)$ in the category of sets. We begin by classifying the trivial theories.

5.1 Lemma. *Let $\varnothing$ be the empty set and let* 1 *denote a one-element set. Then*

1. *The unique function $1T \longrightarrow 1$ is a* **T**-*algebra.*
2. *$\varnothing$ is a* **T**-*algebra in at most one way and this occurs if and only if* $\varnothing T = \varnothing$ (cf. the proof of "2.20 implies 2.18").
3. *Up to isomorphism, there exists exactly one algebraic theory* **T** *such that* 1 *is the only* **T**-*algebra; it is characterized by "$XT = 1$ for all sets X."*
4. *Up to isomorphism, there exists exactly one algebraic theory* **T** *such that* 1 *and $\varnothing$ are the only* **T**-*algebras; it is characterized by "$XT = 1$ for all nonempty sets X and $\varnothing T = \varnothing$."* The proof is safely left as an exercise. □

For obvious reasons, let us call the two algebraic theories of 5.1 *trivial*, and all other algebraic theories of sets *nontrivial*. We now further characterize the nontrivial theories. Notice that the second condition in the proposition below expresses that no equation of form "$v_i = v_j$" for distinct variables v_i and v_j can be deduced in a nontrivial theory.

5.2 Proposition. *The following conditions on* **T** *are equivalent:*

1. **T** *is nontrivial.*
2. *For each set* X, $X\eta: X \longrightarrow XT$ *is injective.*
3. T: **Set** $\longrightarrow$ **Set** *is a faithful functor, that is, whenever* $f, g: X \longrightarrow Y$ *are distinct functions then* $fT, gT: XT \longrightarrow YT$ *are again distinct functions.*

Proof. *1 implies 2.* By hypothesis, some **T**-algebra has at least two elements. By forming a suitably large cartesian power (as in 4.27), for each set X we can construct a **T**-algebra (Y, θ) and an injective function $f: X \longrightarrow Y$. From the naturality square (3.11) we have $X\eta.(fT.\theta) = f.Y\eta.\theta = f.\mathrm{id}_Y = f$

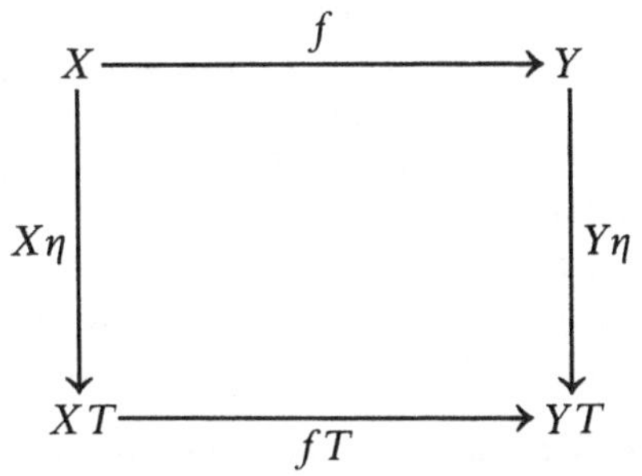

is injective. As $X\eta$ is injective followed by some other function, $X\eta$ is itself surely injective.

2 implies 3. If $f \neq g$ and $Y\eta$ is injective then $f.Y\eta \neq g.Y\eta$, so $X\eta.fT \neq X\eta.gT$. Since fT and gT are **T**-homomorphisms, $fT \neq gT$.

3 implies 1. This is clear, since neither of the two functors involved in the trivial theories are faithful. □

5.3 Definition *Let X be a set.* For finitary **T** (4.18), Theorem 4.25 allows us to treat elements of XT as "E-equivalence classes of **T**-terms"; or "symbolic operations." We view this as a linguistic or a syntactic concept. In general, *let us call elements of* XT *syntactic operations in* X (*with respect to* **T**). For example, $312{+}21{+}{+}{+}$ is a syntactic operation in $X = \{1, 2, 3\}$ with respect to to the theory of abelian groups. Such a symbol induces a semantic operation of abelian group theory in the sense that given any abelian group $(Y, +)$ we get an actual function

$$Y^X \longrightarrow Y:(y_1, y_2, y_3) \longmapsto y_3y_1y_2+y_2y_1+++$$

We have in fact explored the passage from syntactic to semantic operations quite generally in the proof of 4.25 (specifically, 4.26 and the formula "$\delta_\omega = X\hat{\omega}.\xi$"). Let us try to axiomatize semantic operations in their own right. At the very least, such an operation α must assign to each **T**-algebra (Y, θ) a function $(Y, \theta)\alpha: Y^X \longrightarrow Y$. Since homomorphisms are expected to commute with all operations, we should also require the commutative square

$$\begin{array}{ccc} Y^X & \xrightarrow{\;f^X\;} & Y'^X \\ {\scriptstyle (Y,\,\theta)\alpha}\downarrow & & \downarrow{\scriptstyle (Y',\,\theta')\alpha} \\ Y & \xrightarrow[\;f\;]{} & Y' \end{array} \tag{5.4}$$

whenever $f:(Y, \theta) \longrightarrow (Y', \theta')$ is a $\mathbf{T}$-homomorphism (and f^X sends the X-tuple $y:X \longrightarrow Y$ in Y to the X-tuple $y.f:X \longrightarrow Y'$ in Y', that is f^X sends the X-tuple $(y_x : x \in X)$ in Y to the X-tuple $(y_x f : x \in X)$ in Y'). Let us notice that "raising to the Xth power" is a functor $(\;)^X: \mathbf{Set} \longrightarrow \mathbf{Set}$. Let us denote $U^{\mathbf{T}}:\mathbf{Set}^{\mathbf{T}} \longrightarrow \mathbf{Set}$ (as in 4.8) by U for short, and the composite functor $U.(\;)^X:\mathbf{Set}^{\mathbf{T}} \longrightarrow \mathbf{Set}$ by U^X. Then what we have stipulated about α may be summed up by saying: "α is a natural transformation from U^X to U." Let this property define a *semantic operation in X (with respect to* $\mathbf{T}$). To give credibility to this new point of view—that the operations may be defined *after* the homomorphisms are—we prove the following theorem:

5.5 Theorem. *Let* $\mathbf{T}$ *be an algebraic theory in* **Set** *and let* X *be a set. Let* $\mathcal{O}_X(T)$ *be the set of natural transformations from the functor* $(\;)^X$ *(as defined in 5.4) to* T. *Defining* U *and* U^X *as in 5.4, let* $\mathcal{O}_X(\mathbf{T})$ *be the set of natural transformations from* U^X *to* U *(that is, semantic operations in* X*). Then the passage*

$$\begin{array}{c} XT \longrightarrow \mathcal{O}_X(T) \\ \omega \longmapsto (\;)^X \xrightarrow{\;\hat{\omega}\;} T \\ \langle X \xrightarrow{\;f\;} Y,\, Y\hat{\omega}\rangle = \langle \omega,\, XT \xrightarrow{\;fT\;} YT\rangle \end{array} \tag{5.6}$$

is bijective, with inverse

$$\begin{array}{c} \mathcal{O}_X(T) \longrightarrow XT \\ \alpha \longmapsto \langle \mathrm{id}_X,\, X\alpha\rangle \end{array} \tag{5.7}$$

Further, the passage

$$\begin{array}{c} XT \longrightarrow \mathcal{O}_X(\mathbf{T}) \\ \omega \longmapsto U^X \xrightarrow{\;\tilde{\omega}\;} U \\ (Y, \theta)\tilde{\omega} = Y^X \xrightarrow{\;Y\hat{\omega}\;} YT \xrightarrow{\;\theta\;} Y \end{array} \tag{5.8}$$

is bijective, with inverse

$$\begin{array}{c} \mathcal{O}_X(\mathbf{T}) \longrightarrow XT \\ \alpha \longmapsto \langle X\eta,\, (XT, X\mu)\alpha\rangle \end{array} \tag{5.9}$$

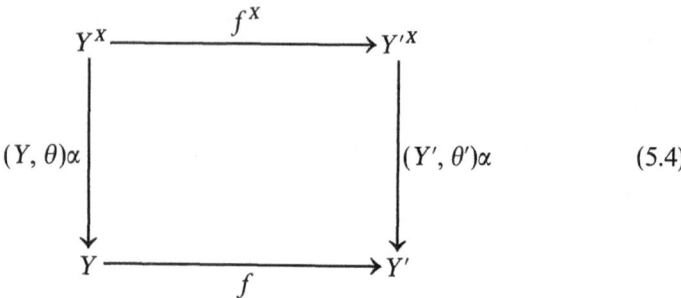

Proof. The passage 5.6 is well defined, that is $\hat{\omega}$ is a natural transformation, precisely because given $f:X \longrightarrow Y$ and $g:Y \longrightarrow Z$, $(f.g)T = fT.gT$. It then follows from the definition of a $\mathbf{T}$-homomorphism that 5.8 is well defined. Let us check that 5.6 and 5.7 are inverse. Starting with ω, we have $\langle \mathrm{id}_X, X\hat{\omega}\rangle = \langle \omega, (\mathrm{id}_X)T\rangle = \omega$; starting with α, for each $f:X \longrightarrow Y$ we have the naturality equation $X\alpha.fT = f^X.Y\alpha$, so that, defining $\omega = \langle \mathrm{id}_X, X\alpha \rangle$, we have $\langle f, Y\hat{\omega} \rangle = \langle\langle \mathrm{id}_X, X\alpha \rangle, fT \rangle = \langle \mathrm{id}_X, f^X.Y\alpha \rangle = \langle f, Y\alpha \rangle$, that is $\hat{\omega} = \alpha$. Let us turn to 5.8 and 5.9. Starting with ω, $\langle X\eta, (XT, X\mu)\tilde{\omega} \rangle = \langle X\eta, XT\hat{\omega}.X\mu \rangle = \langle \omega, X\eta T.X\mu \rangle = \omega$ (by 3.16). Starting with α, for each $f: X \longrightarrow Y$ and each **T**-algebra structure θ on Y we have the **T**-homomorphic extension $f^{\#} = fT.\theta: (XT, X\mu) \longrightarrow (Y, \theta)$ of 4.13, and hence the naturality square

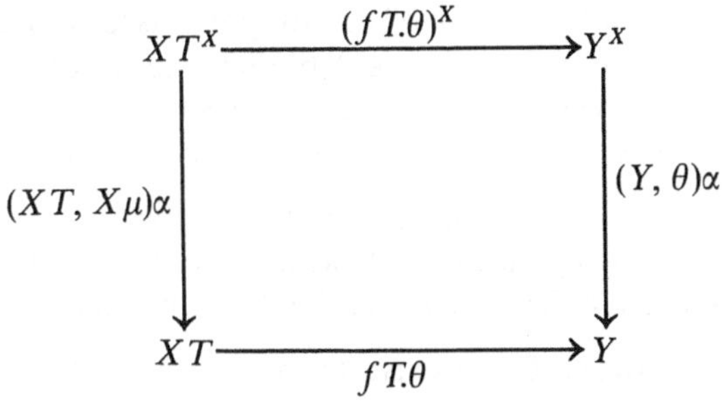

Setting $\omega = \langle X\eta, (XT, X\mu)\alpha \rangle$, we have $\langle f, (Y, \theta)\tilde{\omega} \rangle = \langle f, Y\hat{\omega}.\theta \rangle = \langle X\eta, (XT, X\mu)\alpha.fT.\theta \rangle = \langle X\eta, (fT.\theta)^X.(Y, \theta)\alpha \rangle = \langle X\eta.fT.\theta, (Y, \theta)\alpha \rangle = \langle f.Y\eta.\theta, (Y, \theta)\alpha \rangle = \langle f, (Y, \theta)\alpha \rangle$. $\square$

Passages (5.7) and (5.9) say that naturality is a very powerful constraint, for the natural transformations involved are determined by the value on just one element of just one of the components!

5.10 Definitions. *Let $\omega \in XT$ be a syntactic operation in X. The arity of ω is defined to be the smallest cardinal number "of the set of variables of a formula representing ω" or, more precisely,* $\mathrm{ar}(\omega) = \mathrm{Min}$ *(n : n is a cardinal and there exists $f: n \longrightarrow X$ such that ω is in the image of $fT: nT \longrightarrow XT$).* Thus, 4.18 says that **T** is finitary if and only if every syntactic operation has finite arity. For example, with the help of the unique map $f: \varnothing \longrightarrow X$, we see that the syntactic operation $[xx^{-1}]$ of group theory has arity 0. What is the arity of a semantic operation? Let us first consider a function $\psi: A^X \longrightarrow A$. It may happen that ψ is independent of some of the arguments in X. More precisely, given $S \subset X$, let $\mathrm{res}: A^X \longrightarrow A^S$ denote the restriction map sending $f: X \longrightarrow A$ to its S-restriction (which is just the inclusion map of S composed with f); then there exists at most one factorization

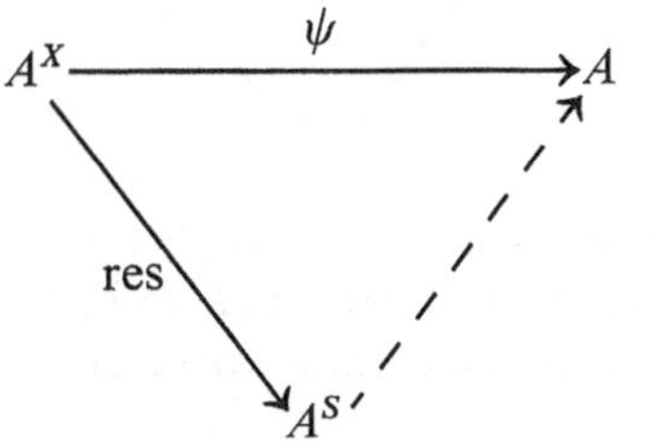

S is a *support of* ψ if such a factorization exists (and ψ is independent of the elements of X not in S). A subset S of X is a *support of* the semantic operation $\alpha: U^X \longrightarrow U$ if S is a support of $(Y, \theta)\alpha: Y^X \longrightarrow Y$ for every **T**-algebra (Y, θ). *The arity of* $\alpha: U^X \longrightarrow U$ *is defined by* "ar(α) = Min(n:n *is a cardinal and there exists a support of* α *of cardinal* n)." For example, let x and x' be distinct elements of X. Then sending $f: X \longrightarrow Y$ to $xf + x'f$ is a semantic operation in X with respect to abelian group theory whose arity is 2. S is a support if and only if $\{x, x'\} \subset S$. We must not infer from this example that the intersection of all supports is a support, however. For the ultrafilter theory of 3.21, if $\mathscr{U} \in X\beta$, then the set of supports of the semantic operation is precisely $\mathscr{U}$! (See exercise 3.)

We now show that syntactic arity and semantic arity coincide:

5.11 Theorem. *Let* **T** *be an algebraic theory of sets, let* X *be a set, and let* $\omega \in XT$ *be a syntactic operation in* X *with corresponding semantic operation* $\tilde{\omega}: U^X \longrightarrow U$ *as in 5.5. Then* ω *and* $\tilde{\omega}$ *have the same arity.*

Proof. ar($\tilde{\omega}$) $\leqslant$ ar(ω). There exist f:ar(ω) $\longrightarrow X$ and $h \in (\text{ar}(\omega))T$ with $\langle h, fT \rangle = \omega$. Let $S = \{uf : u \in \text{ar}(\omega)\}$ be the image of f with inclusion map $i: S \longrightarrow X$ and define p:ar(ω) $\longrightarrow S$ by $up = uf$, so that $f = p.i$. Set $\rho = \langle h, pT \rangle \in ST$. Because the restriction map is a natural transformation res:$(\ )^X \longrightarrow (\ )^S$, it follows from 5.7 that we have

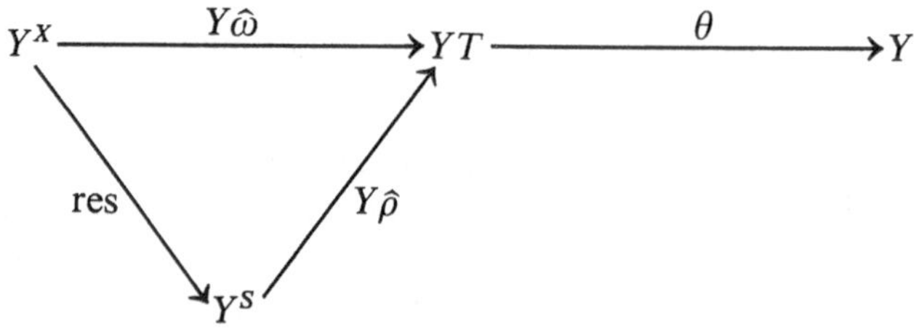

for every **T**-algebra (Y, θ), which shows that S, whose cardinal is at most ar(ω), is a support of $\tilde{\omega}$.

ar(ω) $\leqslant$ ar($\tilde{\omega}$). First suppose ar($\tilde{\omega}$) > 0, so that there exists a nonempty subset S of X of cardinal ar($\tilde{\omega}$) which is a support of $\tilde{\omega}$. Since S is not empty,

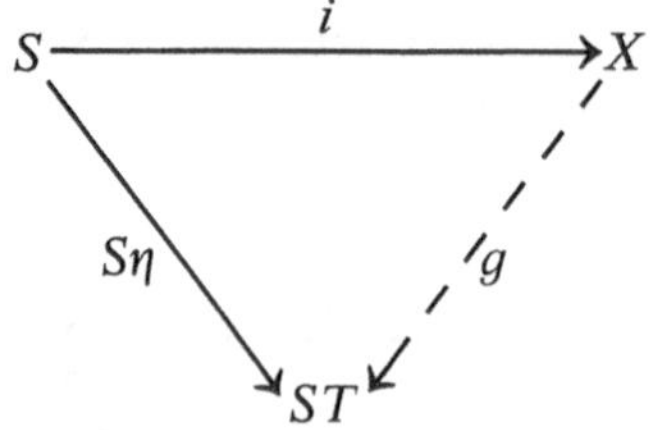

$S\eta$ admits at least one extension g through the inclusion map i as shown above. Because S is a support of $\tilde{\omega}$ and iT:$(ST, S\mu) \longrightarrow (XT, X\mu)$ is a **T**-homomorphism, we have the commutative diagram

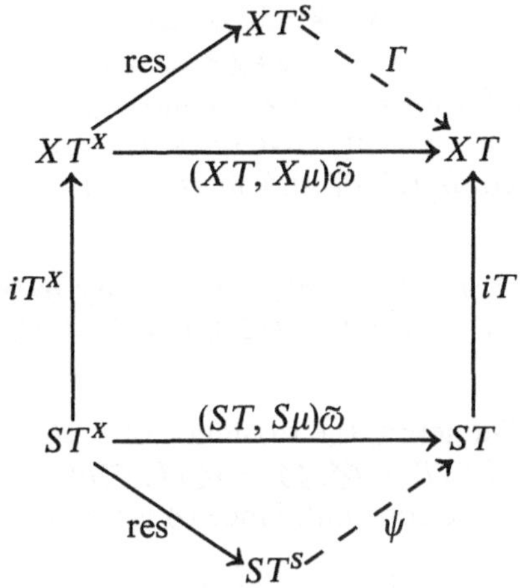

Define $\rho = \langle S\eta, \psi \rangle \in ST$. Then $\langle \rho, iT \rangle = \langle S\eta, \psi.iT \rangle = \langle i.g, \psi.iT \rangle = \langle g, (ST, S\mu)\tilde{\omega}.iT \rangle = \langle g, iT^X.(XT, X\mu)\tilde{\omega} \rangle = \langle g.iT, (XT, X\mu)\tilde{\omega} \rangle = \langle i.g.iT, \Gamma \rangle = \langle S\eta.iT, \Gamma \rangle = \langle i.X\eta, \Gamma \rangle = \langle X\eta, (XT, X\mu)\tilde{\omega} \rangle = \omega$. Now suppose that $\mathrm{ar}(\tilde{\omega}) = 0$. The above argument is still valid—that is, g still exists—providing $ST \neq \varnothing$ (where, now, $S = \varnothing$). Otherwise, $\varnothing$ is an algebra and there exists a factorization

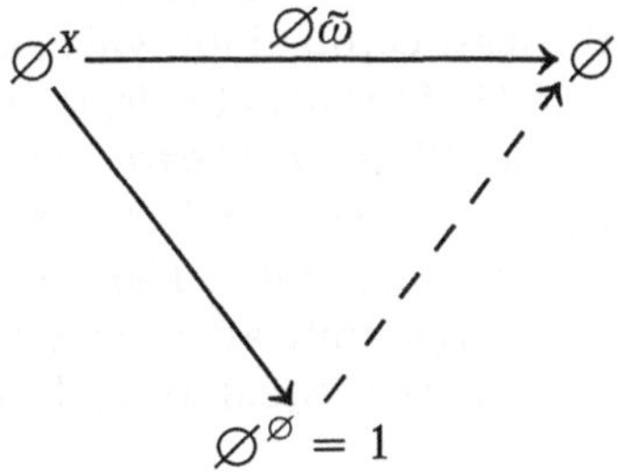

which is a contradiction. □

5.12 Example. Let **T** be the algebraic theory obtained from the equational presentation of semigroups in 4.17 by adjoining the additional equation $\{v_1v_2v_3{**}, v_1v_2{*}\}$. Let $X = \{v_1, v_2\}$, and set $\omega = v_1v_2{*}$. It is clear that $\mathrm{ar}(\omega) \leqslant 2$. The following model (which is actually the free **T**-algebra on two generators)

	x	y	xy	yx	xx	yx
x	xx	xy	xx	xy	xx	xy
y	yx	yy	yx	yy	yx	yy
xy	xy	xy	xy	xy	xy	xy
yx	yx	yx	yx	yx	yx	yx
xx	xx	xx	xx	xx	xx	xx
yy	yy	yy	yy	yy	yy	yy

shows that $\mathrm{ar}(\tilde{\omega}) \geqslant 2$. By 5.11, $\mathrm{ar}(\omega) = \mathrm{ar}(\tilde{\omega}) = 2$.

Let us pause to consider two kinds of constants in universal algebra which appeared incognito in the last stages of the proof of 5.11. A function $f:X \longrightarrow Y$ is *constant* if for all x, x' in X we have $xf = x'f$; this must be the case if X has at most one element. Given $\omega \in nT$, ω is *constant* if $(X, \xi)\tilde{\omega}$ is constant for all $\mathbf{T}$-algebras (X, ξ) and ω is a *true constant* if $\mathrm{ar}(\omega) = 0$. We have:

5.13 Proposition. *Let* $\mathbf{T}$ *be an algebraic theory in* **Set**. *Then* $\varnothing$ *is a* $\mathbf{T}$-*algebra if and only if* $\varnothing T = \varnothing$ *(cf. 5.1(2)). If* $\varnothing T \neq \varnothing$, *every constant is a true constant.*

Proof. If $\varnothing$ is a $\mathbf{T}$-algebra, the existence of $\xi:\varnothing T \longrightarrow \varnothing$ guarantees $\varnothing T = \varnothing$. Conversely, if $\varnothing T = \varnothing$, $\varnothing = (\varnothing T, \varnothing\mu)$ is a $\mathbf{T}$-algebra. Assume $\varnothing T \neq \varnothing$ and let $\omega \in nT$ be constant. Since $\varnothing$ is not a $\mathbf{T}$-algebra, $\mathrm{ar}(\tilde{\omega}) = 0$. By 5.11, $\mathrm{ar}(\omega) = 0$. □

It is possible to be constant without being true. Let $\Omega_1 = \{u\}$, $\Omega_n = \varnothing$ for $n \neq 1$ and let E have the single equation $\{v_1 u, v_2 u\}$. Then an (Ω, E)-algebra is (X, δ) where $\delta:X \longrightarrow X$ is constant. $\varnothing$ is an (Ω, E)-algebra. If $\omega = xu \in \{x\}T$, ω is constant but $\mathrm{ar}(\omega) = 1$.

5.14 Definition. *Let* $\mathbf{T}$ *be an algebraic theory in* **Set**. *Say that* $\mathbf{T}$ *is bounded or that* $\mathbf{T}$ *has rank if there exists a cardinal* N *for which every syntactic operation has arity less than* N, i.e. for all sets X and for all $\omega \in XT$, $\mathrm{ar}(\omega) < N$. If $\mathbf{T}$ is bounded, set M to be the least cardinal for which all syntactic operations have arity less than M. By 5.11, M is also the least cardinal such that every semantic operation has arity less than M. The *rank of* $\mathbf{T}$ is defined to be $M-1$ if M is finite, and M otherwise. A bounded theory whose rank is $\leqslant \aleph_0$ is finitary (4.18). Any other theory is *infinitary*. The algebraic theory for groups has operations of arbitrarily large finite arity such as "$x_1 \cdot \cdots \cdot x_n$" of arity n; therefore, the rank of this theory is equal to $\aleph_0$. In Example 5.12, the rank is 2.

We turn now to describing some interesting algebraic theories of infinite rank.

5.15 Example. Complete and Partially Complete Semilattices. A *finitely complete semilattice* is a partially ordered set $(X, \leqslant)$ in which every finite subset has a supremum (and in particular the empty supremum 0, which is the least element). To be a *homomorphism* $f:(X, \leqslant) \longrightarrow (X', \leqslant')$ of semilattices, we require that f preserves all finite suprema (which is strictly stronger than requiring f to be order preserving). It is well known that semilattices and their homomorphisms have an equational presentation as follows. Let (Ω, E) be the equational presentation for monoids $(X, +, 0)$ ("$+$" $\in \Omega_2$, "0" $\in \Omega_0$ with equations $(x + y) + z = x + (y + z)$, $x + 0 = x = 0 + x$) which are abelian (add the equation $x + y = y + x$) and idempotent (add the equation $x + x = x$). On the one hand, each semilattice with least element 0 and binary supremum operation $+$ is an (Ω, E)-algebra. Then again, given an (Ω, E)-algebra $(X, +, 0)$, we are forced to define $x \leqslant y$ if and only if the supremum of x and y is y i.e., $x + y = y$. It is easy to check that these passages are mutually inverse in such a way that semilattice homomorphisms are just Ω-homomorphisms. More generally, let M be a fixed infinite cardinal. An *M-complete semilattice* is a semilattice $(X, \leqslant)$ for which every subset of X of cardinal less than M has a supremum. The M-complete homomorphisms are required to preserve all n-ary suprema for $n < M$. Finitely complete semilattices are $\aleph_0$-complete semilattices. A *complete semilattice* is a semilattice $(X, \leqslant)$ in which every subset of X has a supremum; the homomorphisms preserve all suprema. The Boolean σ-rings used in measure theory are, in part, $\aleph_1$-complete semilattices. In a complete lattice every subset A has an infimum also, namely $\mathrm{Inf}(A) = \mathrm{Sup}(x : x \leqslant a \text{ for all } a \in A)$. If X is the set of all open subsets of the real numbers (in the usual topology) and $\leqslant$ is inclusion, then $(X, \leqslant)$ is a complete semilattice where suprema are ordinary unions, but the infimum of a family of open sets is the interior of the set-theoretic intersection. The inclusion map of $(X, \leqslant)$ into the complete semilattice of all subsets of the real numbers is a complete semilattice homomorphism which does not preserve the infimum of the countable family $[(-1/n, 1/n) : n = 1, 2, 3, \ldots]$. It is because of the homomorphisms that we distinguish between complete semilattices and complete lattices.

It is easy to check that if (Ω, E) describes semilattices as above, there is a bijective correspondence between Ω-terms in X and finite subsets of X (note that the words of 4.17 reduce to subsets since order and repetition no longer matter, and we must add the empty set to accommodate the true constant 0). The structure map (4.2) of a semilattice $(X, +, 0)$ is the function which assigns to each finite subset of X its $(X, +, 0)$-supremum. Can we create an algebraic theory $\mathbf{T}$ such that XT is the set of all subsets of X and the typical $\mathbf{T}$-structure map $\xi : XT \longrightarrow X$ describes the supremum map of a complete semilattice? We can. Let $\mathbf{T}$ be the algebraic theory of 3.5 (and, as was mentioned in 3.13+, $\mu : TT \longrightarrow T$ is the union map, whereas we showed in 3.10+ that $fT : XT \longrightarrow YT$ is the direct image map). If $\sup : XT \longrightarrow X$ is an arbitrary function then the $\mathbf{T}$-algebra equations 4.9 and 4.10 are clearly equivalent to

1. $\sup\{x\} = x$ for all $x \in X$
2. $\sup(\cup \mathscr{A}) = \sup(\sup A : A \in \mathscr{A})$ for all families of subsets $\mathscr{A} \in XTT$

whereas a $\mathbf{T}$-homomorphism $f : (X, \sup) \longrightarrow (X', \sup')$ preserves sup:

3. $(\sup(A))f = \sup(af : a \in A)$ for all subsets $A \in XT$.

We have at once that the passage from a complete semilattice to its supremum map is a well-defined injection into the $\mathbf{T}$-algebra structures and that homomorphisms are the same on both sides of the fence. The problem is to prove that if $(X, \sup)$ satisfies (1) and (2) above then, via $x \leqslant y$ if and only if $\sup\{x, y\} = y$, X becomes a complete semilattice whose supremum map is sup. We first check reflexivity, antisymmetry, and transitivity. $x \leqslant x$ by (1) and if $x \leqslant y$ and $y \leqslant x$ then $x = \sup\{x, y\} = y$. Suppose $x \leqslant y$ and $y \leqslant z$. Using (1) and (2), $\sup\{x, z\} = \sup\{\sup\{x\}, \sup\{y, z\}\} = \sup(\{x\} \cup \{y, z\}) = \sup\{x, y, z\} = \sup(\{x, y\} \cup \{z\}) = \sup\{\sup\{x, y\}, z\} = \sup\{y, z\} =$

z, that is $x \leqslant z$. Let A be a subset of X. For all $a \in A$, $\sup\{a, \sup(A)\} = \sup(\{a\} \cup A) = \sup(A)$ proving that $\sup(A)$ is an upper bound of A. If x is another upper bound of A then $\sup\{\sup(A), x\} = \sup(A \cup \{x\}) = \sup(\cup(\{a,x\}:a \in A)) = \sup\{\sup\{a,x\}:a \in A\} \sup\{\sup\{x\}\} = x$, so $\sup(A) \leqslant x$. This completes the verification that **T**-algebras are coextensive with complete semilattices.

To deal with M-complete semilattices it seems natural to "truncate" **T** at M by defining $XT_M = \{A \subset X : A \text{ has cardinal } < M\}$. Is T_M a subtheory of **T** in the sense of 3.20? Since every singleton subset is in XT_M the condition on η is true. There is a problem, however, with staying closed under composition. Given $\alpha: A \longrightarrow BT$ and $\beta: B \longrightarrow CT$ then $(\alpha \circ \beta)_a = \cup\{b\beta: b \in a\alpha\}$. If α factors through BT_M (i.e., $a\alpha$ has cardinal $<M$) and β factors through CT_M (i.e., each $b\beta$ has cardinal $<M$) we would hope that $(\alpha \circ \beta)_a$ also has cardinal $<M$. A moment's thought shows that this condition amounts to a rewording of the definition (see the primer on set theory at the end of this section) of a regular cardinal. We formalize with:

(*5.16*) *M is a regular cardinal if and only if T_M is a subtheory of* **T**, *where* **T** *is the theory of 3.5 and XT_M consists of those subsets of X of cardinal $<M$.* By essentially the same proof as in the complete case, the $\mathbf{T}_M$-algebras (where M is an infinite cardinal) are just the M-complete semilattices.

5.17 Example. Complete Atomic Boolean Algebras. A *Boolean algebra* is a commutative ring with unit $(X, +, 0, \text{jux}, 1)$ (where "jux" indicates that we will write multiplication by juxtaposition) in which multiplication is idempotent: $xx = x$. See [Halmos '63]. The standard example is the set of all subsets of a set A where $+$ is symmetric difference, 0 is the empty set, jux is intersection, and 1 is A. By the well-known theorem of Stone ([Stone '36, Theorem 70]) every Boolean algebra is isomorphic to a Boolean subalgebra of subsets of some set. By the discussion in 5.15, we know $(X, \text{jux}, 1)$ is a semilattice. In view of the Stone theorem, it is more natural to define $x \leqslant y$ if and only if $xy = x$ which looks more like "$A \subset B$ if and only if $A \cap B = A$." In any case, a Boolean algebra is a partially ordered set. A *complete Boolean algebra* is a Boolean algebra which has all suprema and infima. A *homomorphism* of complete Boolean algebras must not only be a ring homomorphism, but must preserve as well all infima and suprema. After 5.15, one might expect that complete Boolean algebras arise as the algebras over some theory. This is not the case, as we prove in 5.48. This changes if we impose further restrictions, however. In any partially ordered set $(X, \leqslant)$ $x \in X$ is an *atom* if x is not the least element and if $y < x$ implies y is the least element; that is, the atoms are the minimal elements of $X - \{0\}$. $(X, \leqslant)$ is *atomic* if every element is the supremum of the atoms beneath it. In particular, we know what a complete atomic Boolean algebra is. The set of all subsets of A is a complete atomic Boolean algebra (A is the union of its singleton subsets). It is possible to prove that, up to isomorphism, these are the only complete atomic Boolean algebras.

We turn now to the proof that the complete atomic Boolean algebras may be identified with the algebras over the double power-set theory of 3.19. Unlike the situation of 5.15, we do not know how to interpret the structure map. So let us begin by seeing in what sense an element $\mathscr{A} \in XT$ looks like a syntactic formula in X. It actually is. First, we record what most readers already know (and the rest should check for themselves): any Boolean ring $(X, +, 0, \text{jux}, 1)$ is a lattice with binary infima $x \wedge y = xy$, binary suprema $x \vee y = x + y + xy$ and unique complements (that is, for all x there is unique x' with $x \wedge x' = 0$ and $x \vee x' = 1$) namely $x' = x + 1$. According to the definition of η in 3.19 the element $x \in X$ is the "variable" $\text{prin}(x) \in XT$. For $A \subset X$, $\cap(\text{prin}(x): x \in A) = \{B \subset X : A \subset B\}$. Since $(\text{prin}(x))' = \{B \subset X: x \notin B\}$, we have $\cap(((\text{prin}(x))': x \notin A) = \{B \subset X : B \subset A\}$. It follows at once that $\mathscr{A}$ is the syntactic formula:

$$\mathscr{A} = \bigcup_{A \in \mathscr{A}} \cap(\{\text{prin}(x): x \in A\} \cup \{((\text{prin}(x))': x \notin A\})$$

This immediately forces us to define

5.18 Definition. *If X is a complete atomic Boolean algebra, the structure map of X is defined by*

$$XT \xrightarrow{\ \xi\ } X$$
$$\mathscr{A} \longmapsto \text{Sup}(A^{\not c}: A \in \mathscr{A})$$

where $A^{\not c} = \text{Inf}(A \cup \{x' : x \notin A\}$.

5.19 Proposition. *If X is a complete atomic Boolean algebra and if $A \subset X$ then*

$$A^{\not c} = \begin{cases} x_0 & \text{if } A = \{x : x \geqslant x_0\} \text{ and} \\ & x_0 \text{ is an atom} \\ 0 & \text{otherwise} \end{cases}$$

Proof. Set $B = \{x : x \geqslant A^{\not c}\}$. That $A \subset B$ is obvious. Now suppose that $x \notin A$, so that $A^{\not c} \leqslant x'$. If also $x \in B$ then $A^{\not c} = 0$ or, contrapositively speaking, $A^{\not c} = \mathrm{B}$ whenever $A^{\not c} \neq 0$. If $0 \leqslant x < A^{\not c}$ then $x \notin B = A$, $x < A^{\not c} \leqslant x'$ and $x < x'$ which is possible only if $x = 0$, and this establishes that $A^{\not c}$ is an atom. Let x_0 be any atom and set $A = \{x : x \geqslant x_0\}$. If $x \notin A$, that is if $x \wedge x_0 \neq x_0$, we must have $x \wedge x_0 = 0$ and so $x_0 = (x \wedge x_0) \vee (x' \wedge x_0) = x' \wedge x_0$ proving that $x' \in A$. Therefore $A^{\not c} = \text{Inf}(A) = x_0$ as desired. □

For any set X and subset $A \subset X$, define $A^* \subset XT$ by $A^* = \{\mathscr{A} : A \in \mathscr{A}\}$. We then have the following:

5.20 Proposition. *Let X be a complete atomic Boolean algebra with structure map $\xi: XT \longrightarrow X$ as in 5.18. Then for every subset $B \subset X$ with $B^{\not c} \neq 0$, $B^* = B\xi^{-1}$.*

Proof. By 5.19, if $B^{\not c} \neq 0$ then $B^{\not c}$ is an atom and $B = \{x : x \geqslant B^{\not c}\}$. $B\xi^{-1} = \{\mathscr{A} : \text{Sup}(A : A \in \mathscr{A}) \geqslant B^{\not c}\}$. But for $\mathscr{A} \in B\xi^{-1}$, $\text{Sup}(A^{\not c}: A \in \mathscr{A}) > 0$ so that there exists $A \in \mathscr{A}$ with $A^{\not c} > 0$. As $A^{\not c}$ is an atom, $A^{\not c} = B^{\not c}$ and, by 5.19, in fact $A = B$. Therefore $B\xi^{-1}$ simplifies to B^*. □

It is now clear that the structure map $\xi: XT \longrightarrow X$ of the complete atomic Boolean algebra X does satisfy the **T**-algebra laws 4.9 and 4.10 which amount to

$$\begin{aligned} \mathrm{Sup}(A^{\not c}: x \in A) &= x \qquad \text{for all } x \in X \\ \mathrm{Sup}(A^{\not c}: A^* \in \bar{\mathscr{A}}) &= \mathrm{Sup}(A^{\not c}: A\xi^{-1} \in \bar{\mathscr{A}}) \qquad \text{for all } \bar{\mathscr{A}} \in XTT. \end{aligned} \tag{5.21}$$

The first law, by 5.19, is the statement that every x is the supremum of the atoms beneath it. The second law follows at once from 5.20.

If we start with a **T**-algebra (X, ξ), X becomes a complete atomic Boolean algebra by

$$\begin{aligned} x + y &= (\mathrm{prin}(x) + \mathrm{prin}(y))\xi \qquad \text{(the second "+" is symmetric difference)} \\ xy &= (\mathrm{prin}(x) \cap \mathrm{prin}(y))\xi \\ 0 &= \varnothing\xi \qquad (\varnothing \text{ is the empty family}) \\ 1 &= (2^X)\xi \qquad (2^X \text{ is the set of all subsets of } X) \\ \mathrm{Sup}\, A &= (\mathscr{A}_A)\xi \qquad \text{where} \quad \mathscr{A}_A = \{B \subset X : B \cap A \neq \varnothing\} \\ \mathrm{Inf}\, A &= (\mathscr{A}^A)\xi \qquad \text{where} \quad \mathscr{A}^A = \{B \subset X : A \subset B\} \\ A^{\not c} &= \{A\}\xi \end{aligned} \tag{5.22}$$

The **T**-homomorphism condition 4.11 reads as

$$(\mathrm{Sup}\,(A^{\not c}: A \in \mathscr{A}))f = \mathrm{Sup}(B^{\not c}: Bf^{-1} \in \mathscr{A}) \qquad \text{for all } \mathscr{A} \in XT. \tag{5.23}$$

One needs to prove that (5.22) is indeed a complete atomic Boolean algebra structure whose structure map is ξ, that every complete atomic Boolean algebra satisfies 5.22 with respect to its structure map, and that 5.23 is equivalent to preserving +, jux, 1, Sup, and Inf. All this can be done with the proper choices of $\bar{\mathscr{A}}$'s in 5.21 and $\mathscr{A}$'s in 5.23, and we leave it as a challenging exercise to the reader.

5.24 Compact Hausdorff Spaces. Topological spaces can be studied from the point of view of knowing which ultrafilters converge where, as described below. A topological space X is compact Hausdorff precisely when each ultrafilter converges uniquely, giving rise to a function $\xi: X\beta \longrightarrow X$ which makes it not entirely surprising that these spaces are the same thing as **β**-algebras (3.21). For use later as well as now we set down the theory rather completely.

5.25 Characterization of Ultrafilters. Let $\mathscr{U}$ be a collection of subsets of a set X having the finite intersection property (3.22). Then the following five conditions on $\mathscr{U}$ are equivalent and make $\mathscr{U}$ an *ultrafilter on X*.

1. For all $A \subset X$ either A or its complement $X - A$ belongs to $\mathscr{U}$ (this is 3.23).
2. If $\{A_1, \ldots, A_n\}$ is a finite partition of X then exactly one A_i belongs to $\mathscr{U}$.
3. $X \in \mathscr{U}$; and, if $A_1, \ldots, A_n$ are subsets of X whose union belongs to $\mathscr{U}$ then at least one A_i belongs to $\mathscr{U}$.

4. $\mathscr{U}$ cannot be extended to a larger collection with the finite intersection property and $\mathscr{U}$ is nonempty.

5. If A is a subset of X and $A \notin \mathscr{U}$ then A has empty intersection with some element of $\mathscr{U}$.

Proof Hints. Clearly (3) implies (2), (2) implies (1), and we have (1) implying (3) since $(X - A_1), \ldots (X - A_n)$, $(A_1 \cup \cdots \cup A_n)$ have empty intersection. (1) implies (4): If $F \in \mathscr{F} \supset \mathscr{U}$ and $F \notin \mathscr{U}$ then $X - F \in \mathscr{U}$ so that $\mathscr{F}$ doesn't have the finite intersection property. (4) implies (5): If $A \notin \mathscr{U}$ then $\mathscr{U} \cup \{A\}$ doesn't have the finite intersection property and $U_1 \cap \cdots \cap U_n \cap A = \varnothing$ with $U_i \in \mathscr{U}$. If $U = U_1 \cap \cdots \cap U_n \notin \mathscr{U}$ then similarly $U \cap V_1 \cap \cdots \cap V_m = \varnothing$ with $V_i \in \mathscr{U}$, a contradiction. (5) implies (1): If $A \cap U = \varnothing$ and $(X - A) \cap V = \varnothing$ with $U, V \in \mathscr{U}$ then $U \cap V = \varnothing$, a contradiction. □

If $\mathscr{F} \subset 2^X$ and $A \subset X$, say that A is *close to* $\mathscr{F}$ if there exists $n > 0$ and $F_1, \ldots, F_n \in \mathscr{F}$ with $A \supset F_1 \cap \cdots \cap F_n$. $\mathscr{F}$ is a *filter on* X if $\mathscr{F} \neq \varnothing$, $\varnothing \notin \mathscr{F}$ and every set close to $\mathscr{F}$ is in $\mathscr{F}$. An ultrafilter $\mathscr{U}$ is a filter since $\mathscr{U} \cup \{A\}$ has the finite intersection property if A is close to $\mathscr{U}$.

5.26 Characterization of Principal Ultrafilters. The following conditions on an ultrafilter $\mathscr{U}$ on X are equivalent.

1. $\mathscr{U} = \text{prin}(x)$ (as in 3.19) for some $x \in X$.
2. $\cap\mathscr{U} \neq \varnothing$.
3. Some finite subset of X belongs to $\mathscr{U}$.

The x in (1) is unique. $\mathscr{U}$ is called a *principal ultrafilter*, and the *principal ultrafilter* on x.

Proof. (2) implies (3): if $x \in \cap\mathscr{U}$, $\mathscr{U} \cup \{x\}$ has the finite intersection property. (3) implies (1): there exists $x \in X$ with $\{x\} \in \mathscr{U}$ by 5.25 (3); as $\mathscr{U}$ is a filter, $\text{prin}(x) \subset \mathscr{U}$; as $\text{prin}(x)$ is maximal, $\text{prin}(x) = \mathscr{U}$. □

Since every ultrafilter on X contains X, the only ultrafilters on a finite set are the principal ones. No concrete example of a nonprincipal ultrafilter is known (see the notes at the end of this section). The next theorem uses Zorn's lemma to prove that nonprincipal ultrafilters must exist.

5.27 Plenitude of Ultrafilters; Characterization of Filters. For $\mathscr{F} \subset 2^X$ the following are equivalent.

1. $\mathscr{F}$ is a filter.
2. $\mathscr{F} \neq \varnothing$, $\varnothing \notin \mathscr{F}$, every superset of an element of $\mathscr{F}$ is in $\mathscr{F}$ and for all $n > 0$ and subsets $A_1, \ldots, A_n$ of X, $A_1 \cap \cdots \cap A_n \in \mathscr{F}$ if and only if each $A_i \in \mathscr{F}$.
3. $\mathscr{F}$ is the intersection of a nonempty family of ultrafilters.
4. $\mathscr{F} = \cap(\mathscr{U} : \mathscr{U}$ is an ultrafilter and $\mathscr{F} \subset \mathscr{U})$.

Proof Hints. To show that (1) implies (4) we must prove that for $A \notin \mathscr{F}$ there exists an ultrafilter $\mathscr{U}$ with $A \notin \mathscr{U}$ but $\mathscr{F} \subset \mathscr{U}$. As A is not close to $\mathscr{F}$, $\mathscr{F} \cup \{X - A\}$ has the finite intersection property. By Zorn's lemma, $\mathscr{F} \cup \{X - A\} \subset \mathscr{U}$ for some ultrafilter $\mathscr{U}$. □

Let $\mathcal{T}$ be a topology of open sets on X. It is entirely in the right spirit to say that an ultrafilter $\mathcal{U}$ on X is close to a point x with respect to $\mathcal{T}$ if every neighborhood of x is close to (and therefore in) $\mathcal{U}$. Formally, an ultrafilter $\mathcal{U}$ *converges to x with respect to* $\mathcal{T}$, written $\mathcal{U} \xrightarrow[\mathcal{T}]{} x$ (or just $\mathcal{U} \longrightarrow x$) if $\mathcal{U} \supset \mathfrak{N}_x$ (where $\mathfrak{N}_x = \{N \subset X : x \in N^0\}$ is the set of $\mathcal{T}$-neighborhoods of x). Since $\mathfrak{N}_x$ is a filter and a subset is open if and only if it is a neighborhood of each of its points, we have A is open if and only if $A \in \cap(\cap(\mathcal{U}:\mathcal{U} \longrightarrow x)$: $x \in A)$, that is

(5.28) *A is open if and only if A belongs to each ultrafilter which converges inside it. In particular, the convergence relation determines the topology.*

The following theorem can be found in many standard texts, or can be taken as definition:

5.29 Theorem. *A topological space is compact if every ultrafilter converges to at least one point. A topological space is Hausdorff if every ultrafilter converges to at most one point.* □

Since $\boldsymbol{\beta}$ is a subtheory of the double power-set theory of 3.19, $f : X \longrightarrow Y$ induces $f\beta : X\beta \longrightarrow Y\beta$ via $\langle \mathcal{U}, f\beta \rangle = \{B \subset Y : Bf^{-1} \in \mathcal{U}\}$. It is easy to see that $\langle \mathcal{U}, f\beta \rangle$ is also $\{B \subset Y : B \supset Af \text{ for some } A \in \mathcal{U}\}$.

5.30 Characterization of Continuity. Let $(X, \mathcal{T})$ and $(X', \mathcal{T}')$ be topological spaces and let $f : X \longrightarrow Y$ be a function. Then f is continuous if and only if whenever $\mathcal{U} \longrightarrow x$, $\langle \mathcal{U}, f\beta \rangle \longrightarrow xf$.

Proof. Suppose f is continuous, $\mathcal{U} \longrightarrow x$ and $V \in \mathfrak{N}_{xf}$. As $Vf^{-1} \in \mathfrak{N}_x \subset \mathcal{U}$, $V \in \langle \mathcal{U}, f\beta \rangle$. Conversely, noting that the inverse image of an ultrafilter is again an ultrafilter, we have $\mathcal{U} = \langle \mathcal{U}, f\beta \rangle f^{-1} \supset (\mathfrak{N}_{xf})f^{-1}$ whenever $\mathcal{U} \longrightarrow x$, so that $\mathfrak{N}_x \supset (\mathfrak{N}_{xf})f^{-1}$ and f is continuous. □

We have presented enough background material to establish the interesting result that compact Hausdorff spaces are the same thing as $\boldsymbol{\beta}$-algebras. Let us relativize our notation in 5.20– to the subtheory $\boldsymbol{\beta}$ and define $A^* = \{\mathcal{U} \in X\beta : A \in \mathcal{U}\}$. The algebra laws 4.9 and 4.10 condense to

$$\begin{aligned} (\text{prin}(x))\xi &= x \text{ for all } x \in X \\ \{A \subset X : A^* \in \mathcal{A}\}\xi &= \{A \subset X : A\xi^{-1} \in \mathcal{A}\}\xi \qquad \text{for all } \mathcal{A} \in X\beta\beta. \end{aligned} \tag{5.31}$$

Suppose $(X, \mathcal{T})$ is a compact Hausdorff space with convergence map $\xi : X\beta \longrightarrow X$. Since x belongs to each of its neighborhoods, it is clear that $(\text{prin}(x))\xi = x$. Now let $\mathcal{A} \in X\beta\beta$ and set $x = \{A \subset X : A^* \in \mathcal{A}\}\xi$ and $y = \{A \subset X : A\xi^{-1} \in \mathcal{A}\}\xi$. Suppose $x \neq y$. Then there exist disjoint open sets A, B with $x \in A$ and $y \in B$; (for the reader who is using 5.29 as the definition of Hausdorff: otherwise, $\mathfrak{N}_x \cup \mathfrak{N}_y$ has the finite intersection property and extends to an ultrafilter). By definition, $A^* \in \mathcal{A}$ and $B\xi^{-1} \in \mathcal{A}$. By 5.28, C is open if and only if $C\xi^{-1} \subset C^*$. As B is open and $\mathcal{A}$ is closed under supersets, $B^* \in \mathcal{A}$. Therefore, $(A \cap B)^* = A^* \cap B^* \neq \varnothing$, which is the desired contradiction. This proves that the passage from $\mathcal{T}$ to ξ is well defined; it is injective by 5.28 and homomorphisms are the same on both sides by 5.30. We must show that given an abstract $\boldsymbol{\beta}$-algebra (X, ξ) there is a compact

Hausdorff topology $\mathscr{T}$ whose convergence map is ξ. As discussed above, if what we say is true then $A \in \mathscr{T}$ if and only if $A\xi^{-1} \subset A^*$. It turns out that there is an equivalent definition whose immediate properties are more useful.

5.32 Definition. *Given an abstract* $\boldsymbol{\beta}$*-algebra* (X, ξ) *and* $A \subset X$, *A is open if for all* $\mathscr{U} \in X\beta$ *with* $\mathscr{U}\xi \in A$ *there exists* $U \in \mathscr{U}$ *such that* $U^*\xi \subset A$.

It is obvious that $\varnothing$, X are open and that finite intersections and unions of open sets are open; therefore the collection $\mathscr{T}$ of all open sets forms a topology on X. Suppose $\mathscr{U} \in X\beta$ and $N \in \mathfrak{N}_{\mathscr{U}\xi}$. There exists $U \in \mathscr{U}$ with $U^*\xi \subset N^0$. Since $\text{prin}(u) \in U^*$ and $(\text{prin}(u))\xi = u$ (by the first algebra axiom) for every $u \in U$, $U \subset N$ and $N \in \mathscr{U}$. This proves that $\mathscr{U} \xrightarrow[\mathscr{T}]{} \mathscr{U}\xi$.

5.33 Lemma. *For any subset B of X, the* $\mathscr{T}$*-closure* B^- *of B is contained in* $B^*\xi$.

Proof. This is where we really use the algebra axioms. First of all, if $i: B \longrightarrow X$ is the inclusion map then clearly B^* is just the image of $i\beta: B\beta \longrightarrow X\beta$, so that $B^*\xi$ is just $\langle B \rangle$ as in 4.31. In particular, $B \subset B^*\xi$ and $(B^*\xi)^*\xi = B^*\xi$. To show $B^*\xi$ is $\mathscr{T}$-closed, we must show, given $\mathscr{U}\xi \notin B^*\xi$, that there exists $U \in \mathscr{U}$ with $U^*\xi \cap B^*\xi = \varnothing$. Suppose not. Then $\{U^*: U \in \mathscr{U}\} \cup \{B^*\xi\xi^{-1}\}$ has the finite intersection property and is contained in some ultrafilter $\mathscr{A} \in X\beta\beta$. By 5.25(4), $\{A \subset X: A^* \in \mathscr{A}\} = \mathscr{U}$. Set $\mathscr{V} = \{A \subset X: A\xi^{-1} \in \mathscr{A}\}$. Then $B^*\xi \in \mathscr{V}$. By the second algebra axiom, $\mathscr{U}\xi = \mathscr{V}\xi \in (B^*\xi)^*\xi = B^*\xi$, the desired contradiction. □

We can make quick work of the remaining details. Let $\mathscr{U} \in X\beta$ and suppose $x \in X$ with $\mathscr{U}\xi \neq x$. Since $\{x\}^- \subset \{x\}^*\xi = \{x\}$, $X - \{x\}$ is open and there exists $U \in \mathscr{U}$ with $U^*\xi \subset X - \{x\}$. By 5.33, $x \notin U^-$ and there exists $N \in \mathfrak{N}_x$ with $N \cap U = \varnothing$. In particular, $N \notin \mathscr{U}$, so $\mathscr{U}$ does not converge to x. We have proved that the convergence relation of the topology $\mathscr{T}$ is ξ. As ξ is a function, $\mathscr{T}$ is compact Hausdorff. □

We are now ready to extend the definitions of section 1 to the infinitary case.

5.34 Definitions. *Extending 1.4, an operator domain is a disjoint family of sets,* $\Omega = (\Omega_n: n$ *is a cardinal). As before, an* Ω*-algebra is a pair* (X, δ) *where X is set and* δ *assigns to each* $\omega \in \Omega_n$ *an n-ary operation* $\delta_\omega: X^n \longrightarrow X$. Ω*-homomorphisms are defined exactly as in 1.5 giving rise to the category,* Ω*-alg, of* Ω*-algebras and an obvious underlying set functor* $U: \Omega\text{-alg} \longrightarrow$ **Set**. *Paralleling 5.3, for each cardinal n an n-ary operation of* Ω *is a natural transformation* $\alpha: U^n \longrightarrow U$. *An n-ary* Ω*-equation is a doubleton* $\{\alpha, \alpha'\}$ *where* α *and* α' *are n-ary operations of* Ω. *An* Ω*-equation is an n-ary* Ω*-equation for some n. An equational presentation is a pair* (Ω, E) *where* Ω *is an operator domain as just defined and E is a class of* Ω*-equations.* We are specifically permitting E to be a large set (see the "primer" at the end of this section). *An* Ω*-algebra* (X, δ) *satisfies the* Ω*-equation* $\{\alpha, \alpha'\}$ *just in case* $(X, \delta)\alpha = (X, \delta)\alpha'$. *An* (Ω, E)*-algebra is an* Ω*-algebra which satisfies every equation in E. This defines the category* (Ω, E)*-alg whose objects are the* (Ω, E)*-algebras and whose morphisms are the* Ω*-homomorphisms. Let* $U_E: (\Omega, E)\text{-alg} \longrightarrow$ **Set** *denote the underlying set functor.* By Theorem 5.5, the finitary definition of 1.19 can be recaptured by setting $\Omega_n = \varnothing$ when n is infinite and using only $\aleph_0$-ary equations. *The category of* (Ω, E)*-algebras is called an equationally-definable class.*

5.35 The Lawvere Theory. Let $\mathbf{T}$ be an algebraic theory in **Set**. If $\mathbf{T}$ is bounded, the *regular rank of* $\mathbf{T}$ is the smallest infinite regular cardinal greater than or equal to the rank of $\mathbf{T}$. The *Lawvere theory of* $\mathbf{T}$ is the category Law($\mathbf{T}$) whose objects are {all cardinal numbers} {all cardinal numbers less than the regular rank of $\mathbf{T}$} accordingly as $\mathbf{T}$ is unbounded or bounded. Thus, $\mathrm{Sup}(m_i \in n) \in \mathrm{Law}(\mathbf{T})$ whenever all m_i and n are. A morphism $\alpha : m \rightharpoondown n$ in Law($\mathbf{T}$) is a morphism $\alpha : n \rightharpoonup m$ in $\mathbf{Set}_\mathbf{T}$ (see 3.2), that is a function $\alpha : n \longrightarrow mT$ (Notice how we use different types of arrows to identify which category we mean). Composition and identities are defined just as in $\mathbf{Set}_\mathbf{T}$, specifically

$$
\begin{aligned}
(m \overset{\alpha}{\rightharpoondown} n) * (n \overset{\beta}{\rightharpoondown} p) &= (p \overset{\beta}{\rightharpoonup} n) \circ (n \overset{\alpha}{\rightharpoonup} m) \\
&= p \xrightarrow{\ \beta\ } nT \xrightarrow{\ \alpha T\ } mTT \xrightarrow{\ m\mu\ } mT
\end{aligned}
$$

(where $*$ denotes the composition operation in Law($\mathbf{T}$)), and $n\eta : n \rightharpoondown n$ provides the identities. Since $\mathbf{Set}_\mathbf{T}$ is a category, so is Law($\mathbf{T}$).

Our definition of a $\mathbf{T}$-algebra has so far stressed the monoid form (T, η, μ) of $\mathbf{T}$. We now show how to express 4.9 and 4.10 referring only to the clone form $(T, \eta, \circ)$ of $\mathbf{T}$; also, see exercise 11 of section 4.

5.36 Lemma. *Let* $\mathbf{T}$ *be an algebraic theory in* **Set**. *For each set* X *and function* $\xi : XT \longrightarrow X$ *we may attempt to define a functor* $M_\xi : \mathrm{Law}(\mathbf{T}) \longrightarrow$ **Set** *as follows. On objects,* $nM_\xi = X^n$. *Let* $\alpha : n \rightharpoondown m \in \mathrm{Law}(\mathbf{T})$. *For each* $i \in m$ *we have* $\alpha_i \in nT$ *and so, using 5.6, a function* $X\hat{\alpha}_i : X^n \longrightarrow XT$. *Collecting this m-tuple together, we have a single function* $(X\hat{\alpha}_i : i \in m) : X^n \longrightarrow XT^m$. *Define* $\alpha M_\xi : nM_\xi \longrightarrow mM_\xi$ *by*

$$
\alpha M_\xi = X^n \xrightarrow{(X\hat{\alpha}_i : i \in m)} XT^m \xrightarrow{\ \xi^m\ } X^m
$$

Then M_ξ *is a functor if and only if* (X, ξ) *is a* $\mathbf{T}$*-algebra.*

Proof. All finite cardinals and in particular 1 are in Law($\mathbf{T}$). M_ξ preserves identities if and only if $X\eta^n \cdot \xi^n : X^n \longrightarrow X^n$ is the identity of X^n and this is equivalent to 4.9. We will show that M_ξ preserves composition if and only if ξ satisfies 4.10. First two remarks:

5.37 Remark. Given $\beta : 1 \rightharpoonup n$ (i.e., $\beta \in nT$) and $\alpha : n \rightharpoonup m$ then the following diagram of natural transformations $(\ )^m \longrightarrow T$ is commutative:

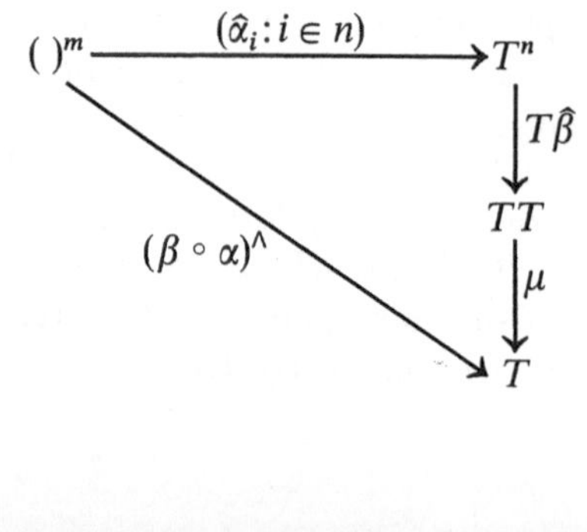

To prove this, 5.7 tells us we have only to check equality of the values assigned by the mth components to id_m. Indeed, $\langle \mathrm{id}_m, (m\hat{\alpha}_i : i \in n).mT\hat{\beta}.m\mu \rangle = \langle \alpha : n \longrightarrow mT, mT\hat{\beta}.m\mu \rangle = \langle \beta, \alpha T.m\mu \rangle = \beta \circ \alpha = \langle \mathrm{id}_m, m(\beta \circ \alpha)^\wedge \rangle$.

5.38 Remark. For any function $\xi : XT \longrightarrow X$, 4.10 is commutative restricted to the elements in the image of $(\varnothing \longrightarrow XT)T$ (that is, restricted to the interpretation in $(XTT, XT\mu)$ of the true constants). To prove this, one need only observe that $(\varnothing \longrightarrow XT)T.X\mu$ and $(\varnothing \longrightarrow XT).\xi T$ are already equal, both being the unique **T**-homomorphism $(\varnothing T, \varnothing\mu) \longrightarrow (XT, X\mu)$.

Now consider the following diagram induced by $\alpha : m \longrightarrow n$ and $\beta : n \longrightarrow p$:

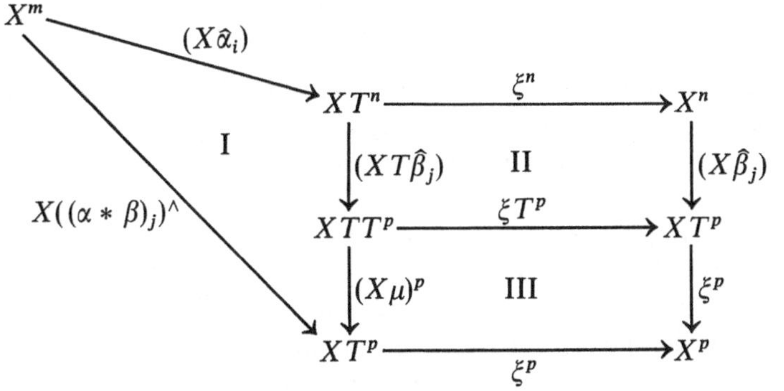

(5.39)

The two boundary paths from X^m to X^p are equal precisely when M_ξ preserves composition whereas III is equivalent to 4.10 (as p can equal 1). I and II always commute (by 5.37 and the naturality of β_j). It is now clear that if 4.10 holds then M_ξ preserves composition. Using 5.38, to prove the converse it is sufficient to show that given $\bar{x} \in XTT$ not in the image of $(\varnothing \longrightarrow XT)T$ there exist $m, n \in \mathrm{Law}(\mathbf{T})$, $\alpha : m \longrightarrow n$, $\beta : n \longrightarrow 1$ and $h : m \longrightarrow X$ such that $\langle h, (X\hat{\alpha}_i : i \in n).XT\hat{\beta} \rangle = \bar{x}$. This amounts to "unravelling $\bar{x}$ as a word of words" and can be done as follows: Since $\bar{x} \in (XT)T$ there exists $n \in \mathrm{Law}(\mathbf{T})$ and $f : n \longrightarrow XT$ such that $\bar{x}$ is in the image of fT, that is $\langle f, XT\hat{\beta} \rangle = \langle \beta, fT \rangle = \bar{x}$ for some $\beta \in nT$. By our hypothesis on $\bar{x}$, we may assume $n > 0$. For each $i \in n$ we can find, similarly, $m_i \in \mathrm{Law}(\mathbf{T})$, $g_i : m_i \longrightarrow X$ and $\gamma_i \in m_i T$ with $\langle g_i, X\hat{\gamma}_i \rangle = \langle \gamma_i, g_i T \rangle = f_i \in XT$. By the definition of the objects of $\mathrm{Law}(\mathbf{T})$, $m = \mathrm{Sup}(m_i : i \in n) \in \mathrm{Law}(\mathbf{T})$. Since m is equipotent with the disjoint union of the m_i there exists a family $(in_i : i \in n)$ of injections $in_i : m_i \longrightarrow m$ with disjoint images and such that the union of the images is m. One checks routinely that

$$(\ )^m \xrightarrow{in_i.-} (\ )^{m_i} \xrightarrow{\hat{\gamma}_i} T$$

is a natural transformation and hence has form $\hat{\alpha}_i$ for unique $\alpha_i \in mT$. This defines $\alpha : m \longrightarrow n$. There exists unique $h : m \longrightarrow X$ such that $in_i.h = g_i$ for all $i \in n$. We have $\langle h, (X\hat{\alpha}_i).XT\hat{\beta} \rangle = (\langle g_i, X\hat{\gamma}_i \rangle : i \in n)XT\hat{\beta} = (f_i : i \in n)XT\hat{\beta} = \langle f, XT\hat{\beta} \rangle = \bar{x}$. $\square$

We are now ready to prove the extension of 4.25 to infinitary algebraic theories. Rather than relying on a Birkhoff variety argument, we will present the operations and equations explicitly and prove that they work. In particular, this provides a different (and more informative) proof of 4.25.

5.40 Theorem. *Let* **T** *be an algebraic theory in* **Set**. *Then there exists an equational presentation* (Ω, E) *as in 5.34, with* $\Omega_n = \varnothing$ *if* $n \notin \mathrm{Law}(\mathbf{T})$ *(see 5.35), and every equation n-ary for some* $n \in \mathrm{Law}(\mathbf{T})$, *such that* **T**-*algebras and* (Ω, E)-*algebras are isomorphic as categories of sets with structure (as defined in the statement of 4.15).*

Proof. Define (Ω, E) as follows:

$$\begin{aligned} \Omega_n &= \{n\} \times nT & &\text{if } n \in \mathrm{Law}(\mathbf{T}) \\ \Omega_n &= \varnothing & &\text{if } n \notin \mathrm{Law}(\mathbf{T}). \end{aligned} \tag{5.41}$$

Denoting the underlying set functor from Ω-alg by U, each $\alpha \in nT$ becomes an n-ary operation $\hat{\alpha}: U^n \longrightarrow U$ of Ω by $(X, \delta)\hat{\alpha} = \delta_\alpha: X^n \longrightarrow X$ (where we write $\alpha \in \Omega_n$ for the more cumbersome (n, α)). Define the equations E by $E = E_1 \cup E_2 \cup E_3$ where

E_1 is the class of all equations $\{U^p \xrightarrow{a.-} U^n \xrightarrow{\hat{\alpha}} U, U^p \xrightarrow{b.-} U^m \xrightarrow{\hat{\beta}} U\}$ corresponding to $(p, n, m, a, b, \alpha, \beta)$ such that p, n, $m \in \mathrm{Law}(\mathbf{T})$, $a: n \longrightarrow p$ and $b: m \longrightarrow p$ are functions, $\alpha \in nT$, $\beta \in mT$ subject to the conditions that p is the union of the image of a and the image of b and $\langle \alpha, aT \rangle = \langle \beta, bT \rangle \in pT$.

E_2 is the single equation $\{\widehat{1\eta}: U \longrightarrow U, \mathrm{id}: U \longrightarrow U\}$ (thinking of $1\eta: 1 \longrightarrow 1T$ as an element of Ω_1).

E_3 consists of the class of equations $\{\widehat{\alpha * \beta}: U^m \longrightarrow U, U^m \xrightarrow{(\hat{\alpha}_i)} U^n \xrightarrow{\hat{\beta}} U\}$ corresponding to all (m, n, α, β) with $\alpha: m \rightharpoondown n$ and $\beta: n \rightharpoondown 1$ in $\mathrm{Law}(\mathbf{T})$.

5.42 Proposition. *If* $h: A \longrightarrow B$ *is injective, so is* $hT: AT \longrightarrow BT$.

Proof. We consider three cases: If A is nonempty we can extend the identity function of A to s as shown below:

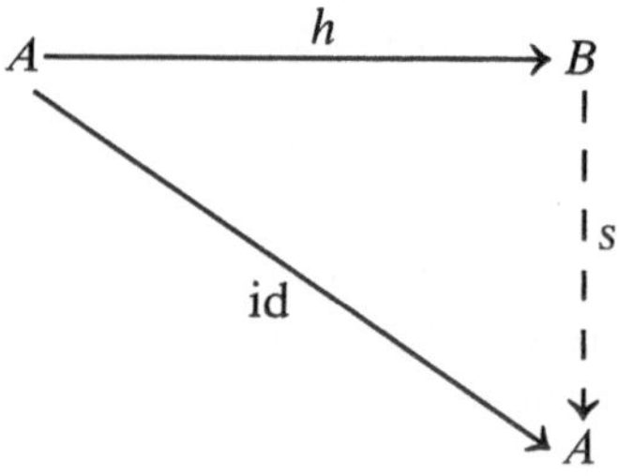

Then, since $hT.sT = \mathrm{id}_{AT}$, hT is injective. If A is empty and AT is also empty then hT is injective (as is any function from the empty set). Finally, consider $A = \varnothing$ but $\varnothing T \neq \varnothing$. Then there exists a function $s: B \longrightarrow \varnothing T$. As $hT.s^\#$ is the unique **T**-homomorphism from $(\varnothing T, \varnothing\mu)$ to itself, $hT.s^\# = \mathrm{id}_{\varnothing T}$ and hT is injective.

5.43 Proposition. *An Ω-algebra (X, δ) satisfies E_1 if and only if there exists a (necessarily unique) function $\xi : XT \longrightarrow X$ such that for all $\alpha \in \Omega_n$*

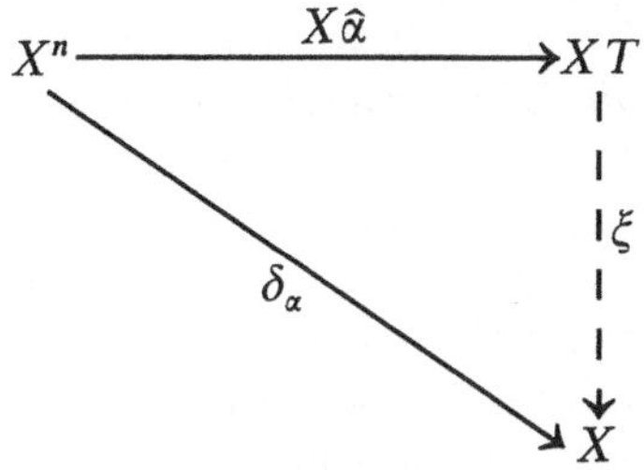

Proof. If $\bar{x} \in XT$ then there exists $n \in \mathrm{Law}(\mathbf{T})$, $f : n \longrightarrow X$ and $\alpha \in nT$ such that $\langle \alpha, fT \rangle = \bar{x}$, and we are forced to define $\bar{x}\xi = \langle f, \delta_\alpha \rangle$. Suppose now that $m \in \mathrm{Law}(\mathbf{T})$, $g : m \longrightarrow X$ and $\beta \in mT$ are again such that $\langle \beta, gT \rangle = \bar{x}$. For ξ to be well defined, we must be able to prove that $\langle f, \delta_\alpha \rangle = \langle g, \delta_\beta \rangle$. Let $S \subset X$ be the union of the images of f and g. There exists $p \in \mathrm{Law}(\mathbf{T})$ and a bijection $\psi : p \longrightarrow S$. Define $a : n \longrightarrow p$ and $b : m \longrightarrow p$ by $a = \bar{f}.\psi^{-1}$ and $b = \bar{g}.\psi^{-1}$, where $\bar{f} : n \longrightarrow S$ and $\bar{g} : m \longrightarrow S$ are defined by the diagram

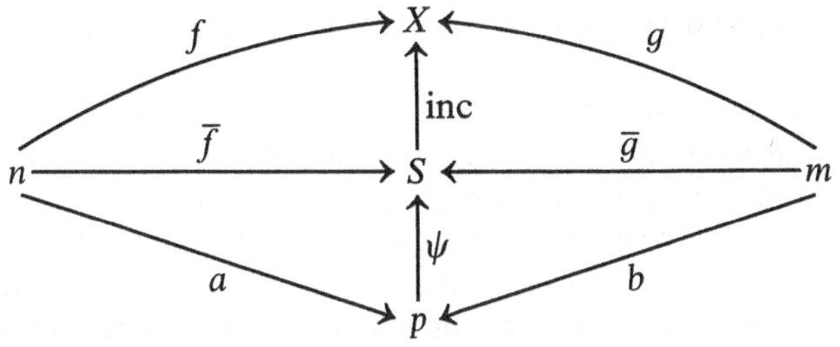

above. Define $h : p \longrightarrow X = \psi.\mathrm{inc}$. Then $\langle \alpha, aT \rangle hT = \langle \alpha, fT \rangle = \langle \beta, gT \rangle = \langle \beta, bT \rangle hT$. Since hT is injective by 5.42, $\langle \alpha, aT \rangle = \langle \beta, bT \rangle$. Therefore, there is an equation in E_1 corresponding to $(p, n, m, a, b, \alpha, \beta)$. If (X, δ) satisfies this equation, $\langle f, \delta_\alpha : X^n \longrightarrow X \rangle = \langle a.h, \delta_\alpha \rangle = \langle h, (a \cdot -).\delta_\alpha : X^p \longrightarrow X \rangle = \langle h, (b \cdot -).\delta_\beta : X^p \longrightarrow X \rangle = \langle g, \delta_\beta : X^m \longrightarrow X \rangle$. Hence ξ is well defined if (X, δ) satisfies E_1. Conversely, suppose ξ is well defined and we have an equation in E_1 corresponding to $(p, n, m, a, b, \alpha, \beta)$. Then for every $h : p \longrightarrow X$ we have $\langle h, (a \cdot -).\delta_\alpha : X^p \longrightarrow X) = \langle a.h, X\hat{\alpha}.\xi : X^n \longrightarrow X \rangle = \langle \alpha, (a.h)T \rangle \xi = \langle \alpha, aT \rangle hT.\xi = \langle \beta, bT \rangle hT.\xi = \langle h, (b \cdot -).\delta_\beta : X^p \longrightarrow X \rangle$, and (X, δ) satisfies all equations in E_1. This completes the proof of 5.43.

The remaining details are old hat. If (X, δ) satisfies E_1 giving rise to $\xi : XT \longrightarrow X$, then the precise content of 5.36 is that (X, δ) satisfies E_2 and E_3 if and only if (X, ξ) is a $\mathbf{T}$-algebra (although one must notice that the proof of 5.36 made it clear that to prove M_ξ preserves composition one could always assume $p = 1$). The inverse passage from (X, ξ) to (X, δ) and the proof that the two sorts of homomorphism are the same is achieved exactly as in 4.25. □

5.44 Definitions. *Let $\mathscr{A}$ be an arbitrary category and let $U: \mathscr{A} \longrightarrow$* **Set** *be a functor. As in 5.3, define $U^X: \mathscr{A} \longrightarrow$* **Set** *for each set X by $U^X = U.(\ )^X$. If n is a cardinal, U is tractable at n providing the class of all natural transformations from U^n to U is a small set (see the "primer" at the end of this section). U is tractable if U is tractable at n for every cardinal n. An equational presentation (Ω, E) (as in 5.34) is tractable just in case U_E: (Ω, E)-alg $\longrightarrow$* **Set** *is tractable.*

If **T** is an algebraic theory in **Set**, $U^{\mathbf{T}}$ is tractable by Theorem 5.5. It follows that any (Ω,E) which presents $\mathbf{Set}^{\mathbf{T}}$ (e.g. as in 5.41) is tractable, which imposes a necessary condition on (Ω, E) in order that it present $\mathbf{Set}^{\mathbf{T}}$ for some **T**. Happily, tractability is sufficient:

5.45 Theorem. *Let (Ω, E) be an equational presentation as in 5.34. Providing (Ω, E) is tractable, there exists an algebraic theory* **T** *in* **Set** *such that* **T**-*algebras and (Ω, E)-algebras are isomorphic as categories of sets with structure (as defined in the statement of 4.15).*

Informal comments in lieu of proof. For each (small) set A define a (perhaps large) set $A\Omega$ "inductively" by

$$a \in A\Omega \text{ for all } a \in A$$

whenever $\omega \in \Omega_n$ and $(p_i : i \in n)$ is an n-tuple in $A\Omega$ then $(p_i : i \in n)\omega \in A\Omega$ (cf. 1.8, 1.9).

Despite the highly intuitive appeal, the proof that we have a principle of algebraic recursion (cf. 1.14) and, in particular, the proof that $A\Omega$ makes any sense is difficult and will not be given in this book (see the notes at the end of this section). Accepting algebraic recursion makes the rest proceed smoothly. Define E_A as in 2.1 and then set $AT = A\Omega/E_A$. At some stage in the argument it becomes clear that the tractability assumption forces AT to be a small set. The infinitary versions of 2.2 and the universal property 2.5 are established using the old proofs, and (as was remarked in 2.16–) the ability to build an algebraic theory around T is a formal consequence of this universal property. The remaining details are a straight-forward rehash of 4.1 and 4.15. □

We will offer a rigorous (but different sort of) proof of 5.45 in Chapter 3 (see 3.1.26).

We close the section with two examples of nontractable equational presentations.

5.46 Complete Lattices Are Not Tractable at 3. The class of complete lattices and homomorphisms which simultaneously preserve supremum and infimum is equationally presentable but not tractable at 3. Let us observe first that complete semilattices (as in 5.15) are equational as a consequence of 5.5 and 5.41; in fact, a study of that construction allows us to throw away some of the operations and equations, and the following is an equational presentation of complete semilattices:

$$\begin{aligned} \Omega_n &= \{\mathrm{Sup}_n\} \qquad \text{for each cardinal } n. \\ E &= E_1 \cup E_2 \cup E_3 \end{aligned} \tag{5.47}$$

where E_1 consists of the equations $\mathrm{Sup}_n(af) = \mathrm{Sup}_n(bf)$ indexed by each instance of a pair $a, b: n \longrightarrow m$ of surjections; E_2 is the single equation $\mathrm{Sup}_1(x) = x$; and E_3 consists of the equations $\mathrm{Sup}_p(\mathrm{Sup}_m(\Gamma_i f): i \in p) = \mathrm{Sup}_n(f)$, where $(m \xrightarrow{\Gamma_i} n: i \in p)$ is a family of functions such that the union of the images covers all of n. (Of course, each side in any of the above equations is a natural transformation in accordance with the Definition 5.34; E_1 is trying to say (and does!) that $\mathrm{Sup}_n(f) = \mathrm{Sup}_n(g)$ whenever $f, g: n \longrightarrow X$ have the same image).

Define a new operator domain Ω' by $\Omega'_n = \{\mathrm{Inf}_n\}$ and define Ω'-equations E' by substituting "Inf" for "Sup" in each of the equations in E above. Since supremum in a partially ordered set $(X, \leqslant)$ is the same thing as infimum in the partially ordered set $(X, \geqslant)$, our interpretation of the operators in 5.47 was biased, and we may be comfortable in viewing complete semilattices as (Ω', E')-algebras as well. It is now clear that complete lattices is a full subcategory of $(\Omega \cup \Omega', E \cup E')$-algebras, and all that is missing is the guarantee that the Sup operators induce the same partial order as the Inf operators, that is that "$\mathrm{Sup}_2(x, y) = y$ if and only if $\mathrm{Inf}_2(x, y) = x$." The equational way to say this is the well known *absorptive laws*:

$$\mathrm{Inf}_2(x, \mathrm{Sup}_2(x, y)) = x$$
$$\mathrm{Sup}_2(x, \mathrm{Inf}_2(x, y)) = x$$

Now that we have seen that complete lattices are equational, let us explore the tractability properties of the underlying set functor U. U has only six 2-ary operations namely the true constants $0 = \mathrm{Sup}_{\varnothing}$, $1 = \mathrm{Inf}_{\varnothing}$ the two projections and Sup_2, Inf_2 (as is easy to check directly). It is at least mildly surprising that by adding one new variable we get not only infinitely many operations, but a large set of them. A proof of this can be found in [Hales '64, section 3]. While Hales' proof is too involved to present here, the construction is quite simple. For each ordinal i define a 3-ary operation Γ_i as follows:

$$\Gamma_0(x, y, z) = x$$
$$\Gamma_{i+1} = \mathrm{Sup}_2(x, \mathrm{Inf}_2(y, \mathrm{Sup}_2(z, \mathrm{Inf}_2(x, \mathrm{Sup}_2(y, \mathrm{Inf}_2(z, \Gamma_i))))))$$
$$\Gamma_i = \mathrm{Sup}_n(\psi\Gamma_j : j < i) \qquad \text{if } i \text{ is a limit ordinal (and } \psi \text{ is a conveniently prechosen bijection with the cardinal } n).$$

What Hales did was to construct for each pair of ordinals an example of a complete lattice on which the corresponding operations differ. The same proof shows that there are infinitely many 3-ary operations even for just finitely-complete lattices.

5.48 Complete Boolean Algebras Are Not Tractable at $\aleph_0$. The class of complete Boolean algebras and homomorphisms which simultaneously preserve supremum and complement (and hence everything else such as infima, + etc.) is equationally presentable but not tractable at $\aleph_0$. The reader may provide her own favorite equational presentation; for us, it is easiest to view a complete Boolean algebra as a complete lattice which is distributive and complemented. To this end, adjoin a single unary operator, c, to the operator domain of 5.47 and impose the equations of 5.47 together with four new ones:

$$\begin{aligned}
\mathrm{Inf}_2(\mathrm{Sup}_2(x, y), z) &= \mathrm{Sup}_2(\mathrm{Inf}_2(x, z), \mathrm{Inf}_2(y, z)) \\
\mathrm{Sup}_2(\mathrm{Inf}_2(x, y), z) &= \mathrm{Inf}_2(\mathrm{Sup}_2(x, z), \mathrm{Sup}_2(y, z)) \\
\mathrm{Sup}_2(x, cx) &= \mathrm{Inf}_\varnothing \\
\mathrm{Inf}_2(x, cx) &= \mathrm{Sup}_\varnothing
\end{aligned}$$

To check that we really have a Boolean algebra here, sceptics should note that $x + y = \mathrm{Inf}_2(\mathrm{Sup}_2(x, y), c\,\mathrm{Inf}_2(x, y))$.

With the help of exercise 19, it is not hard to prove, as is certainly suggested by 5.5, that the class of n-ary operations of any equational presentation contains a subalgebra freely generated by n, i.e., with the universal property of 4.12 with respect to algebras and homomorphisms (although is not necessarily a genuine algebra because it may be built on a large set). It suffices to show, then, that the assumption that we have a complete Boolean algebra F and a function $\eta:\aleph_0 \longrightarrow F$ with the universal property

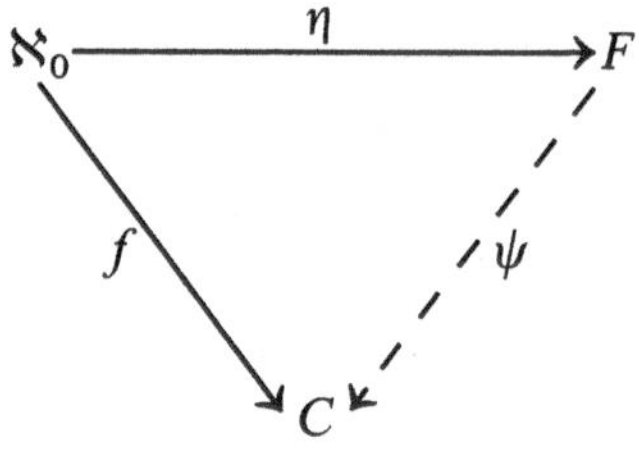

that every function $f:\aleph_0 \longrightarrow C$ to a complete Boolean algebra C extends uniquely to a complete Boolean homomorphism $\psi:F \longrightarrow C$, leads to the conclusion that F admits a surjection to every ordinal number. In fact, by 4.12 and 4.14 this proves immediately that complete Boolean algebras is not **T**-presentable.

Since the image of ψ (as above) is a complete Boolean subalgebra of C containing the image of the sequence f, it is sufficient to construct, given an ordinal α, a complete Boolean algebra C possessing a subset equipotent with α and a sequence $f:\aleph_0 \longrightarrow C$ such that no proper subalgebra of C contains the image of f.

Let X be an arbitrary topological space. For each subset U of X let $U^\perp$ denote the complement of the closure of U. U is a *regular open set* if U is the interior of its closure, that is $U = U^{\perp\perp}$. An important result (see [Halmos '63, §4, §7]) is that the set of all regular open sets forms a complete Boolean algebra with

$$\begin{aligned}
\mathrm{Sup}_n(U_i : i \in n) &= \mathrm{Int\ Cls}(\cup(U_i : i \in n)) \\
\mathrm{Inf}_n(U_i : i \in n) &= \mathrm{Int\ Cls}(\cap(U_i : i \in n)) \\
c(U) &= U^\perp
\end{aligned}$$

Fix an infinite ordinal α and provide the set α with the discrete topology. Let $\mathbf{N} = \{0, 1, 2, \ldots\}$ be the set of natural numbers, and let $X = \alpha^{\mathbf{N}}$ have the product topology. If $(S_n : n \in \mathbf{N})$ is a sequence of subsets of α then $\{f \in X : f_n \in S_n\}$ is closed since it is the cartesian product of the closed sets S_n. For $n \in \mathbf{N}$ and $S \subset \alpha$, $A_{n,S} = \{f \in X : f_n \in S\}$ is a typical subbasic open set; as mentioned above, it is also closed. As all clopen sets are regular open sets, $A_{n,S}$ is a regular open set. In particular, for $n \in \mathbf{N}$ and $\beta < \alpha$, $A_{n,\beta} = \{f \in X : f_n = \beta\}$ is a regular open set. For n, $m \in \mathbf{N}$, $B_{n,m} = \{f \in X : f_n \leqslant f_m\}$ is closed (use nets) as is its complement (use nets again), so $B_{n,m}$ is a regular open set. Let $\mathscr{B}$ be the complete Boolean subalgebra of regular open sets generated by the countable family of all $B_{n,m}$'s and all $A_{n,\beta}$'s (for finite ordinals β), where "subalgebra generated by" means "the intersection of all containing subalgebras"). No proper subalgebra of $\mathscr{B}$ contains this countable family. Since the $A_{0,\beta}$ are pairwise distinct, the proof is completed by proving that all $A_{0,\beta} \in \mathscr{B}$. To do this we will establish "for all $n \in \mathbf{N}$, $A_{n,\beta} \in \mathscr{B}$" for each $\beta < \alpha$ by transfinite induction. We already know this for finite β, which provides the basis. Now suppose this is true for $\beta < \gamma$; we must prove it for γ.

Fix $n \in \mathbf{N}$. As $\cup(A_{n,\beta} : \beta < \gamma) = \{f \in X : f_n < \gamma\}$ is clopen, $\mathrm{Sup}(A_{n,\beta} : \beta < \gamma) = \cup(A_{n,\beta} : \beta < \gamma)$ which proves that $\{f \in X : f_n < \gamma\} \in \mathscr{B}$. It follows that the set $C_{n,m} = \{f \in X : f_n \leqslant f_m \text{ or } f_m < \gamma\} = \mathrm{Sup}(B_{n,m}, \{f \in X : f_m < \gamma\})$ belongs to $\mathscr{B}$. Set $C_n = \mathrm{Inf}(C_{n,m} : m \in \mathbf{N}) = \mathrm{Int}(\cap(C_{n,m} : m \in \mathbf{N}))$. We claim that $C_n = \{f \in X : f_n \leqslant \gamma\}$. On the one hand, given $g \in X$ with $g_n \leqslant \gamma$ then for all $m \in \mathbf{N}$ either $g_n \leqslant g_m$ or else $g_m < g_n \leqslant \gamma$ so we have $g \in C_{n,m}$. Thus, $\{f \in X : f_n \leqslant \gamma\}$ is a subset of $\cap(C_{n,m} : m \in \mathbf{N})$ and is (subbasic) open, so is contained in C_n. Conversely, suppose given $f \in C_n$. Then some basic open neighborhood of f is contained in $\cap(C_{n,m} : m \in \mathbf{N})$, that is there exists a nonempty finite subset F of $\mathbf{N}$ such that every function agreeing with f on F is in every $C_{n,m}$. Define $g \in X$ by

$$g_m = \begin{cases} f_m & \text{if } m \in F \text{ or } m = n \\ \gamma & \text{otherwise} \end{cases}$$

There exists $m' \in \mathbf{N}$ with $g(m') = \gamma$. Since $g \in C_{n,m'}$ and "$g(m') < \gamma$" is false, we have $f_n = g_n \leqslant g_{m'} = \gamma$ as desired. This proves that $\{f \in X : f_n \leqslant \gamma\}$ belongs to $\mathscr{B}$. Noting that $\{f \in X : f_n \leqslant \gamma\}$ and $\{f \in X : f_n < \gamma\}$ are both clopen and in $\mathscr{B}$ we conclude that $A_{n,\gamma} = \{f \in X : f_n \leqslant \gamma\} - \{f \in X : f_n < \gamma\} = \mathrm{Inf}(\{f \in X : f_n \leqslant \gamma\}, \{f \in X : f_n < \gamma\}^{\perp}) \in \mathscr{B}$ and we are done. □

Primer on Set Theory

We outline a few concepts from set theory which were needed in this section. The outline is easily filled in by consulting [Monk '69] and is somewhat expanded in the exercises. See also [Mac Lane '71, I.6]. Another primer appears at the end of section 3.1.

Ordinals, as defined below, are sets of sets. The smallest ordinal, denoted **0**, is the empty set. If x is an ordinal, the next biggest ordinal is its *successor* $s(x) = x \cup \{x\}$. Thus each integer n induces the ordinal $\mathbf{n} = s^n(0)$. In normal usage we write n rather than $\mathbf{n}$. The first infinite ordinal $\boldsymbol{\omega}$ is the union of the chain $0 \subset 1 \subset 2 \cdots$. The next ordinals are the $s^n(\boldsymbol{\omega})$ and $\cup s^n(\boldsymbol{\omega})$. More formally, a set of sets x is $\in$-*transitive* if whenever $y \in x$ and $z \in y$ then $z \in x$. An *ordinal* is a sets of sets x such that x is $\in$-transitive and such that for all $y \in x$, y is $\in$-transitive. If x, y are ordinals then (using the "axiom of regularity") exactly one of "$x \in y$," "$x = y$," "$y \in x$" occurs ([Monk 9.9]) so that the class Ord of all ordinals is linearly ordered via $x \leqslant y$ if $x = y$ or $x \in y$. If X is a nonempty set of ordinals then $\cap\{x : x \in X\}$ is an ordinal and is in X ([Monk 9.10]); in particular, X has a least element. Further, for ordinals x, y, $x \leqslant y$ holds if and only if $x \subset y$.

Every ordinal x satisfies $x = \{y : y \text{ is an ordinal and } y < x\}$ ([Monk 9.13 (iii)]). This establishes the (*first*) *principle of transfinite induction* [Monk 10.1]: to define a function on an ordinal x it suffices to define $f(y)$ for all $y < x$. We used this in 5.48. An ordinal x is a *successor ordinal* if x has form $s(y)$ (i.e., "x–1 exists") and x is a *limit ordinal* if $x \neq 0$ and x is not a successor ordinal. ω is the smallest limit ordinal. The (*second*) *principle of transfinite induction* [Monk 10.4] asserts: to define a function f on an ordinal x it suffices to define $f(0)$, to define $f(y + 1)$ in terms of $f(y)$ whenever $y + 1 < x$ and to define $f(y)$ in terms of $\{f(z) : z < y\}$ whenever y is a limit ordinal and $y < x$. Cf. the construction of Γ_i in 5.46. The "algebraic recursion" of the proof comments of 5.45 generalizes transfinite induction.

A *cardinal* is an ordinal which is not equipotent ("equipotent" means "in bijective correspondence with") with a smaller ordinal. The finite ordinals n are cardinals. ω is also a cardinal but *qua* cardinal it is customary to call it $\aleph_0$. $s(\omega)$ is not a cardinal. Given any set A there exists a unique cardinal x (using the axiom of choice) such that A and x are equipotent ([Monk 18.3]); x is the *cardinality of* A, $x = \text{card}(A)$. A is *uncountable* if $\text{card}(A) > \aleph_0$; otherwise, A *is countable*. If A admits an injection into or a surjection from B then $\text{card}(A) \leqslant \text{card}(B)$. As discussed above, the cardinals constitute a linearly ordered class such that every nonempty subset has a least element. If x is a cardinal, x^+ denotes the next largest cardinal. There is no largest cardinal, that is, x^+ always exists ([Monk 18.13]). If $(X_i : i \in I)$ is a family of cardinals, their *sum* $\sum x_i$ is the cardinality of the disjoint union $\{(y, i) : y \in x_i\}$ of the sets x_i ([Monk 20.1]). A cardinal x is *regular* ([Monk 21.18]) if x is infinite and if for every family $(x_i : i \in I)$ of cardinals with each $x_i < x$ and $\text{card}(I) < x$, it is the case that $\sum x_i < x$. Starting with $\aleph_0$ and defining $\aleph_{n+1} = (\aleph_n)^+$, $\sum \aleph_n$ is the smallest infinite cardinal which is not regular. For any infinite cardinal x, x^+ is regular ([Monk 21.14]).

A *chain* in a partially ordered set $(X, \leqslant)$ is a subset C of X such that whenever $x, y \in C$ either $x \leqslant y$ or $y \leqslant x$. m is a *maximal element of* $(X, \leqslant)$ if for every $x \in X$ it is false that $x > m$. *Zorn's lemma* asserts: if every chain in $(X, \leqslant)$ has an upper bound (i.e., there exists u in X, not necessarily in C, such that $u \geqslant c$ for every c in C) then $(X, \leqslant)$ has at least one maximal element. The theorem works if $(X, \leqslant)$ is empty since the empty set is a chain with no upper bound. It is well known that Zorn's lemma is equivalent to the axiom of choice ([Monk section 16]). In the context of the proof hints to 5.27, X is the set of all families $\mathscr{F}$ with the finite intersection property, $\mathscr{F} \leqslant \mathscr{G}$ means $\mathscr{F} \subset \mathscr{G}$ and a maximal element is an ultrafilter by 5.25 (iv).

Paradox: it is easy to prove that the set Ord is an ordinal ([Monk 9.7]). Thus Ord $\in$ Ord, i.e. Ord $<$ Ord, which is impossible. Also, the *Russell Paradox*: let R be the set of all sets of sets x such that $x \notin x$; then R is a set of sets, but if $R \in R$ then $R \notin R$ whereas if $R \notin R$ then $R \in R$. The way these paradoxes are resolved is by insisting that certain "classes" such as Ord and R above are "impredicatively defined" (that is, there is obvious "self-definition" in phrases such as "the ordinal of all ordinals" and "the set of all sets of sets") and are not *bona fide* sets. For example, Ord is not really an ordinal because it is not a set.

To truly resolve the crises of the preceding paragraph would lead us far afield from the subject matter of this book. We refer the reader to [Mac Lane '71, I.6] for a description of some related problems and of the "one universe" set theory that is adequate for our needs. (See also [Fraenkel, Bar Hillel and Levy '73, II.7].) In brief, there is a "universe" U modelling the "set of all sets." A *small set* is an element of U. A subset of U which is not an element of U is a *large set*. Thus Ord and R above are large sets. Any set which admits an injection from or a surjection to every cardinal is a large set. A cartesian product of a family of small sets indexed by a small set is a small set; on the other hand, the cartesian product of all small sets is a large set. *A priori*, there is no reason why $\mathcal{O}_X(\mathbf{T})$ or $\mathcal{O}_X(T)$ need be small sets, but this is proved in 5.5. The class of operations of an operator domain as in 5.34 will be a large set unless there exists a cardinal N such that $\Omega_n = \varnothing$ whenever $n > N$. The intended meaning of "tractable" in 5.44 is that "$A\Omega/E_A$" in the spirit of 2.1 is a small set for all A even though $A\Omega$ may be large. 5.43 asserts that the free lattice on 3 generators is a large set. The category **Set** of sets and functions has as objects small sets.

Notes for Section 5

Our trivial algebraic theories were dubbed "inconsistent" by [Lawvere '63, page 51]. The idea that algebraic operations are natural transformations (our semantic operations) is due to Lawvere (see [Lawvere '63, page 69, Theorem 1]) and was emphasized by Linton (see [Linton, '66, '69]). Lawvere defined algebras as set-valued functors as in 5.36. The inverse passages 5.6 and 5.7 is an instance of the well-known Yoneda lemma ([Yoneda '54]) of category theory; see [Mac Lane '71]. [Lawvere '63, pages 52–53] called our constants *definable constants* and our true constants *expressible constants*; our 5.13 is his proposition 5.

Examples 5.15, 5.17 were well known to the Zürich school. It was also known ([Linton '66, section 5]) that compact Hausdorff spaces were representable as the algebras of a suitable theory; according to Barr (personal communication), he, Beck, and Linton convinced themselves, in 1965, that the constructions of 5.24 could be given (and this appeared in [Manes '67, '69]). See also [Paré '71] (presented in [Mac Lane '71, VI. 9]). [Semadeni '74–A] and cf. [Gonshor '74].

It was pointed out to us by M. H. Stone that a perfectly modern definition of nonprincipal ultrafilter was given in 1908! by F. Riesz ([Riesz '08, p. 23]).

His five axioms for a collection of subsets $\mathscr{U}$ of a set to be an "ideale verdichtungstelle" are: (1) every superset of an element of $\mathscr{U}$ is again in $\mathscr{U}$; (2) if A, B are disjoint and if their union is in $\mathscr{U}$ then $A \in \mathscr{U}$ or $B \in \mathscr{U}$; (3) the intersection of two elements of $\mathscr{U}$ is again in $\mathscr{U}$; (4) $\mathscr{U}$ is maximal with respect to properties (1)–(3); (5) the intersection of all elements of $\mathscr{U}$ is empty. See [Bell and Slomson '71] for the use of ultrafilters in model theory (and Chapters 5, 6 there for details concerning the structure of ultrafilters *per se*). The axiom of choice, which is equivalent to Zorn's lemma, was used to prove the "ultrafilter theorem" (every filter is contained in an ultrafilter) used in 5.27. There exist models of (Zermelo-Frankel) set theory in which (1) the ordinal ω has no nonprincipal ultrafilters or (2) every infinite set has at least one nonprincipal ultrafilter but the ultrafilter theorem fails or (3) the ultrafilter theorem holds but the axiom of choice fails. See [Jech '73, page 82, page 132, Theorem 7.1]. Since one expects any "actual construction" of a nonprincipal ultrafilter to build one on ω in any model of set theory, it is popular to assert that "it is impossible to construct *any* example of a non-principal ultrafilter"; at this writing, however, this assertion is only a conjecture.

The characterization of topological concepts in the language of ultrafilter convergence can be found in [Choquet '48]; we thank H. R. Fischer and O. Wyler for pointing out this reference.

Infinitary universal algebra begins with the founding paper [Birkhoff '35] where algebraic recursion is assumed without comment. A number of works have been devoted to a rigorous construction of free Ω-algebras when Ω is infinitary but bounded: [Diener '66], [Felscher '65, '72], [Harzheim '66], [Henkin, Monk and Tarski '71], [Kerkhoff '65], [Lowig '52, 57], and [Słomiński '59]. Słomiński's monograph provides (among other things) a treatment along the lines of section 1. Kerkhoff's construction of free Ω-algebras, quite similar to 1.1.7, strikes us as being the simplest: given a set A, let B be the set of all subsets of the set of all (finite) words on the set

$$A + \cup\{n : \Omega_n \neq \varnothing\} + \cup\Omega_n$$

observe that B is an Ω-algebra via

$$\begin{array}{ccc} B^n & \xrightarrow{\delta_\omega} & B \\ (S_\alpha : \alpha < n) & \longmapsto & \{n\} \cup \{n\alpha w : \alpha < n, w \in S_\alpha\} \end{array}$$

and set $A\Omega$ to be the Ω-subalgebra of B generated by $\{\{a\} : a \in A\}$.

Unbounded universal algebra was first recognized by [Linton '66, '69] and 5.35–5.45 is an adaptation of Linton's work. Felscher proved that (Ω, E)-algebras are coextensive with functors on the Lawvere theory in [Felscher '69, '72 3.2].

[Birkhoff '35, Theorem 27] proved that the free finitely-complete lattice on 3 generators is infinite, attributing the question to [Klein '34]. The proof of 5.48 was given independently by [Gaifmann '64] and [Hales '64]. The simpler proof we presented is from [Solovay '66].

Exercises for Section 5

1. Recall that "characteristic function" establishes a bijection between subsets of X and functions $X \longrightarrow 2$, where $2 = \{0, 1\}$ so that, in particular, a collection of subsets of X is a function $2^X \longrightarrow 2$. Also recall that an I-indexed family of functions $(f_i : X \longrightarrow Y)$ corresponds to the single function $f : X \longrightarrow Y^I$.
   (a) Let $\mathbf{T}$ be the double power-set theory of 3.19. Thinking of XT as the set of functions from 2^X to 2, show that η sends x to the xth projection $pr_x : 2^X \longrightarrow 2$ and that the characteristic function of $x(\alpha \circ \beta)$ is given by

$$2^Z \xrightarrow{(\chi_{y\beta}: y \in Y)} 2^Y \xrightarrow{\chi_{x\alpha}} 2$$

   (b) (Kock-Lawvere) Let $\mathbf{S}$ be any algebraic theory in **Set** and let $(2, \xi)$ be an $\mathbf{S}$-algebra structure on 2. Define XS_ξ to be the set of $\mathbf{S}$-homomorphisms from the (cartesian power) algebra $(2, \xi)^X$ to $(2, \xi)$. Show that $\mathbf{S}$ is a subtheory of the double power-set theory as in (a). See also exercise 11 of 2.3.
2. In this exercise we abstract the structure of the set of supports of a function $\psi : A^X \longrightarrow A$. In case $A = \varnothing$, $X \neq \varnothing$ and the set of supports of ψ is the set of all nonempty subsets of X. This case is singular and we concentrate on the case with $A \neq \varnothing$. A *quasifilter on* a set X is a nonempty collection $\mathscr{F}$ of subsets of X satisfying (i) every superset of an element of $\mathscr{F}$ is in $\mathscr{F}$, and (ii) the intersection of two elements of $\mathscr{F}$ is again in $\mathscr{F}$.
   (a) If $\psi : A^X \longrightarrow A$ is any function, with A nonempty, show that the set of supports of ψ is a quasifilter on X. [Hint: if F, G have empty intersection, prove directly that ψ is constant; otherwise define functions e, f, g as shown below with ae, cf, dg identities; also, see exercise 21 of 2.1.]

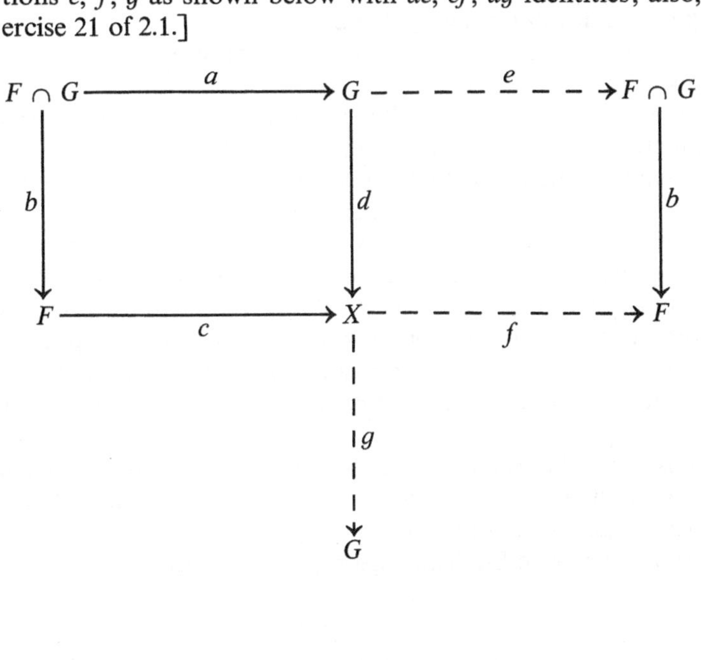

(b) If $\mathscr{F}$ is any quasifilter on X, prove that $\mathscr{F}$ is the set of supports of its characteristic function $2^X \longrightarrow 2$.
(c) Prove that $\mathscr{F}$ is a quasifilter on X if and only if (iii) $X \in \mathscr{F}$, and (iv) for all $F, G \subset X$, $F \cap G \in \mathscr{F}$ if and only if $F, G \in \mathscr{F}$.
(d) Let $\Omega_0 = \{e\}$, $\Omega_1 = \{\cdot\}$, $\Omega_n = \varnothing$ otherwise. $2 = \{0, 1\}$ is an Ω-algebra with $\delta_e = 1$ and $\delta_{\cdot}$ defined by

	1	0
1	1	0
0	0	0

Show that $\mathscr{F}$ is a quasifilter on X if and only if its characteristic function $(2, \xi)^X \longrightarrow (2, \xi)$ is an Ω-homomorphism.
(e) Let $AT = \{\mathscr{F} : \mathscr{F} \text{ is a quasifilter on } X\}$. Show that AT is a subtheory of the double power-set theory. [Hint: use (d) and exercise 1.]
(f) [Day '75]. A partially-ordered set is *directed* if each two elements have an upper bound. Let **T** be the quasifilter theory of (e). Show that $\mathbf{Set}^{\mathbf{T}}$ may be identified with the category whose objects are complete lattices satisfying

$$\mathrm{Inf}(\mathrm{Sup}\, A_i : i \in I)) = \mathrm{Sup}(\mathrm{Inf}(a_i : i \in I) : (a_i) \in \Pi A_i)$$

for each family (A_i) of directed subsets, and whose morphisms preserve all infima and all suprema of directed subsets. [Hint: cf. 5.17; if $\mathscr{F} \in XT$, we have

$$\mathscr{F} = \bigcup_{F \in \mathscr{F}} \bigcap_{x \in F} (\mathrm{prin}(x)).]$$

3. Let $\boldsymbol{\beta}$ be the ultrafilter theory of 3.21.
(a) For $\mathscr{U} \in X\boldsymbol{\beta}$, show that $\mathscr{U}$ is the set of supports of the semantic operation $\tilde{\mathscr{U}}$.
(b) Show that $\mathscr{U}$ is an ultrafilter on X if and only if its characteristic function is a Boolean ring homomorphism $2^X \longrightarrow 2$.
(c) Recapture the result that $\boldsymbol{\beta}$ is a subtheory of the double power-set theory by combining (b) and exercise 1.
4. Let $AT = \{\mathscr{F} : \mathscr{F} \text{ is a filter on } X\}$. Show that T is a subtheory of the double power-set theory. [Hint: in the context of exercise 2d, add the true constant e' and let $\delta_{e'} = 0$.] Wyler [to appear] presents the **T**-algebras as "interval-like" complete semilattices with compact topology.
5. Why is "$x_1 x_2 m x_3 m x_4 \cdots$" not a valid infinitary operation?
6. Prove that $\boldsymbol{\beta}$ is unbounded. [Hint: look up the definition of "uniform ultrafilter".]
7. Establish that the Boolean algebra of Lebesgue-measureable subsets of (say) the unit interval is countably complete and atomic but that the quotient algebra modulo "equal almost everywhere," while still a countably-complete Boolean algebra, is nonatomic.

8. As mentioned in the proof of 5.48, the regular open sets of any topological space form a complete Boolean algebra. What conditions on the space force this algebra to be atomic?
9. With respect to the axioms for an ultrafilter given by Riesz in the notes, show that his axioms are equivalent to ours but that (4) is implied by the other axioms.
10. Prove that a uniform space is complete if and only if every Cauchy ultrafilter converges. [Hint: prove that if two convergent filters do not converge to exactly the same sets of points then their intersection is not Cauchy; then use the ultrafilter theorem.]
11. Let $\mathscr{A}$ be a collection of subsets of X and define $\mathscr{A}^\bullet = \{\mathscr{U} \in X\beta$: for all $U \in \mathscr{U}$ there exists $A \in \mathscr{A}$ with $A \subset U\}$. Show that $\{\mathscr{A}^\bullet : \mathscr{A} \subset 2^X\}$ is the set of all closed subsets of the topological space $(X\beta, X\mu)$.
12. Prove that any variety (cf. exercise 6+ of section 4) of $\boldsymbol{\beta}$-algebras which contains at least one algebra having two or more elements must be the variety of all $\boldsymbol{\beta}$-algebras. [Hint: let 2 be the unique two-element $\boldsymbol{\beta}$-algebra; show that the inclusion $X\beta \subset 2^{(2^X)}$ is a $\boldsymbol{\beta}$-subalgebra.]
13. Show that if **T** is not trivial then the syntactic operation $1\eta : 1 \longrightarrow 1T$ has arity 1.
14. In the context of 5.36, show that a **T**-homomorphism from (X, ξ) to (Y, θ) is the same thing as a natural transformation from M_ξ to M_θ.
15. Let $\omega \in XT$ and let S be a subset of X with inclusion map $i : S \longrightarrow X$. Show that S is a support of $\tilde{\omega}$ if and only if ω is in the image of $iT : ST \longrightarrow XT$. [Hint: study the proof of 5.11.]
16. Say that **T** is *atomic* if every operation has a minimal support, that is, for all $\omega \in XT$ there exists a support S of $\tilde{\omega}$ such that no proper subset of S is a support of $\tilde{\omega}$. Prove that every finitary theory and the power-set theory are atomic but that the ultrafilter theory is not atomic.
17. Prove that the double power-set theory is not atomic. [Hint: for $\mathscr{B} \in XT$ and $S \subset X$ define $\mathscr{B}^S = \{A \subset X$: there exists $B \in \mathscr{B}$ such that $A \cap S = B \cap S\}$; then $\mathscr{B} \subset \mathscr{B}^S$; show that S is a support of $\mathscr{B}$ if $\mathscr{B} = \mathscr{B}^S$.]
18. Our discussion of the supremum-infimum duality in 5.48 should not be misconstrued as effecting changes in the algebraic theory itself; in fact, for T the power-set functor, show that intersection is *not* a natural transformation $TT \longrightarrow T$.
19. Let (Ω, E) be an equational presentation and let $U : \Omega\text{-alg} \longrightarrow$ **Set**, $V : (\Omega, E)\text{-alg} \longrightarrow$ **Set** be the underlying set functors.
    (a) For each set n let $A(n)$ be the class of natural transformations from U^n to U. For each ω in Ω_m, $\tilde{\omega} : U^m \longrightarrow U$ induces the structure of an Ω-algebra on $A(n)$ by $(p_i : i \in m)\delta_\omega = (p_i)\tilde{\omega}$. Define $n\eta : n \longrightarrow A(n)$ by $\langle i, n\eta \rangle = pr_i : U^n \longrightarrow U$. Show that the intersection of all subalgebras of $A(n)$ containing the image of n has the universal property of 4.12 that all Ω-algebra valued functions from n admit a unique extension to an Ω-homomorphism.
    (b) Similarly, construct the perhaps large free (Ω, E)-algebra by repeating the construction of (a) for the class $B(n)$ of natural transfor-

mations from V^n to V [Hint: to prove that $B(n)$ satisfies E show that $B(n)$ may be identified with $A(n)/E_n$ in a sense similar to that of 2.1].

20. In this exercise we indicate how compact abelian groups arise as the algebras over a theory in **Set**. We begin by reviewing some of the theory of character groups (see e.g. [Hewitt and Ross '63], [Pontrjagin '46]). Let S denote the circle group, that is, the compact metric abelian group of complex numbers of modulus 1 with complex multiplication as group operation. For each locally compact abelian group C, the character group $C^\wedge$ of C is the locally compact abelian group of all continuous homomorphisms from C to S with neighborhood basis at the origin $\{U(F, n) : F \text{ is a compact subset of } C \text{ and } n = 1, 2, 3 \ldots\}$ where $U(F, n)$ is the set of all characters χ in $C^\wedge$ such that $|x\chi - 1| < 1/n$ for all x in F. If C, D are locally compact abelian groups then the passage from $f : C \longrightarrow D$ to $f^\wedge : D^\wedge \longrightarrow C^\wedge$, where $\chi f^\wedge = f.\chi$, establishes a bijection between the two sets of continuous homomorphisms. The map $C \longrightarrow (C^\wedge)^\wedge$ which sends c to "evaluate at c" is a topological isomorphism, so "$(C^\wedge)^\wedge = C$." C is compact if and only if $C^\wedge$ is discrete.
    (a) For each set A, consider the discrete group (ignore the natural product topology!) S^A of functions from A to S and let AT denote the underlying set of the compact abelian group $(\mathsf{S}^A)^\wedge$. Let $A\eta : A \longrightarrow AT$ send a to the projection $pr_a : \mathsf{S}^A \longrightarrow \mathsf{S}$. Given $\beta : B \longrightarrow CT$, define the homomorphism $\psi : \mathsf{S}^C \longrightarrow \mathsf{S}^B$ by $f\psi = (b \longmapsto f\beta_b)$ and hence the map $\beta^\# : BT \longrightarrow CT$ by $\beta^\# = \psi^\wedge$. Prove that $(T, \eta, \circ)$ is an algebraic theory in **Set** where $\alpha \circ \beta = \alpha.\beta^\#$.
    (b) Let C be a compact abelian group with underlying set $|C|$. Define a function $\xi : |C|T \longrightarrow |C|$ by $\chi\xi = c$ where c is the unique element of C such that the restriction of $\chi : \mathsf{S}^{|C|} \longrightarrow \mathsf{S}$ to $C^\wedge$ is "evaluate at c." Prove that $(|C|, \xi)$ is a **T**-algebra.
    (c) Complete the proof that "compact abelian groups and continuous homomorphisms" = "**T**-algebras and **T**-homomorphisms."

The next two exercises provide insight into the algebra definition of exercise 4.11 for theories in **Set**.

21. A cardinal n is a *generating cardinal* for the algebraic theory **T** in **Set** if every p in XT admits a factorization

$$p = (1 = A_1) \xrightarrow{\alpha_1} A_2 \longrightarrow \cdots \longrightarrow A_k \xrightarrow{\alpha_k} X$$

such that for all $1 \leqslant i \leqslant k$ and $a \in A_i$, $ar(a\alpha_i) < n^\bullet$, where

$$n^\bullet = \begin{cases} n + 1 & \text{(if } n \text{ is finite)} \\ n & \text{(if } n \text{ is infinite)} \end{cases}$$

If **T** has a generating cardinal, the least one is the *generating rank* of **T**.
    (a) Show that **T** has generating rank 2 if $\mathbf{Set}^\mathbf{T}$ = monoids.
    (b) A *comparison algebra* [Kennison '75] is an (Ω, E)-algebra where Ω has a single operation C of arity 4 and E consists of the following five equations:

$$\begin{aligned} &aaxyC = x && abxxC = x \\ &abxyC = baxyC && ababC = b \\ &abtuvwC\hat{t}\hat{u}\hat{v}\hat{w}CC = abt\hat{t}Cabu\hat{u}Cabv\hat{v}Cabw\hat{w}CC \end{aligned}$$

(see exercise 3.3.8). Show that the corresponding algebraic theory has generating rank 3. [Hint: show $abxyC = yabyWabxWW$ if $tuvW = tuvtC$.]

(c) Prove that "generating rank $\leqslant$ rank."

(d) Let n be a generating cardinal for **T** and let $\xi: XT \longrightarrow X$ satisfy "$X\eta.\xi = \mathrm{id}_X$" and "for all $\alpha, \beta: A \longrightarrow XT$ with $\mathrm{card}(A) < n^\bullet$, if $\alpha.\xi = \beta.\xi$ then $\alpha^{\#}.\xi = \beta^{\#}.\xi$". Prove that (X, ξ) is a **T**-algebra. [Hint: use exercise 4.11; let AT_k be the subset of those elements of AT which, in the definition of "generating cardinal" above, admit a factorization of size k; prove "for arbitrary A and $\alpha, \beta: A \longrightarrow XT$, if $\alpha.\xi = \beta.\xi$ then $\langle p, \alpha^{\#}.\xi\rangle = \langle p, \beta^{\#}.\xi\rangle$ for all $p \in AT_k$" by induction on k; the assumptions on ξ provide the basis; if $p \in AT_{k+1}$, the basis argument also proves $(\alpha_{k+1} \circ \alpha).\xi = (\alpha_{k+1} \circ \beta).\xi$; now use the induction hypothesis.]

22. Let **T** be an algebraic theory in **Set**, let $\mathscr{A}$ be an arbitrary class of sets and let $\xi: XT \longrightarrow X$ satisfy $X\eta.\xi = \mathrm{id}_X$. Prove that the following two conditions are equivalent:

(a) For every $\alpha, \beta: A \longrightarrow XT$ with $A \in \mathscr{A}$, if $\alpha.\xi = \beta.\xi$ then $\alpha^{\#}.\xi = \beta^{\#}.\xi$.

(b) "ξ commutes with $\mathscr{A}$-ary operations", i.e., for every $A \in \mathscr{A}$ and $\omega \in AT$, the following diagram commutes:

$$\begin{array}{ccccc} XT^A & & \xrightarrow{\quad (XT,\, X\mu)\tilde{\omega} \quad} & & XT \\ \Big\downarrow{\scriptstyle \xi^A} & & & & \Big\downarrow{\scriptstyle \xi} \\ X^A & \xrightarrow[\quad X\hat{\omega} \quad]{} & XT & \xrightarrow[\quad \xi \quad]{} & X \end{array}$$

[Hint: to prove (b) from (a), for arbitrary α consider $\beta = \alpha.\xi.X\eta$.]

23. [Michael '51]. Let $\mathscr{K}$ be the category of compact Hausdorff spaces and continuous maps. For A in $\mathscr{K}$, topologize the set, AT, of closed subsets of A by designating as open subbasis the family $\{U^\star: U \text{ open in } A\}$ where $U^\star = \{S \in AT: S \cap U \neq \varnothing\}$. Show that AT is compact Hausdorff, that $A\eta: A \longrightarrow AT$, $a \longmapsto \{a\}$ is continuous, that if $\alpha: A \longrightarrow BT$ is continuous then

$$\alpha^{\#}: AT \longrightarrow BT, \qquad S \longmapsto \bigcup\{a\alpha: a \in S\}$$

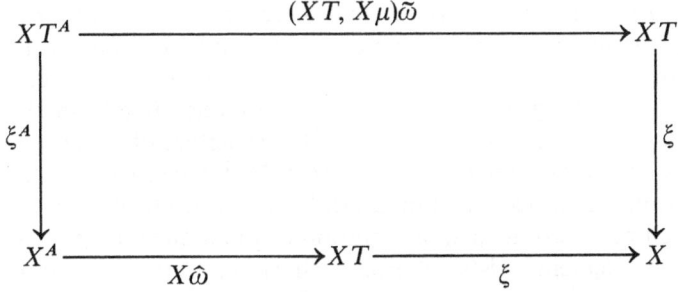

is well-defined and continuous and hence that $(T, \eta, \circ)$, $\alpha \circ \beta = \alpha.\beta^{\#}$, is an algebraic theory in $\mathscr{K}$. [Hint: the only detail not readily available in Michael's paper is the continuity of $\alpha^{\#}$; to this end, it suffices for open U in B to check that $U^{\star}(\alpha^{\#})^{-1} = (U\alpha^{-1})^{\star}$.] We do not know how to interpret the **T**-algebras, but they clearly have to do with "continuous selections."

24. Let α be a regular cardinal, let **T** have rank $\leqslant \alpha$ and let (X, ξ) be **T**-algebra possessing an inclusion-minimal set A of generators with $\operatorname{card}(A) = \alpha$. Show that every set generating (X, ξ) has cardinal $\geqslant \alpha$ (and hence that any two minimal sets of generators have the same (cardinal). [Hint: if $\langle B \rangle = X$ then for b in B choose $S_b \subset A$ with $\operatorname{card}(S_b) < \alpha$ and $b \in \langle S_b \rangle$; as $\bigcup S_b = A$, $\operatorname{card}(B) \geqslant \alpha$.]

Exercises for the Primer on Set Theory

A. A left-to-right picture of the first few ordinals is

$$0\ 1\ 2 \cdots n \cdots \omega\ s(\omega)\ s^2(\omega) \cdots$$

Extend the picture to the right as far as you can. Then prove that the set of all countable ordinals is uncountable. [Hint: consider the least uncountable ordinal.]

B. Prove *Cantor's theorem*: for any set A, $\operatorname{card}(2^A) > \operatorname{card}(A)$ (where 2^A means the set of all subsets of A). [Proof outline: no $f: A \longrightarrow 2^A$ can be surjective since (cf. Russell's Paradox) the set $\{a \in A : a \notin af\}$ is not in the image of f.] Conclude that, for any cardinal x, x^+ exists and $x < x^+ \leqslant 2^x$. Look up the "generalized continuum hypothesis" which asserts that $x^+ = 2^x$.

C. Use Zorn's lemma to prove that there exists a family $\mathscr{F}$ of circles (boundary and interior) in the plane such that no two intersect (meaning, also, that no two are tangent) and yet so "densely distributed" that any circle not already in $\mathscr{F}$ must intersect a circle already in $\mathscr{F}$. Observe that your argument makes no use of the structure of circles.

D. Zorn's lemma was introduced in [Zorn '35] where its equivalence with the axiom of choice was noted (and proved in [Kneser '50]). The following proof outline is an adaptation of Banaschewski '53. [Assume no maximal element exists. (i) For each chain C there exists $u(C)$ in X strictly greater than each element of C. The sole use of the axiom of choice is the existence of this choice function u from chains to elements; if you believe u really exists then Zorn's lemma is "true." (ii) Define $\mathbf{C}$ to be the set of all chains C such that (a) every nonempty subset of C has a least element and (b) for every x in C, $x = u(C_x)$ where C_x is the chain of all y in C with $y < x$. (iii) Prove that for every C in $\mathbf{C}$ the chain $C^+ = C \cup \{u(C)\}$ is again in $\mathbf{C}$. (iv) If $C, D \in \mathbf{C}$ exactly one of "$C = D$," "$C \subset D, C \neq D$, and $C = D_x$ for some x in D," "$D \subset C, D \neq C, D = C_x$ for some x in C" holds (this takes some work). (v) $W = \cup \mathbf{C} \in \mathbf{C}$. (vi) $u(W) \in W$, the desired contradiction.]

E. Use Zorn's lemma to prove that for any two sets X, Y, either X admits an injection into Y or Y admits an injection into X. [Proof outline: use the partially ordered set of all (A, f) with $A \subset X$ and $f: A \longrightarrow Y$ injective; show that a maximal (A, f) is such that f is onto or $A = X$.]

F. Selfreference is not really a set-theoretic phenomenon. Consider the truth value of the sentence "this sentence is false."

Chapter 2

Trade Secrets of Category Theory

This chapter provides a selfcontained introduction to some elementary topics in category theory. Familiar constructions of set theory are generalized to an arbitrary category. "Sets with structure" generalizes to "objects with structure," providing a universe in which to discuss "algebraic structure."

1. The Base Category

A proviso such as "all spaces are assumed Hausdorff and all maps are assumed continuous" is the mathematical author's way of saying "let Hausdorff topological spaces and continuous maps be the base category." The most familiar base category is the category **Set** of sets and functions. In this section, we explore how some familiar constructions involving sets and functions can be described in more arbitrary categories.

1.1 Assumption. *For the balance of this section, fix a category $\mathscr{K}$. We will assume that $\mathscr{K}$ is locally small (also: $\mathscr{K}$ has small hom-sets) in the sense that for each pair (A, B) the class $\mathscr{K}(A, B)$ is a small set (as defined in the primer on set theory of section 1.5).*

Set is certainly locally small. In a category of "structured sets" morphisms from X to Y are determined as functions from the underlying set of X to that of Y; therefore all categories of structured sets are locally small. In pure category theory one is interested in "functor categories" (see exercises 1, 2.9 and 3.2.5) such as the category whose objects are functors from **Set** to itself and whose morphisms are natural transformations (obviously an important category with regard to the material in Chapter 1); this category is not locally small. We are therefore making some concessions to everyday mathematics in insisting that $\mathscr{K}$ be locally small. The numerous examples we will draw upon in this chapter are by and large restricted to categories of sets with structure. Just to see what is abstractly possible, let us note two different sorts of category.

1.2 Example. If $\mathscr{K}$ has only one object, then the set M of all morphisms of $\mathscr{K}$ consists solely of endomorphisms of the unique object (where, in any category, an *endomorphism* is a morphism whose domain and codomain coincide). Composition is an everywhere-defined associative operation on M and the identity map of the unique object is a two sided unit. $\mathscr{K}$ is the same thing as a monoid.

1.3 Definitions. *A preordered class is a (perhaps large) class C equipped with a reflexive and transitive binary relation. A preordered category is a category in which there is at most one morphism from A to B for each pair (A, B)*

of objects. For all practical purposes the two notions are the same (define $A \leqslant B$ to mean there exists a morphism from A to B). Henceforth we will treat preordered classes as if they were categories.

1.4 Isomorphisms in a Category. It is traditionally clear when two structured sets of the same sort are "abstractly the same." This occurs just in case there is a "structure-preserving" bijection, called an isomorphism, between them; in actual context, the following definition of "isomorphism" usually coincides with the intuitively most natural concept of "structure-preserving relabelling." The morphism $f: A \longrightarrow B$ is defined to be an *isomorphism in* $\mathscr{K}$ just in case there exists a morphism $f^{-1}: B \longrightarrow A$ in $\mathscr{K}$ such that $f.f^{-1} = \mathrm{id}_A$ and $f^{-1}.f = \mathrm{id}_B$. In **Set**, f is an isomorphism if and only if f is a bijection. For $\mathscr{K}$ = groups and homomorphisms, f is an isomorphism if and only if f is a bijective homomorphism. For $\mathscr{K}$ = topological spaces and continuous maps, f is an isomorphism if and only if f is a homeomorphism (it is not sufficient that f be a continuous bijection). Two objects A, B in $\mathscr{K}$ are *isomorphic in* $\mathscr{K}$ if there exists an isomorphism $f: A \longrightarrow B$ in $\mathscr{K}$. Isomorphism is an equivalence relation, often written "$\cong$", on the objects of $\mathscr{K}$. One of the most fundamental philosophical principles in category theory is: *isomorphic objects are abstractly the same*. Since the constructions of category theory are, as a rule, unique only "up to isomorphism," we should never lose sight of what categorical isomorphism means. For example, in the category of metric spaces and continuous maps, "isomorphic" only means "homeomorphic" which may or may not be adequate, depending on context.

A monoid is a group if and only if all of its morphisms are isomorphisms. A preordered class is partially ordered (that is, $\leqslant$ is antisymmetric as well as reflexive and transitive) if and only if every isomorphism is an identity morphism.

1.5 Products. In set theory, the product of a family $(A_i : i \in I)$ of sets is the set A of all I-tuples $(a_i : i \in I)$ with each $a_i \in A_i$. The function $A \longrightarrow A_j$ which sends (a_i) to a_j is called the jth projection function. If A' is a set and if we are given functions $f_i : A' \longrightarrow A_i$ then there exists a unique function $f: A' \longrightarrow A$ such that $f.p_i = f_i$ for all $i \in I$; f is defined by $af = (af_i : i \in I)$.

Given a family of objects $(A_i : i \in I)$ in a category $\mathscr{K}$, a *product of* $(A_i : i \in I)$ *with respect to* $\mathscr{K}$ is an object A of $\mathscr{K}$ and an I-tuple of $\mathscr{K}$-morphisms of form $p_i : A \longrightarrow A_i$ possessing the "universal property" that whenever A' is an object of $\mathscr{K}$ similarly equipped with an I-tuple of $\mathscr{K}$-morphisms of form

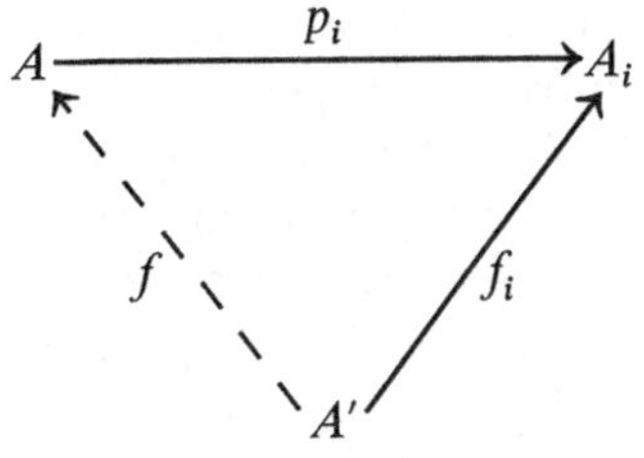

$f_i: A' \longrightarrow A_i$ there exists a unique $\mathscr{K}$-morphism $f: A' \longrightarrow A$ such that $f.p_i = f_i$ for all $i \in I$. ("Universal property" is a vague term which refers to a construction whose central feature is the existence of a unique morphism subject to a categorical property.) The morphisms p_i are called *projections.*

1.6 Proposition. *Any two products of* $(A_i : i \in I)$ *are isomorphic.*

Proof. Suppose $p_i: A \longrightarrow A_i$ and $q_i: B \longrightarrow A_i$ are both products of (A_i). Consider the unique induced maps as shown below:

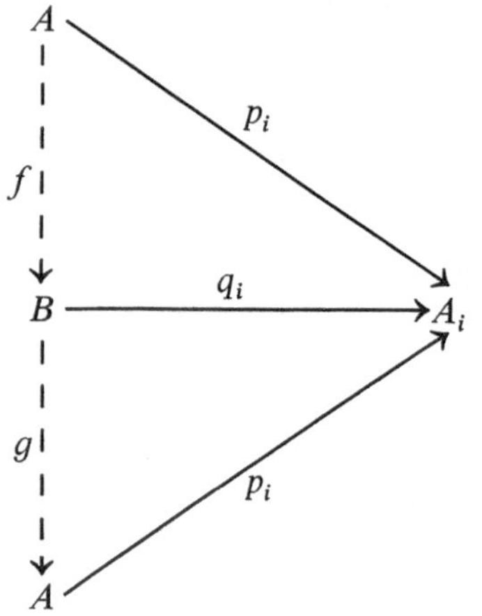

Then $(fg)p_i = f(gp_i) = fq_i = p_i = (\mathrm{id}_A)p_i$ for all $i \in I$, which proves $fg = \mathrm{id}_A$. Similarly, $gf = \mathrm{id}_B$. Therefore f is an isomorphism transforming one set of projections into the other (which would seem to be as isomorphic as two products could possibly be without being equal). □

Two things are worth noticing in the above proof: it works for all universal properties; it required two of the three category axioms.

Because of 1.6, we can think in terms of *the* product of (A_i) and write it as $\prod A_i$. In practice, "$\prod A_i$" is either any convenient choice of—or the isomorphism class of all—I-tuples $p_i: A \longrightarrow A_i$ with the universal property; for most categorical purposes, these distinctions do not matter. In some contexts, the notation "$\prod A_i$" means just the object A, e.g. as in "consider p_j: $\prod A_i \longrightarrow A_j$" which is synonymous with "let $p_i: A \longrightarrow A_i$ be a product of (A_i)".

The "size" of a product is the size of I. The smallest product is the empty product (i.e., I is empty), which it is standard to call a *terminal object of* $\mathscr{K}$, often denoted by the symbol 1. A terminal object is the same thing as an object A possessing the universal property that for all objects A' there exists a unique map $A' \longrightarrow A$. In **Set**, a terminal object is a 1-element set, which explains the notation "1." The category of all sets which do not have exactly one element and functions does not have a terminal object. Unary products always exist ($\mathrm{id}_A : A \longrightarrow A$ is one.) Binary products are better written $A_1 \times A_2$. $\mathscr{K}$ *has small products* if $\prod A_i$ exists for every family $(A_i : i \in I)$ of objects of $\mathscr{K}$ with I a small set. Similarly, $\mathscr{K}$ *has finite products, has countable products* and so on accordingly as $\prod A_i$ exists for finite families, for countable families, et cetera.

Several examples follow and more appear in the exercises.

1.7 Example. **Set** has products, the usual ones. The category of finite sets and functions has finite products (the usual ones). The fact that an infinite product (in the usual sense) of finite sets is not necessarily finite strongly suggests but does not prove that finite sets does not have products. To justify our intuition, notice that morphisms $1 \longrightarrow X$ are essentially the same thing as elements of X. Using the universal property of a product, this proves that the elements of $\prod A_i$ are indeed in bijective correspondence with the elements of the usual product.

1.8 Example. The category of topological spaces and continuous functions has products; one provides the usual cartesian product set with what is normally called the product topology, or the topology of pointwise convergence. A net $(a_{i,\alpha})$ converges to (a_i) in $\prod A_i$ if and only if for all i, $a_{i,\alpha}$ converges to a_i. This statement both characterizes the product topology and amounts to the universal property.

1.9 Example. The category of metric spaces and distance-decreasing maps (we call the function $f:(X, d) \longrightarrow (X', d')$ *distance decreasing* if for all $x, y \in X$, $d'(xf, yf) \leqslant d(x, y)$) has finite products and many other—but not all—products. Since all constant functions are distance decreasing, we can use the one-element metric space as in 1.7 to argue that if $(A, d) = \prod(A_i, d_i)$ A must be the usual product of the sets A_i. Because projections must be distance decreasing, we must have $d((a_i), (b_i)) \leqslant \text{Sup}(d_i(a_i, b_i) : i \in I)$. It is now easy to prove that $\prod(A_i, d_i)$ exists if and only if $\text{Sup}(d_i(a_i, b_i) : i \in I)$ is finite for every pair (a_i), (b_i) in the usual product set A; and then (A, d) is the product where d is this supremum. For any fixed M, any family of metric spaces of diameter $\leqslant M$ has a product which is itself of diameter $\leqslant M$.

(1.10) In a preordered class, products are exactly the same thing as infima.

1.11 Proposition. *Let $\mathscr{K}$ have products and let* **T** *be an algebraic theory in $\mathscr{K}$. Then the category $\mathscr{K}^{\mathbf{T}}$ of* **T**-*algebras has products.*

Proof. Let (A_i, ξ_i) be a family of **T**-algebras and let $p_i : A \longrightarrow A_i$ be a product diagram in $\mathscr{K}$. By the universal property, there exists a unique morphism $\xi : AT \longrightarrow A$ such that

$$\begin{array}{ccc} AT & \xrightarrow{(p_i)T} & A_iT \\ {\scriptstyle \xi}\big\downarrow & & \big\downarrow{\scriptstyle \xi_i} \\ A & \xrightarrow[p_i]{} & A_i \end{array}$$

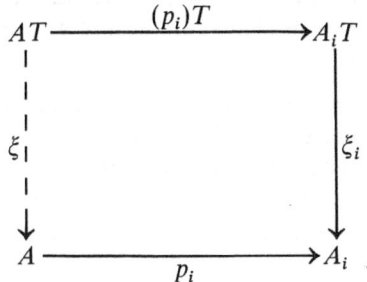

Although 1.4.27 was nominally restricted to **Set**, the reasoning there is perfectly general and proves that (A, ξ) is a **T**-algebra. It remains to establish the universal property. Suppose we have given **T**-homomorphisms $f_i:(B, \theta) \longrightarrow (A_i, \xi_i)$. There exists a unique $\mathscr{K}$-morphism $f:B \longrightarrow A$ such that $f.p_i = f_i$ for all i. In the diagram below, we must show that the

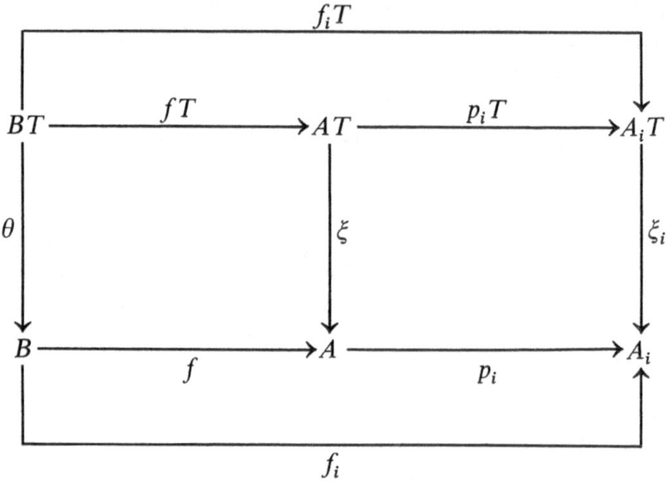

leftmost square commutes given that all the outer rectangles do. But this follows immediately from the universal property, since the leftmost square is commutative followed by each p_i. □

The above proposition is our first encounter with "categorical universal algebra."

In all of the examples of products so far, there is no evidence that the underlying set of the product structured set, when it exists, is not always the usual product set. The following is such a counterexample.

1.12 Example. Consider the category whose objects are metric spaces with base point $(X, d, \bar{x})$ (the "base point" $\bar{x}$ is simply an arbitrary element of X) and distance-decreasing base-point preserving (i.e., $xf = x'$) functions. Every family $(X_i, d_i, \bar{x}_i)$ has a product (X, d, x) where X is the subset of the usual product of the X_i consisting of all tuples (x_i) with the property that $\mathrm{Sup}(d_i(x_i, \bar{x}_i))$ is finite. As in 1.9, d is defined by $d((x_i), (y_i)) = \mathrm{Sup}(d_i(x_i, y_i)$: $i \in I)$ which is guaranteed to be finite by the definition of X and the fact that each d_i satisfies the triangle inequality. $(\bar{x}_i)$ provides the base point. The projections are the restrictions of the usual ones. In general, X is a proper subset of the usual product set.

1.13 Equalizers. Given sets A_1, A_2 and two functions $f, g:A_1 \longrightarrow A_2$ the inclusion map i of the subset $A = \{x \in A_1 : xf = xg\}$ on which f and g agree can be characterized up to isomorphism by the following universal property: for every function $i':A' \longrightarrow A_1$ such that $i'f = i'g$ there exists a unique function $h:A' \longrightarrow A$ such that $hi = i'$ (since the image of i' is contained in A, h is defined by $a'h = a'i'$). Given morphisms $f, g: A_1 \longrightarrow A_2$ in a category $\mathcal{K}$, an *equalizer of* (f, g) is an object E and a morphism $i: E \longrightarrow A$ with the following universal property:

(1) $i.f = i.g$

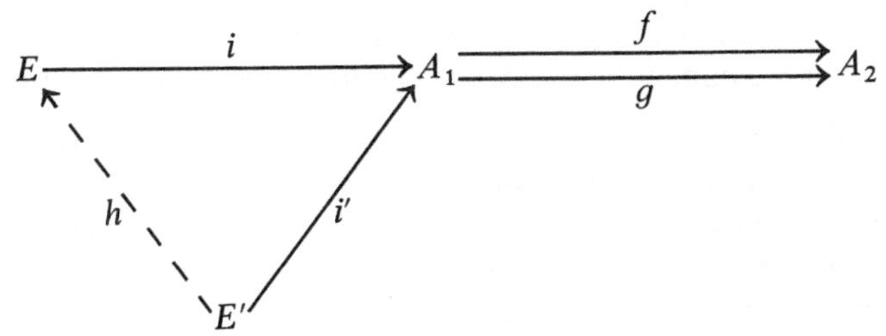

(2) Given i' with $i'.f = i'.g$, there exists unique h with $h.i = i'$.

By the same sort of reasoning as in 1.6, equalizers are unique up to isomorphism. We speak of *the* equalizer of f and g and write eq(f, g) to denote any convenient representative of—or the entire isomorphism class of—all equalizers $i: E \longrightarrow A$ of $f, g: A_1 \longrightarrow A_2$. $\mathcal{K}$ *has equalizers* if eq(f, g) exists for every pair $f, g: A_1 \longrightarrow A_2$.

Most categories of sets with structure have equalizers via the appropriate "substructure" on the subset of points on which f and g agree. For topological spaces, use the relative topology. For metric spaces (either 1.9 or 1.12) just restrict the metric to the subset. For groups and homomorphisms, the subset in question is a subgroup. The latter, or course, is another instance of categorical universal algebra.

1.14 Proposition. *Let* $\mathcal{K}$ *have equalizers and let* $\mathbf{T}$ *be an algebraic theory in* $\mathcal{K}$. *Then the category* $\mathcal{K}^{\mathbf{T}}$ *of* $\mathbf{T}$*-algebras has equalizers.*

Proof. Let $f, g: (A, \xi) \longrightarrow (B, \theta)$ be $\mathbf{T}$-homomorphisms and let $i: E \longrightarrow A = \text{eq}(f, g)$ in $\mathcal{K}$. Consider the diagram:

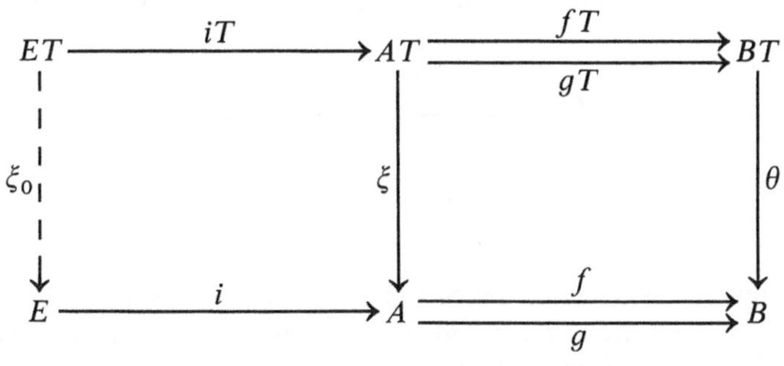

We have $(iT.\xi).f = iT.(\xi.f) = iT.fT.\theta = (i.f)T.\theta = (i.g)T.\theta = iT.\xi.g$. From the universal property, we obtain a unique ξ_0 with $\xi_0.i = iT.\xi$. We need to show on the one hand that (E, ξ_0) is a $\mathbf{T}$-algebra (since then surely i becomes a $\mathbf{T}$-homomorphism) and on the other hand that the universal property is satisfied (which amounts to the assertion that if (E', γ) is a **T**-algebra and if $f: E' \longrightarrow E$ is a $\mathscr{K}$-morphism then $f:(E', \gamma) \longrightarrow (E, \xi_0)$ is a **T**-homomorphism providing $f.i: (E', \gamma) \longrightarrow (A, \xi)$ is). Consideration of the details of why this worked for products (see 1.4.27 and 1.11), shows that the result was a formal consequence of the fact that when two maps into a product are the same followed by each projection the two maps were already equal. Applying the same concepts, the proof is completed by observing

1.15 Proposition. *Whenever $i: E \longrightarrow A$ is the equalizer of two morphisms $f, g: A \longrightarrow B$ in $\mathscr{K}$ and whenever two morphisms $t, u: E' \longrightarrow E$ have the property that $t.i = u.i$, then $t = u$.* The proof of 1.15 is clear, since if i' denotes the common value of $t.i$ and $u.i$, $i'.f = i'.g$ and t and u are both the unique morphism induced by i'. □

1.16 Example. The category of nonempty sets and functions does not have equalizers. For let $f, g: A \longrightarrow B$ be a pair of functions between nonempty sets which do not agree on any element of A: if $i: E \longrightarrow A$ satisfied $i.f = i.g$ then, since all objects are nonempty sets, there exists $x \in E$ and f and g agree on xi, a contradiction. A category theorist believes that a category without equalizers is "incomplete" and regards with suspicion statements such as "all sets will be assumed nonempty" which preface many books and papers; to her, this is like assuming that all complex numbers are nonzero.

1.17 Example. (Suggested by M. Barr.) Let $\mathscr{K}$ be the category whose objects are abelian groups which have no elements of order 2 and which are *2-divisible* (i.e., for all x there exists y with $2y = x$), and whose morphisms are group homomorphisms. Given $f, g: A \longrightarrow B$ in $\mathscr{K}$, let E_0 be the subgroup $\{a \in A : af = ag\}$, define $E_{k+1} = 2E_k$ and set E to be the intersection of all E_k. Then E is 2-divisible (if x is in E let y_k in E_0 satisfy $2^k y_k = x$; as A has no elements of order 2, $2y_{k+1} = y_k$, so that $2y_1 = x$ with y_1 in E). It is then clear that the inclusion map of E is the equalizer of f, g in $\mathscr{K}$. For a specific case, let **Q** be the additive group of rational numbers, let **Z** be the subgroup of integers and consider the canonical projection and zero map:

$$\mathbf{Q} \underset{0}{\overset{\theta}{\rightrightarrows}} \mathbf{Q}/\mathbf{Z}$$

Here $E_0 = Z$ whereas $E = 0$. Thus $\mathscr{K}$ has equalizers but they are not constructed at the level **Set**.

1.18 Limits. A *diagram scheme* Δ is given by a set $N(\Delta)$ of *nodes* and a specification to each ordered pair (i, j) of nodes a set $\Delta(i, j)$ of *edges from i to j*, satisfying the axiom that $\Delta(i, j)$ is disjoint from $\Delta(i', j')$ if $(i, j) \neq (i', j')$. For example, any category is a diagram scheme where the objects are the nodes and the morphisms are the edges. A *diagram in* a category $\mathscr{K}$ is a pair (Δ, D) where Δ is a diagram scheme and D assigns to each node $i \in N(\Delta)$ an object D_i of $\mathscr{K}$, and then D assigns to each edge $\alpha \in \Delta(i, j)$ a $\mathscr{K}$-morphism

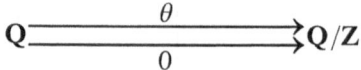

$D_\alpha : D_i \longrightarrow D_j$. A *lower bound* of a diagram (Δ, D) is a pair (L, ψ) where L is an object of $\mathscr{K}$ and ψ assigns to each node $i \in N(\Delta)$ a $\mathscr{K}$-morphism ψ_i of form $\psi_i : L \longrightarrow D_i$ such that

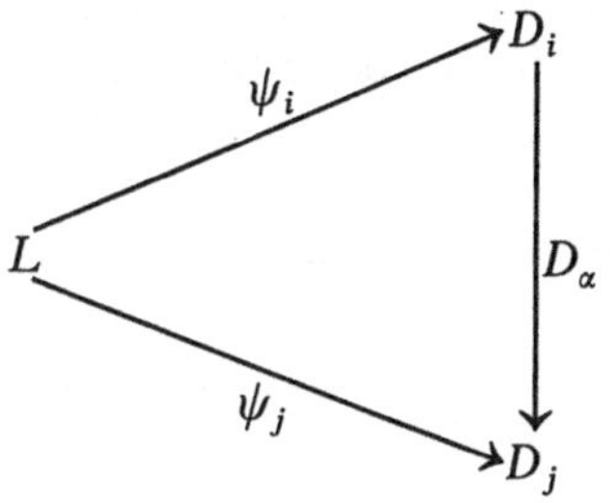

commutes for every edge $\alpha \in \Delta(i, j)$. A *limit* of the diagram (Δ, D) is a lower bound (L, ψ) with the universal property that whenever (L', ψ') is another lower bound there exists a unique $\mathscr{K}$-morphism $f : L' \longrightarrow L$ such that

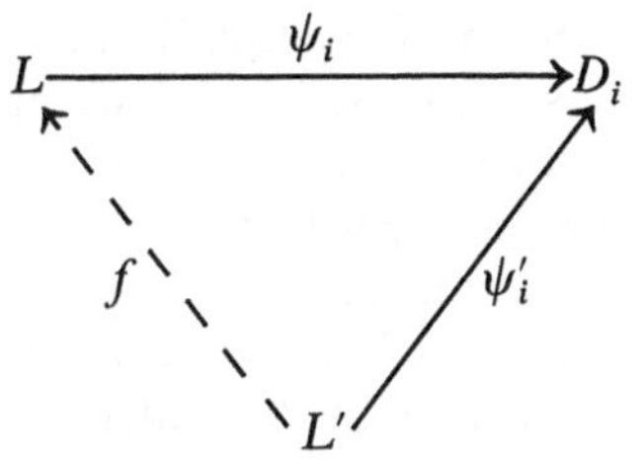

is commutative for all nodes i.

(1.19) If $\mathscr{K}$ is a preordered category, a lower bound of (Δ, D) is an object L with $L \leqslant D_i$ for all i. L is a limit if and only if $L = \mathrm{Inf}(D_i)$

(1.20) Let I be an arbitrary set, and define Δ by $N(\Delta) = I$, $\Delta(i, j) = \varnothing$ for all $i, j \in I$. For each category $\mathscr{K}$, a diagram of form (Δ, D) is the same thing as an I-indexed family of objects of $\mathscr{K}$. For such diagrams, "limit" means "product."

There is an alternate equivalent definition of lower bounds which is useful in practice. A subset F of $N(\Delta)$, for a given diagram scheme Δ, is called *final* if for all $j \in N(\Delta)$ with $j \notin F$ there exists $i \in F$ with $\Delta(i, j) \neq \varnothing$. Given (Δ, D), a *lower bound relative to* F is a pair (L, ψ) where L is an object of $\mathscr{K}$ and ψ assigns to each $i \in F$ a $\mathscr{K}$-morphism ψ_i of form $\psi_i : L \longrightarrow D_i$ subject to the conditions that the outer (solid) square of the diagram

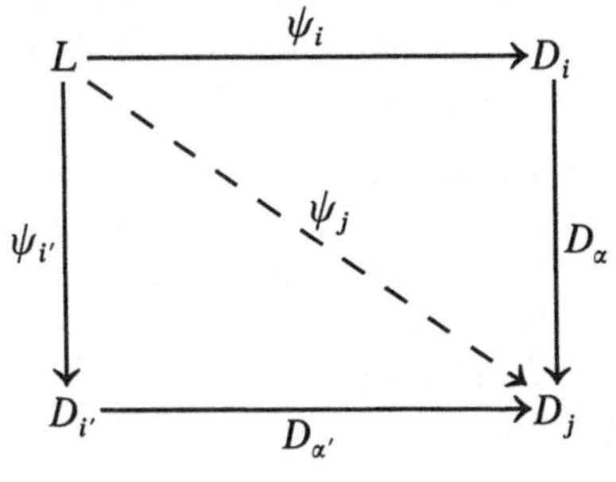

is commutative for all $i, i' \in F$, $j \in N(\Delta)$, $\alpha \in \Delta(i, j)$, $\alpha' \in \Delta(i', j)$ and that the triangle $\psi_i . D_\alpha = \psi_j$ is commutative whenever $i, j \in F$, $\alpha \in \Delta(i, j)$. So long as F is final, it is obvious that "restricting to F" is a bijective passage from lower bounds of (Δ, D) to lower bounds relative to F of (Δ, D) whose inverse is obtained by defining ψ_j as indicated in the diagram above. For all practical purposes, lower bounds relativized to a final set are the same as lower bounds.

(1.21) Let Δ have two nodes and two edges as shown:

$$t \overset{\alpha}{\underset{\beta}{\rightrightarrows}} u$$

A diagram (Δ, D) in $\mathscr{K}$ is any pair of $\mathscr{K}$-morphisms with the same domain and codomain. Given such a diagram $f, g: A \longrightarrow B$, a lower bound relative to the final set $\{t\}$ is a $\mathscr{K}$-morphism $i: E \longrightarrow A$ such that $i.f = i.g$. Limits of such diagrams are the same thing as equalizers.

By exactly the same reasoning used in 1.6, we have that any two limits of the same diagram are isomorphic. We speak of *the* limit of (Δ, D) and write $\lim D$ to denote any convenient representative of—or the equivalence class of all—limits of (Δ, D). A diagram scheme Δ is *small* if $N(\Delta)$ is a small set and if $\Delta(i, j)$ is a small set for all i, j. *$\mathscr{K}$ has small limits* and $\mathscr{K}$ is *small complete* if for every diagram (Δ, D) in $\mathscr{K}$ with Δ small, (Δ, D) has a limit. A category may have some large limits. For example, if Δ has two nodes i, j with $\Delta(j, i) = \varnothing$ but $\Delta(i, j)$ a large set and if (Δ, D) is a diagram such that only two distinct morphisms $D_i \longrightarrow D_j$ are among the D_α, $\lim D$ is just the same as the equalizer of the two morphisms involved. On the other hand, if $\mathscr{K}$ is not preordered, so that there exist objects A, B admitting at least two distinct morphisms $f, g: A \longrightarrow B$ then, for each class I, there are 2^I distinct I-tuples of morphisms from A to B. Therefore, if the product P of I copies of B exists, there are at least 2^I $\mathscr{K}$-morphisms from A to P; since $\mathscr{K}$ is locally small, no such P can exist if I is a large set. With the exception of preordered classes, locally small categories never have large products.

The following quite remarkable theorem guarantees that most familiar categories have small limits and justifies our preoccupation with products and equalizers.

1.22 Theorem. *$\mathscr{K}$ has small limits if and only if $\mathscr{K}$ has small products and equalizers; in that case, any small limit can be represented as the equalizer of a pair of maps between two products.*

Proof. Let (Δ, D) be a diagram in $\mathscr{K}$ with Δ small. The following diagram defines (using the universal property of the rightmost product) the morphisms f and g. Let $(E, k) = \mathrm{eq}(f, g)$. Defining ψ_i as above, $(E, \psi) = \lim D$. The proof follows immediately, since

(1) For any object E', an assignment ψ' of $\psi'_i: E \longrightarrow D_i$ for all $i \in N(\Delta)$ corresponds to a single morphism $\psi': E \longrightarrow \prod(D_i : i \in N(\Delta))$.

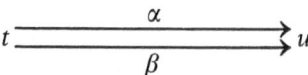

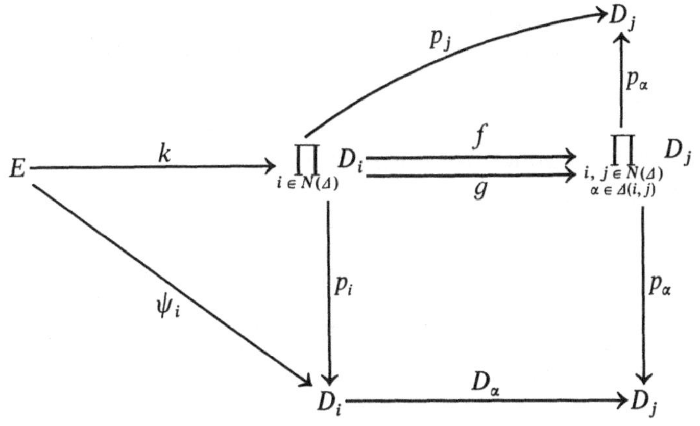

(2) (E', ψ') as in (1) is a lower bound of (Δ, D) if and only if $\psi'.f = \psi'.g$. □

1.23 Duality. The *dual* or *opposite* of a monoid M is the monoid M^{op} sharing the same set of elements with M but whose composition $*$ is defined by $x * y = yx$. Again, the *dual* or *opposite* of a preordered class $(X, \leqslant)$ is the preordered class $(X, \leqslant)^{op} = (X, \geqslant)$ (where, of course, $x \geqslant y$ means $y \leqslant x$). Both of these are instances of a more general construction. The *dual* or *opposite* of a category $\mathscr{K}$ is the category $\mathscr{K}^{op}$ defined as follows. The objects of $\mathscr{K}^{op}$ are the same as the objects of $\mathscr{K}$. We define morphisms by $\mathscr{K}^{op}(A, B) = \mathscr{K}(B, A)$. As in the case of clones (1.2.7), two different categories share the same objects and some notational distinction is in order: let us write $f: A \longrightarrow\!\!\!< B$ to mean $f: B \longrightarrow A$ in $\mathscr{K}$. Then composition in $\mathscr{K}^{op}$ is defined by

$$(A \xrightarrow{f}\!\!< B)(B \xrightarrow{g}\!\!< C) = (C \xrightarrow{g} B)(B \xrightarrow{f} A)$$

The identities are provided by $\mathrm{id}_A: A \longrightarrow\!\!\!< A = \mathrm{id}_A: A \longrightarrow A$. The category axioms are clear and it is obvious that $\mathscr{K}^{op}$ is locally small if $\mathscr{K}$ is. Moreover, $(\mathscr{K}^{op})^{op} = \mathscr{K}$, so $\mathscr{K}^{op}$ is a typical category. If $S(\mathscr{K})$ is a statement about an arbitrary category $\mathscr{K}$, S^{op} is the statement defined by $S^{op}(\mathscr{K}) = S(\mathscr{K}^{op})$. For example, consider the statement

$S(\mathscr{K})$: *Given* $f: A \longrightarrow B$ and $g: B \longrightarrow C$ in $\mathscr{K}$,
if f and g have right inverses, so does $f.g$

The statement is true in every $\mathscr{K}$ since if $f_1: B \longrightarrow A$ and $g_1: C \longrightarrow B$ with $f.f_1 = \mathrm{id}_A$ and $g.g_1 = \mathrm{id}_B$ then $(f.g)(g_1.f_1) = \mathrm{id}_A$. We deduce that S^{op} (which has the same domain as S) is also universally true. From the point of view of $\mathscr{K}$, S^{op} asserts that the composition of maps having a left inverse again has a left inverse. Notice that if $\mathscr{K}$ and $\mathscr{L}$ are dual to each other the situation is abstractly symmetric; we do not know if "$A \longrightarrow B$" refers to $\mathscr{K}$ or to $\mathscr{L}$. From the "base category" point of view, one of the categories is "real" and the other "abstract" (e.g. consider **Set** and **Set**^{op}). This contextual asymmetry is one of the reasons duality is useful. The *principle of categorical duality* is: *S^{op} is universally true if S is.* Duality cuts the work in half, as will be illustrated with frequency hereafter. Even when we are interested in specific "concrete" categories, duality forces us to look at some rather abstract ones.

1.24 Co-Concepts. The dual of "freeble" is *co*freeble. For example, if $f:A \longrightarrow B$ in $\mathscr{K}$ then B is the domain of f in $\mathscr{K}^{\mathrm{op}}$ and hence is the *co*domain of f in $\mathscr{K}^{\mathrm{op\,op}} = \mathscr{K}$ which is consistent with our original terminology. A more precise algorithm is

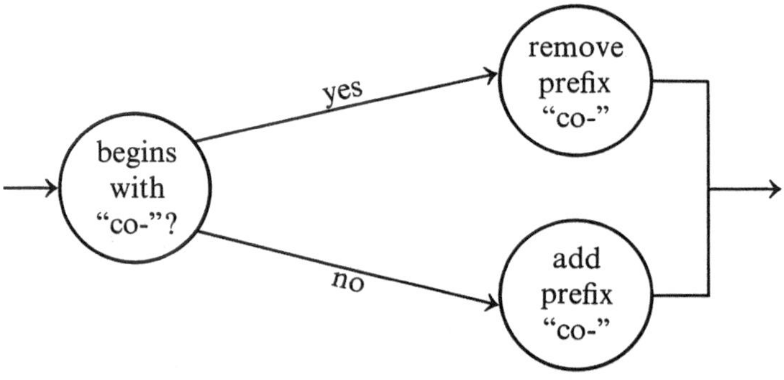

and one usually doesn't bother with the hyphens (but note: co-complete, co-optimal).

1.25 Coproducts. A diagram (we can now use that term with aplomb) $\mathrm{in}_i : A_i \longrightarrow A$ is a *coproduct* in $\mathscr{K}$ if, of course, $\mathrm{in}_i : A \longleftarrow A_i$ is a product in $\mathscr{K}^{\mathrm{op}}$. It follows at once from the dual of 1.6 that coproducts are unique up to isomorphism (note: coisomorphisms are isomorphisms).

The notation for coproducts is $\coprod A_i$. Binary coproducts are better written $A + B$. The empty coproduct is a coterminal object which it is more standard to call an *initial object*. A common symbol for an initial object is 0.

Abstractly, product and coproduct are the same concept. But let us explore what coproducts look like in some familiar contexts.

(1.26) In **Set**, $\coprod A_i$ exists and is the disjoint union $\{(i, a) : i \in I \text{ and } a \in A_i\}$. The ith injection (it is standard to call the coproduct coprojections *injections*) in_i sends $a \in A_i$ to (i, a). The universal property is easy:

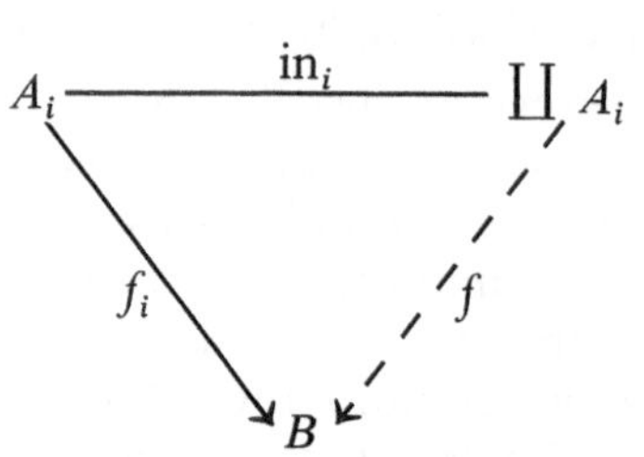

The unique f is defined by $(i, a)f = af_i$. 0 is the empty set.

(1.27) The category of topological spaces and continuous functions has coproducts. Provide the disjoint union at the level of sets with the largest topology making all the injection functions continuous (a set in $\coprod A_i$ is open if and only if its intersection with each A_i was already open).

Unlike the situation for products, familiar categories of sets with structure tend to have coproducts but the underlying set of the coproduct is different from the disjoint union.

(1.28) The category of abelian groups has coproducts. $\coprod A_i$ is usually called the *direct sum*, and is written $\oplus A_i$; it consists of the subgroup of $\prod A_i$ of all tuples (a_i) such that $a_i = 0$ for all but at most finitely many i. The ith injection map sends a to the I-tuple $(\delta^i_j a : j \in I)$ where "δ" denotes the Kronecker delta. For the universal property, f is defined by $f = \sum f_i$ (where

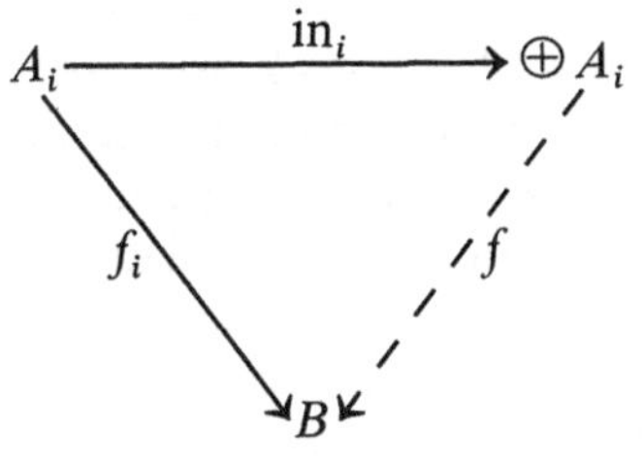

the sum is pointwise finite).

1.29 Coequalizers. A diagram

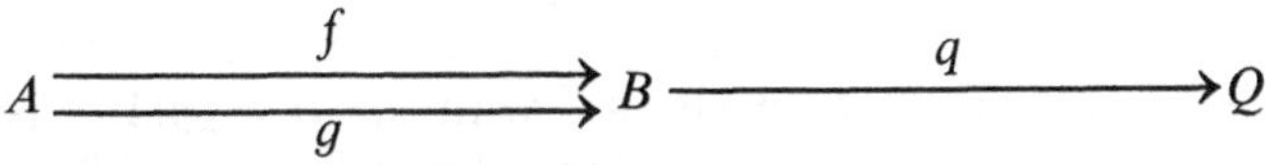

in $\mathcal{K}$ is a coequalizer if, of course, $q = \text{eq}(f, g)$ in $\mathcal{K}^{\text{op}}$.

(1.30) **Set** has coequalizers. Given $f, g : A \longrightarrow B$ let R be the intersection of all equivalence relations on B containing $\{(af, ag) : a \in A\}$. Set $Q = B/R$ with canonical projection $q : B \longrightarrow Q$. Then $q = \text{coeq}(f, g)$. Clearly, $f.q = g.q$.

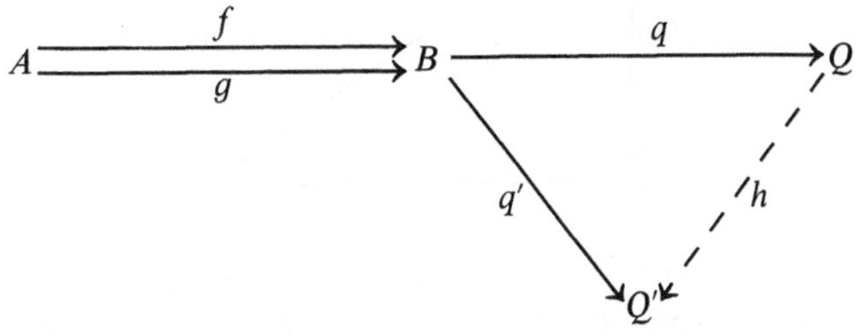

Now suppose given q' as shown above with $f.q' = g.q'$. Then $S = \{(b_1, b_2) : b_1 q' = b_2 q'\}$ is an equivalence relation on B containing R. h is uniquely defined by $(bR)h = bq'$.

(1.31) The category of topological spaces and continuous functions has coequalizers. Construct the coequalizer at the level **Set** as in 1.30 and assign Q the quotient topology.

(1.32) The category of abelian groups and homomorphisms has coequalizers. Given $f, g: A \longrightarrow B$, define $q: B \longrightarrow Q$ to be the canonical projection $B \longrightarrow B/\mathrm{Im}(f - g)$. If $\mathbf{Z}$ is the additive group of integers and $\mathbf{Z}_2$ is the 2-element group,

$$\mathbf{Z} \underset{2}{\overset{0}{\rightrightarrows}} \mathbf{Z} \xrightarrow{\quad q \quad} \mathbf{Z}_2$$

is the coequalizer in "abelian groups" although the coequalizer in **Set** is infinite (see exercise 22). An abelian group is *torsion-free* if each of its elements is of infinite order. Then $0, 2: \mathbf{Z} \longrightarrow \mathbf{Z}$ is in the category of torsion-free abelian groups although $\mathbf{Z}_2$ is not; but, in general, the category of torsion-free abelian groups does have coequalizers (namely $B \longrightarrow Q \longrightarrow Q/T$, where Q is the coequalizer in the category of abelian groups and T is the *torsion subgroup* of Q consisting of all elements of finite order). Thus, $\mathrm{coeq}(0, 2) = 0$ in the category of torsion-free abelian groups.

(1.33) In a preordered class, given $f, g: A \longrightarrow B$, $f = g$ so that $\mathrm{id}_B = \mathrm{coeq}(f, g)$.

1.34 Colimits. Given a diagram scheme Δ define the *dual* scheme Δ^{op} by $N(\Delta^{\mathrm{op}}) = N(\Delta)$, $\Delta^{\mathrm{op}}(i, j) = \Delta(j, i)$. If (Δ, D) is a diagram in $\mathscr{K}$, define the diagram $(\Delta^{\mathrm{op}}, D^{\mathrm{op}})$ in $\mathscr{K}^{\mathrm{op}}$ by $(D^{\mathrm{op}})_i = D_i$ and, for $\alpha \in \Delta^{\mathrm{op}}(i, j)$, $(D^{\mathrm{op}})_\alpha = D_\alpha: D_i \longleftarrow D_j$. A subset C of $N(\Delta)$ is *cofinal* if C is final qua subset of $N(\Delta^{\mathrm{op}})$, i.e., for all $j \in N(\Delta)$, $j \notin C$, there exists $i \in C$ with $\Delta(j, i) \neq \varnothing$. If C is cofinal, an *upper bound of* (Δ, D) *relative to* C is a lower bound of $(\Delta^{\mathrm{op}}, D^{\mathrm{op}})$ relative to C, that is a pair (L, ψ) where L is an object of $\mathscr{K}$ and ψ assigns a morphism $\psi_i: D_i \longrightarrow L$ to each $i \in C$ in such a way that the square (and triangle $*$)

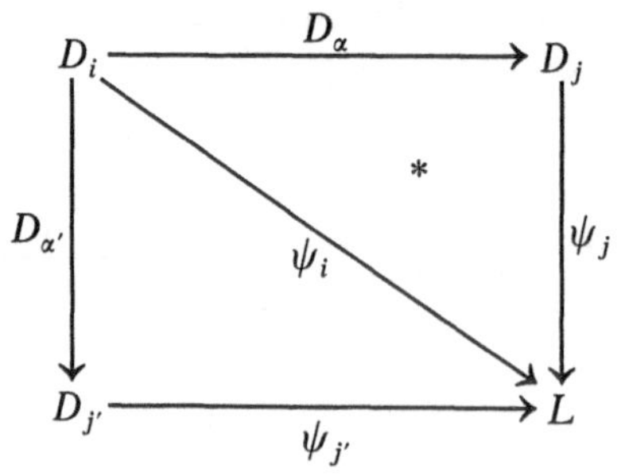

is commutative for all $j, j' \in C$, for all $i \in N(\Delta)$ and for all $\alpha \in \Delta(i, j)$, $\alpha' \in \Delta(i, j')$ (and $*$ holds when also $i \in C$). An upper bound of (Δ, D) is an upper bound of (Δ, D) relative to $C = N(\Delta)$. An upper bound (L, ψ) is a colimit of (Δ, D) just in case it is a limit of $(\Delta^{\mathrm{op}}, D^{\mathrm{op}})$, i.e., just in case it is an upper bound with the universal property displayed below with respect to other upper bounds (L', ψ'):

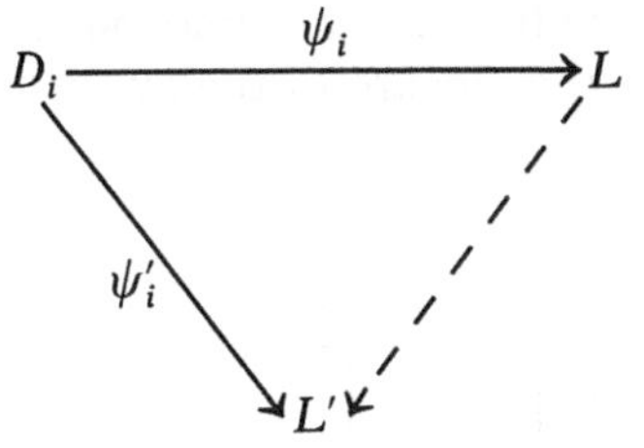

The notation for colimits is colim D. By the dual of 1.22, a category is small cocomplete if and only if it has coproducts and coequalizers.

1.35 Various Epimorphisms. "Surjective" is an important property of functions. As it turns out, there are numerous categorical definitions which characterize "surjective" in **Set**; we will content ourselves with three of them.

Let $f:A \longrightarrow B$ in $\mathscr{K}$. f is *split epi* or f is a *split epimorphism* if f has a left inverse, that is if there exists $d:B \longrightarrow A$ in $\mathscr{K}$ with $d.f = \mathrm{id}_B$. f is a *coequalizer* if $f = \mathrm{coeq}(g_1, g_2)$ for some pair $g_1, g_2:C \longrightarrow A$. f is *epi* or f is an *epimorphism* if for all pairs $t, u:B \longrightarrow T$ such that $f.t = f.u$, $t = u$.

1.36 Hierarchy Theorem for Epimorphisms. *A split epimorphism is a coequalizer and a coequalizer is an epimorphism.*

Proof. If $f:A \longrightarrow B$ is split epi with $d.f = \mathrm{id}_B$ then $f = \mathrm{coeq}(\mathrm{id}_A, f.d)$. That coequalizers are epi is dual to 1.15. □

Epimorphisms are important in diagram chasing. A typical situation is shown below. Assume that we wish to prove that (?) commutes given that

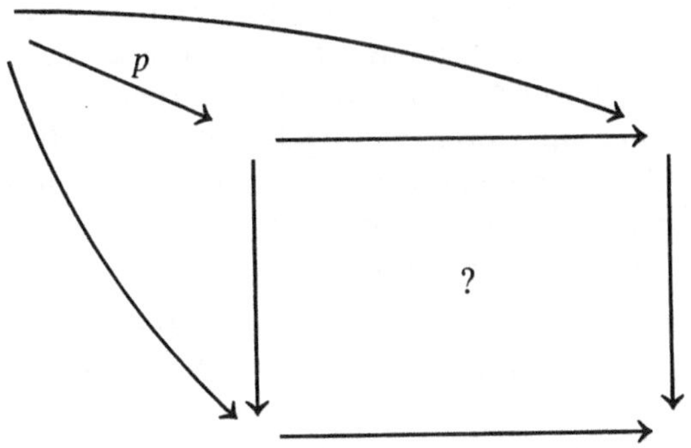

the peripheral diagram and the two triangles do. There is no problem if we know p is an epimorphism.

(1.37) In **Set**, all epimorphisms are split and all three concepts mean "surjective." The axiom of choice says that surjections are split epi. Since any function $f:A \longrightarrow B$ composes equally with the characteristic functions $\chi_B, \chi_{\mathrm{Im}(f)}:B \longrightarrow \{0, 1\}$ it is clear that epimorphisms are surjective.

(1.38) In the category of Hausdorff spaces and continuous maps $f:A \longrightarrow B$ is a coequalizer if and only if f is surjective and B has the quotient topology induced by f, and f is epi if and only if the image of f is a dense subset of B. The first statement and half of the second are referred to exercise 11. We must show that every epimorphism has a dense image. For any $f:A \longrightarrow B$, let I be the closure of the image of f and form $(B + B)/I$ as shown below:

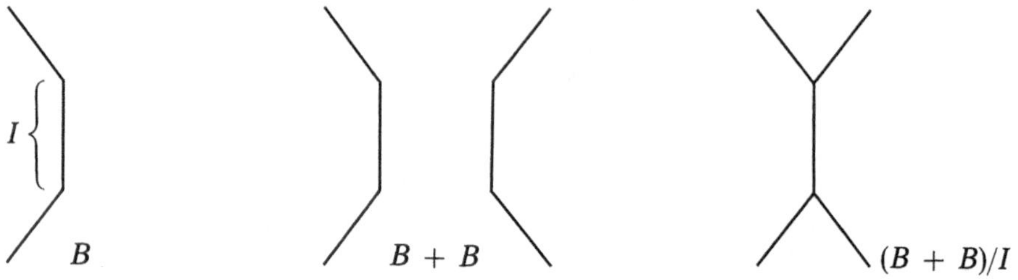

There are injections $\text{in}_1, \text{in}_2: B \longrightarrow B + B$ and a canonical projection $p: B + B \longrightarrow (B + B)/I$. Since $f.(\text{in}_1.p) = f.(\text{in}_2.p)$, we have that $I = B$ when f is epi. (We leave it as an exercise to prove that $(B + B)/I$ is Hausdorff.)

(1.39) In the category of abelian groups and homomorphisms, epimorphisms are the same thing as surjective homomorphisms. Surely surjective homomorphisms are epi. Given a homomorphism $f:A \longrightarrow B$, let I be the image of f and let $p:B \longrightarrow B/I$ be the canonical projection. Then $f.p = f.0$. If f is epi, $p = 0$ which implies that $I = B$.

(1.40) Epimorphisms need not be surjective in $\mathbf{Set}^{\mathbf{T}}$. Consider the inclusion map $i:\mathbf{N} \longrightarrow \mathbf{Z}$ of the natural numbers into the integers as a homomorphism of rings or of monoids, take your choice; in either category, i is epi. To prove it, observe that any monoid homomorphism, f, defined on $\mathbf{Z}$ satisfies $(-n)f \cdot (n)f = e$ for every $n \in \mathbf{N}$, thereby forcing $(-n)f = (nf)^{-1}$.

(1.41) A *Boolean σ-algebra* is a Boolean algebra with countable suprema and countable infima (see 1.5.17). A fundamental structure in measure theory is a set together with a sub Boolean σ-algebra of subsets. By a *homomorphism* of Boolean σ-algebras we mean a Boolean algebra homomorphism which preserves the countable suprema and infima. It is an open question whether epimorphisms are surjective in this category.

1.42 Proposition. *Given $f:A \longrightarrow B$ and $g:B \longrightarrow C$ in $\mathscr{K}$, then (1) If f, g are epi, so is $f.g$; if f, g are split epi, so is $f.g$; (2) If $f.g$ is epi, so is g; if $f.g$ is split epi, so is g.*

Proof. (1) The split version was illustrated in 1.23. For epimorphisms, if $(f.g).t = (f.g).u$ then $g.t = g.u$ and $t = u$. (2) For epimorphisms, if $g.t = g.u$ then surely $(f.g).t = (f.g).u$ so that $t = u$; for split epimorphisms, if $d.(f.g) = \text{id}$, surely $(d.f).g = \text{id}$. □

The analog of 1.42 for coequalizers is not always true (see 1.57 and 1.58).

1.43 Various Monomorphisms. The dual concepts to split epi, coequalizer, and epi are *split mono* (or *split monomorphism*), *equalizer*, and *mono* (or *monomorphism*). Thus, $f:A \longrightarrow B$ is split mono if there exists $s:B \longrightarrow A$ with $f.s = \text{id}_A$, f is an equalizer if $f = \text{eq}(g_1, g_2)$ for some pair $g_1, g_2: B \longrightarrow C$ and f is mono if for all pairs $t, u: T \longrightarrow A$ such that $t.f = u.f$ we have $t = u$. Dual to 1.36 and 1.42 we have at once.

1.44 Proposition. *Split monos are equalizers and equalizers are monos. If f and g are mono or split mono, so is $f.g$. If $f.g$ is mono or split mono, so is f.* □

(1.45) In **Set**, monos are the same thing as injective functions (use constant functions for t and u). Given any subset A of a set X, it is easy to construct functions $f, g: X \longrightarrow Y$ with $A = \mathrm{eq}(f, g)$, and this makes it clear that all monos are equalizers. In fact, if $f: A \longrightarrow B$ is mono and A is nonempty, then f is split mono (if $I = \mathrm{Im}(f)$, let $s = f^{-1}$ on I and any element of A elsewhere). The inclusion map of the empty set into a nonempty one is mono, but never split mono.

1.46 Proposition. *Let* **T** *be an algebraic theory in $\mathscr{K}$. A* **T***-homomorphism $f:(A, \xi) \longrightarrow (B, \theta)$ is a monomorphism in $\mathscr{K}^{\mathbf{T}}$ if and only if $f: A \longrightarrow B$ is mono in $\mathscr{K}$.*

Proof. Clearly, mono in $\mathscr{K}$ implies mono in $\mathscr{K}^{\mathbf{T}}$. For the converse, let $t, u: C \longrightarrow A$ be arbitrary morphisms in $\mathscr{K}$. Let $t^{\#}$, $u^{\#}$ be the unique homomorphic extensions of t, u as provided by the universal property of $(CT, C\mu)$ (see 1.4.12) and shown below (the breaks in the arrows denotes that the morphisms are only in $\mathscr{K}$, not in $\mathscr{K}^{\mathbf{T}}$).

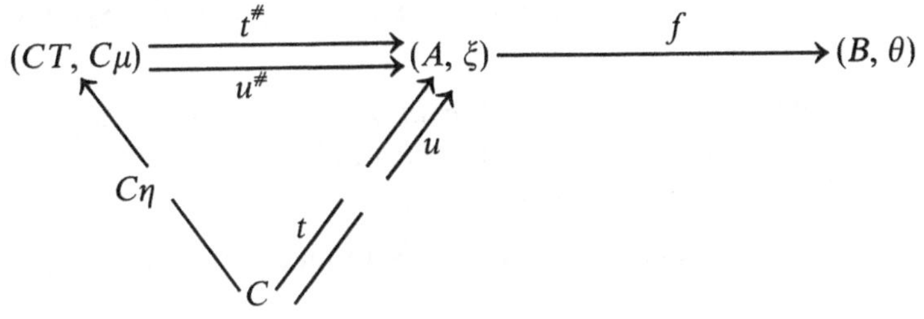

If $t.f = u.f$ then $t^{\#}.f = u^{\#}.f$ (as they are both homomorphisms and agree on the generators) so that $t^{\#} = u^{\#}$, and $t = u$. □

Comparing 1.46 with 1.40 and 1.41 we see that $\mathbf{Set}^{\mathbf{T}}$ behaves more predictably with respect to monos than with epimorphisms.

It is clear that if $\mathscr{K}$ is a "category of sets with structure", injective morphisms will be mono in $\mathscr{K}$. The following example shows that sometimes monos are not injective functions.

(1.47) Let $\mathscr{K}$ be the category of 2-divisible abelian groups as in 1.17. Let A be the multiplicative group of non-zero real numbers and set $f: A \longrightarrow A$ to be the the squaring homomorphism $xf = x^2$. Since f identifies x and $-x$, f is not injective. But f is mono. For let $t, u: T \longrightarrow A$ with $t.f = u.f$. Let $x \in T$. Then $y^2 = x$ for some y, and $xt = y^2t = (yt)^2 = ytf = yuf = xu$, as desired.

1.48 Image Factorization. The categorical view of the image of a function f is as a factorization $f = p.i$ with p surjective and i injective. An axiomatic theory appears in 3.4.1. We present here two of the many possible such theories in an arbitrary category. Given a morphism $f: A \longrightarrow B$ in $\mathscr{K}$, a *coequalizer-mono factorization of f* is a factorization $f = p.i$ with p a coequalizer in $\mathscr{K}$ and i a monomorphism in $\mathscr{K}$. The dual concept is *epi-equalizer factorization*, that is $f = p.i$ with p an epimorphism and i an equalizer.

1.49 Proposition. *Coequalizer-mono factorizations are unique up to isomorphism.*

Proof. Suppose $p.i = f = p'.i'$ with p, p' coequalizers and i, i' mono-

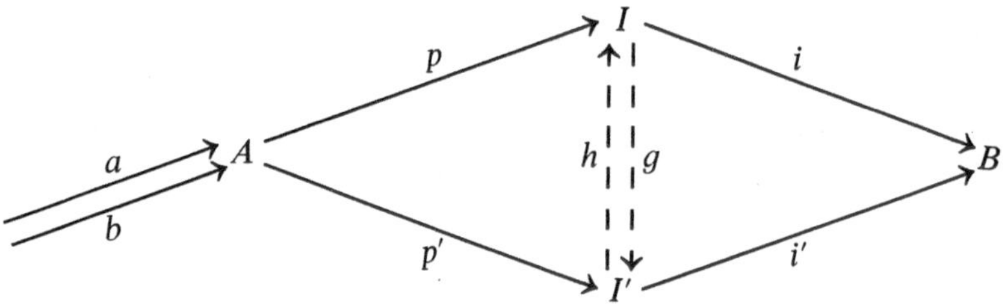

morphisms. There exists a, b with $p = \text{coeq}(a, b)$. As $a.p'.i' = a.p.i = b.p.i. = b.p'.i'$ and i' is mono, $a.p' = b.p'$ so there exists a unique g such that $p.g = p'$. Since p is epi, $g.i' = i$. Symmetrically, there exists $h{:}I' \longrightarrow I$ with $p'.h = p$ and $h.i = i'$. Either because p is epi or i is mono, $g.h = \text{id}_I$. Symmetrically, $h.g = \text{id}_{I'}$. □

1.50 Corollary. *Given $f{:}A \longrightarrow B$ in $\mathscr{K}$, the following three conditions on f are equivalent*: (1) *f is an isomorphism*; (2) *f is a coequalizer and f is mono*; *and* (3) *f is an equalizer and f is epi.*

Proof. If f is a coequalizer and f is mono, $f.\text{id} = f = \text{id}.f$ are two

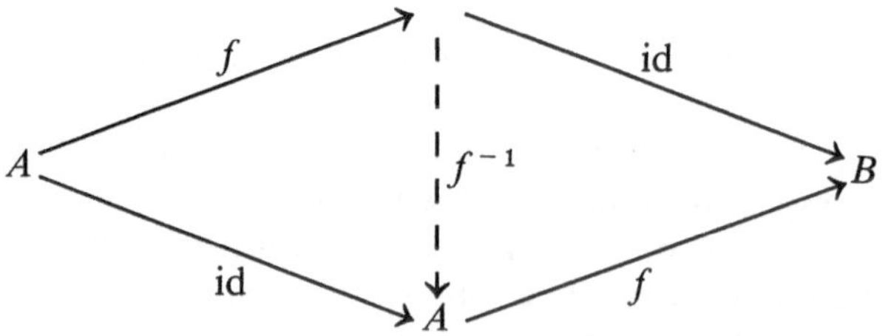

coequalizer-mono factorizations of f, giving rise to f^{-1} as above. □

Examples such as 1.38 and 1.40 show that a morphism which is epi and mono need not be an isomorphism.

$\mathscr{K}$ *has coequalizer-mono factorizations* if every morphism in $\mathscr{K}$ has a coequalizer-mono factorization. For example, **Set** has coequalizer-mono factorizations and epi-equalizer factorizations and they both coincide with surjective-injective factorizations.

(1.51) The category of topological spaces and continuous maps has coequalizer-mono factorizations and epi-equalizer factorizations. Given a continuous map $f{:}A \longrightarrow B$ with image factorization $f = p.i$ at the level of sets we can provide $\text{Im}(f)$ with the quotient topology induced by p, in which case i is continuous and (p, i) is a coequalizer-mono factorization of f, or we can provide Im(f) with the subspace topology induced by i, in which case p is continuous and (p, i) is an epi-equalizer factorization. The details are left as exercises.

A well-known way to construct the image factorization of a function $f: A \longrightarrow B$ is to divide out by the *equivalence relation, E, of f* (where E is the equivalence relation on A given by $E = \{(x, y): xf = yf\}$). Let p be the

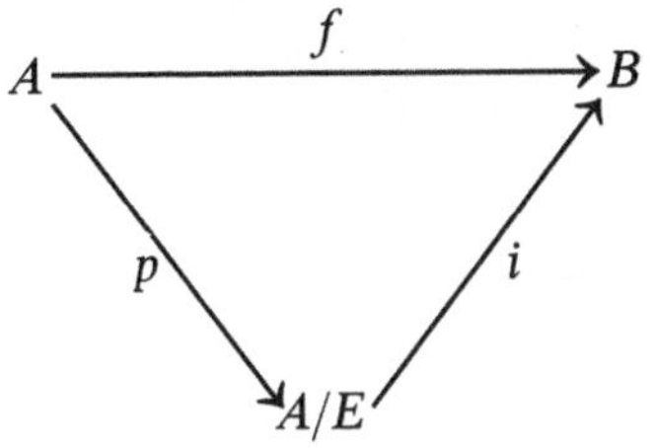

canonical projection as shown. Since xEy if and only if $xf = yf$, $(xE)i = xf$ is a well-defined injection. Thus $f = p.i$ is a coequalizer-mono factorization of f. We now explore the possibility that this construction can be imitated in $\mathscr{K}$.

Consider a diagram scheme Δ with three nodes i, j, k and just two edges, $\alpha \in \Delta\ (i, k)$ and $\beta \in \Delta\ (j, k)$. A diagram (Δ, D) in $\mathscr{K}$ looks like

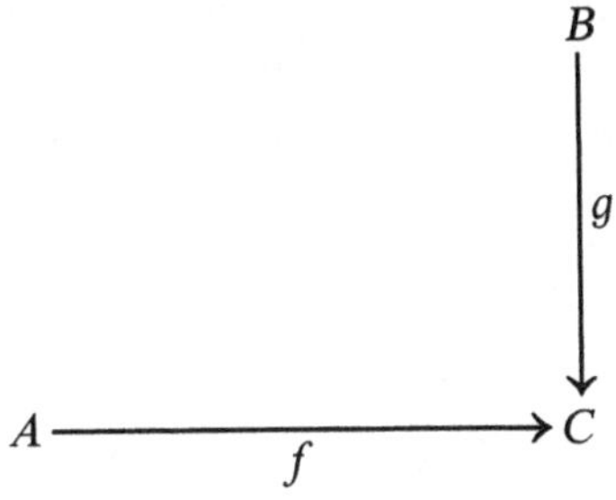

(i.e., $D_\alpha = f$ and $D_\beta = g$). Since $\{i, j\}$ is final, lim D is an object P of $\mathscr{K}$ equipped with two morphisms $a: P \longrightarrow A$ and $b: P \longrightarrow B$ such that $a.f = b.g$, and universal with this property as shown below:

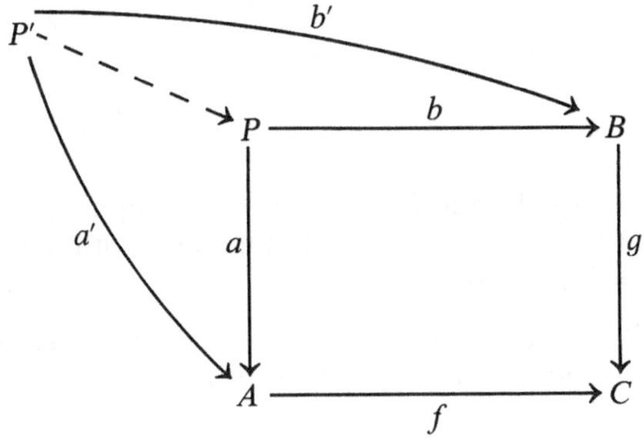

The square $a.f = b.g$ is called a *pullback square*, (P, a, b) is the *pullback of* (f, g, C) and b is the *pullback of* f *along* g. (The dual concept is called a *pushout*.) In **Set**, P is constructed as the subset $\{(a, b) : af = bg\}$ of $A \times B$ with a and b the restrictions of the coordinate projections. In particular, if $f = g$, P is the equivalence relation of f. In this case, a glance at 1.30 shows that the canonical projection $C \longrightarrow C/P$ is the coequalizer of (a, b).

Consider a morphism $f : A \longrightarrow B$ in $\mathscr{K}$. The *kernel pair of* f is the pullback (E, a, b) of (f, f, B). Let us assume that this kernel pair exists and that $p : A \longrightarrow C = \operatorname{coeq}(a, b)$ also exists. Define $i : C \longrightarrow B$ as shown below by the universal property of a coequalizer

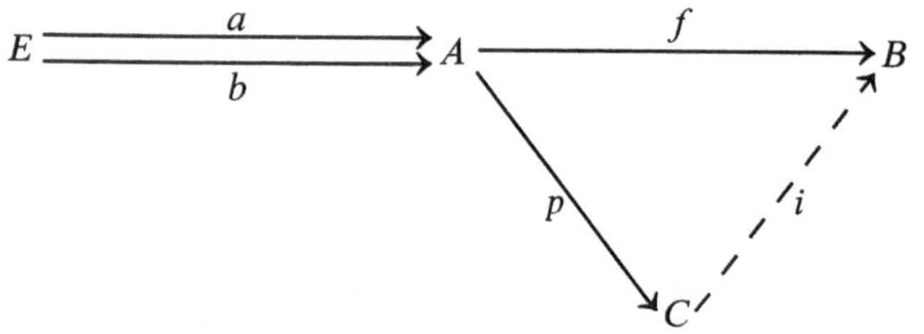

(1.52)

(since $a.f = b.f$). As we see shortly, the factorization $f = p.i$ is a very good candidate for the coequalizer-mono factorization of f.

1.53 Lemma. *Let $p : A \longrightarrow I$ be a coequalizer in $\mathscr{K}$. Then if the kernel pair of p exists, p is the coequalizer of its kernel pair.*

Proof. We assume that $p = \operatorname{coeq}(a', b')$ for some $a', b' : E' \longrightarrow A$.

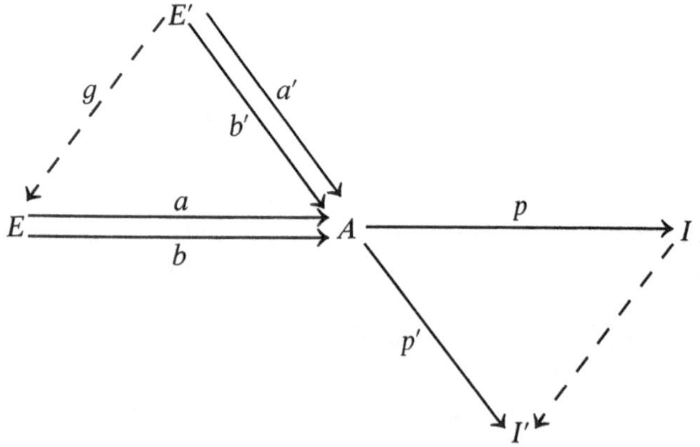

Let $a, b : E \longrightarrow A$ be the kernel pair of p. Suppose p' is given with $a.p' = b.p'$. Since $a'.p = b'.p$ there exists unique g with $g.a = a'$, $g.b = b'$. Therefore, $a'.p' = b'.p'$, as desired. $\square$

1.54 Proposition. *Let $f : A \longrightarrow B$ in $\mathscr{K}$. If f has a kernel pair with a coequalizer, then the factorization 1.52 is the only candidate for a coequalizer-*

mono factorization of f; that is, if f has a coequalizer-mono factorization it is isomorphic to the factorization of 1.52.

Proof. Let $f = p.i$ be a coequalizer-mono factorization of f, and let $a, b: E \longrightarrow A$ be the kernel pair of f. Because i is mono, a pair a', b':

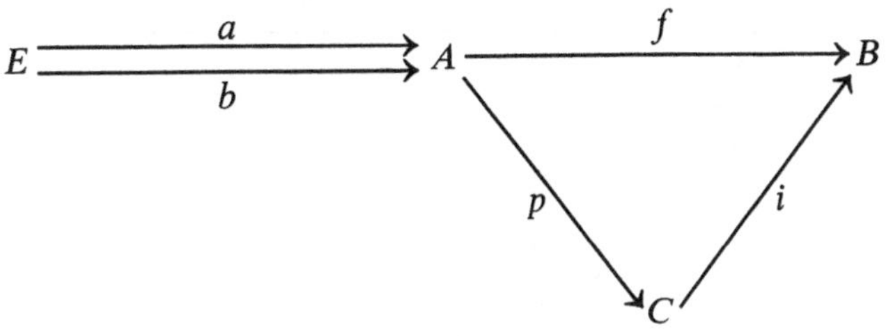

$E' \rightrightarrows A$ satisfies $a'.f = b'.f$ if and only if it satisfies $a'.p = b'.p$. It follows that (a, b) is also the kernel pair of p. By 1.53, $p = \operatorname{coeq}(a, b)$. Since the morphism $i: C \longrightarrow B$ of 1.52 is unique, the proof is complete. □

1.55 Proposition. *Let* **T** *be an algebraic theory in* **Set**. *Then* $\mathbf{Set}^{\mathbf{T}}$ *has coequalizer-mono factorizations and they are constructed at the level of sets.*

Proof. Let $f:(A, \alpha) \longrightarrow (B, \beta)$ be a **T**-homomorphism. Let $f = (p: A \longrightarrow C).(i: C \longrightarrow B)$ be the usual image factorization at the level of sets. By 1.4.31, there exists a unique **T**-algebra structure $\gamma: CT \longrightarrow C$ on C such that p and i become **T**-homomorphisms. Let $a, b: E \longrightarrow A$ be the

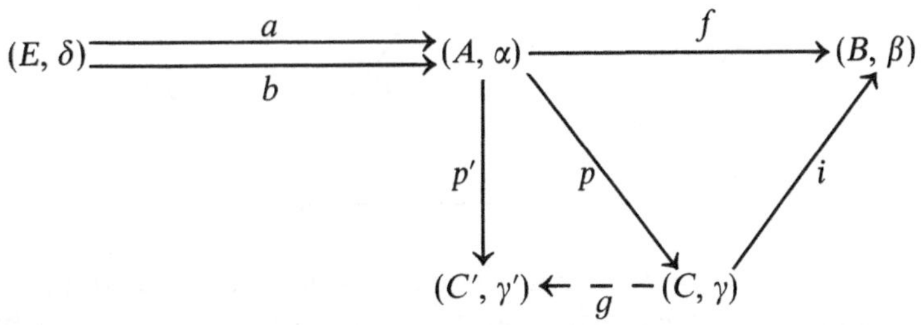

kernel pair of f in **Set**. By analyzing the proof of 1.22 in the context of 1.11 and 1.14, it is clear that there exists a unique $\delta: ET \longrightarrow E$ by virtue of which a and b become **T**-homomorphisms, (and then $a, b:(E, \delta) \longrightarrow (A, \alpha)$ is in fact the kernel pair of f in $\mathbf{Set}^{\mathbf{T}}$, a fact we do not need to use here). By 1.54, it will suffice to show that $p = \operatorname{coeq}(a, b)$ in $\mathbf{Set}^{\mathbf{T}}$. Suppose $p':(A, \alpha) \longrightarrow (C', \gamma')$ with $a.p' = b.p'$. Since $p = \operatorname{coeq}(a, b)$ in **Set**, there exists a unique function $g: C \longrightarrow C'$ with $p.g = p'$. We must show that g is a **T**-homomorphism. This amounts to a slightly updated version of 1.2.6:

(*1.56*) *Given an algebraic theory in* **Set**, *a surjective* **T**-*homomorphism* $p:(A, \alpha) \longrightarrow (C, \gamma)$, *a* **T**-*algebra* (C', γ') *and a function* $g: C \longrightarrow C'$ *such that* $p.g$ *is a* **T**-*homomorphism, then* $g:(C, \gamma) \longrightarrow (C', \gamma')$ *is again a* **T**-*homomorphism.*

To prove 1.56, we use just the sort of diagram that appeared in the advertisement for epimorphisms of 1.36+, namely:

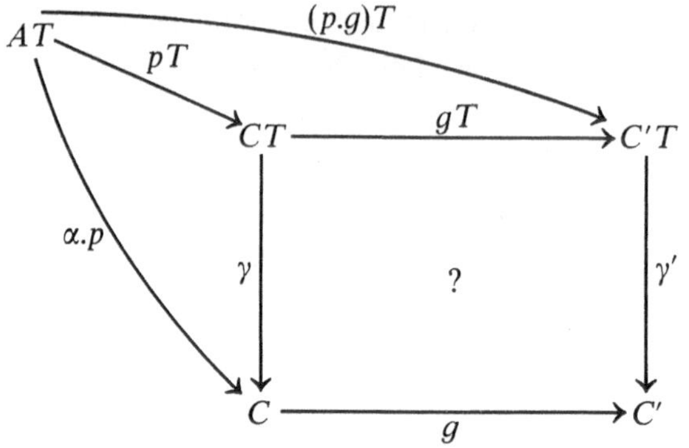

Crucial is the use of 1.4.29 which guarantees that pT is epi. □

1.57 Proposition. *Let $\mathscr{K}$ have coequalizer-mono factorizations. Then 1.42 is true for coequalizers, that is, given $f: A \longrightarrow B$ and $g: B \longrightarrow C$ in $\mathscr{K}$ we have*

(1) *if f, g are coequalizers then so is $f.g$, and*
(2) *if $f.g$ is a coequalizer, so is g.*

Proof. We prove (2) first. Let $g = p.i$ be a coequalizer-mono factoriza-

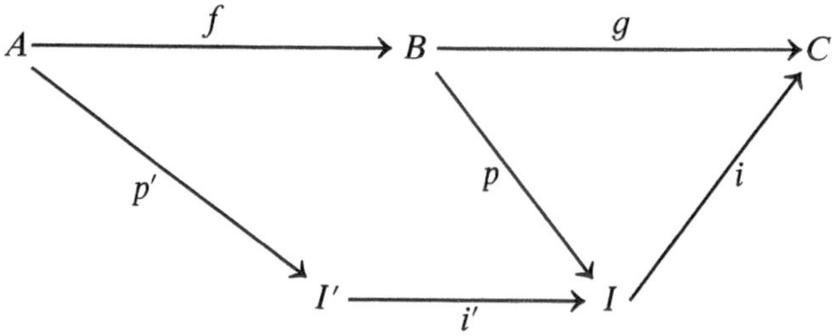

tion, and then let $f.p = p'.i'$ be a coequalizer-mono factorization. As $p'(i'.i)$ is a coequalizer-mono factorization of $f.g$, $i'.i$ is an isomorphism by 1.49. Since i is both split epi and mono, it follows from 1.50 that i is an isomorphism.

To prove (1), let $f = \text{coeq}(a, b)$ and let $p.i$ be a coequalizer-mono factorization of $f.g$. Since i is mono we have $a.p = b.p$ which induces unique h with

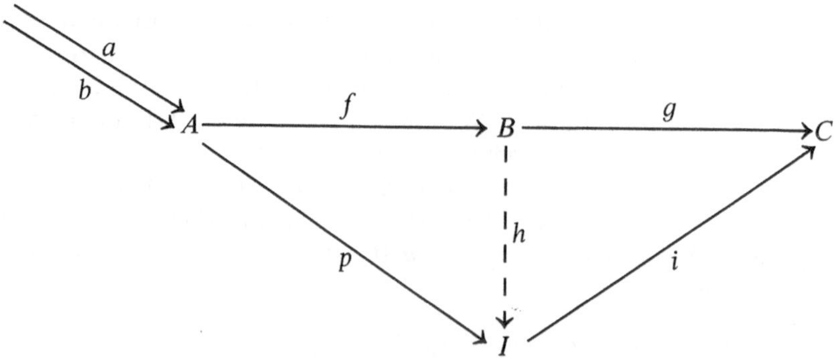

with $f.h = p$. As f is epi, $h.i = g$. By (2), i is a coequalizer and hence, by 1.50, an isomorphism. □

The following example shows that even a category with all small limits and colimits (and all factorizations as in 1.52 in particular) need not necessarily have coequalizer-mono factorizations.

(1.58) Let $\mathscr{K}$ be the category of abelian groups with no element of order 4 (that is $4x = 0$ implies $2x = 0$) and group homomorphisms. Products, equalizers, and coproducts (1.28) are formed just as they are in the category of all abelian groups. The proof that this category has coequalizers will be postponed until 3.7.13. Therefore, $\mathscr{K}$ has all small limits and colimits. Let $\mathbf{Z}$, $\mathbf{Z}_2$, and $\mathbf{Z}_4$ denote, respectively, the abelian groups of integers, integers modulo 2, and integers modulo 4. The homomorphism $2{:}\mathbf{Z} \longrightarrow \mathbf{Z}$ (sending x to $2x$) is the equalizer of the morphisms $p, 0{:}\mathbf{Z} \longrightarrow \mathbf{Z}_2$ in $\mathscr{K}$ (p is the canonical projection). Composing $2{:}\mathbf{Z} \longrightarrow \mathbf{Z}$ with itself gives the $\mathscr{K}$-morphism $4{:}\mathbf{Z} \longrightarrow \mathbf{Z}$. While $4{:}\mathbf{Z} \longrightarrow \mathbf{Z}$ is the equalizer of $p, 0{:}\mathbf{Z} \longrightarrow \mathbf{Z}_4$ in the category of all abelian groups, $\mathbf{Z}_4$ is not in $\mathscr{K}$. Suppose that there exists f, $g{:}\mathbf{Z} \longrightarrow A$ in $\mathscr{K}$ with $4 = \mathrm{eq}(f, g)$. As $4f(1) = 4g(1)$ and A is in $\mathscr{K}$, $2(f(1) - g(1)) = 0$, that is, $2.f = 2.g$ and there exists a unique $h{:}\mathbf{Z} \longrightarrow \mathbf{Z}$ with $h.4 = 2$. Since $4n = 2$ has no solution in $\mathbf{Z}$, we get a contradiction to the assertion that $4{:}\mathbf{Z} \longrightarrow \mathbf{Z}$ was an equalizer. By the dual of 1.57, it follows that $\mathscr{K}$ does not have epi-equalizer factorizations. $\mathscr{K}^{\mathrm{op}}$, then, is a category with small limits and colimits in which the composition of coequalizers is not a coequalizer and hence which does not have coequalizer-mono factorizations.

1.59 Generators and Cogenerators. A fixed object, G, of $\mathscr{K}$ can be used to make actual sets out of $\mathscr{K}$-objects and actual functions out of $\mathscr{K}$-morphisms. If A is a $\mathscr{K}$-object write "$a \in A$" just in case $a{:}G \longrightarrow A$ in $\mathscr{K}$. Each $\mathscr{K}$-morphism $f{:}A \longrightarrow B$ acts functionally on elements since if $a \in A$ we literally have $af \in B$. Notice that $a(\mathrm{id}_A) = a$ and $a(fg) = (af)g$. G is a *generator* if morphisms are distinguished by their functional action; more precisely, given $f, g{:}A \longrightarrow B$ in $\mathscr{K}$ with $f \neq g$ there exists $a{:}G \longrightarrow A \in A$ with $af \neq ag$.

The prototype example of a generator is $\mathscr{K} = \mathbf{Set}$ and $G = 1$. Actually, every nonempty set is a generator in $\mathbf{Set}$. More generally, if $\mathbf{T}$ is an algebraic theory in $\mathbf{Set}$ and A is a nonempty set then the free algebra $(AT, A\mu)$ is a generator in $\mathbf{Set}^{\mathbf{T}}$. In the category of topological spaces and continuous maps or the category of metric spaces and distance-decreasing maps (1.9) 1 is again a generator.

An object C of $\mathscr{K}$ is a *cogenerator in* $\mathscr{K}$ if, of course, C is a generator in $\mathscr{K}^{\mathrm{op}}$, that is if, when $f, g{:}A \longrightarrow B$ with $f \neq g$, there exists $h{:}B \longrightarrow C$ with $fh \neq gh$.

If I is a set and A is an object of $\mathscr{K}$, the *Ith power of A*, A^I, is the product of I copies of A.

1.60 Proposition. *Let C be an object of $\mathscr{K}$ and assume that the power C^I exists for all small sets I. Then C is a cogenerator if and only if for all $\mathscr{K}$-objects B the evaluation map, ev_B, of B is a monomorphism, where ev_B is* *defined by the diagram*:

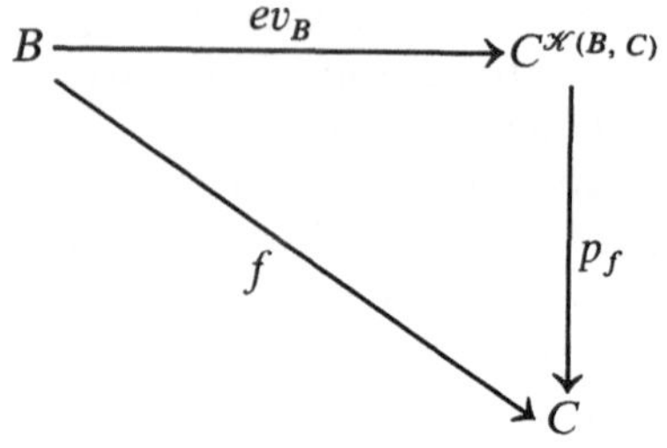

Proof. This is an easy exercise; note that $\mathscr{K}$ must be locally small. □

(1.61) The circle group S (i.e., the complex numbers of modulus 1) is a cogenerator in the category of abelian groups. While this result is well known (the phrase is "there exist enough homomorphisms to the circle"), it is not particularly easy to prove. It follows from 1.60, that every abelian group is isomorphic to a subgroup of a product of circles.

(1.62) In **Set**, every set with at least two elements is a cogenerator.

(1.63) Let $\mathscr{K}$ be the category of complete semilattices and supremum-preserving functions (1.5.15). Let C be the two-element lattice $\{0, 1\}$ with $0 \leqslant 1$. Then C is a cogenerator. It suffices to show that whenever $a \neq b$ in the complete semilattice X then there exists $h: X \longrightarrow C$ with $ah \neq bh$. If $a < b$, define h by $xh = 0$ if $x \leqslant a$, and $xh = 1$ otherwise. If $a \nless b$, define h by $xh = 0$ if $x \leqslant b$, and $xh = 1$ otherwise.

(1.64) The category of groups and homomorphisms does not have a cogenerator. Given any set X, consider those bijections from X to itself which leave fixed all but finitely many elements; such bijections, which are essentially permutations of a finite set, have parity, that is, are either even or odd. It is well known that the group of all even permutations of the set X forms a simple group if X has any number of elements greater than 4. In particular, if C is a would-be cogenerator there exists a simple group S of cardinal larger than that of C. Since $\mathrm{id}_S \neq 0$, there exists a nonzero homomorphism $f: S \longrightarrow C$. Since S is simple, the kernel of f is either zero or all of S, and either is impossible.

1.65 Monosubobjects. Let $i: A \longrightarrow X$, $j: B \longrightarrow X$ be monomorphisms with codomain X. If there exists $t: A \longrightarrow B$ with $tj = i$ then t is unique

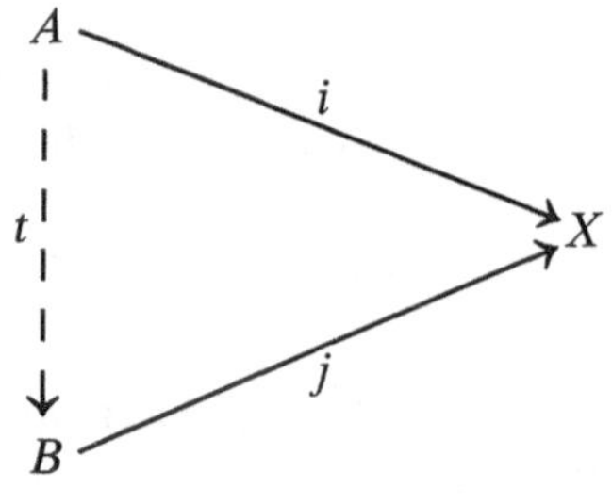

(as j is mono) and t is a monomorphism (as i is mono) and we write: $i \leqslant j$. This defines a reflexive and transitive relation on the class of monomorphisms with codomain X. i and j are *isomorphic* if $i \leqslant j$ and $j \leqslant i$; note that t as above is indeed an isomorphism and that "isomorphic" is an equivalence relation. An isomorphism class of X-valued monomorphisms is called a *monosubobject of* X. Writing $[i]$ for the isomorphism class of the mono $i:A \longrightarrow X$, "$[i] \leqslant [j]$ if $i \leqslant j$" is well defined and defines a partial ordering ("inclusion") on the class of monosubobjects of X.

(1.66) In **Set**, the passage from $[i]$ to the image of i is well defined and establishes a bijection from the set of monosubobjects of X to the set of subsets of X.

1.67 Direct Images. If $\mathscr{K}$ has coequalizer-mono factorizations, each morphism $f:A \longrightarrow B$ induces the "direct image map" $[i]f = [j]$ where (p, j) is any coequalizer-mono factorization of $i.f$. The diagram below shows

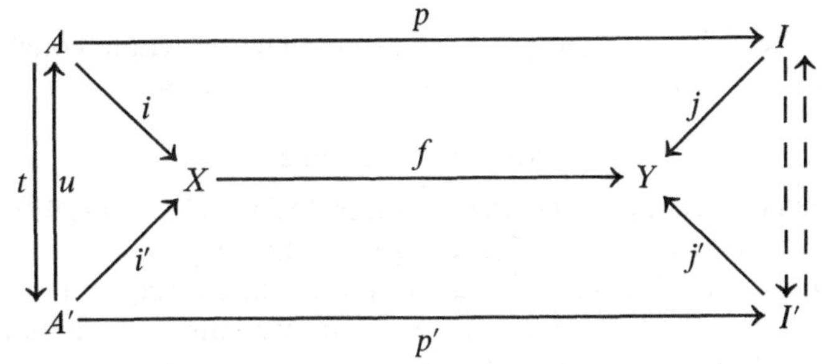

that if $[i] = [i']$ and (p', j') is a coequalizer-mono factorization of $i'.f$ then $[j] = [j']$ (i.e., as u is an isomorphism, (up, j) is another coequalizer-mono factorization of $i'.f$).

1.68 Inverse Images. If $\mathscr{K}$ has pullbacks, every morphism $f:X \longrightarrow Y$ induces the "inverse image map" $[i]f^{-1} = [j]$ where j is a pullback of i

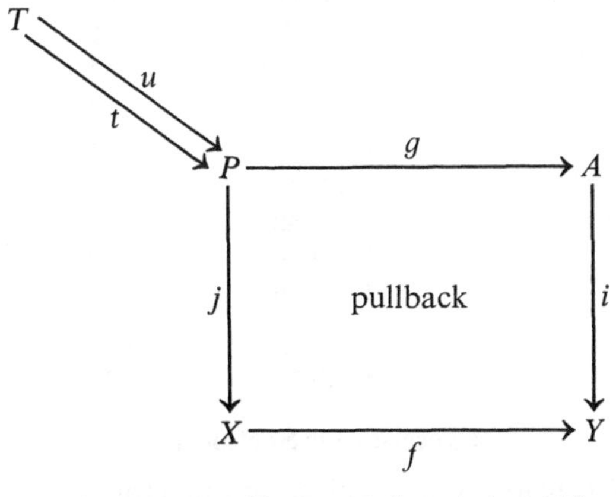

along f. To see that j is mono, if $tj = uj$ then $tgi = tjf = ujf = ugi$ and, since i is mono, $tg = ug$. Therefore, $t = v = u$ where v is the unique map

induced by

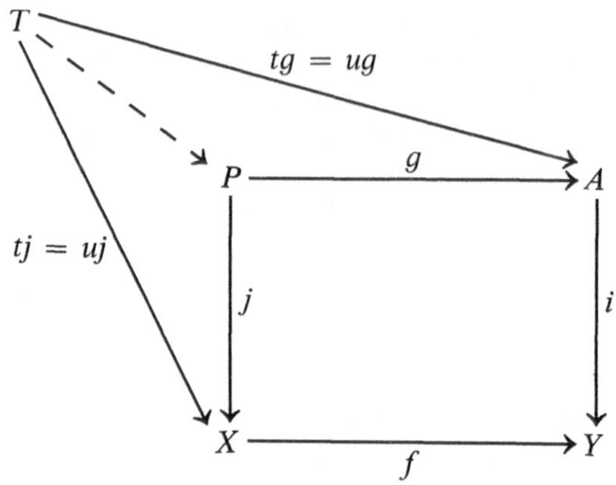

The proof that $[j] = [j']$ if $[i] = [i']$ and that the whole construction is independent of the choice of pullback is an easy exercise.

Notes for Section 1

The open question of (1.41) was raised by [Linton '66, page 93]. Example 1.58 is due to [Isbell '64, page 7]; see also [Kelly '69].

The following table should aid the reader in seeking out some of the expository literature of category theory. The last three columns record, in rough terms, the number of pages devoted to the subject matter covered, respectively, in sections 1, 2, and 3 of this chapter.

Author	Language	Pages	1	2	3
Arbib and Manes '75	English	160	70	40	15
Bucur and Deleanu '68	English	224	30	30	2
Brinkmann and Puppe '66	German	107	30	6	0
Ehresmann '65	French	358	60	20	10
Felscher '65–A	German	65	15	10	4
Freyd '64	English	164	26	20	0
Goguen *et al.* '75	English	85	18	16	0
Hasse and Michler '66	German	358	30	0	40
Herrlich and Strecker '74	English	400	102	93	9
Mac Lane '71	English	262	57	44	1
Mitchell '65	English	273	24	35	0
Pareigis '70	English	268	40	50	0
Schubert '72	English	385	49	26	0

Exercises for Section 1

1. Let $\mathscr{K}$ be the category whose objects are functors from the category of complete Boolean algebras to the category of sets and whose morphisms

are natural transformations. Use 1.5.5 and 1.5.48 to prove that $\mathscr{K}$ is not locally small.

2. Given $f: A \longrightarrow B$ in $\mathscr{K}$, prove that f is an isomorphism if and only if there exist $g: B \longrightarrow A$ and $h: B \longrightarrow A$ such that $fg = \mathrm{id}_A$ and $hg = \mathrm{id}_B$.
3. Show that a monoid M has binary products (*qua* category) if and only if M is isomorphic to $M \times M$. For infinite X, show that $(2^X, \cap)$ is such a monoid.
4. Analyze the following categories for completeness and cocompleteness.
   (a) monoids and monoid homomorphisms
   (b) groups and group homomorphisms
   (c) vector spaces over a field and linear maps
   (d) topological abelian groups and continuous homomorphisms
   (e) rings and ring homomorphisms
   (f) fields and ring homomorphisms
5. Let $\mathscr{K}$ be the category of metric spaces and distance-decreasing maps (as in 1.9) and, similarly, let $\mathscr{L}$ be the category of metric spaces of diameter at most M and distance-decreasing maps. Show that $\mathscr{K}$ fails to have binary coproducts whereas $\mathscr{L}$ has all coproducts. Show that both categories have coequalizers. [Hint: set $\mathrm{coeq}(f, g) = Y/R$ where $R = \{(y, y'): yh = y'h$ for all h in the category with $fh = gh\}$ with metric $d(yR, zR) = \mathrm{Sup}(d(ah, bh): fh = gh$ as above and $a \in yR, b \in zR)$.]
6. Let $\mathscr{K}$ be the complete poset $2 = \{0, 1\}$ with $0 < 1$ *qua* category. Prove that 2 is large complete and large cocomplete.
7. Give explicit constructions for limits and colimits in **Set** based on 1.22.
8. This exercise provides a reason why limits are more intuitive than colimits. Let $\mathscr{K}$ be a (locally small) category and let (Δ, D) be a diagram in $\mathscr{K}$. Each object A induces the set-valued functor $\mathscr{K}(A, -)$: $\mathscr{K} \longrightarrow$ **Set** *represented by A* whose value on the object X is the set $\mathscr{K}(A, X)$ and whose value on the morphism $f: X \longrightarrow Y$ is the function $-.f: \mathscr{K}(A, X) \longrightarrow \mathscr{K}(A, Y)$ defined by $g \longrightarrow g.f$ (cf. 1.59). Via $-.D_\alpha: \mathscr{K}(A, D_i) \longrightarrow \mathscr{K}(A, D_j)$, $(\Delta, \mathscr{K}(A, D))$ is a diagram in **Set**. If (L, ψ) is a lower bound of D, the triangle

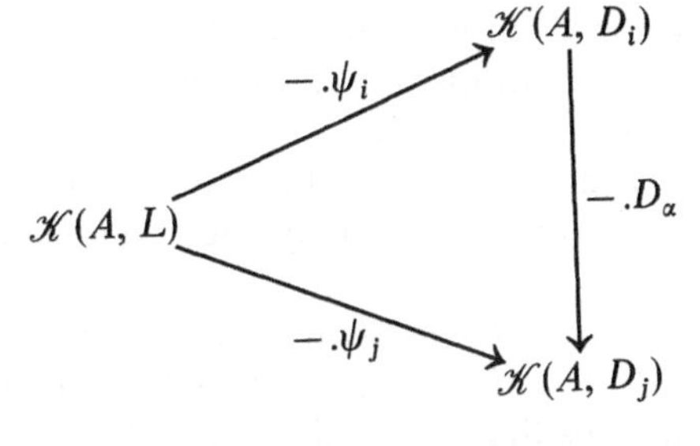

is commutative so that $(\mathscr{K}(A, L), -.\psi)$ is a lower bound of $\mathscr{K}(A, D)$.

(a) Prove that (L, ψ) is a limit of D if and only if for every object A of $\mathscr{K}$, $(\mathscr{K}(A, L), -.\psi)$ is a limit of $\mathscr{K}(A, D)$ in **Set**.

(b) For each A in $\mathscr{K}$, the functor $\mathscr{K}(-, A):\mathscr{K}^{op} \longrightarrow \mathbf{Set}$ *corepresented by* A is the functor $\mathscr{K}^{op}(A, -)$. Prove that (L, ψ) is a colimit of D if and only if for every object A of $\mathscr{K}$, $(\mathscr{K}(L, A), \psi.-)$ is a *limit* of $\mathscr{K}(D, A)$ in **Set**.

9. A "disjoint union" of categories is defined in the obvious way. Show that $\mathscr{K} = \mathscr{K}^{op}$ holds if and only if $\mathscr{K}$ is a disjoint union of abelian monoids. [Hint: take note of Axiom 3 in the definition of a category.]
10. Let $\mathscr{K}$ be the category whose objects are sets and whose morphisms $f:X \longrightarrow Y$ are relations, that is, the Kleisli category of the theory of 1.3.5. Show that the identity function on objects and the passage from a relation to its inverse defines a functor from $\mathscr{K}$ to $\mathscr{K}^{op}$ which establishes an isomorphism of categories (see the next section) $\mathscr{K} \cong \mathscr{K}^{op}$.
11. Let $\mathscr{K}$ be the category of topological spaces and continuous maps and let $\mathscr{L}$ be the category of Hausdorff topological spaces and continuous maps. Given $f:X \longrightarrow Y$ in $\mathscr{K}$ show that f is a coequalizer if and only if f is surjective and Y has the quotient topology induced by f (i.e., a subset B of Y is open if and only if Bf^{-1} is open in X); establish the same result in $\mathscr{L}$. Show that in both categories, eq(f, g) is the subset on which f and g agree provided with the subspace topology; and show that eq(f, g) is a closed subset in the Hausdorff case.
12. Say that $f:A \longrightarrow B$ in $\mathscr{K}$ is a *regular epimorphism* if for every $h:A \longrightarrow C$ satisfying "for every pair $t, u:T \longrightarrow A$, if $tf = uf$ then $th = uh$" there exists unique $\bar{h}:B \longrightarrow C$ with $f\bar{h} = h$.

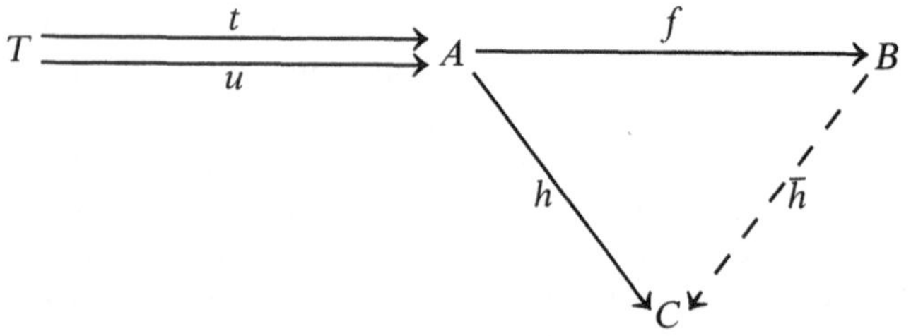

(a) Prove that every coequalizer is a regular epimorphism and that every regular epimorphism is an epimorphism.
(b) Prove that f is an isomorphism if and only if f is a regular epimorphism and f is a monomorphism.
(c) Prove that if f is a regular epimorphism and if the kernel pair of f exists then f is a coequalizer (namely the coequalizer of its kernel pair).

13. A category $\mathscr{K}$ satisfies the *axiom of choice* if every epimorphism in $\mathscr{K}$ is split epi. (Cf. exercise 1 of section 1.2.) Prove that the category of vector spaces over a field satisfies the axiom of choice. [Hint: by the axiom of choice in **Set**, every vector space has a basis.]
14. Given arbitrary morphisms $i:A \longrightarrow X$, $j:B \longrightarrow X$ say that i and j are *equivalent* if there exist factorizations:

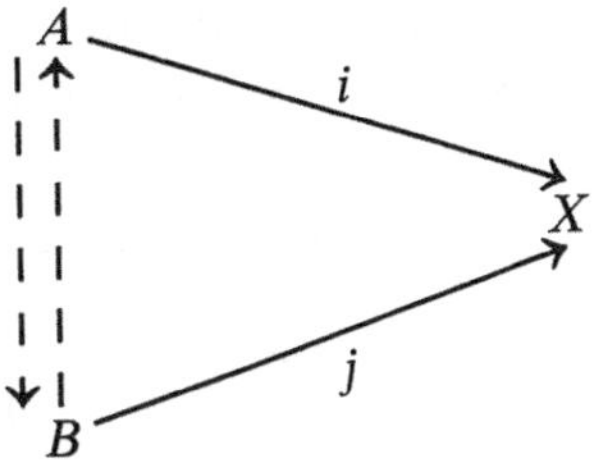

The equivalence class $\langle i \rangle$ is a *generalized subobject of* X. Show that, in any category $\mathscr{K}$, the passage $[i] \longmapsto \langle i \rangle$ from monosubobjects to generalized subobjects is well defined and injective but that this passage is surjective if and only if every morphism in $\mathscr{K}$ factors as a split epi followed by a mono.

15. Say that a pair $p, q : E \longrightarrow X$ of functions is an *equivalence relation* if the induced function $E \longrightarrow X \times X$ is injective and is such that its image is an equivalence relation on X. In the spirit of exercise 8, say that a pair $p, q : E \longrightarrow X$ of morphisms in a category $\mathscr{K}$ is an *equivalence relation* if for every object A, the pair of functions

$$\mathscr{K}(A, E) \underset{-.q}{\overset{-.p}{\rightrightarrows}} \mathscr{K}(A, X)$$

is an equivalence relation.
   (a) Translate the four conditions on $-.p$, $-.q$ into simple diagrammatic statements about p, q in $\mathscr{K}$.
   (b) Prove that the kernel pair of a morphism is always an equivalence relation.
   (c) Construct an example in the category of topological spaces and continuous maps of an equivalence relation which is not a kernel pair. [Hint: "discretize" a kernel pair.]

16. Consider the commutative diagram

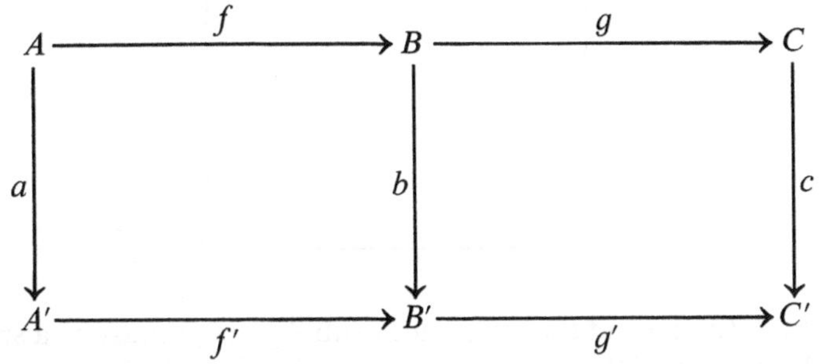

   (a) Show that if both squares are pullbacks then the outer rectangle is a pullback.
   (b) Prove that if the rightmost square and the outer rectangle are pullbacks then the leftmost square is a pullback.

17. An object P in a category $\mathscr{K}$ is *projective* if for every epimorphism $e: A \longrightarrow B$ and morphism $f: P \longrightarrow B$ there exists $g: P \longrightarrow B$ with

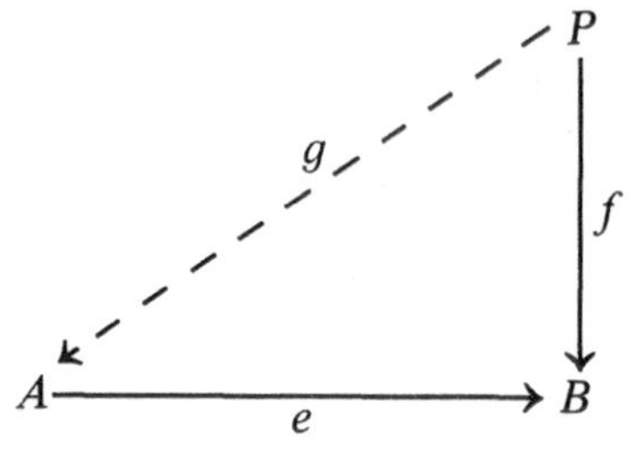

$g.e = f$.
   (a) Let $\mathbf{T}$ be an algebraic theory in **Set**. Prove that every free $\mathbf{T}$-algebra is projective in $\mathbf{Set}^{\mathbf{T}}$.
   (b) Let $\mathscr{K}$ be any category which has a projective generator. Prove that the pullback of an epimorphism along any morphism is again an epimorphism.
18. A *Kleisli algebra* is a pair $(M, \mathbf{T})$ where M is a monoid (*qua* one-object category) and $\mathbf{T}$ is an algebraic theory in M.
   (a) Show that a Kleisli algebra is the same thing as a quintuple $(M, \cdot, e; \circ, \eta)$ where $(M, \cdot, e)$ and $(M, \circ, \eta)$ are monoids satisfying the law $(x \cdot \eta) \circ y = x \cdot y$ for all x, y. [Hint: clone form!]
   (b) Show that $(M, +, 0; \circ, \eta)$ is a Kleisli algebra if $(M, +, 0)$ is an abelian monoid, $\eta + \eta = 0$ and $x \circ y = x + y + \eta$. In this case, show that T is the identity functor and that $\mu = \eta$.
19. A *subobject classifier in* a category $\mathscr{K}$ possessing a terminal object 1 is a pair (Ω, t) where Ω is an object of $\mathscr{K}$ and $t: 1 \longrightarrow \Omega$ is (necessarily) a monomorphism such that the pullback of t along any morphism exists and, in fact, the passage from $\chi: X \longrightarrow \Omega$ to $[t]\chi^{-1}$ establishes a bijection

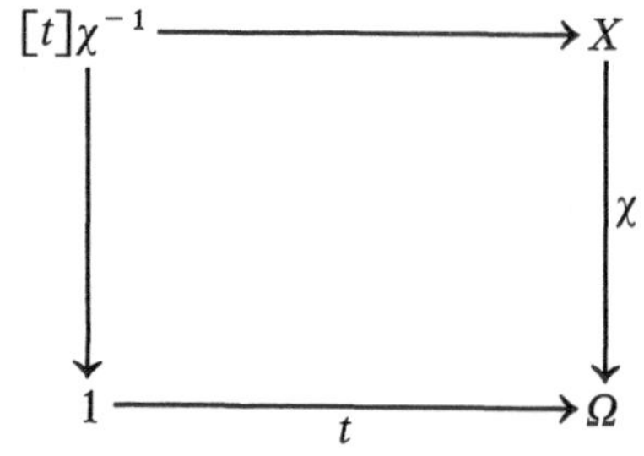

between $\mathscr{K}(X, \Omega)$ and the set of monosubobjects (it must be a small set if $\mathscr{K}$ is locally small!) of X. If A is a subobject of X the corresponding $\chi_A: X \longrightarrow \Omega$ is the "characteristic morphism" of A.
   (a) Prove that all subobject classifiers are unique up to isomorphism [Hint: use exercise 16.]
   (b) Prove that 2 is a subobject classifier in **Set**.
   (c) Let $\mathscr{K}$ be the category whose objects are topological spaces and

whose morphisms are functions which are continuous, open, and closed. Prove that the two-element discrete space is a subobject classifier.

(d) Let M be a monoid. An *M-set* is a pair (X, ξ) where $\xi: X \times M \longrightarrow X$, $(x, m) \longmapsto xm$ satisfies $xe = x$ and $x(mm') = (xm)m'$. If (X, ξ) and (Y, θ) are *M-sets*, and *equivariant* map $f: (X, \xi) \longrightarrow (Y, \theta)$ satisfies $(xf)m = (xm)f$. Set $\Omega = \{A \subset M: AM \subset A\}$ and define an M-set structure on Ω by $Am = \{x \in M: xm \in A\}$. Show that Ω is a subobject classifier in the category of M-sets and equivariant maps.

20. Given a metric space (X, d) and an element x of X show that $d(x, -): (X, d) \longrightarrow \mathbf{R}$ (where $\mathbf{R}$ has the usual metric) is distance decreasing. Conclude that $\mathbf{R}$ is a cogenerator in the category (1.9) of metric spaces. An object K of (any category) $\mathscr{K}$ is *injective* if K is projective in $\mathscr{K}^{\mathrm{op}}$ (as defined in exercise 17) that is, if for every $f: A \longrightarrow K$

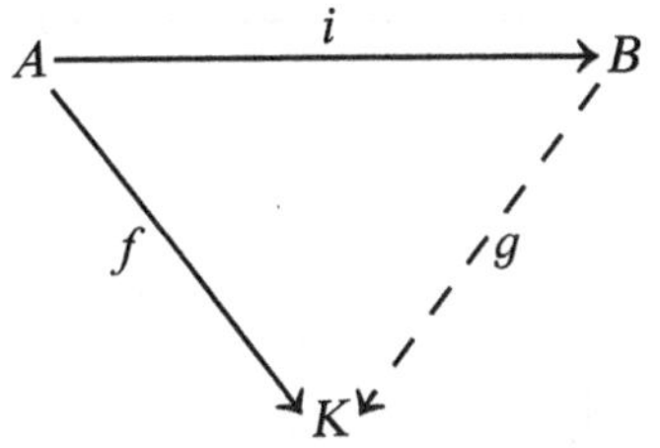

and mono $i: A \longrightarrow B$ there exists an extension $g: B \longrightarrow K$ with $i.g = f$. Prove *McShane's theorem* ([McShane '34, Theorem 1]): $\mathbf{R}$ is injective in the category of metric spaces. [Hint: if f is defined and distance decreasing on a subset A, $xg = \mathrm{Sup}\{af - d(x, a): a \in A\}$ is a suitable extension.]

21. Given the commutative diagram in $\mathscr{K}$ shown below and an arbitrary

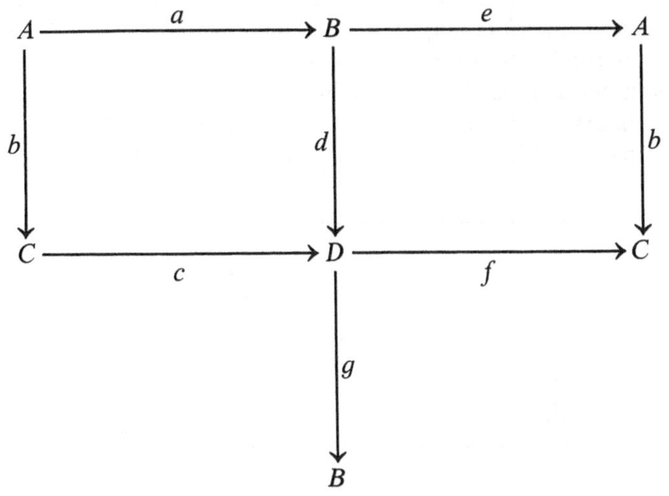

$$ae = \mathrm{id}_A$$
$$cf = \mathrm{id}_C$$
$$dg = \mathrm{id}_B$$

functor $H:\mathscr{K} \longrightarrow \mathscr{L}$ prove that

$$\begin{array}{ccc} AH & \xrightarrow{aH} & BH \\ {\scriptstyle bH}\downarrow & & \downarrow{\scriptstyle dH} \\ CH & \xrightarrow[cH]{} & DH \end{array}$$

is a pullback diagram in $\mathscr{L}$. [Hint: if $t:T \longrightarrow C$, $u:T \longrightarrow B$ satisfies $tc = ud$ then $(ue)b = udf = tcf = t$ and $(ue)a = ueadg = uebcg = tcg = udg = u$.] Reconsider the hint in 1.5 exercise 2(a) with H: **Set** $\longrightarrow$ **Set**^{op} defined by $BH = A^B$, using the result above to show that

$$\begin{array}{ccc} A^X & \longrightarrow & A^F \\ \downarrow & & \downarrow \\ A^G & \longrightarrow & A^{F \cap G} \end{array}$$

is a pushout in **Set** if $F \cap G$ is nonempty.

22. Let **T** be a finitary theory in **Set** and let $f, g:(X, \xi) \longrightarrow (Y, \theta)$ be a pair of **T**-homomorphisms whose image S in $Y \times Y$ contains the diagonal. Prove that the coequalizer of f and g exists and is constructed at the level **Set**. [Hint: the equivalence relation generated by S is the union of the chain $(SS^{-1})^n$ of subalgebras and so is itself a subalgebra.]
23. Investigate "isomorphic" in the following categories of metric spaces: Metric spaces and *Lipschitz* maps (i.e. for some fixed $M > 0$, $d(xf, yf) \leqslant M\, d(x, y)$); metrizeable topological spaces and continuous maps; metric spaces of diameter at most 1 and Lipschitz maps.
24. Let **T** be a theory in **Set** and let **S** be a subtheory of **T**. Show that rank(S) $\leqslant$ rank(T). [Hint: it suffices to prove that the square is a

$$\begin{array}{ccc} nS & \xrightarrow{iS} & AS \\ \downarrow & & \downarrow \\ nT & \xrightarrow[iT]{} & AT \end{array}$$

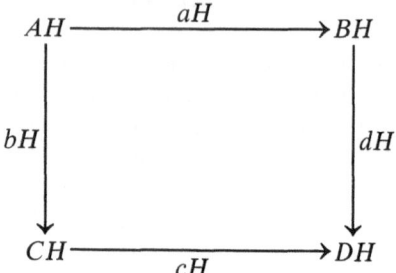

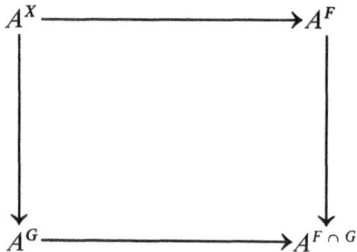

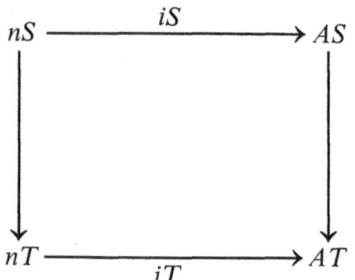

pullback whenever n is a nonempty subset of A; use exercise 21.]

In the following three exercises we describe algebraic theories in **Set** in a style close to the original formulation by [Lawvere '63]. See exercise 3.2.7 for a generalization to arbitrary categories.

25. (For **Set** only.) An *algebraic theory in coproduct form* is a pair $(\mathscr{L}, \eta)$ where $\mathscr{L}$ is a category with the same objects as **Set** (we let $\alpha: A \longrightarrow B$ mean $\alpha \in \mathscr{L}(A, B)$) and η provides specified coproduct diagrams

$$(1 \xrightarrow{(A\eta)_a} A : a \in A)$$

in $\mathscr{L}$ for each set A.

(a) Given (L, η), show that $(T, \eta, (-)^{\#})$ is an algebraic theory in extension form (Exercise 1.3.12) if $AT = \mathscr{L}(1, A)$, $\langle a, A\eta\rangle = (A\eta)_a$ and if

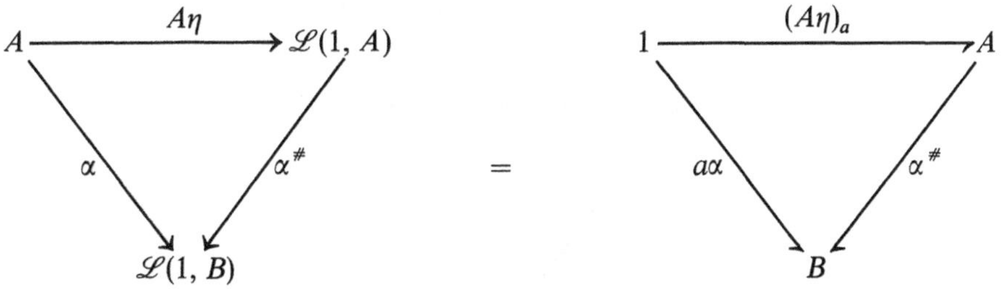

(b) Given $(T, \eta, (-)^{\#})$, show that $(\mathscr{L}, \eta)$ is a theory in coproduct form if $\mathscr{L}$ is the Kleisli category of $(T, \eta, \circ)$ $(\alpha \circ \beta = \alpha.\beta^{\#})$ and $(A\eta)_a = \langle a, A\eta\rangle : 1 \longrightarrow AT$.

(c) Clarify and prove: the passages of (a) and (b) are mutually inverse up to isomorphism.

26. A *finitary Lawvere theory* is a pair $(\mathscr{L}, \eta)$ where $\mathscr{L}$ is a category with $\mathrm{Ob}(\mathscr{L}) = \mathbf{N}$ (we write $\alpha: n \longrightarrow m$ if $\alpha \in \mathscr{L}(m, n)$) and η assigns coproduct diagrams

$$(1 \xrightarrow{(n\eta)_i} n : i \in n)$$

in $\mathscr{L}$ for every n in $\mathbf{N}$.

(a) Let $(\mathscr{L}, \eta)$ be a finitary Lawvere theory. For each set A let Δ^A be the diagram scheme with nodes all (n, f) with n in $\mathbf{N}$ and $f: n \longrightarrow A$ and with edges h shown below, and let D^A be the Δ^A-diagram in **Set**:

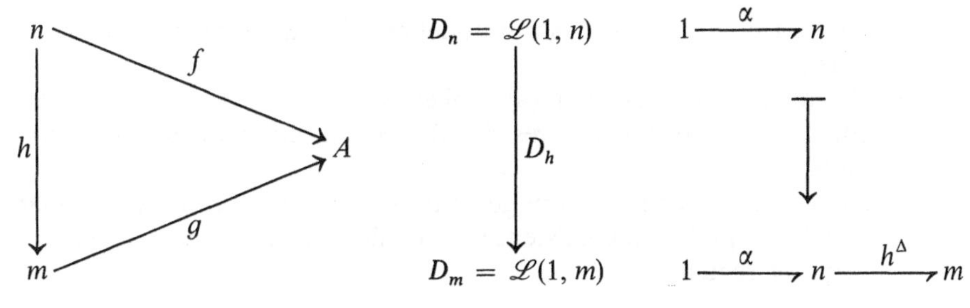

where h^Δ is defined by the coproduct property

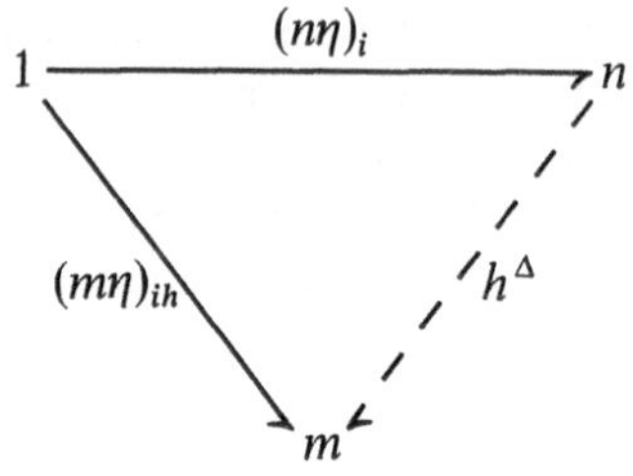

Define $AT = \text{colim } D^A$ with canonical injections $\Gamma^A_{(n,f)}: D_n \longrightarrow AT$. Define $A\eta: A \longrightarrow AT$ by $\langle a, A\eta\rangle = \langle 1\eta, \Gamma^A_{(1,a)}\rangle$. Given $\alpha: A \longrightarrow BT$, define $\alpha^\#: AT \longrightarrow BT$ by the colimit property

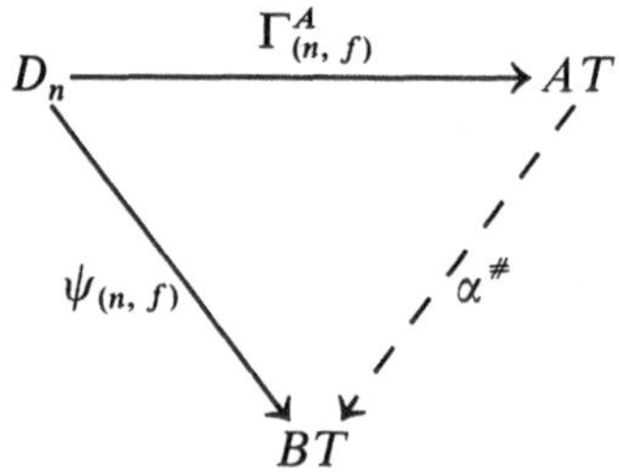

where, using the finiteness of n to choose a factorization

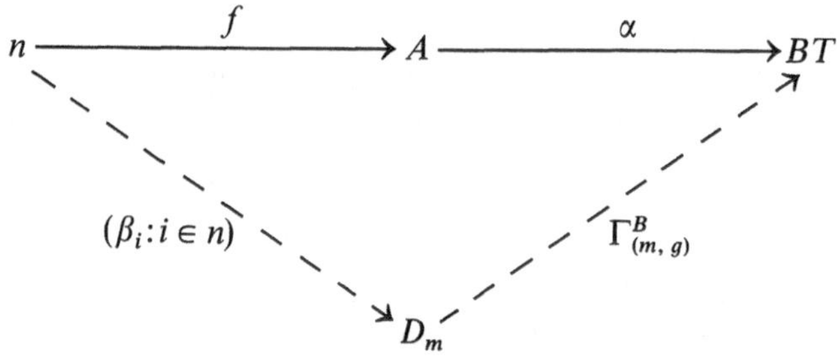

we define

$$(1 \xrightarrow{\tau} n)\psi_{(n,f)} = (1 \xrightarrow{\tau} n \xrightarrow{\beta} m)\Gamma^B_{(m,g)}$$

Prove that $(T, \eta, (-)^\#)$ is a well-defined finitary algebraic theory in extension form.

(b) Show that the passage in (a) is bijective up to isomorphism. [Hint: for the inverse passage, restrict the construction of exercise 25(b) to **N**.]

(c) State and prove the obvious generalization of (b) for arbitrary regular cardinals. [Hint: the existence of the factorization $f.\alpha = (\beta_i).\Gamma^B_{(m,g)}$ requires regularity.]

27. Let X be a fixed set. Verify that $(\mathscr{L}, \eta)$ is a finitary Lawvere theory if $\mathscr{L}(n, m)$ = partial functions from $X \times n$ to $X \times m$ and $\langle x, (n\eta)_i \rangle = (x, i)$. Observe that this theory is much more naturally described this way than in $(T, \eta, \circ)$ form. This theory is named $[X, 1]$ in [Elgot '75, Section 2.3]. A hint of the connection with computer science lies in observing that the interpretations of schemes as in exercise 1.1.8 are morphisms in $[X, 1]$. An open problem is to formulate the role played by the algebras of the theories that arise in the context of the paper of Elgot mentioned above.

2. Free Objects

The free **T**-algebras of 1.4.12 can be described in terms of the underlying $\mathscr{K}$-object functor $U^{\mathbf{T}}: \mathscr{K}^{\mathbf{T}} \longrightarrow \mathscr{K}$ and, hence, abstracted to the level of any functor $U: \mathscr{A} \longrightarrow \mathscr{K}$; we will say that such functors "have a left adjoint." Examples and standard existence theorems are given. Paralleling the development in section 1.2, we observe that every adjointness induces an algebraic theory.

2.1 Definition. *Let $U: \mathscr{A} \longrightarrow \mathscr{K}$ be a functor and let K be an object of $\mathscr{K}$. A free $\mathscr{A}$-object over K with respect to U is a pair (F, η) where F is an object of $\mathscr{A}$ and $\eta: K \longrightarrow FU$ is a morphism in $\mathscr{K}$ subject to the universal property that whenever (A, f) is another such pair (that is, A is an object in $\mathscr{A}$ and $f: K \longrightarrow AU$ in $\mathscr{K}$) there exists a unique $\mathscr{A}$-morphism $f^{\#}: F \longrightarrow A$*

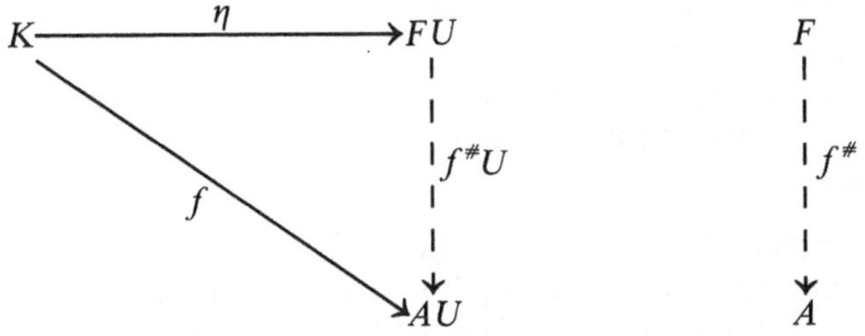

such that $\eta . f^{\#} U = f$. For us, the fundamental example is 2.3 below; our first such encounter was 1.1.20. For general heuristics, think of $\mathscr{K}$ as the base category and $\mathscr{A}$ as a category of $\mathscr{K}$-objects with additional structure with "forgetful" functor U (for more on this, see section 3). F is the object "freely generated by the generators K" and η represents "inclusion of the generators". The universal property says that "a $\mathscr{K}$-morphism on K with values in an $\mathscr{A}$-object admits a unique $\mathscr{A}$-morphic extension."

2.2 Proposition. *Free objects are unique up to isomorphism.*

Proof. Suppose that (F, η) and (F', η') are both free over K with respect to U. Consider the unique $\mathscr{A}$-morphisms g and h as shown in the diagram on the following page:

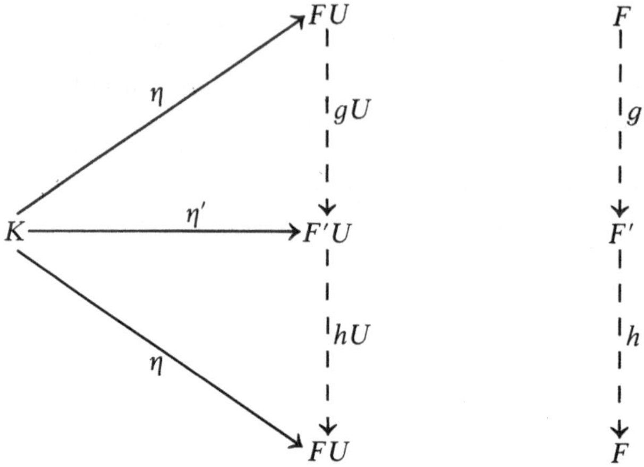

Since $\eta.(gh)U = \eta.gU.hU = \eta = \eta.(\mathrm{id}_F)U$, $gh = \mathrm{id}_F$. Symmetrically, $hg = \mathrm{id}_{F'}$. Therefore, g and h are mutually inverse isomorphisms the U of which converts each η into the other; it seems hard to imagine a more stringent isomorphism statement! □

The reader might compare 2.2 with 1.4.14.

2.3 Example. If $\mathbf{T}$ is an algebraic theory in the category $\mathscr{K}$ then, for each K in $\mathscr{K}$, $(KT, K\mu; K\eta)$ is a free $\mathscr{K}^{\mathbf{T}}$-object over K with respect to $U^{\mathbf{T}}:\mathscr{K}^{\mathbf{T}} \longrightarrow \mathscr{K}$. This is the content of 1.4.12. Notice that the two uses of the symbol "η" are coherent.

2.4 Example. Let U be the forgetful functor from the category of topological spaces and continuous maps to the category of sets and functions. For each set n let F_n be the set n provided with its discrete topology. Let $\eta:n \longrightarrow F_nU$ be the identity function of n. Then (F_n, η) is free over n with respect to U.

2.5 Example. Let $\mathscr{A}$ be the category whose objects are fields and whose morphisms are unit-preserving ring homomorphisms. Let n be a set with two elements. Then there exists no free field (F, η) over n with respect to the forgetful functor $U:\mathscr{A} \longrightarrow \mathbf{Set}$. For suppose such (F, η) existed. Let $\mathbf{Z}_2$ be the two element field and let $f:n \longrightarrow \mathbf{Z}_2U$ be a bijection. From the first diagram below and the dual of 1.42 we see that η is injective. Now consider

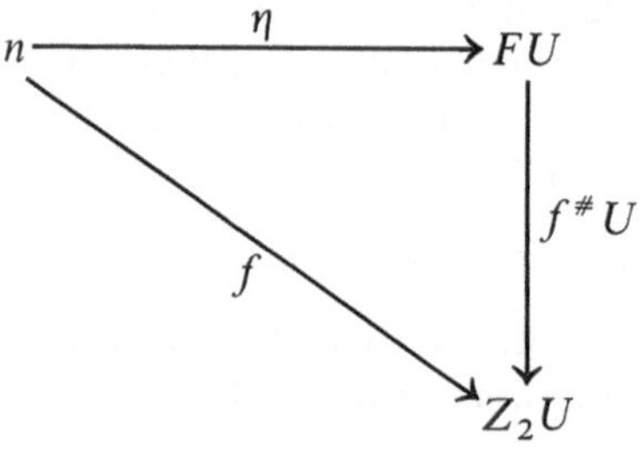

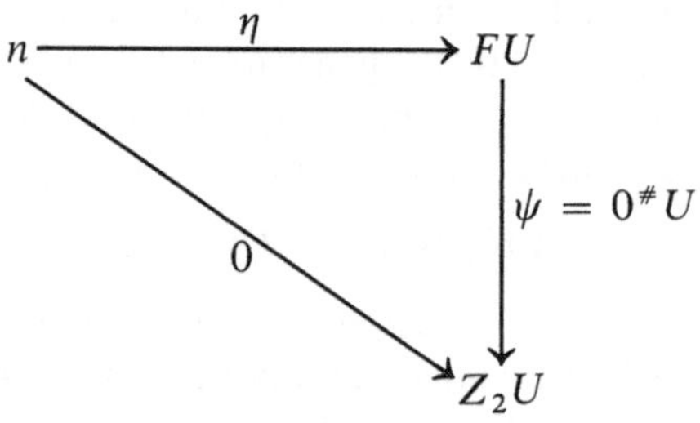

the extension ψ of the zero map 0 as shown. Since η is injective, the kernel of ψ is not 0. Since ψ maps 1 to 1, the kernel of ψ is not F. This contradicts the well-known fact that no field has a nontrivial ideal.

Fields come very close to being presentable by finitary operations and equations; the problem is that "multiplicative inverse" is not a unary operation since it is not defined at 0. Example 2.5 says that, from the categorical point of view, field theory is far from being part of universal algebra.

2.6 Various Subcategories. Let $\mathscr{C}$ be the category of categories and functors. A *subcategory of* $\mathscr{K}$ is a $\mathscr{C}$-monosubobject of $\mathscr{K}$ as defined in 1.65. Consideration of the three morphism category

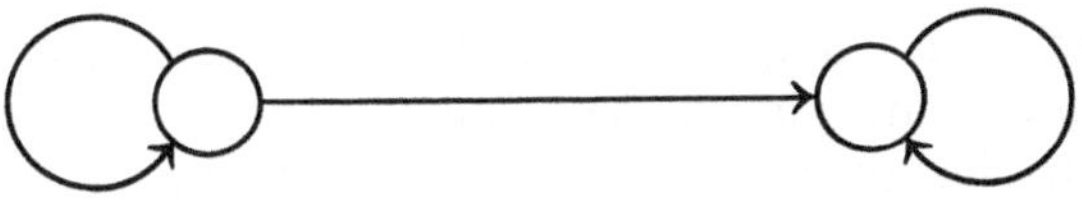

makes it clear that $H: \mathscr{L} \longrightarrow \mathscr{K}$ is a monomorphism if and only if H is injective on morphisms or, equivalently, the associated functions

$$\mathscr{L}(L_1, L_2) \xrightarrow{L \mapsto LH} \mathscr{K}(L_1H, L_2H) \qquad \text{for } L_1, L_2 \text{ in } \mathscr{L}$$

are injective.

By a *literal subcategory* of $\mathscr{K}$ we mean a category $\mathscr{L}$ with $\text{Obj}(\mathscr{L}) \subset \text{Obj}(\mathscr{K})$ and $\mathscr{L}(L_1, L_2) \subset \mathscr{K}(L_1, L_2)$, i.e., a subclass of $\mathscr{K}$-morphisms closed under identities and composition. The arbitrary subcategory $[H]$ is represented by the inclusion functor of the literal subcategory

$$\{L_1H \xrightarrow{fH} L_2H : L_1 \xrightarrow{f} L_2 \in \mathscr{L}\}$$

In the future, the term "literal" will not be used even though it may have been intended. This parallels the conventions used in **Set** where, almost always, subobjects are intended to be literal subsets.

A subcategory $\mathscr{L}$ of $\mathscr{K}$ is *full* if for every L_1, L_2 in $\mathscr{L}$ and $f: L_1 \longrightarrow L_2$ in $\mathscr{K}$, f is in $\mathscr{L}$; or, in the fussy "nonliteral" language, given L_1, L_2 in $\mathscr{L}$ and $f: L_1H \longrightarrow L_2H$ in $\mathscr{K}$, there exists (necessarily unique) $\bar{f}: L_1 \longrightarrow L_2$ in $\mathscr{L}$ with $\bar{f}H = f$. Thus, full subcategories of $\mathscr{K}$ are in bijective correspondence with subclasses of $\text{Obj}(\mathscr{K})$. For example, if $\mathscr{K}$ is the category of topological spaces and continuous maps, "Hausdorff spaces" defines a full subcategory whereas "Hausdorff spaces and continuous open maps" defines a subcategory which is not full.

Let $\mathscr{L}$ be a subcategory of $\mathscr{K}$ and let K be an object in $\mathscr{K}$. A *reflection of K in* $\mathscr{L}$ is a free $\mathscr{L}$-object over K with respect to the inclusion functor $U: \mathscr{L} \longrightarrow \mathscr{K}$. It is common to delete U from the picture of the universal property, as shown on the following page, but we must keep in mind the

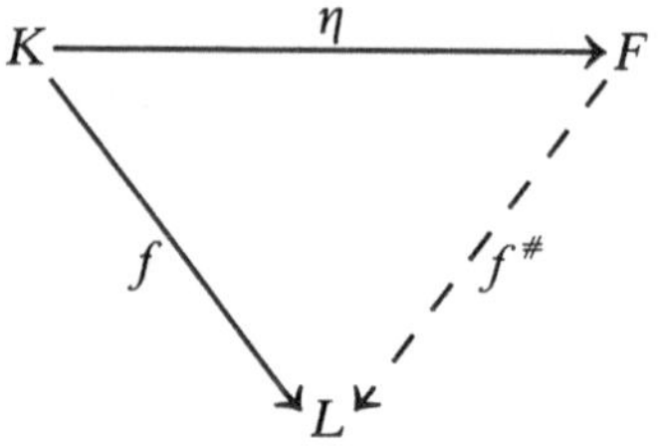

proviso that F and L are objects in $\mathscr{L}$ and that (in case $\mathscr{L}$ is not full) $f^{\#}$ is again required to be in $\mathscr{L}$. In this context, η is called the *reflection* map. $\mathscr{L}$ is a *reflective* subcategory of $\mathscr{K}$ just in case every object of K has a reflection in $\mathscr{L}$.

2.7 Abelianizing a Group. Abelian groups is a full reflective subcategory of groups and homomorphisms. The failure of a group G to be abelian is measured by each instance of "$ab \neq ba$," that is by the set $C = \{ab(ba)^{-1}: a, b \in G\}$ of commutators. If N is the normal subgroup generated by C, then G/N is abelian because $aN\,bN\,(bN\,aN)^{-1} = (ab(ba)^{-1})N = N$. Let $\eta: G \longrightarrow G/N$ be the canonical projection. To check that η is the reflection map, we

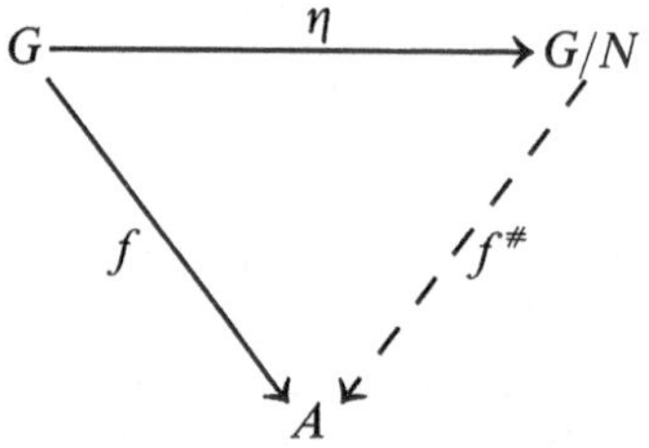

are forced to define $(aN)f^{\#} = af$. This is well defined since the kernel of f is a normal subgroup of G which, because A is abelian, contains C.

2.8 The β-Compactification of a Topological Space. Compact Hausdorff spaces form a full reflective subcategory of topological spaces and continuous maps. We will prove that an arbitrary topological space has a compact Hausdorff reflection by making use of the algebraic theory $\boldsymbol{\beta}$ of 1.3.21.

The reader should first refresh herself on the notations used in 1.5.24. One change: to reserve η for the current reflection map of interest, we write $\text{prin}: X \longrightarrow X\beta$ for the principal ultrafilter map. The secret is the following commutative diagram associated with a set X, a compact Hausdorff space ($=$ $\boldsymbol{\beta}$-algebra) (C, ξ) and a continuous map $\psi:(X\beta, X\mu) \longrightarrow (C, \xi)$. What we can read at once from this diagram is:

(2.9) For each ultrafilter $\mathscr{U}$ on X, $\langle \mathscr{U}, f\beta \rangle \longrightarrow \mathscr{U}\psi$. In particular, by 1.5.30, if $\mathscr{T}$ is a topology on X then $f:(X, \mathscr{T}) \longrightarrow (C, \xi)$ is continuous if and only if ($\mathscr{U} \underset{\mathscr{T}}{\longrightarrow} x$ implies $\mathscr{U}\psi = xf$).

Fix a topological space $(X, \mathscr{T})$. Define a binary relation R_0 on the set $X\beta$ by $\mathscr{U}R_0\mathscr{V}$ if and only if $\mathscr{U}$ and $\mathscr{V}$ converge to a common point with

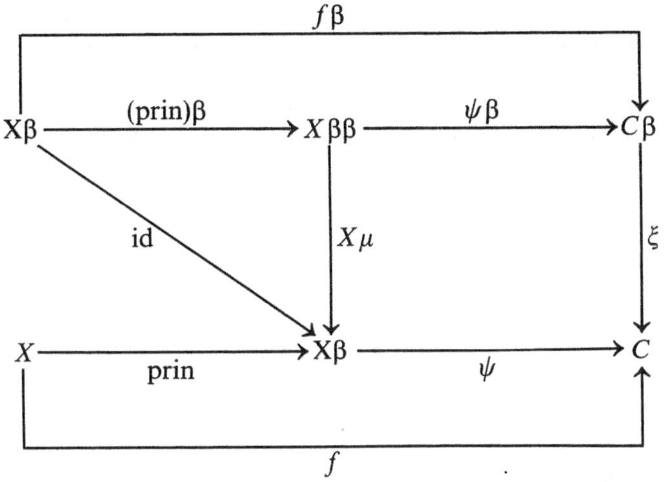

respect to $\mathscr{T}$. Let R be the smallest closed equivalence relation on $(X\beta, X\mu)$ containing R_0. Let $\theta: X \longrightarrow X\beta/R$ be the canonical projection. As is well known from topology (but see 3.1.13 where it turns out to be an important structural fact in universal algebra) the space $F = X\beta/R$ in the quotient topology induced by $(X\beta, X\mu)$ and θ is again compact Hausdorff. We will show that $\eta = \text{prin}.\theta:(X, \mathscr{T}) \longrightarrow F$ is the desired reflection map. To prove that η is continuous, we use 2.9 with $C = F$, $\psi = \theta$ and $f = \eta$. Let $\mathscr{U} \xrightarrow[\mathscr{T}]{} x$. Since also $\text{prin}(x) \xrightarrow[\mathscr{T}]{} x$, $\mathscr{U} R_0 \text{prin}(x)$, and $\mathscr{U}\theta = \text{prin}(x)\theta = x\eta$ as we desired. To establish the universal property, let $f:(X, \mathscr{T}) \longrightarrow (C, \xi)$ be continuous with (C, ξ) a $\boldsymbol{\beta}$-algebra. It takes only 1.4.12 to produce the unique continuous ψ shown below:

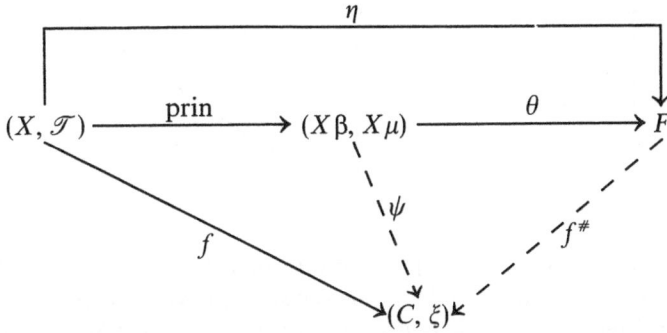

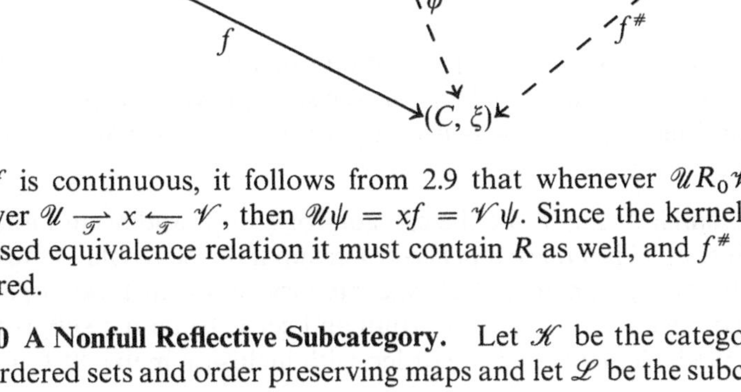

Since f is continuous, it follows from 2.9 that whenever $\mathscr{U} R_0 \mathscr{V}$, that is, whenever $\mathscr{U} \xrightarrow[\mathscr{T}]{} x \xleftarrow[\mathscr{T}]{} \mathscr{V}$, then $\mathscr{U}\psi = xf = \mathscr{V}\psi$. Since the kernel pair of ψ is a closed equivalence relation it must contain R as well, and $f^{\#}$ is induced as desired.

2.10 A Nonfull Reflective Subcategory. Let $\mathscr{K}$ be the category of partially ordered sets and order preserving maps and let $\mathscr{L}$ be the subcategory of complete partially ordered sets and supremum-preserving maps. Examples abound to show $\mathscr{L}$ is not full. Let $(X, \leqslant)$ be an arbitrary partially ordered set and set $F = \{A \subset X\colon$ whenever $x \leqslant a$ and $a \in A$ then $x \in A\}$. With inclusion as partial order, F is complete and in fact ordinary intersection and union provide the infima and suprema. To establish the universal prop-

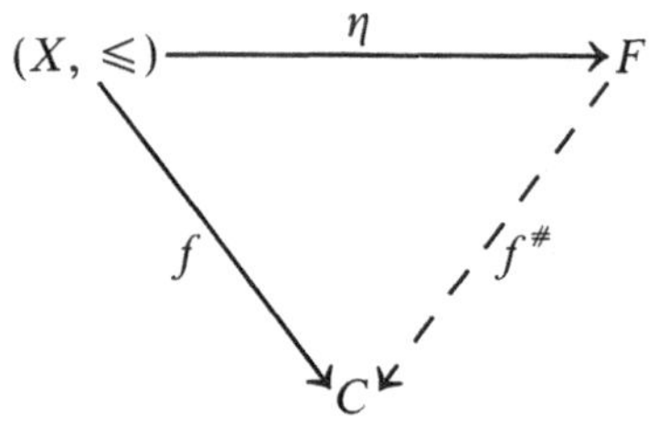

erty, define $x\eta = \{y \in X : y \leqslant x\}$ (which is clearly order preserving) and define $Af^{\#} = \mathrm{Sup}(af : a \in A)$, which makes sense since C is complete. That $f^{\#}$ is supremum preserving and that $f \leqslant \eta.f^{\#}$ is true for any function f. To prove that $f \geqslant \eta.f^{\#}$, we need to know that f was order preserving.

2.11 Cofree Objects. The duality theory of 1.23 extends quite well to functors, essentially as a special case of the dual of a diagram as discussed in 1.34. Given a functor $U\colon \mathscr{A} \longrightarrow \mathscr{K}$ define the functor $U^{\mathrm{op}}\colon \mathscr{A}^{\mathrm{op}} \longrightarrow \mathscr{K}^{\mathrm{op}}$ by $AU^{\mathrm{op}} = AU$ and $(f\colon A \longleftarrow B)U^{\mathrm{op}} = fU\colon AU^{\mathrm{op}} \longleftarrow BU^{\mathrm{op}}$. Given an object K of $\mathscr{K}$, a *cofree $\mathscr{A}$-object over K with respect to U* is a free $\mathscr{A}^{\mathrm{op}}$-object over K with respect to U^{op}. If (C, ε) is cofree over K with respect to U, the picture of the universal property is

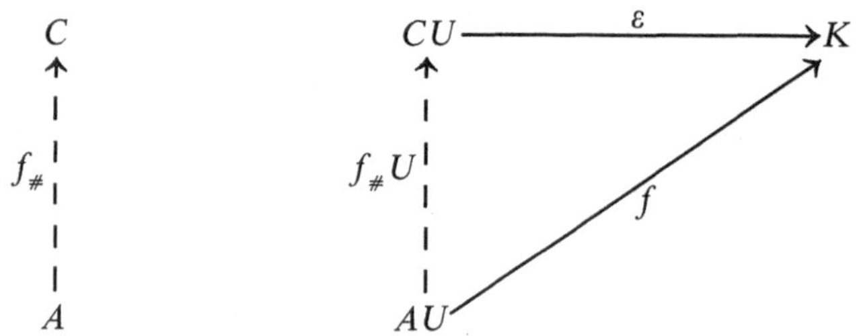

2.12 Example. Let U be the forgetful functor from topological spaces to sets as in 2.4. For each set n let C_n be the set n provided with its indiscrete topology and let $\varepsilon\colon C_nU \longrightarrow n = \mathrm{id}_n$. Then (C_n, ε) is cofree over n with respect to U.

2.13 Example. Let $\mathscr{K}$ be the category of groups and homomorphisms. Say that a group G is a *torsion-generated* group if the elements of finite order generate the group. Then the full subcategory $\mathscr{L}$ of all torsion-generated groups is coreflective. Let G be a group and set C to be the subgroup of G generated by all elements of finite order, with inclusion map ε. If $f\colon T \longrightarrow G$

and T is torsion-generated, then it is clear that $\mathrm{Im}(f) \subset C$, which defines $f_\#$.

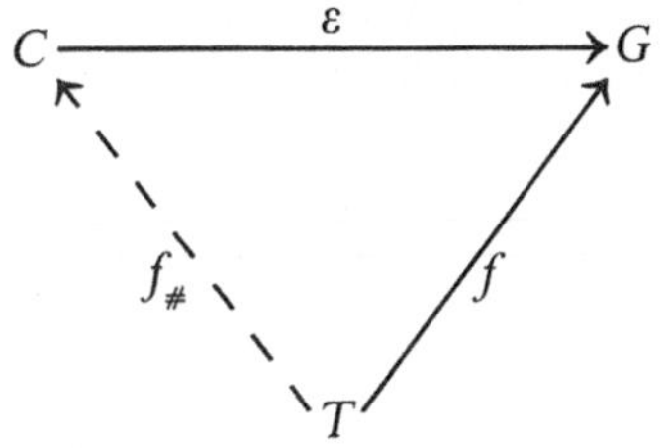

2.14 Example. Let X_0 be a set and consider the functor $- \times X_0$: **Set** $\longrightarrow$ **Set** which sends $f: A \longrightarrow B$ to $f \times \mathrm{id}: A \times X_0 \longrightarrow B \times X_0$. Then for any set Y, $(Y^{X_0}, \varepsilon: Y^{X_0} \times X_0 \longrightarrow Y)$ is the cofree set over Y with respect to $- \times X_0$, where ε is the evaluation map, $(f, x)\varepsilon = xf$. It

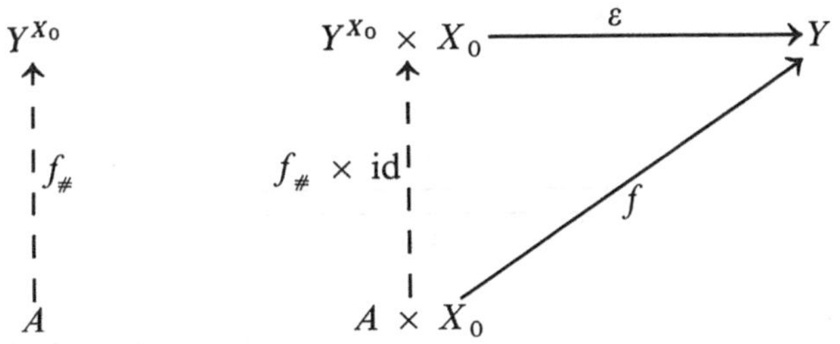

is clear that $f_\#$ exists uniquely via $\langle x, af_\# \rangle = (a, x)f$.

2.15 Adjointness. An *adjointness* is a 6-tuple $(\mathscr{A}, \mathscr{K}, U, F, \eta, \varepsilon)$ where $\mathscr{A}$ and $\mathscr{K}$ are categories, $U: \mathscr{A} \longrightarrow \mathscr{K}$ and $F: \mathscr{K} \longrightarrow \mathscr{A}$ are functors and $\eta: \mathrm{id}_{\mathscr{K}} \longrightarrow FU$ and $\varepsilon: UF \longrightarrow \mathrm{id}_{\mathscr{A}}$ are natural transformations, subject to the so called "triangular identities"

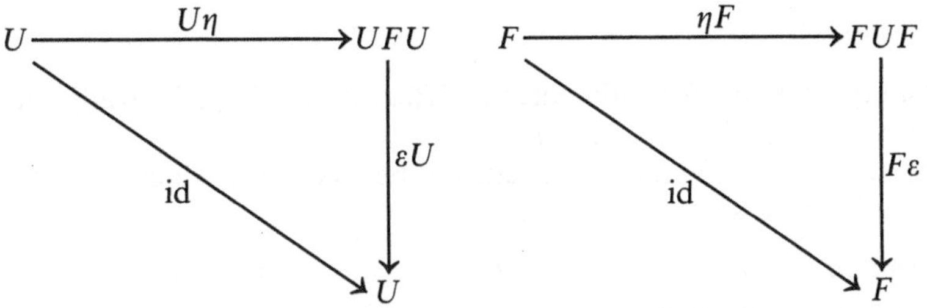

(2.16)

The diagrams of 2.16 are "objectwise." Thus, the first triangle asserts that $AU\eta: AU \longrightarrow AUFU$ (which uses the action of U on objects) when composed with $A\varepsilon U: AUFU \longrightarrow AU$ (which uses the action of U on morphisms) is $\mathrm{id}_{AU}: AU \longrightarrow AU$ for every object A in $\mathscr{A}$. With the period notation: $AU\eta . A\varepsilon U = \mathrm{id}_{AU}$. Some of the formalism concerned with the ways functors and natural transformations interact is developed at the end of this section.

There is an immediate duality theory for adjointness. First consider the picture of the adjointness $(\mathscr{A}, \mathscr{K}, U, F, \eta, \varepsilon)$

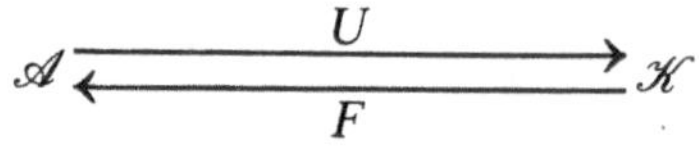

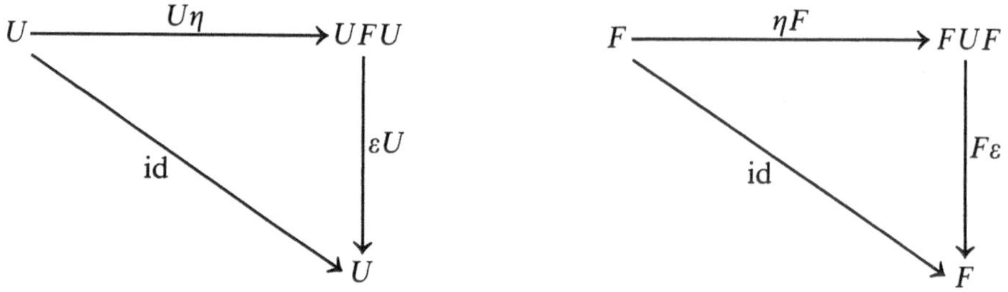

Now, stare at:

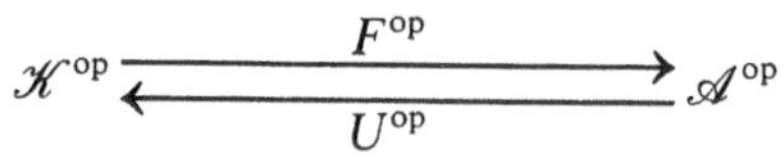

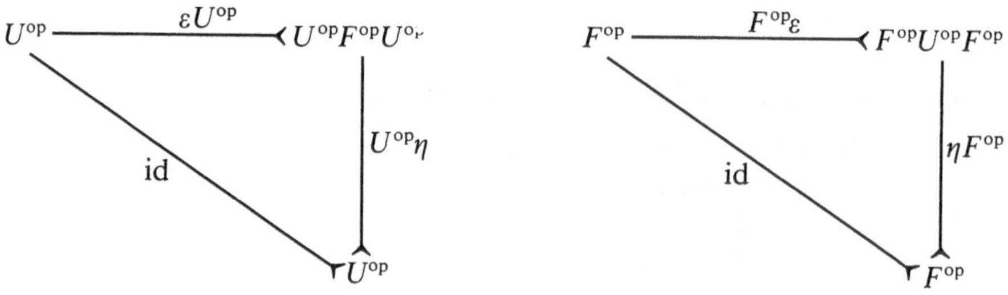

We have at once

2.17 Duality Principle for Adjointness. If $(\mathscr{A}, \mathscr{K}, U, F, \eta, \varepsilon)$ is an adjointness then so is $(\mathscr{K}^{\mathrm{op}}, \mathscr{A}^{\mathrm{op}}, F^{\mathrm{op}}, U^{\mathrm{op}}, \varepsilon, \eta)$.

We now justify the overlap in notations between free objects, cofree objects, and adjointness:

2.18 Theorem. *Let $U:\mathscr{A} \longrightarrow \mathscr{K}$ be a functor. Then the following two conditions on U are equivalent:*

1. *For every object K in $\mathscr{K}$ there exists a free object over K with respect to U.*
2. *There exists an adjointness of form $(\mathscr{A}, \mathscr{K}, U, F, \eta, \varepsilon)$.*

Proof. *(1) implies (2).* Let $(KF, K\eta)$ be free over K with respect to U. Given $f:K \longrightarrow L$ in $\mathscr{K}$, the universal property forces the definition of

$fF: KF \longrightarrow LF$ so as to make $\eta: \mathrm{id}_{\mathscr{K}} \longrightarrow FU$ a natural transformation:

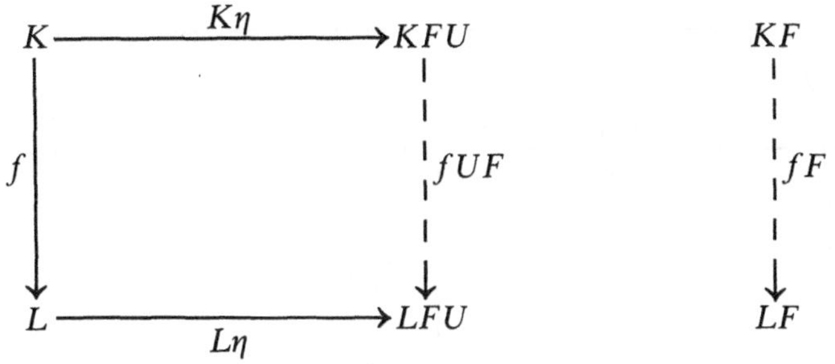

The functorial equations, $(f.g)F = fF.gF$ and $(\mathrm{id}_K)F = \mathrm{id}_{KF}$, are immediate consequences of the universal property. The first triangle in 2.16 forces us to define ε by $A\varepsilon = (\mathrm{id}_{AU})^{\#}$:

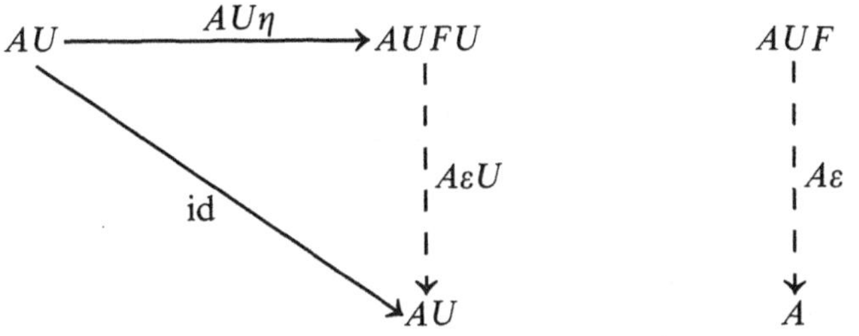

Of general interest, is the formula

(2.19) *For all* $f: K \longrightarrow AU$, $f^{\#} = fF.A\varepsilon$, which is seen from

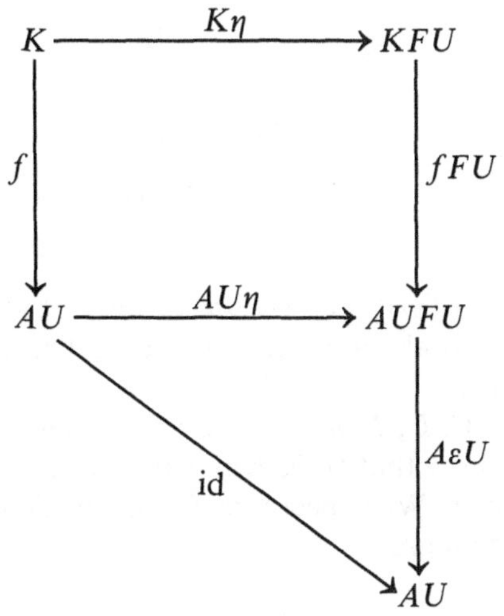

Finally, look at the two diagrams:

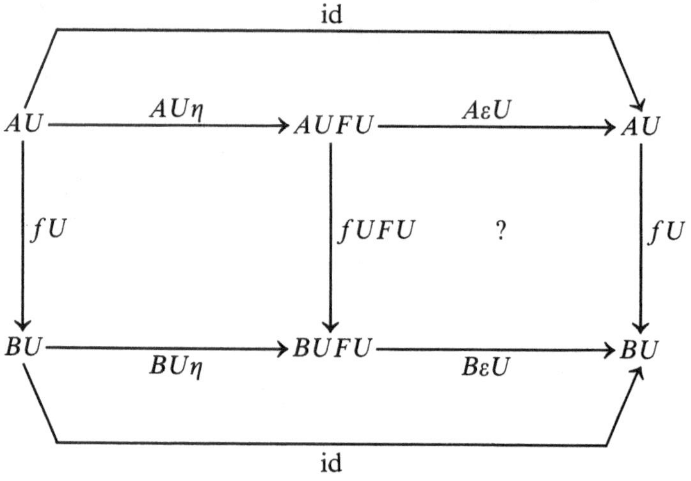

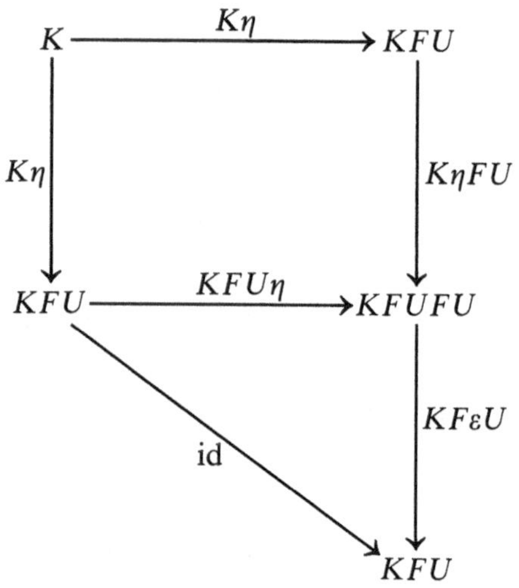

The first shows that $A\varepsilon.f = (fU)^{\#} = fUF.B\varepsilon$ for all $f\colon A \longrightarrow B$ in $\mathscr{A}$, that is, that ε is natural; the second establishes the second triangle in 2.16 by showing that $\eta\mathrm{F}.\mathrm{F}\varepsilon = (\mathrm{id}_{FU})^{\#}$.

(*2*) *implies* (*1*). Let $(\mathscr{A}, \mathscr{K}, U, F, \eta, \varepsilon)$ be an adjointness and let K be an object in $\mathscr{K}$. We wish to show that $(KF, K\eta)$ is free over K with respect to U. Let $f\colon K \longrightarrow AU$ be given. We expect that $f^{\#}$ will be defined as in (2.19), so define $f^{\#}$ that way. The diagram

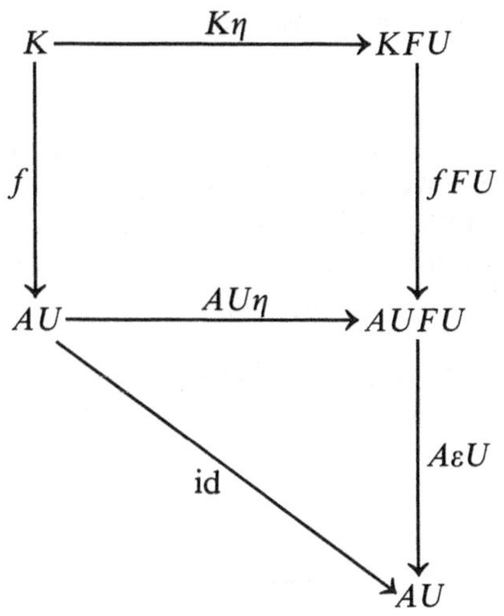

proves that $K\eta.f^{\#} = f$. Suppose also, $\psi : KF \longrightarrow A$ satisfies $K\eta.\psi U = f$. Then

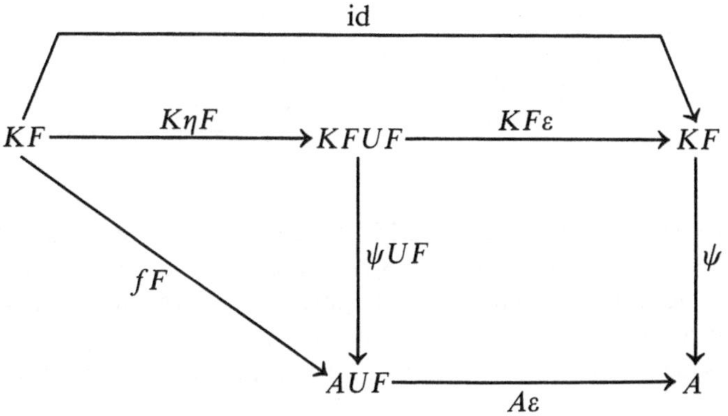

proves that $\psi = f^{\#}$. □

With the help of 2.17 we see that the dual of 2.18 reads as follows:

2.19 Theorem (Dual to 2.18). *Let $F : \mathscr{K} \longrightarrow \mathscr{A}$ be a functor. Then there exists an adjointness of form $(\mathscr{A}, \mathscr{K}, U, F, \eta, \varepsilon)$ if and only if for every object A in $\mathscr{A}$ there exists a cofree $\mathscr{K}$-object over A with respect to F.* □

If $(\mathscr{A}, \mathscr{K}, U, F, \eta, \varepsilon)$ is an adjointness, there are bijections

$$\mathscr{A}(KF, A) \cong \mathscr{K}(K, AU)$$

for all K in $\mathscr{K}$ and A in $\mathscr{A}$. This looks quite like the definition of the adjoint

U^* of the operator U between Hilbert spaces:

$$(xU^*, y) = (x, yU)$$

This analogy is responsible for the "adjoint" in adjointness. Indeed, given an adjointness as above we say that F is a *left adjoint to* U and that U *has* F *as a left adjoint*; symmetrically, U is a *right adjoint to* F and F *has* U *as a right adjoint*. We also rephrase either of the two equivalent statements about U in 2.18 by saying that U *has a left adjoint*.

2.20 The Algebraic Theory of an Adjointness. Let $(\mathscr{A}, \mathscr{K}, U, F, \eta, \varepsilon)$ be an adjointness. For the very general reasons discussed in 2.31, $T = FU$: $\mathscr{K} \longrightarrow \mathscr{K}$ is a functor and

$$\mathrm{id}_{\mathscr{K}} \xrightarrow{\ \eta\ } T \qquad TT = F(UF)U \xrightarrow{\ \mu = F\varepsilon U\ } FU = T$$

are natural transformations. For each object A of $\mathscr{A}$, the diagrams

$$\begin{array}{ccc} AU & \xrightarrow{\ AU\eta\ } & AUT \\ & {\scriptstyle \mathrm{id}_{AU}} \searrow & \big\downarrow {\scriptstyle A\varepsilon U} \\ & & AU \end{array} \qquad \begin{array}{ccc} AUTT & \xrightarrow{\ A\varepsilon UT\ } & AUT \\ {\scriptstyle AU\mu} \big\downarrow & & \big\downarrow {\scriptstyle A\varepsilon U} \\ AUT & \xrightarrow[\ A\varepsilon U\ ]{} & AU \end{array}$$

are immediate consequences of 2.15, which is surely reminiscent of 1.4.9 and 1.4.10. In fact, by setting $A = KF$, and directly observing that $\eta T.\mu = (\eta F.F\varepsilon)U = \mathrm{id}_T$, we have all of the diagrams of 1.3.16, that is $\mathbf{T} = (T, \eta, \mu)$ is an algebraic theory in $\mathscr{K}$. $\mathbf{T}$ is called the *algebraic theory induced by* the adjointness $(\mathscr{A}, \mathscr{K}, U, F, \eta, \varepsilon)$. Noting that U transforms the naturality square

$$\begin{array}{ccc} AUF & \xrightarrow{\ fUF\ } & BUF \\ {\scriptstyle A\varepsilon} \big\downarrow & & \big\downarrow {\scriptstyle B\varepsilon} \\ A & \xrightarrow[\ f\ ]{} & B \end{array} \quad \text{into} \quad \begin{array}{ccc} AUT & \xrightarrow{\ fUT\ } & BUT \\ {\scriptstyle A\varepsilon U} \big\downarrow & & \big\downarrow {\scriptstyle B\varepsilon U} \\ AU & \xrightarrow[\ fU\ ]{} & BU \end{array}$$

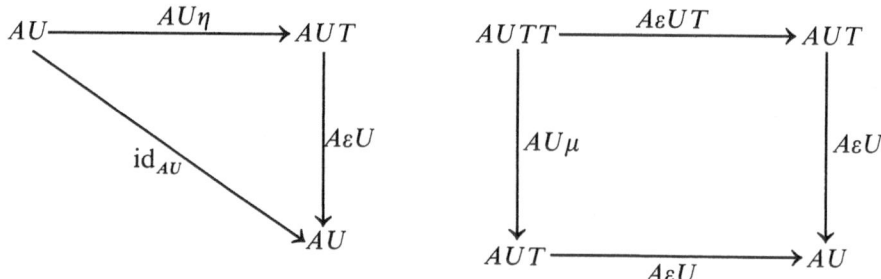

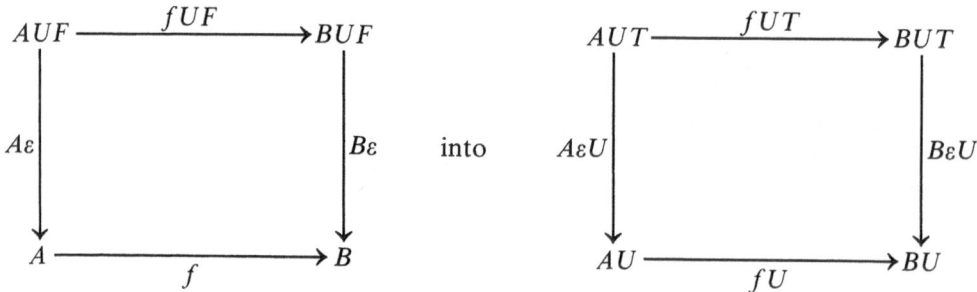

the passage from A to $(AU, A\varepsilon U)$ describes a functor $\Phi: \mathscr{A} \longrightarrow \mathscr{K}^{\mathbf{T}}$ such that the following diagram of functors commutes:

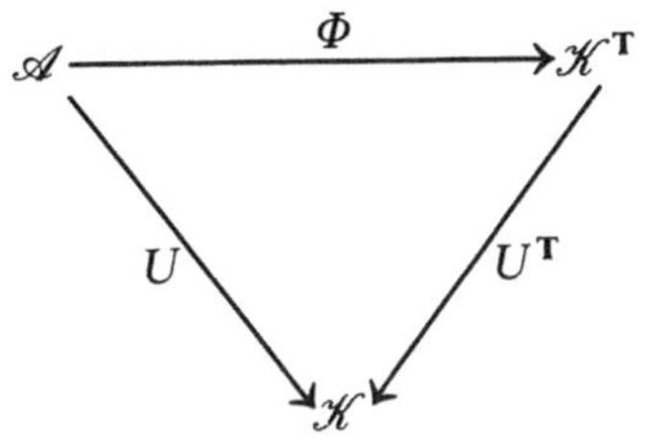

(2.21)

Notice that (2.21) forces the definition of Φ on morphisms. Φ is called the *semantics comparison functor of* the adjointness $(\mathscr{A}, \mathscr{K}, U, F, \eta, \varepsilon)$. Roughly speaking, Φ measures the extent to which the object A in $\mathscr{A}$ is the $\mathscr{K}$-object AU "with algebraic structure" (see 3.2 exercise 9). For example, if $\mathbf{T}$ is an algebraic theory in $\mathscr{K}$ and if $U = U^{\mathbf{T}}$ we have (from the proofs of 1.4.12 and 2.18) a canonical adjointness $(\mathscr{K}^{\mathbf{T}}, \mathscr{K}, U^{\mathbf{T}}, F^{\mathbf{T}}, \eta, \varepsilon)$ where $KF^{\mathbf{T}} = (KT, K\mu)$ and $(K, \xi)\varepsilon = \xi:(KT, K\mu) \longrightarrow (K, \xi)$; it is obvious that the algebraic theory of this adjointness is exactly $\mathbf{T}$ on the nose and that the semantics comparison functor is the identity functor.

We turn our attention now to the problem of characterizing when an arbitrary functor has a left adjoint. We begin with an important necessary condition.

2.22 Proposition. *Let $U:\mathscr{A} \longrightarrow \mathscr{K}$ have a left adjoint. Then U preserves limits; that is for every diagram (Δ, D) in $\mathscr{A}$ (see 1.18) and for every limit (L, ψ) of (Δ, D), $(LU, \psi U)$ is a limit of the $\mathscr{K}$-diagram (Δ, DU).*

Proof. Let $(\mathscr{A}, \mathscr{K}, U, F, \eta, \varepsilon)$ be an adjointness. Let (K, Γ) be a lower bound of (Δ, DU). For $\alpha \in \Delta(i, j)$ we have

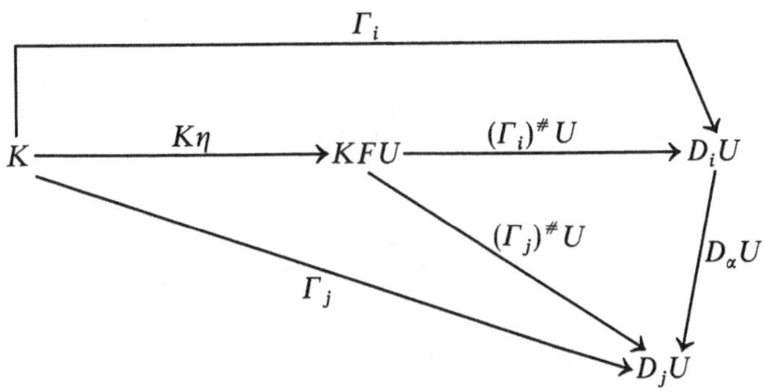

which shows that $\Gamma_i^{\#}.D_\alpha = \Gamma_j^{\#}$, that is, that $(KF, \Gamma^{\#})$ is a lower bound of

(Δ, D). This induces a unique map $f: KF \longrightarrow L$ with $f.\psi_i = \Gamma_i^{\#}$ for all i.

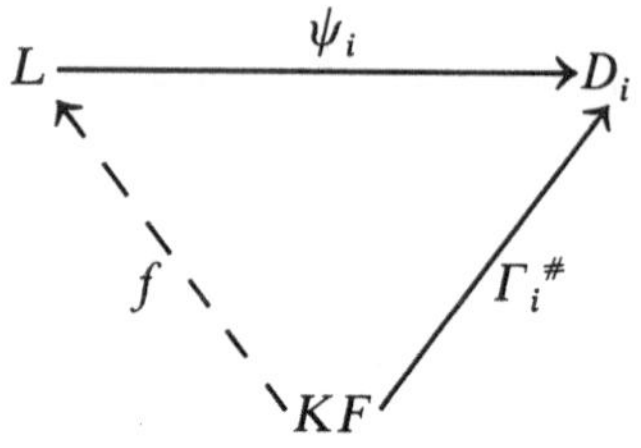

As shown below, $f_{\#} = K\eta.fU: K \longrightarrow LU$ satisfies $f_{\#}.\psi_i U = \Gamma_i$ for

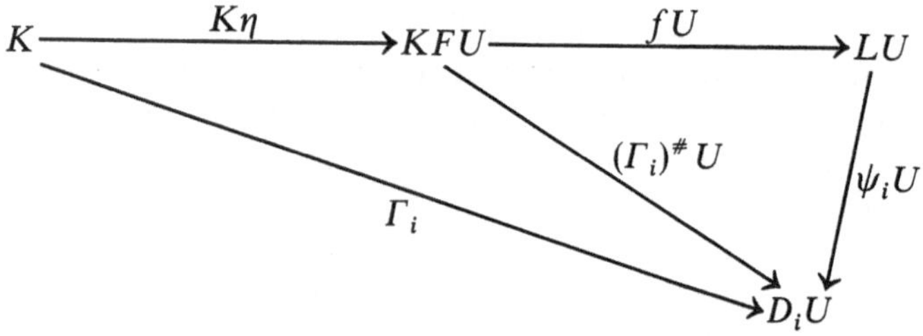

all i. If also $g: K \longrightarrow LU$ is such that $g.\psi_i U = \Gamma_i$ for all i, then the diagrams

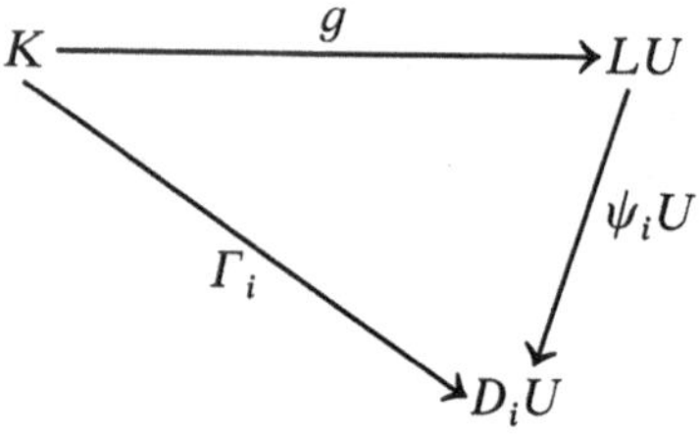

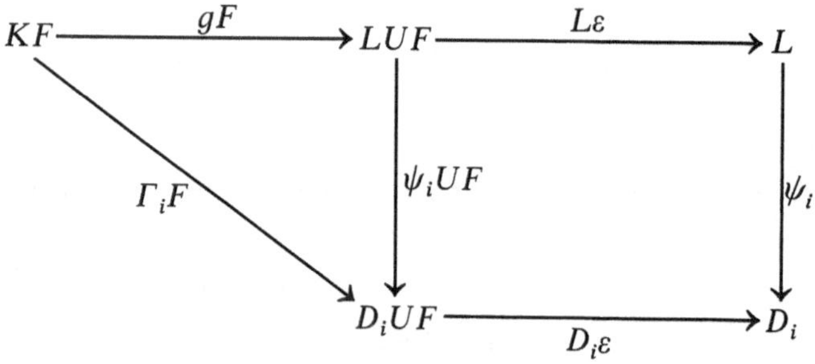

prove that $g^{\#}.\psi_i = \Gamma_i^{\#}$ for all i, so that $g^{\#} = f$, and $f_{\#} = g$ as desired. □

2.23 The Solution Set Condition. Let $U: \mathscr{A} \longrightarrow \mathscr{K}$ be a functor and let K be an object in $\mathscr{K}$. *U satisfies the solution set condition at K* providing there exists a small set $\mathscr{S}_K$ of pairs (S, s) with S in $\mathscr{A}$ and $s: K \longrightarrow SU$ having the property that for all pairs (A, f) with $f: K \longrightarrow AU$ there exists (S, s) in $\mathscr{S}_K$ and $\psi: S \longrightarrow A$ with $s.\psi U = f$. Of course the set of all pairs (A, f) has

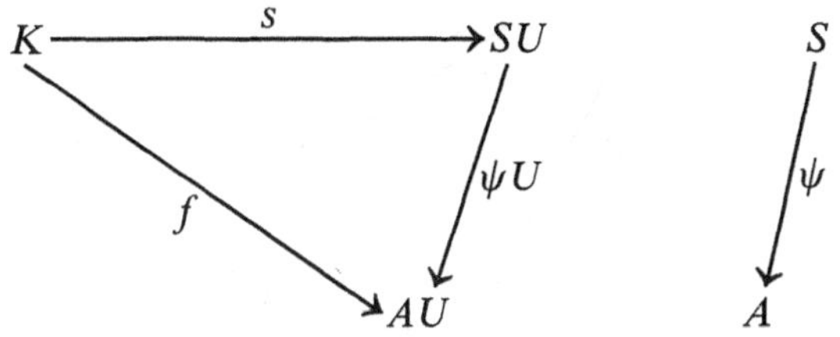

such a property, but it is not necessarily a small set. If U has a left adjoint, on the other hand, $\{(KF, K\eta)\}$ is a one-element solution set for K.

2.24 General Adjoint Functor Theorem (P. Freyd). *Assume that $\mathscr{A}$ is a locally small (1.1) category which has small limits (1.21+). Let $U: \mathscr{A} \longrightarrow \mathscr{K}$ be a functor. Then necessary and sufficient for the functor U to have a left adjoint are the three conditions:*

1. *U preserves small products (i.e., if $p_i: A \longrightarrow A_i$ is a product in $\mathscr{A}$ and i ranges over a small set then $p_i U: AU \longrightarrow A_i U$ is a product in $\mathscr{K}$).*
2. *U preserves equalizers (i.e., if $h = \mathrm{eq}(f, g)$ in $\mathscr{A}$ then $hU = \mathrm{eq}(fU, gU)$ in $\mathscr{K}$).*
3. *U satisfies the solution set condition at K for every K in $\mathscr{K}$.*

Proof. We have already observed the necessity of these conditions, so we concentrate on the sufficiency. Let us point out at once that U preserves all small limits because $\mathscr{A}$ has all small limits and because of the construction used to prove 1.22. Fix K in $\mathscr{K}$. Our object is to produce a free $\mathscr{A}$-object (F, η) over K with respect to U. For later use it is helpful to break the proof into two steps. First, we will construct a *weakly free $\mathscr{A}$-object over K with respect to U*, this being a pair (F', η') with the weakened universal property, shown below, that for all (A, f) there exists at least one $\psi: F' \longrightarrow A$ in $\mathscr{A}$

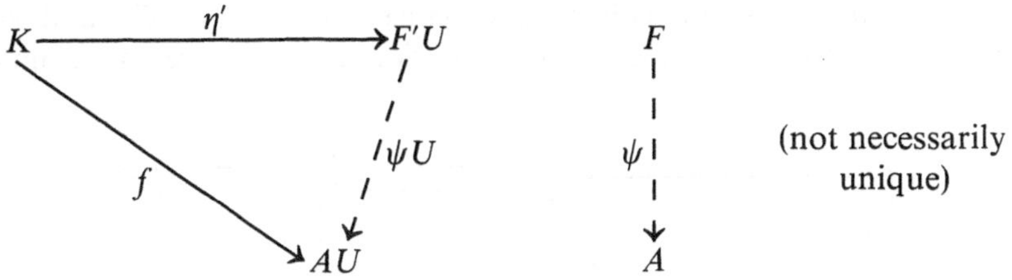

with $\eta'.\psi U = f$. To this end, let $\mathscr{S}_K$ be a solution set for K as in 2.23; in essence, we are trying to find a one-element solution set (F', η') to replace $\mathscr{S}_K$. Set $F' = \prod(S:(S, s) \in \mathscr{S}_K)$ in $\mathscr{A}$. Of course the existence of F' depends crucially on the smallness condition on $\mathscr{S}_K$. Since U preserves products, $p_s U: F'U \longrightarrow SU$ is a product in $\mathscr{K}$ giving rise to a unique map η' with

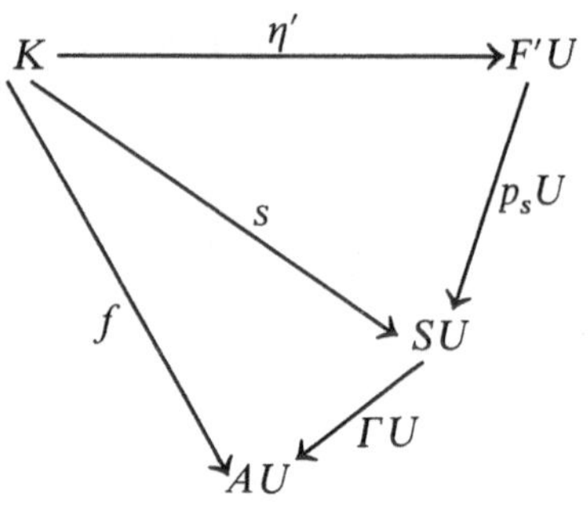

$\eta'.p_sU = s$ for all $(S, s) \in \mathscr{S}_K$. If (A, f) is an arbitrary pair then, by the definition of $\mathscr{S}_K$, there exists $(S, s) \in \mathscr{S}_K$ and $\Gamma: S \longrightarrow A$ in $\mathscr{A}$ with $s.\Gamma U = f$. Setting $\psi: F' \longrightarrow A = p_s.\Gamma$, we have $\eta'.\psi U = f$ as desired.

Extending slightly the definition of 1.13, say that a *collective equalizer* of a set C of morphisms from A_1 to A_2 is a map $i: E \longrightarrow A_1$ universal with the property that $ix = iy$ whenever $x, y \in C$; that is $ix = iy$ for all $x, y \in C$ and whenever $i'x = i'y$ for all $x, y \in C$ then there exists unique h with $h.i = i'$.

The second of the promised two steps is the proof is the following lemma:

2.25 Lemma. *Let $U: \mathscr{A} \longrightarrow \mathscr{K}$ be a functor, let K be an object in $\mathscr{K}$, and let (F', η') be a weakly free $\mathscr{A}$-object over K with respect to U with the property that the set $C = \{F' \xrightarrow{x} F' : \eta'.xU = \eta'\}$ has a collective equalizer $i: F \longrightarrow F'$ which is preserved by U (that is iU is the collective equalizer of $\{xU : x \in C\}$). Assume further that every pair of $\mathscr{A}$-morphisms has an equalizer and that U preserves these equalizers. Then, for suitable η, (F, η) is free over K with respect to U.*

Before proving 2.25, let us record that its hypotheses are available in the context of 2.24. We have already constructed a weakly free (F', η'). Since $\mathscr{A}$ is locally small, any set of endomorphisms of F' will have a collective equalizer (this being just a limit by the obvious extension of 1.21) which is preserved by U. Thus, the proof of 2.25 completes as well the proof of 2.24.

Proof. As iU is the collective equalizer of $\{xU : x \in C\}$ and since $\eta'.xU = \eta'.yU$ for all $x, y \in C$ (the definition of C), there exists unique

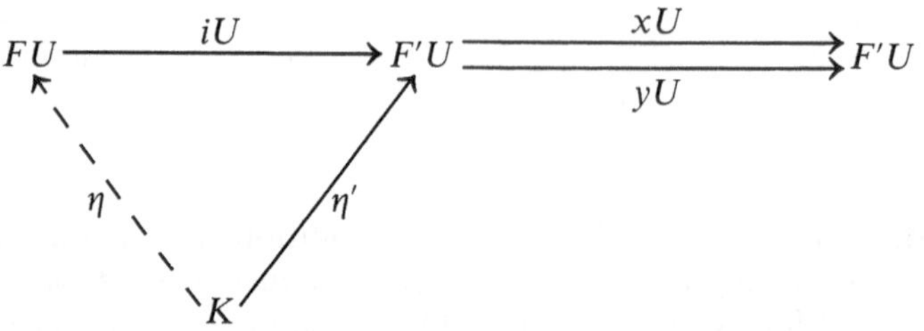

η with $\eta.iU = \eta'$. It is obvious that (F, η) is also weakly free over K with respect to U. What we must prove is that whenever $\psi, \psi': F \longrightarrow A$ are

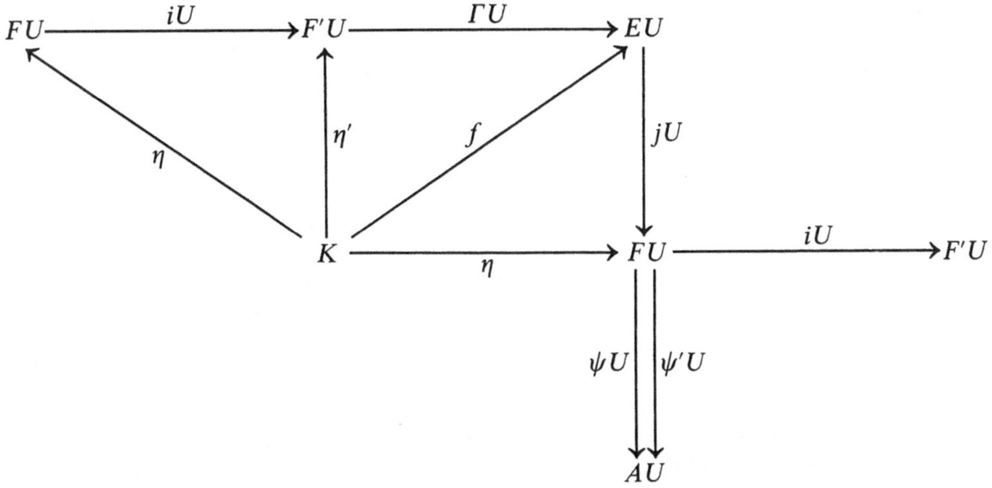

$\mathscr{A}$-morphisms such that $\eta.\psi U = \eta.\psi' U$, then $\psi = \psi'$. Form $j = \text{eq}(\psi, \psi')$ in $\mathscr{A}$ so that $jU = \text{eq}(\psi U, \psi' U)$ in $\mathscr{K}$. Since $\eta.\psi U = \eta.\psi' U$, there exists unique f with $f.jU = \eta$. Since (F', η') is weakly free, there exists $\Gamma: F' \longrightarrow E$ with $\eta'.\Gamma U = f$. Set $x: F' \longrightarrow F' = \Gamma.j.i$. Reading off the diagram above, we see that $\eta'.xU = \eta.iU = \eta'$, and $x \in C$. As $\text{id}_{F'} \in C$, $i.x = i$, that is we have $i.\Gamma.j.i = \text{id}_F.i$. Since i is a monomorphism, $i.\Gamma.j = \text{id}_F$. It follows that $\psi = i.\Gamma.j.\psi = i.\Gamma.j.\psi' = \psi'$. We are done. □

2.26 Example. Let $\mathscr{A}$ be the category of topological groups and continuous group homomorphisms and let $U: \mathscr{A} \longrightarrow \mathscr{K}$ be the forgetful functor to topological spaces and continuous maps. Then U has a left adjoint. The proof illustrates the use of 2.24. Clearly $\mathscr{A}$ is locally small. The way one constructs products in $\mathscr{A}$ is to provide the product group with the product topology. With regard to equalizers, the subspace topology on a subgroup makes for a topological subgroup. Therefore all side conditions on $\mathscr{A}$ and conditions (1) and (2) in 2.24 are obvious at a glance. To prove (3), fix a topological space K, set $\mathscr{S}$ to be the set of all cardinals less than or equal to the cardinality of the free group generated by K, and define $\mathscr{S}_K = \{(S, s): S \in \mathscr{A}, s: K \longrightarrow SU \in \mathscr{K}$ and the underlying set of S belongs to $\mathscr{S}\}$. $\mathscr{S}_K$ is a small set. Consider arbitrary (A, f). Let I be the algebraic subgroup

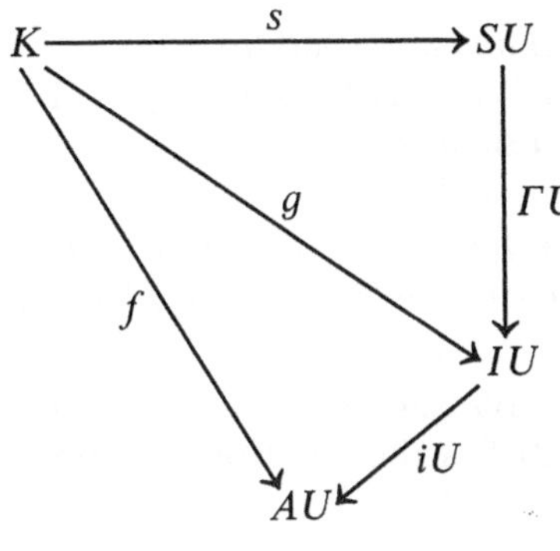

of A generated by the image of f made into a topological group by using the subspace topology. Then the inclusion map $i:I \longrightarrow A$ is a morphism in $\mathscr{A}$. Since the cardinal of I is in $\mathscr{S}$ (cf. 1.4.31) there exists $S \in \mathscr{S}$ and a bijection $\Gamma:S \longrightarrow I$. It is clear that there exists a unique topological group structure on S making Γ a topological group isomorphism (specifically, a subset of S is open just in case Γ of it was open and $s * s' = (s\Gamma \cdot s'\Gamma)\Gamma^{-1}$). Finally, define $s = g.(\Gamma U)$ (in the diagram above).

2.27 Example. Let $U:\mathscr{A} \longrightarrow \mathscr{K}$ be the forgetful functor from the category of metric spaces and distance-decreasing maps (1.9) to the category of topological spaces and continuous maps. Then U does not have a left adjoint. An easy way to see this is to know that it is rare for an uncountable product of metrizeable topological spaces to be metrizeable. For a specific example, we may cite [Hewitt and Ross '62, Theorem 8.11] which asserts if I is an uncountable set and if K is a denumerably infinite discrete space, then K^I is not normal (and hence, not metrizeable). But K is metrizeable by the discrete metric d which keeps all distinct pairs of points exactly one unit apart. The product $(K, d)^I$ exists in $\mathscr{A}$ (1.9) but is not preserved by U, so we invoke 2.22.

2.28 Example (J. R. Isbell). There exists a set-valued functor $U:\mathscr{A} \longrightarrow$ **Set** from the category of groups and group homomorphisms which preserves limits but which does not have a left adjoint. Let I be the proper class of all cardinals and for each $\alpha \in I$ let S_α be a simple group of cardinality $\geqslant \alpha$ (1.64). Define $GU = \prod(\mathscr{A}(S_\alpha, G):\alpha \in I)$. Then GU is a small set because "zero" is the only homomorphism from S_α to G if the cardinal of S_α exceeds that of G. If $\psi:G \longrightarrow G' \in \mathscr{A}$, define ψU by $\langle (f_\alpha), \psi U\rangle = (f_\alpha.\psi)$. That U is a functor is clear. We leave the proof that U preserves products and equalizers to the reader (see exercise 13). Suppose that (F, η) were free over the 1-element set 1 with respect to U. Fix $\beta \in I$ with S_β of larger cardinal than F. Define $(g_\alpha) \in S_\beta U$ by $g_\alpha = 0$ except when $\alpha = \beta$ when $g_\beta = \mathrm{id}_{S_\beta}$. Then there exists $\psi:F \longrightarrow S_\beta$ with $\eta_\beta.\psi = \mathrm{id}_{S_\beta}$, the desired contradiction.

A comparison of the methods of 1.64 and 2.28 suggests a possible relationship between "not having a cogenerator" and "not satisfying the solution set condition." This is in fact the case:

2.29 Special Adjoint Functor Theorem (P. Freyd). *Let $\mathscr{A}$ be a category satisfying*

1. *$\mathscr{A}$ is locally small and has small limits.*
2. *$\mathscr{A}$ is well-powered; that is for each object A in $\mathscr{A}$ the class of mono-subobjects of A (1.65) is a small set.*
3. *$\mathscr{A}$ has a cogenerator (1.59).*

Let $U:\mathscr{A} \longrightarrow \mathscr{K}$ be a functor to the (locally small) category $\mathscr{K}$. Then a necessary and sufficient condition that U have a left adjoint is that U preserves products and equalizers.

Proof. Fix K in $\mathscr{K}$. By 2.24,, we need only show that the conditions on $\mathscr{A}$ guarantee that U satisfies the solution set condition at K. Let C be a cogenerator in $\mathscr{A}$ and let P be the cartesian power object $C^{\mathscr{K}(K, CU)}$ in $\mathscr{A}$. This is meaningful, since $\mathscr{K}$ is locally small. Since $\mathscr{A}$ is well powered, there exists a small set $\mathscr{S}$ of objects in $\mathscr{A}$ such that every subobject of P is represented by a monomorphism with domain in $\mathscr{S}$. Set $\mathscr{S}_K = \{(S, s): S \in \mathscr{S}$ and $s: K \longrightarrow SU\}$. Then $\mathscr{S}_K$ is a small set, again since $\mathscr{K}$ is locally small. Now consider an arbitrary pair (A, f) with $f: K \longrightarrow AU$. Consider the cartesian power $Q = C^{\mathscr{A}(A, C)}$ which we may since $\mathscr{A}$ is locally small. For each $x: A \longrightarrow C$ in $\mathscr{A}$ we have $f.xU: K \longrightarrow CU$ in $\mathscr{K}$, which induces unique α as shown below:

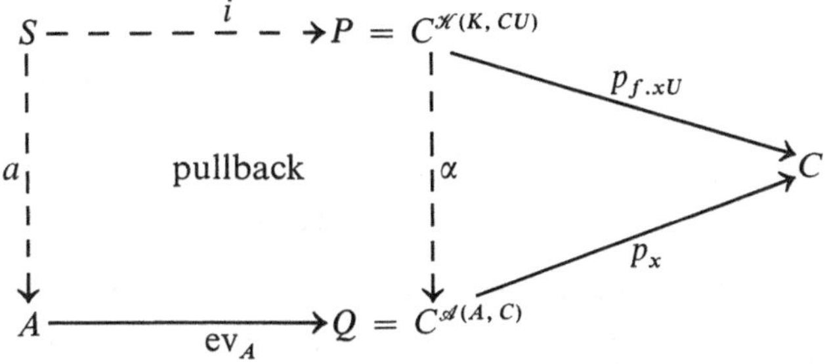

The evaluation map ev_A (1.60) is a monomorphism because C is a cogenerator, that is A may be thought of as a subobject of Q. Because $\mathscr{A}$ has small limits we may take the inverse image of A under α (1.68) which is a subobject of P; that is, there exists a pullback square as shown above with $S \in \mathscr{S}$. Because U preserves small limits, U of the above square is still a pullback and PU

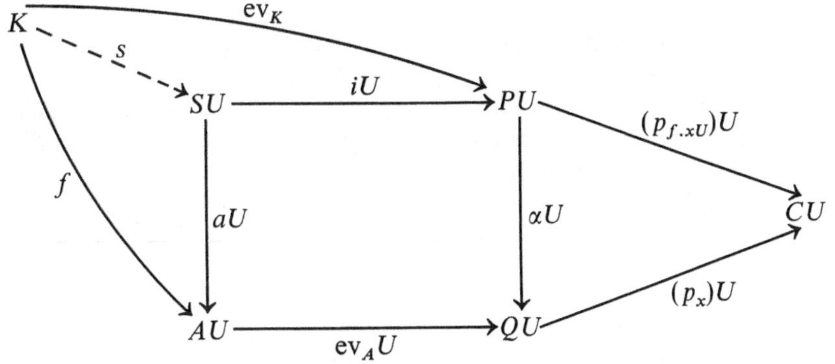

and QU are still powers of CU as shown. In particular, we can define ev_K as in 1.60. For each $x: A \longrightarrow C$ we have $\text{ev}_K.\alpha U.(p_x)U = \text{ev}_K.(p_{f.xU})U = f.xU$ (definition of ev_K) $= f.(\text{ev}_A.p_x)U = f.\text{ev}_A U.(p_x)U$. As $(p_x)U$ is an arbitrary product projection, $\text{ev}_K.\alpha U = f.\text{ev}_A U$. There exists unique s with $s.aU = f$ and $s.iU = ev_K$. In particular, $s.aU = f$. As $(S, s) \in \mathscr{S}_K$, we are done. □

We mention now another property of functorial adjointness reminiscent of operators on Hilbert space.

2.30 Composition Theorem for Adjoints. *Let* $U_1:\mathscr{A} \longrightarrow \mathscr{B}$ *and* $U_2:\mathscr{B} \longrightarrow \mathscr{C}$ *be functors which have a left adjoint. Then* $U_1U_2:\mathscr{A} \longrightarrow \mathscr{C}$ *has a left adjoint; indeed, if* $(\mathscr{A}, \mathscr{B}, U_1, F_1, \eta_1, \varepsilon_1)$ *and* $(\mathscr{B}, \mathscr{C}, U_2, F_2, \eta_2, \varepsilon_2)$ *are adjointnesses, then so is* $(\mathscr{A}, \mathscr{C}, U_1U_2, F_2F_1, \eta_2.F_2\eta_1U_2, U_1\varepsilon_2F_1.\varepsilon_1)$. "$(U_1U_2)^* = U_2^*U_1^*$."

Proof. Fix C in $\mathscr{C}$. We check the universal property for the advertised $C\eta = C\eta_2.CF_2\eta_1U_2$. Let $f:C \longrightarrow AU_1U_2$ be given. There exists unique

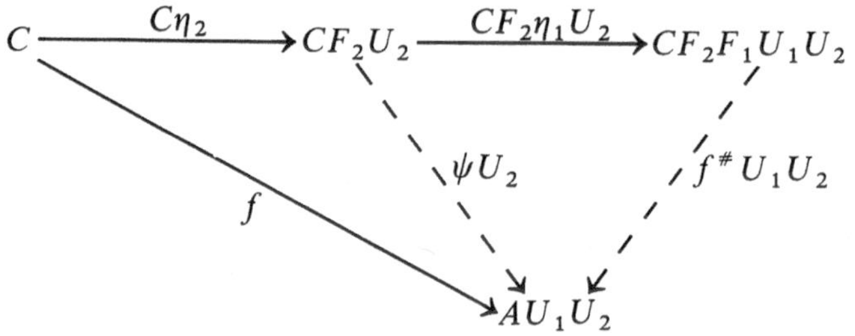

$\psi:CF_2 \longrightarrow AU_1$ with $C\eta_2.\psi U_2 = f$. Thus, there exists unique f : $CF_2F_1 \longrightarrow A$ such that $CF_2\eta_1.f^{\#}U_1 = \psi$. It is now trivial to check that $f^{\#}$ is unique with respect to the property that $C\eta_2.CF_2\eta_1U_2.f^{\#}U_1U_2 = f$. As regards the formulas in the composite adjointness, the construction of 2.18 immediately provides everything except perhaps the formula for ε; but by 2.17, this must be dual to the formula for η. □

As has been clear since 1.3.16, and again in 2.16, there are various ways in which functors and natural transformations interact. We close this section by formalizing some of these interactions. Additional development is provided in 3.2.6.

2.31 The Godement Calculus. Given functors $F_1, F_2, F_3:\mathscr{K} \longrightarrow \mathscr{L}$ and natural transformations $\tau:F_1 \longrightarrow F_2$ and $\sigma:F_2 \longrightarrow F_3$ the *pointwise composition* or "*horizontal composition*" $\tau.\sigma:F_1 \longrightarrow F_3$ is defined, using composition in the category $\mathscr{L}$, by $A(\tau.\sigma) = A\tau.A\sigma$. Naturality is seen at once from the diagram

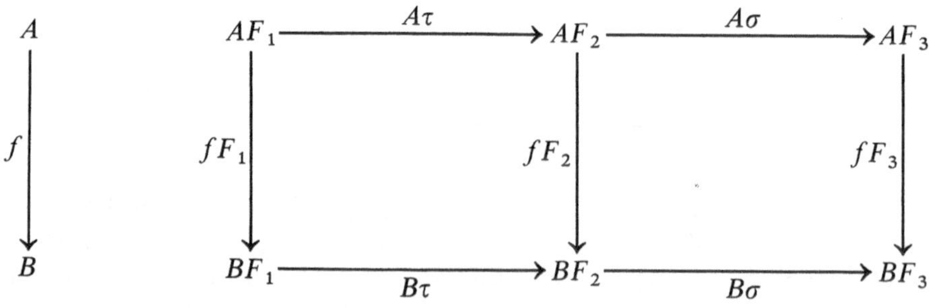

Now consider the functors $F:\mathscr{K} \longrightarrow \mathscr{L}$, $G_1, G_2:\mathscr{L} \longrightarrow \mathscr{M}$, $H:\mathscr{M} \longrightarrow \mathscr{N}$ and let $\tau:G_1 \longrightarrow G_2$ be a natural transformation. Define $F\tau:FG_1 \longrightarrow FG_2$ by $A(F\tau) = (AF)\tau$. The naturality of $F\tau$ is a special case of the naturality of τ:

$$\begin{array}{ccccc} A & & AFG_1 & \xrightarrow{AF\tau} & AFG_2 \\ {\scriptstyle f}\big\downarrow & & {\scriptstyle fFG_1}\big\downarrow & & \big\downarrow{\scriptstyle fFG_2} \\ B & & BFG_1 & \xrightarrow[BF\tau]{} & BFG_2 \end{array}$$

Also, define $\tau H: G_1H \longrightarrow G_2H$, using the action of H on morphisms, by $X(\tau H) = (X\tau)H$. $H\tau$ is natural because the functor H preserves the commutativity of the appropriate naturality square:

$$\begin{array}{ccccc} X & & XG_1H & \xrightarrow{(X\tau)H} & XG_2H \\ {\scriptstyle f}\big\downarrow & & {\scriptstyle fG_1H}\big\downarrow & & \big\downarrow{\scriptstyle fG_2H} \\ Y & & YG_1H & \xrightarrow[(Y\tau)H]{} & YG_2H \end{array}$$

The *Godement rules* are:

Given $\mathscr{K} \xrightarrow{F_1 \xrightarrow{\tau} F_2} \mathscr{L} \xrightarrow{G} \mathscr{M} \xrightarrow{H} \mathscr{N}$

then $(\tau G)H = \tau(GH): F_1GH \longrightarrow F_2GH$. We write τGH.

Given $\mathscr{K} \xrightarrow{F} \mathscr{L} \xrightarrow{G} \mathscr{M} \xrightarrow{H_1 \xrightarrow{\tau} H_2} \mathscr{N}$

then $F(G\tau) = (FG)\tau: FGH_1 \longrightarrow FGH_2$. We write $FG\tau$.

Given $\mathscr{K} \xrightarrow{F} \mathscr{L} \xrightarrow{G_1 \xrightarrow{\tau} G_2} \mathscr{M} \xrightarrow{H} \mathscr{N}$

then $(F\tau)H = F(\tau H): FG_1H \longrightarrow FG_2H$. We write $F\tau H$.

Given $\mathscr{K} \xrightarrow{F} \mathscr{L} \xrightarrow{G_1 \xrightarrow{\tau} G_2 \xrightarrow{\sigma} G_3} \mathscr{M} \xrightarrow{H} \mathscr{N}$

then $F(\tau.\sigma)H = F\tau H.F\sigma H: FG_1H \longrightarrow FG_3H$.

The proof of these assertions is routine and is left to the reader.

Notes for Section 2

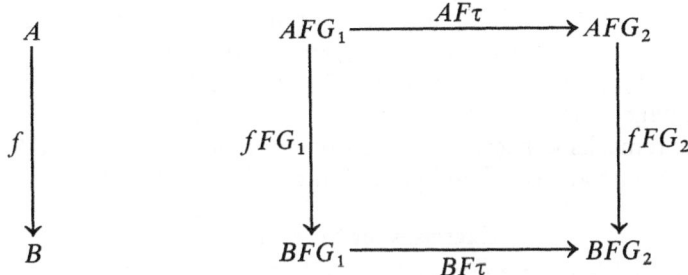

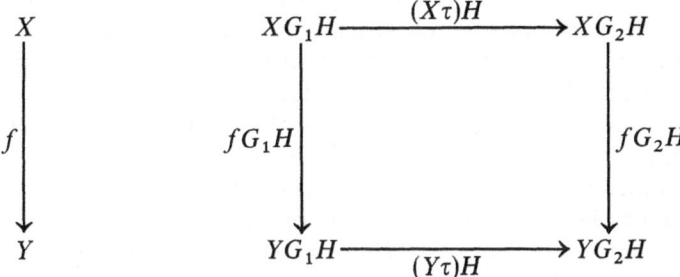

Adjoint functors were introduced by [Kan '58]. The result of 2.8 is well known (see e.g. [Kelley '55, page 153]) although attention is usually restricted to the completely regular case since it is felt, for reasons unclear, that η should be a subspace. The algebraic theory of 2.20 originates with [Huber '61, Theorem 4.2]. The adjoint functor theorems of 2.24, 2.29 were present in Freyd's 1960 Princeton dissertation and appeared as [Freyd '64, Chapter 3, exercises J, M]. Examples similar to 2.28 may be found in [Gabriel and Ulmer '71, page 176].

Further remarks concerning the history of adjointness can be found in [Felscher '65, introduction] and [Mac Lane '71, pages 103, 132].

Exercises for Section 2

1. Retrace the developments in 1.2.5–1.2.14 replacing the underlying set functor from (Ω, E)-algebras with an arbitrary functor which has a left adjoint.
2. Using well-known properties of metric completion, show that complete metric spaces is a full reflective subcategory of the category of metric spaces and distance-decreasing maps as in 1.9.
3. Show that partially ordered sets is a full reflective subcategory of the category of reflexive and transitively ordered sets and order-preserving maps. [Hint: divide out by antisymmetry.] Let X be a set and let Y be the collection of all families of subsets of X with reflexive and transitive ordering $\mathscr{A} \leqslant \mathscr{B}$ if the topology generated by $\mathscr{A}$ (i.e., with $\mathscr{A}$ as subbase) is contained in the topology generated by $\mathscr{B}$; show that the reflection of $(Y, \leqslant)$ may be identified with the set of topologies on X.
4. Let $\mathscr{K}$ be the category of M-sets of exercise 1.19(d). Prove that the underlying set functor from $\mathscr{K}$ has both left and right adjoints. [Hint: for any set A, the free M-set has underlying set $A \times M$ and the cofree M-set has underlying set A^M.]
5. Let $f: M \longrightarrow N$ be a monoid homomorphism, i.e., a functor between one-object categories. Open question: if f has a left adjoint must f be an isomorphism?
6. Let $f: X \longrightarrow Y$ be a function and regard the order-preserving inverse-image map $f^{-1}: 2^Y \longrightarrow 2^X$ as a functor between partially ordered categories. Show that f^{-1} has both left and right adjoints. [Hint: direct image provides the left adjoint; the right adjoint can be discovered using the universal property on singletons.]
7. A category $\mathscr{K}$ is *cartesian closed* if $\mathscr{K}$ has finite products and if, for every object A, the functor $- \times A: \mathscr{K} \longrightarrow \mathscr{K}$ has a right adjoint. Prove that the category of M-sets (exercise 1.19(d)) is cartesian closed (cf. 2.14).
8. A *propositional logic* is a cartesian closed category $\mathscr{K}$ such that for all objects p, q, $\mathscr{K}(p, q)$ has at most one element. The objects of $\mathscr{K}$ are "propositions" and the reflexive and transitive order $p \leqslant q$ determining $\mathscr{K}$ may be interpreted "there exists a proof of q given p." Study the adjunction induced by fixed p noting that ε establishes modus ponens as a rule of inference. Consider the case where $\mathscr{K}$ is the usual Boolean algebra of propositions.

9. If $\mathscr{K}$, $\mathscr{L}$ are categories, the *functor category* $\mathscr{L}^{\mathscr{K}}$ has as objects all functors $\mathscr{K} \longrightarrow \mathscr{L}$, has as morphisms natural transformations and has as composition and identities the pointwise ones. Establish that $\mathscr{L}^{\mathscr{K}}$ is a category. Isomorphisms in $\mathscr{L}^{\mathscr{K}}$ are called *natural equivalences.* Show that the natural transformation $\tau: F \longrightarrow G$ is a natural equivalence if and only if each $A\tau$ is an isomorphism in $\mathscr{L}$. Show that the category of M-sets as in exercise 1.19 is the same thing as $\mathbf{Set}^M$.
10. Let $U: \mathscr{A} \longrightarrow \mathbf{Set}$ be a set-valued functor. U is *representable* if there exists an $\mathscr{A}$-object A such that F is naturally equivalent to $\mathscr{A}(A, -)$ (see exercise 1.8); show that this *representing object* A is unique up to isomorphism. Prove that U is representable if and only if there exists a free object over 1 with respect to U. More generally, for any functor $U: \mathscr{A} \longrightarrow \mathscr{B}$ and object B of $\mathscr{B}$, show that there exists a free object over B with respect to U if and only if $\mathscr{B}(B, (-)U): \mathscr{A} \longrightarrow \mathbf{Set}$ is representable.
11. Prove the *Yoneda lemma*: given $U: \mathscr{A} \longrightarrow \mathbf{Set}$ and A in $\mathscr{A}$, the class of natural transformations from $\mathscr{A}(A, -)$ to U is a small set, indeed is in bijective correspondence with AU. [Hint: generalize 1.5.6, 1.5.7.]
12. Show that $\mathscr{L}^{\mathscr{K}}$ is at least as complete, or as cocomplete, as $\mathscr{L}$ is. [Hint: given a diagram in $\mathscr{L}^{\mathscr{K}}$, to define the needed functor use the obvious pointwise construction on objects and use the universal property in $\mathscr{L}$ to induce the action on morphisms.]
13. If $\mathscr{L}$ is complete (so that exercise 12 establishes the completeness of $\mathscr{L}^{\mathscr{K}}$) show that the full subcategory of $\mathscr{L}^{\mathscr{K}}$ of limit-preserving functors is closed under limits. Use this observation and exercise 1.8 to prove that the functor U of 2.28 is limit preserving. Why is this functor not representable? Show that the underlying set functor from complete Boolean algebras to sets is representable even though it does not have a left adjoint. Thus "left adjoint" implies "representable" and "representable" implies "limit preserving" but neither of these implications is reversible.
14. Say that $\mathscr{A}$ is a *SAFT category* if $\mathscr{A}$ is a locally small, small complete, well-powered category with a cogenerator (so that, by the Special Adjoint Functor Theorem 2.29, every limit preserving functor from $\mathscr{A}$ to a locally small category has a left adjoint). Prove that the category of metric spaces of diameter at most 1 and distance-decreasing maps is a SAFT category. [Hint: modify exercise 1.20.] Show that the category of Banach spaces and linear maps of norm at most 1 is a SAFT category. [Hint: the product is a subset of the product set; use the Hahn-Banach theorem to prove that the scalar field (which may be either real or complex) is an—in fact injective—cogenerator.] Show that the "unit disc" functor from Banach spaces to metric spaces has a left adjoint. Why does the "underlying metric space" functor not have a left adjoint?
15. A *metric space with base point* is (X, d, x_0) where (X, d) is a metric space and the "base point" x_0 is any element of X. Let $\mathscr{K}$ be the category of metric spaces with base point and distance-decreasing base point pre-

serving maps. The following sums up, in more categorical language, some of the constructions of [Arens and Eells '56, section 2].

(a) Show that the forgetful functor from Banach spaces (as in exercise 14) to $\mathscr{K}$ has a left adjoint.

(b) Let (F, η) be the free Banach space over (X, d, x_0) with respect to the functor in (a). Prove that η is an isometry into. [Hint: for each x in X the scalar-valued map $d(x, -) - d(x, x_0)$ is distance-decreasing.]

16. A cartesian closed category with a subobject classifier (see exercises 7, 1.19) is called a *topos*. In previous exercises it has been observed that the category $\mathbf{Set}^M$ of M-sets is a topos. Prove, more generally, that $\mathbf{Set}^{\mathscr{A}}$ is a topos for any small category $\mathscr{A}$. [Hint: use the Yoneda lemma (exercise 11) to define F^G and Ω.]

17. Prove the *truncated adjoint functor theorem*: Let α be any cardinal $\geqslant 1$. Let $U:\mathscr{A} \longrightarrow \mathscr{K}$ be a functor. An *α-small set* is a set of cardinal less than α; and *α-product* in $\mathscr{A}$ is an I-indexed product with I of cardinal less than α. Assume that $\mathscr{A}$ has *α-products* and equalizers of pairs. Then U has a left adjoint if and only if U preserves α-products, U preserves equalizers, and for every K in $\mathscr{K}$ there exists a solution set $\mathscr{S}$ such that $\mathscr{S}$ is α-small and such that, for every $(S, f) \in \mathscr{S}$, the set of all $x:S \longrightarrow S$ with $f.xU = f$ is α-small.

18. Let $\mathscr{A}$ be the category of rings and unit-preserving ring homomorphisms, let $\mathscr{K}$ be the category of abelian groups and homomorphisms, and let $U:\mathscr{A} \longrightarrow \mathscr{K}$ be the forgetful functor. Use the general adjoint functor theorem to prove that U has a left adjoint. Then show that the free ring over an abelian group is given by the well-known *integral group ring* construction.

19. Let $\mathscr{K}$ be the category of topological spaces and continuous maps. If I denotes the unit interval, show that $- \times I:\mathscr{K} \longrightarrow \mathscr{K}$ has a right adjoint. [Hint: use the compact-open topology.] If X is any topological space, show that $- \times X:\mathscr{K} \longrightarrow \mathscr{K}$ preserves coproducts but need not have a right adjoint if X is not locally compact Hausdorff. For any two topological spaces X, Y the *tensor product $X \otimes Y$ of X and Y* is the set $X \times Y$ provided with the largest topology such that the maps $y \longmapsto (x, y)$ (for all x) as well as the maps $x \longmapsto (x, y)$ (for all y) are continuous. Show that $- \otimes X:\mathscr{K} \longrightarrow \mathscr{K}$ has a right adjoint for every space X. [Hint: pointwise convergence]

20. Use the general adjoint functor theorem to prove that the forgetful functor from the category of compact groups to the category of topological spaces as well as the forgetful functor from the category of compact groups to the category of groups have left adjoints.

21. A *weak limit* of a diagram (Δ, D) is a lower bound (L, ψ) such that given any other lower bound (L', ψ') there exists (not necessarily unique) $f:L' \longrightarrow L$ with $f.\psi_i = \psi'_i$ for all i.

(a) Show that a morphism $f:A \longrightarrow B$ is a monomorphism if and only if the commutative square

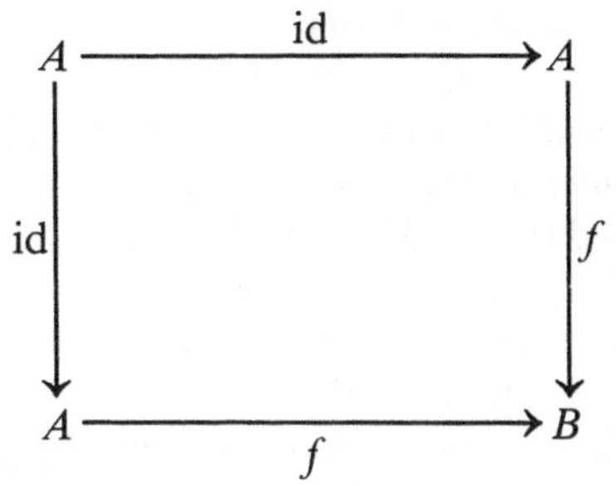

is a weak pullback.

(b) Show that any functor which preserves weak limits must also preserve limits. [Hint: generalize (a).]

22. Prove *Huber's theorem*: Let $(\mathscr{A}, \mathscr{K}, U, F, \eta, \varepsilon)$ be an adjointness and let (T_1, η_1, μ_1) be an algebraic theory in $\mathscr{A}$; then $(FT_1U, \eta.F\eta_1U, FT_1\varepsilon T_1U.F\mu_1U)$ is an algebraic theory in $\mathscr{K}$. [Hint: use 2.20 and 2.30.]

23. [Bergman '75]. Let $\mathscr{K}$ be the category of rings with unit, let $\mathscr{L}$ be the category of groups and let $GL_n:\mathscr{K} \longrightarrow \mathscr{L}$ (for fixed n) be the "general linear group" functor whose action on morphisms sends (a_{ij}) to $(\langle a_{ij}, f\rangle)$. Show that GL_n has a left adjoint.

3. Objects with Structure

In a base category $\mathscr{K}$, whatever "structure" an object K has is implicit and treated as a primitive concept. In this section, we axiomatize "categories of $\mathscr{K}$-objects with additional structure." This provides a suitable universe in which to place categories of $\mathscr{K}$-objects with algebraic structure (that is, $\mathscr{K}^{\mathbf{T}}$). The theory is also interesting in its own right.

3.1 Categories of $\mathscr{K}$-Objects with Structure. Let $\mathscr{K}$ be a (fixed base) category. A *literal category, $\mathscr{C}$, of $\mathscr{K}$-objects with structure* (the more general "category of $\mathscr{K}$-objects with structure" is defined in 3.3) is defined by the following two data and two axioms:

$\mathscr{C}$ assigns to each object K of $\mathscr{K}$ a class $\mathscr{C}(K)$ of *$\mathscr{C}$-structures on K*. A *$\mathscr{C}$-structure* is a pair (K, s) with $s \in \mathscr{C}(K)$.

For each ordered pair $(K, s; L, t)$ of $\mathscr{C}$-structures, $\mathscr{C}$ assigns a subset $\mathscr{C}(s, t)$ of $\mathscr{K}(K, L)$ of *$\mathscr{C}$-admissible $\mathscr{K}$-morphisms from* (K, s) *to* (L, t); to denote that $f:K \longrightarrow L$ is in $\mathscr{C}(s, t)$ we will write $f:(K, s) \longrightarrow (L, t)$ or $f:s \longrightarrow t$ (if necessary, imposing additional decoration should more than one $\mathscr{C}$ be in the picture).

The two axioms are

Axiom of Composition. If $f:s \longrightarrow t$ and $g:t \longrightarrow u$ then $f.g:t \longrightarrow u$.

Structure is Abstract. If $f:K \longrightarrow L$ is an isomorphism in $\mathscr{K}$ then for all $t \in \mathscr{C}(L)$ there exists unique $s \in \mathscr{C}(K)$ such that $f:s \longrightarrow t$ and $f^{-1}:t \longrightarrow s$.

It follows from these axioms that $\mathrm{id}_K : s \longrightarrow s$ for all $s \in \mathscr{C}(K)$. As a result, there is a category—also denoted $\mathscr{C}$—with objects all $\mathscr{C}$-structures and as morphisms all $\mathscr{C}$-admissible $\mathscr{K}$-morphisms. There is also the obvious *underlying $\mathscr{K}$-object functor* $U : \mathscr{C} \longrightarrow \mathscr{K}$ which on objects sends (K, s) to K.

We have not insisted that $\mathscr{C}(K)$ is always a small set; when this is the case we say that $\mathscr{C}$ is *small over* $\mathscr{K}$.

As a typical example, we may regard "topological groups" as a category of groups with structure. Here, $\mathscr{K}$ is the category of groups, $\mathscr{C}(K)$ is the set of all topologies on the set K compatible with the group structure, and a group homomorphism is admissible just in case it is continuous.

3.2 Various Functors. Let $U : \mathscr{A} \longrightarrow \mathscr{B}$ be a functor. U is *faithful* if for each pair (A, A') of objects in $\mathscr{A}$, the passage from $f : A \longrightarrow A'$ to $fU : AU \longrightarrow A'U$ is injective. For any object B in $\mathscr{B}$ there is a *constant functor* $U_B : \mathscr{A} \longrightarrow \mathscr{B}$ which sends every object to B and every morphism in $\mathscr{A}$ to the identity map of B. Notice that no constant functor from the category of groups to the category of sets is faithful. The picture of a

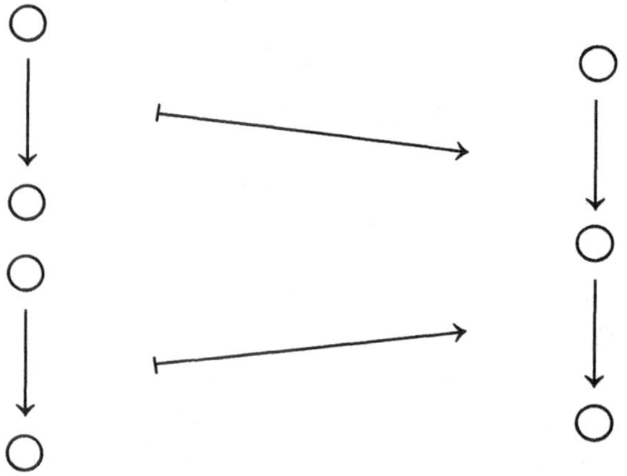

functor from a 4-object category to a 3-object one illustrates the principle that the class of morphisms which constitutes the image of a faithful functor need not be closed under composition; i.e., it is false, even for faithful functors, that "the image of a subcategory is a subcategory." This pathology disappears if we add either the condition that U be injective on objects (so that $[U]$ is a subcategory) or the stipulation that U be full; U is *full* if for every pair (A, A') of $\mathscr{A}$-objects, the passage from $f : A \longrightarrow A'$ to $fU : AU \longrightarrow A'U$ is surjective. Thus the subcategory $[H]$ is full in the sense of 2.6 if and only if H is full. U is an *isomorphism of* categories, and the categories $\mathscr{A}$ and $\mathscr{B}$ are *isomorphic*, if U is bijective on objects, full and faithful. Equivalently, U is an isomorphism in the category of categories and functors. U is a *full representative subcategory of* $\mathcal{B}$ if U is a full subcategory such that every object of $\mathcal{B}$ is isomorphic to an object of the form AU. U is a *full replete subcategory of* $\mathcal{B}$ if every object isomorphic to an object in U is already in U. Thus U is an isomorphism if and only if U is a full, representative, replete subcategory.

There is no doubt that isomorphic categories should be treated as abstractly the same for all categorical purposes. The multiplicity of functorial definitions suggested above allows many concepts of "nearly isomorphic" of which the most fundamental is *equivalence* (see e.g. [Mac Lane '71, page 18]). Although we will avoid a serious discussion of this point, it is worthwhile to observe that with respect to internal properties of a category, as exemplified by the material in section 1, the existence of a full and faithful functor $U:\mathcal{A} \longrightarrow \mathcal{B}$ such that every object in $\mathcal{B}$ is isomorphic to an object of form AU, forces $\mathcal{A}$ and $\mathcal{B}$ to have very similar structure (e.g. with respect to having coequalizer-mono factorizations). On the other hand, consider the picture (below) of a full representative subcategory between two preordered categories:

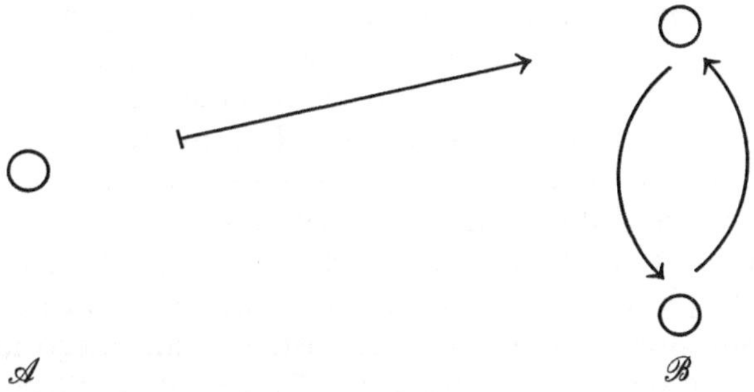

With respect to suitably external properties, $\mathcal{A}$ and $\mathcal{B}$ seem quite different. For example, $\mathcal{A}$ has one endofunctor, but $\mathcal{B}$ has four.

3.3 Concrete Categories. Let $\mathcal{K}$ be a (fixed base) category. A *concrete category over* $\mathcal{K}$ is a pair $(\mathcal{A}, U)$ where $\mathcal{A}$ is a category and $U:\mathcal{A} \longrightarrow \mathcal{K}$ is a faithful functor. A *homomorphism* $H:(\mathcal{A}, U) \longrightarrow (\mathcal{B}, V)$ *of* concrete categories is a functor $H:\mathcal{A} \longrightarrow \mathcal{B}$ which is *over* $\mathcal{K}$ in the sense that the diagram

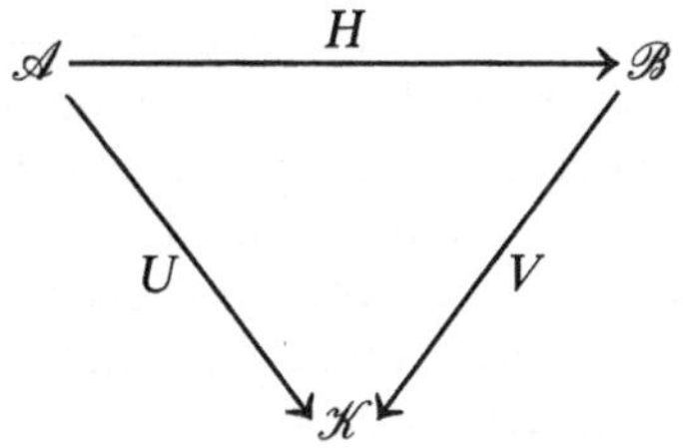

of functors commutes (on objects and on morphisms). Notice that if H is such a homomorphism, $(\mathscr{A}, H)$ is a concrete category over $\mathscr{B}$. The diagram

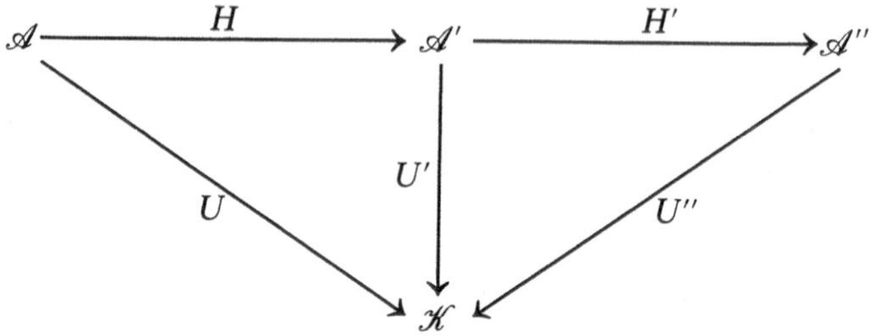

shows that the composition of homomorphisms is again one. With this composition, and using identity functors, we get the category Con($\mathscr{K}$) whose objects are concrete categories and whose morphisms are the homomorphisms defined above. Notice that there is no guarantee that Con($\mathscr{K}$) is locally small if $\mathscr{K}$ is. If $\mathscr{C}$ is a literal category of $\mathscr{K}$-objects with structure with underlying $\mathscr{K}$-object functor U then $(\mathscr{C}, U)$ is an object in Con($\mathscr{K}$). Extending but slightly the definition in 3.1, a concrete category $(\mathscr{A}, U)$ will be called a *category of $\mathscr{K}$-objects with structure* if there exists a literal category of $\mathscr{K}$-objects with structure $\mathscr{C}$ such that $(\mathscr{A}, U)$ is isomorphic to $\mathscr{C}$ in Con($\mathscr{K}$). Notice, incidentally, that if $H:(\mathscr{A}, U) \longrightarrow (\mathscr{B}, V) \in \text{Con}(\mathscr{K})$ then if $H:\mathscr{A} \longrightarrow \mathscr{B}$ is an isomorphism of categories then $H^{-1}:\mathscr{B} \longrightarrow \mathscr{A}$ is automatically over $\mathscr{K}$; that is, for functors over $\mathscr{K}$, there is no difference between isomorphisms in Con($\mathscr{K}$) and isomorphisms of categories.

The full subcategory of Con($\mathscr{K}$) consisting of all categories of sets with structure will be denoted as Struct($\mathscr{K}$). By definition, Struct($\mathscr{K}$) is a full replete subcategory of Con($\mathscr{K}$). This reflects our confidence that the invariants in a given category of structured objects will be implicit in the forgetful functor, a principle which has been with us since 1.2.17. In the sequel we will freely write "let $\mathscr{C} \in \text{Struct}(\mathscr{K})$," and assume notationally that $\mathscr{C}$ is literal (as in 3.1) when the properties under discussion are isomorphism invariant.

Let $(\mathscr{A}, U)$ be an arbitrary concrete category over $\mathscr{K}$. For each object K of $\mathscr{K}$ define $\mathscr{A}_K$ to be the class of all objects A in $\mathscr{A}$ such that $AU = K$. If $\mathscr{C} \in \text{Struct}(\mathscr{K})$, $\mathscr{C}_K$, and $\mathscr{C}(K)$ are, clearly, essentially the same thing. Carrying the analogy further, given $f:K \longrightarrow L$ in $\mathscr{K}$ and $\mathscr{A} \in \mathscr{A}_K$, $B \in \mathscr{A}_L$ say that f is *admissible from A to B* if there exists $\psi:A \longrightarrow B$ with $\psi U = f$; this definition seems in the right spirit because ψ is unique when it exists. For each K there is a canonical preordering (i.e., a reflexive and transitive relation) on $\mathscr{A}_K$ defined by $A \leqslant B$ if and only if $\text{id}_K:A \longrightarrow B$ is admissible. The preordered class $(\mathscr{A}_K, \leqslant)$ is called the *fibre over K*. For $\mathscr{C} \in \text{Struct}(\mathscr{K})$, $(\mathscr{C}(K), \leqslant)$ is even partially ordered (that is, $\leqslant$ is antisymmetric as well) since "structure is abstract" guarantees that if $\text{id}_K:s \longrightarrow t$ and $\text{id}_K:t \longrightarrow s$ then

$s = t$. In general, $(\mathscr{A}, U)$ in Con$(\mathscr{K})$ is *antisymmetric* if all of its fibres are partially ordered classes.

3.4 Antisymmetrization Theorem. *"Antisymmetric" is a full reflective replete subcategory of* Con$(\mathscr{K})$. *The antisymmetric reflection functor of each concrete category is always full, faithful, and surjective on objects.*

Proof. Define the equivalence relation $A \sim B$ on the objects of $\mathscr{A}$ by $A \sim B$ if and only if $AU = K = BU$ and $\mathrm{id}_K : A \longrightarrow B$, $\mathrm{id}_K : B \longrightarrow A$ are admissible, and let $[A]$ denote the equivalence class of A. Extend $\sim$ to an equivalence relation on the class of all morphisms of $\mathscr{A}$ by $\psi : A \longrightarrow B \sim \psi' : A' \longrightarrow B'$ if and only if $[A] = [A']$, $[B] = [B']$ and $\psi U = \psi' U$. Define a category $[\mathscr{A}]$ with objects all $[A]$, morphisms $[\mathscr{A}]([A], [B]) = \{[\psi] | \psi : A \longrightarrow B\}$, $[\psi_1][\psi_2] = [\psi_1\psi_2]$, $\mathrm{id}_{[A]} = [\mathrm{id}_A]$. It is routine to check that all this is well-defined, giving rise to the functor $[\ ] : \mathscr{A} \longrightarrow [\mathscr{A}]$ which is full, faithful and surjective on objects. There exists a unique functor $[U] : [\mathscr{A}] \longrightarrow \mathscr{K}$ such that $[\ ].[U] = U$, and $([\mathscr{A}], [U])$ is an antisymmetric concrete category over $\mathscr{K}$. If $H : (\mathscr{A}, U) \longrightarrow (\mathscr{B}, V) \in$

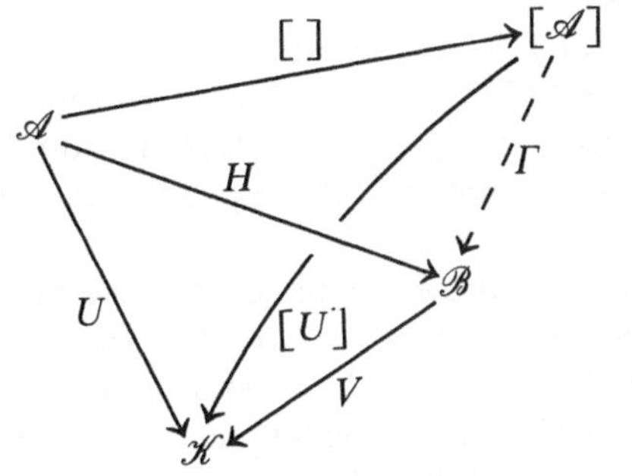

Con$(\mathscr{K})$, and if $(\mathscr{B}, V)$ is antisymmetric, the desired unique Γ is defined by $[A]\Gamma = AH$. $\square$

To illustrate the ideas in 3.4, let $\mathscr{A}$ be the category of metric spaces and continuous maps and let $U : \mathscr{A} \longrightarrow$ **Set** be the underlying set functor. Then $([\mathscr{A}], [U])$ is isomorphic in Con(**Set**) to the category of metrizeable topological spaces and continuous maps.

3.5 Structural Reflection Theorem. *Let $(\mathscr{A}, U)$ be a concrete category over $\mathscr{K}$. The following three statements are true:*

1. *$(\mathscr{A}, U)$ has a reflection in Struct$(\mathscr{K})$.*
2. *If $\Phi : (\mathscr{A}, U) \longrightarrow (\mathscr{C}, V) \in$* Con$(\mathscr{K})$ *with $(\mathscr{C}, V) \in$* Struct$(\mathscr{K})$ *then Φ is a reflection of $(\mathscr{A}, U)$ in* Struct$(\mathscr{K})$ *if and only if Φ is full and every object in $\mathscr{C}$ is isomorphic to an object of form $A\Phi$.*
3. *If $(\mathscr{A}, U)$ is antisymmetric then its reflection map in* Struct$(\mathscr{K})$ *is a full representative subcategory.*

Proof. Consider the situation shown below, where $(\mathscr{C}, V)$ and $(\mathscr{C}', V')$

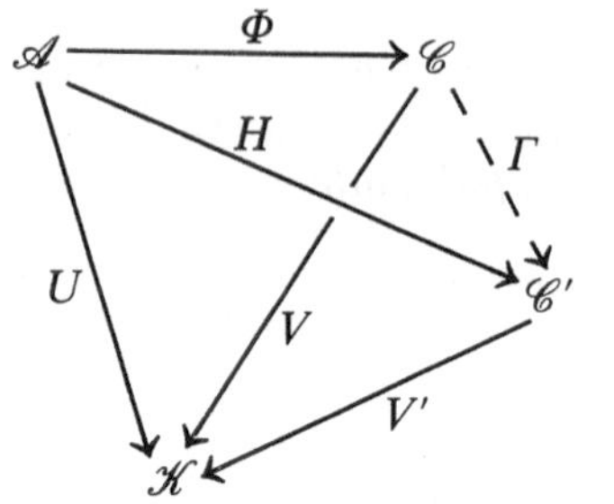

are in Struct($\mathscr{K}$) and Φ is full and representative. Let C be a $\mathscr{C}$-object. By hypothesis, there exists an isomorphism $\bar{f}:A\Phi \longrightarrow C$. Set $f = \bar{f}U$. Since $\mathscr{C}'$ satisfies "structure is abstract" there exists $C\Gamma$ such that $f:AH \longrightarrow C\Gamma$ and its inverse are admissible. To see that Γ is well defined on objects, suppose also that $\bar{g}:A'\Phi \longrightarrow C$ is an isomorphism and that C' is unique such that $g:A'H \longrightarrow C'$ and its inverse are admissible. Since Φ is full and $\bar{f}.(\bar{g})^{-1}:A\Phi \longrightarrow A'\Phi$, $f.g^{-1}$ is admissible from A to A' and hence, via H, from AH to $A'H$. It follows that $\mathrm{id}_{CV} = f^{-1}.(f.g^{-1}).g$ is admissible from $C\Gamma$ to C'; and symmetrically, id_{CV} is admissible from C' to $C\Gamma$ so that they are equal. To complete the "if" of (2) it is necessary to verify that if $f:C_1 \longrightarrow C_2$ is admissible in $\mathscr{C}$ then $f:C_1\Gamma \longrightarrow C_2\Gamma$ is admissible in $\mathscr{C}'$, that $A\Phi\Gamma = AH$ and that Γ is unique with these properties; we leave this as an easy exercise.

By combining 3.4 with the above, we have only to establish that if $(\mathscr{A}, U)$ is antisymmetric then there exists $(\mathscr{C}, V)$ in Struct($\mathscr{K}$) and $\Phi:(\mathscr{A}, U) \longrightarrow (\mathscr{C}, V)$ in Con($\mathscr{K}$) such that Φ is a full representative subcategory. This is done as follows. For fixed K in $\mathscr{K}$, consider the class of all pairs (A, f) with $f:K \longrightarrow AU$ an isomorphism. Given two such pairs (A, f) and (A', f'), say that $(A,f) \leqslant (A',f')$ if for every pair (g, B) with $g:K \longrightarrow BU$, if $f.g:A \longrightarrow B$ is admissible then so is $f'.g:A' \longrightarrow B$. Then $\leqslant$ is reflexive and transitive,

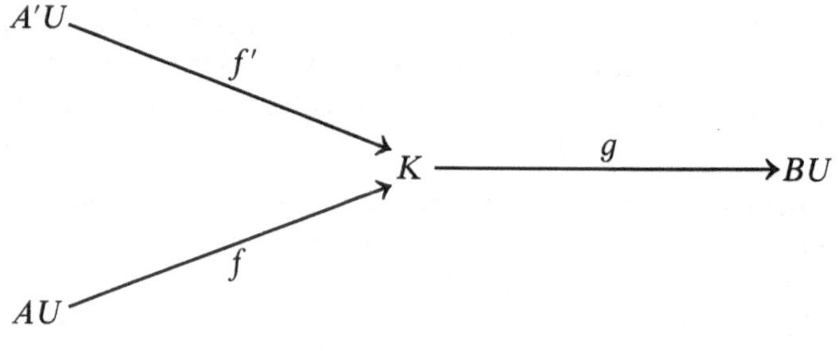

so "$(A,f) \sim (A',f')$ if $(A,f) \leqslant (A',f')$ and $(A',f') \leqslant (A,f)$" is an equivalence relation. Let $\mathscr{C}(K)$ be the class of all $\sim$-equivalence classes $[A, f]$ of pairs

(A, f). If $[A, f] \in \mathscr{C}(K)$, $[B, g] \in \mathscr{C}(L)$ and $h: K \longrightarrow L$ define h to be admissible from $[A, f]$ to $[B, g]$ if and only if $f.h.g^{-1}: A \longrightarrow B$ is admissible in $\mathscr{A}$. This is well defined, since if $(A, f) \sim (A', f')$ and $(B, g) \sim (B', g)$ we argue as follows: as $f.h.g^{-1}: A \longrightarrow B$ and $(A, f) \leqslant (A', f')$, $f'.h.g^{-1}: A' \longrightarrow B$; as $(B', g') \leqslant (B, g)$, $g.(g')^{-1}: B \longrightarrow B'$; therefore $(f'.h.g^{-1}).g.(g')^{-1} = f'.h.(g')^{-1}: A' \longrightarrow B'$ is admissible. $A\Phi$ is defined to be $[A, \mathrm{id}_{AU}]$. The numerous omitted verifications are left as an easy calculation. □

3.6 Summary. We have shown that an arbitrary concrete category over $\mathscr{K}$ admits a full and faithful functor over $\mathscr{K}$ to a category of $\mathscr{K}$-objects with structure whose image contains an isomorph of every object, and that all such functors are isomorphic (2.2!). Our remarks in 3.2 suggest that, for our purposes, a concrete category and its structural reflection have very similar properties.

Here is a typical example of structural reflection. Let $\mathscr{A}$ be the category of groups and homomorphisms, let $\mathbf{Z}$ be the group of integers and define $U: \mathscr{A} \longrightarrow \mathbf{Set}$ as follows. Set $AU = \mathscr{A}(\mathbf{Z}, A)$. For $f: A \longrightarrow B$ in $\mathscr{A}$, set $\langle x, fU \rangle = x.f$. Then $(\mathscr{A}, U)$ is a concrete category over **Set** which is, in fact, antisymmetric. It does not satisfy the existence condition in "structure is abstract." If $U': \mathscr{A} \longrightarrow \mathbf{Set}$ is the usual underlying set functor, then $(\mathscr{A}, U') \in \mathrm{Struct}(\mathbf{Set})$. It is obvious how to make AU into a group isomorphic to A, thereby defining the full representative subcategory $\Phi: (\mathscr{A}, U) \longrightarrow (\mathscr{A}, U')$ which is the structural reflection of $(\mathscr{A}, U)$. This justifies our intuition that "U and U' are practically the same."

3.7 Constructions in Struct($\mathscr{K}$). Regardless of the nature of $\mathscr{K}$, Struct($\mathscr{K}$) has products. $\mathrm{id}_{\mathscr{K}}: \mathscr{K} \longrightarrow \mathscr{K}$ (which is in Struct($\mathscr{K}$) via 1-element fibres and all maps admissible) is the terminal object. More generally, given any family $(\mathscr{C}_i : i \in I)$ in Struct($\mathscr{K}$), the product $\mathscr{C}$ is defined by

$$\mathscr{C}(K) = \prod_{i \in I} \mathscr{C}_i(K)$$

$$f:(s_i) \longrightarrow (t_i) \quad \text{if and only if} \quad f: s_i \longrightarrow t_i \quad \text{for all } i.$$

With the obvious projection homomorphisms $\mathscr{C} \longrightarrow \mathscr{C}_i$ (sending $(K, (s_j))$ to (K, s_i)) it is routine to verify that $\mathscr{C}$ is the categorical product. If $\mathscr{C} \in \mathrm{Struct}(\mathscr{K})$ and if $\mathscr{P}$ is a full replete subcategory of $\mathscr{C}$ then the axioms of 3.1 hold (construct what is needed in $\mathscr{C}$ and observe it is in $\mathscr{P}$) so that $\mathscr{P}$ is in Struct($\mathscr{K}$). Products and full replete subcategories are often used together to create new structures out of old. "Topological groups" is a typical example over $\mathscr{K} = \mathbf{Set}$. Let $\mathscr{C}_1$, $\mathscr{C}_2 \in \mathrm{Struct}(\mathbf{Set})$ be "spaces" and "groups." Then $\mathscr{C}_1 \times \mathscr{C}_2$ is the category of "topologized groups," the objects being sets equipped with (unrelated) topological and group structure. "Topological groups" is then a full replete subcategory of $\mathscr{C}_1 \times \mathscr{C}_2$.

We now begin to classify structure. Of fundamental importance in this book is:

3.8 Algebraic Structure. $\mathscr{C} \in \mathrm{Struct}(\mathscr{K})$ is *algebraic over* $\mathscr{K}$ if there exists an algebraic theory $\mathbf{T}$ in $\mathscr{K}$ such that $\mathscr{C}$ is isomorphic in $\mathrm{Struct}(\mathscr{K})$ to $\mathscr{K}^{\mathbf{T}}$. $(\mathscr{A}, U) \in \mathrm{Con}(\mathscr{K})$ is *weakly algebraic* if its structural reflection (3.5) is algebraic.

3.9 Discretely-Ordered Structure. It is sometimes the case that "bijective admissible maps are isomorphisms"; for example, this is true for groups but false for topological spaces. We now formalize when this is true. A preordered set $(X, \leqslant)$ is *discrete* if $\leqslant$ is the equality relation, that is if $x \leqslant y$ then $x = y$. $\mathscr{C} \in \mathrm{Struct}(\mathscr{K})$ is *discretely ordered* if for every K, the fibre $(\mathscr{C}(K), \leqslant)$ is discrete. This is equivalent to stipulating that whenever $f:(K, s) \longrightarrow (L, t)$ is admissible with $f: K \longrightarrow L$ an isomorphism in $\mathscr{K}$, then also $f^{-1}: t \longrightarrow s$. (Proof: there exists unique t' with $f: s \longrightarrow t'$ and $f^{-1}: t' \longrightarrow s$; since $f^{-1}.f: t' \longrightarrow t$, $t' \leqslant t$ and $t' = t$.) Notice that

(3.10) *Algebraic implies discretely ordered.*

(3.11) *Any product of discretely ordered categories of $\mathscr{K}$-objects with structure is again discretely ordered; any full replete subcategory of a discretely ordered category is again discretely ordered.*

3.12 Duality. The *dual* $(\mathscr{A}, U)^{\mathrm{op}}$ *of* the concrete category $(\mathscr{A}, U: \mathscr{A} \longrightarrow \mathscr{K})$ over $\mathscr{K}$ is the concrete category $(\mathscr{A}^{\mathrm{op}}, U^{\mathrm{op}})$ over $\mathscr{K}^{\mathrm{op}}$ where (cf. 1.34) $U^{\mathrm{op}}: \mathscr{A}^{\mathrm{op}} \longrightarrow \mathscr{K}^{\mathrm{op}}$ is defined by

$$AU^{\mathrm{op}} = AU$$

$$(A \xleftarrow{\ f\ } B)U^{\mathrm{op}} = AU \xleftarrow{\ fU\ } BU$$

Passing from a concrete category to its dual establishes an isomorphism of categories $\mathrm{Con}(\mathscr{K}) \cong \mathrm{Con}(\mathscr{K}^{\mathrm{op}})$, the action on homomorphisms being depicted below:

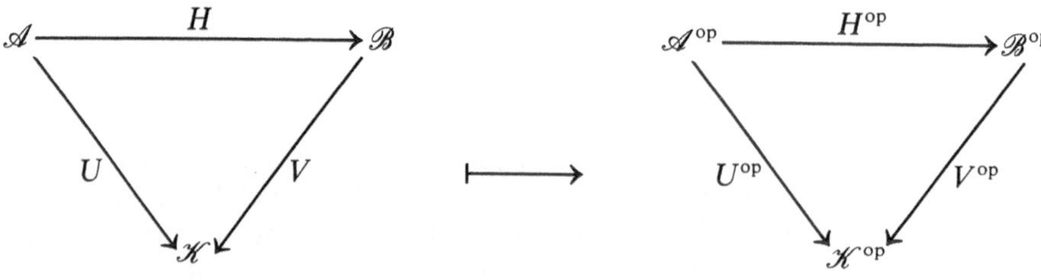

It is clear that, by restriction, duality establishes an isomorphism $\mathrm{Struct}(\mathscr{K}) \cong \mathrm{Struct}(\mathscr{K}^{\mathrm{op}})$. Because of duality, our work in classifying structure is "cut in half"; for even if our original aim was to study only categories of sets with structure, it would have been necessary to consider the much more abstract idea of a category of $\mathbf{Set}^{\mathrm{op}}$-objects with structure in order to take advantage of duality.

"Discretely ordered" is *self-dual* in that $\mathscr{C}$ is discretely ordered over $\mathscr{K}$ if and only if $\mathscr{C}^{op}$ is discretely ordered over $\mathscr{K}^{op}$. On the other hand, "algebraic" is not a self-dual property. Say that $\mathscr{C}$ is *coalgebraic over* $\mathscr{K}$ if, of course, $\mathscr{C}^{op}$ is algebraic over $\mathscr{K}^{op}$.

3.13 Closure Operators and Interior Operators. Let the category $\mathscr{K}$ be a partially ordered class. Consider an algebraic theory $\mathbf{T} = (T, \eta, \mu)$ in $\mathscr{K}$. To say T is a functor is only to say T is an order-preserving endomorphism. The existence of η and μ translate to the properties $x \leqslant xT$ and $xTT = xT$. Therefore T is just what is usually called a closure operator on $\mathscr{K}$ (and it is more standard to write x^- instead of xT); i.e., the other axioms are guaranteed to hold because all diagrams in $\mathscr{K}$ commute. $\mathscr{K}^{\mathbf{T}}$ is just the full subcategory of closed elements, that is all x such that $x^- \leqslant x$. As is well known, if $\mathscr{K}$ is a complete partially-ordered set (including the empty infimum 1) then the full subcategory $\mathscr{A}$ or $\mathscr{K}$ is algebraic over $\mathscr{K}$ if and only if $\mathscr{A}$ is closed under infima (for the inverse passage, define $x^- = \operatorname{Inf}(y : x \leqslant y)$). Dually, $\mathscr{A}$ is coalgebraic over $\mathscr{K}$ if and only if $\mathscr{A}$ is closed under suprema. As discussed further in exercise 10, $\mathscr{A}$ is coalgebraic over $\mathscr{K}$ if and only if $\mathscr{A}$ can be constructed, up to isomorphism, as the coalgebras over an algebraic cotheory. In the partially ordered case, an algebraic cotheory is usually called an interior operator, being an order-preserving endomorphism $\circ$ satisfying $x^\circ \leqslant x$ and $x^{\circ\circ} = x^\circ$. The coalgebras are the open elements.

Well known constructions in topology are "the smallest topology making a family of functions from a set to a bunch of spaces continuous" and, dually, "the largest topology making a family of functions from a bunch of spaces to a set continuous." This is easy to formulate in Struct($\mathscr{K}$):

3.14 Optimal and Co-Optimal Families. Let $\mathscr{C}$ be a literal category of $\mathscr{K}$-objects with structure (for convenience, as the extension of the ideas to arbitrary concrete categories will be obvious). A family (not necessarily small!) of admissible maps of form

$$(K, s) \xrightarrow{\ f\ } (L_i, t_i)$$

is *optimal* if whenever (K', s') is a $\mathscr{C}$-structure and $g : K' \longrightarrow K$ is a $\mathscr{K}$-morphism such that $g.f_i : s' \longrightarrow t_i$ is admissible for all i then $g : s' \longrightarrow s$ is also admissible. An *optimal map* is a one-element optimal family. Consider the situation

$$K \xrightarrow{\ f_i\ } (L_i, t_i)$$

(i.e., $f_i : K \longrightarrow L_i$ is a family of $\mathscr{K}$-morphisms and $t_i \in \mathscr{C}(L_i)$). An *optimal lift of* $(f_i : K \longrightarrow (L_i, t_i))$ is a structure $s \in \mathscr{C}(K)$ such that $f_i : (K, s) \longrightarrow (L_i, t_i)$ is an optimal family. Optimal lifts, when they exist, are unique in view of the antisymmetry of $\mathscr{C}$.

Dually, a family $f_i : (L_i, t_i) \longrightarrow (K, s)$ is *co-optimal* if the family $f_i : (K, s) \longleftarrow (L_i, t_i)$ of admissible morphisms of $\mathscr{C}^{op} \in \text{Struct}(\mathscr{K}^{op})$ is optimal; that is, specifically, if whenever (K', s') is a $\mathscr{C}$-structure and $g: K \longrightarrow K'$ is a $\mathscr{K}$-morphism such that

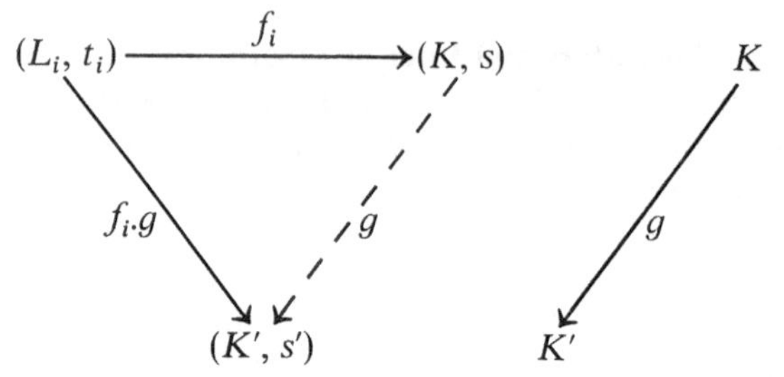

$f_i.g$ is admissible for all i then g is also admissible.

3.15 Example. Let $\mathscr{C} \in$ Struct(**Set**) be topological spaces and continuous maps. Any family (not necessarily small!) $f_i: K \longrightarrow (L_i, \mathscr{T}_i)$ has optimal lift $\mathscr{S} \in \mathscr{C}(K)$, where a subbase for $\mathscr{S}$ is all sets of form Uf_i^{-1} with $U \in \mathscr{T}_i$. This is a well known construction in topology and such $\mathscr{S}$ is conventionally called the "smallest" or "weakest" topology making all f_i continuous since if $\mathscr{S}' \in \mathscr{C}(K)$ makes each f_i continuous then $\mathscr{S}' \supset \mathscr{S}$; from the point of view of the fibre $(\mathscr{C}(K), \leqslant)$ we have $\mathscr{S}_1 \leqslant \mathscr{S}_2$ if and only if $\mathscr{S}_2 \subset \mathscr{S}_1$, which explains why we avoid such terminology. Any family $f_i:(L_i, \mathscr{T}_i) \longrightarrow K$ has a co-optimal lift $\mathscr{S}$, namely $\mathscr{S} = \{U \subset K\colon \text{for all } i,\ Uf_i^{-1} \in \mathscr{T}_i\}$. The definitions "optimal family" and "co-optimal family" make sense when (f_i) is the empty family. The empty optimal lift in $\mathscr{C}(K)$ is the indiscrete topology whereas the empty co-optimal lift is the discrete topology.

3.16 Example. If **T** is an algebraic theory in $\mathscr{K}$, if (Δ, D)—write $D_i = (L_i, \xi_i)$—is a diagram in $\mathscr{K}^{\mathbf{T}}$ and if (K, ψ) is a limit of $(\Delta, DU^{\mathbf{T}})$ in $\mathscr{K}$ then the unique morphism $\xi: KT \longrightarrow K$ such that each ψ_i becomes a **T**-homomorphism renders $(K, \xi; \psi)$ a limit of D in $\mathscr{K}^{\mathbf{T}}$ (the proof is easy and is given in 3.1.19 below). It is true, moreover, that $\psi_i:(K, \xi) \longrightarrow (L_i, \xi_i)$

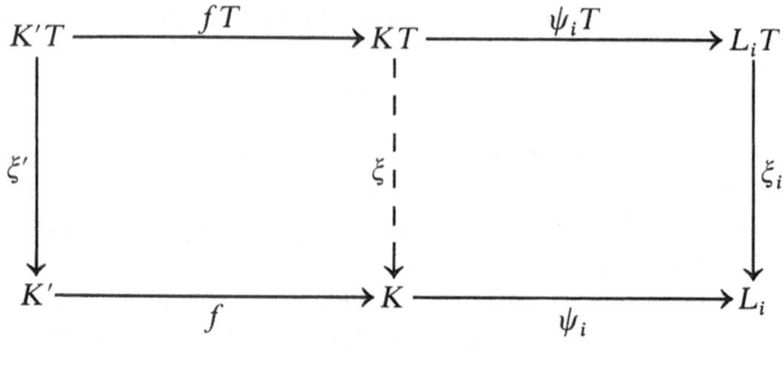

is optimal in $\mathscr{K}^{\mathbf{T}}$. To prove this, observe (as shown above) that if $\xi'.f.\psi_i = fT.\psi_i T.\xi_i$ for all i then $\xi'.f = fT.\xi$ since both maps are induced by the same universal property (cf. 3.1.20).

It is not true that all families in $\mathscr{K}^{\mathbf{T}}$ have optimal lifts. For example, if $\mathscr{K} = \mathbf{Set}$ then an inclusion $A \subset (X, \xi)$ cannot lift unless A is a subalgebra of (X, ξ) (see 1.4.31 which asserts that $A = \langle A \rangle$ if and only if the inclusion of A lifts to a homomorphism).

3.17 Construction of Limits. Let $\mathscr{C} \in \text{Struct}(\mathscr{K})$. The functor $U: \mathscr{C} \longrightarrow \mathscr{K}$ *constructs limits* if for every (not necessarily small!) diagram (Δ, D) in $\mathscr{C}$ and every limit (K, ψ) of (Δ, DU) in $\mathscr{K}$ the family $\psi_i: K \longrightarrow D_i = (L_i, t_i)$ has an optimal lift s. It follows at once that $\psi_i:(K, s) \longrightarrow (L_i, t_i)$ is a limit of D in $\mathscr{C}$; thus, if U constructs limits, then $\mathscr{C}$ has and U preserves whatever sorts of limit $\mathscr{K}$ has. Example 3.16 shows that $U^{\mathbf{T}}: \mathscr{K}^{\mathbf{T}} \longrightarrow \mathscr{K}$ constructs limits.

Dually, $U:\mathscr{C} \longrightarrow \mathscr{K}$ *constructs colimits* if $U^{\text{op}}:\mathscr{C}^{\text{op}} \longrightarrow \mathscr{K}^{\text{op}}$ constructs limits, that is, whenever $\psi_i: D_i \longrightarrow K$ is such that (K, ψ) is a colimit of (Δ, DU), it has a co-optimal lift (which is then a colimit of D in $\mathscr{C}$). $U^{\mathbf{T}}: \mathscr{K}^{\mathbf{T}} \longrightarrow \mathscr{K}$ rarely constructs colimits (e.g. Example 1.28 shows that coproducts of abelian groups are not built on disjoint unions; see also section 7 of Chapter 3). By 3.15, the underlying set functor from topological spaces constructs both limits and colimits.

3.18 Example. Let U be the forgetful functor from the category of Banach spaces and norm-decreasing maps to the category of metric spaces with base point (as described in exercise 2.15). Then U constructs limits but fails to construct infinite coproducts (which in the Banach space category are obtained by completing the weak direct sum with the "sum" norm).

3.19 Example. The underlying set functor from real vector spaces constructs limits but fails to construct coproducts. On the other hand, the underlying abelian group functor from real vector spaces constructs both limits and colimits.

3.20 Optimal Substructures and Co-Optimal Quotient Structures. Let $\mathscr{C} \in \text{Struct}(\mathbf{Set})$. If (K, s) is a $\mathscr{C}$-structure, a subset A of K is an *optimal subset of* (K, s) if the inclusion map $A \longrightarrow (K, s)$ has an optimal lift. Dually, a surjection $f:K \longrightarrow L$ is a *co-optimal quotient of* (K, s) if f has a co-optimal lift. While many such $\mathscr{C}$ admit numerous sorts of "substructure," the optimal ones are almost always an important type. Similarly for quotients. We illustrate with examples, pausing to prove a lemma useful here and in section 3.1:

3.21 Lemma. *Let $\mathscr{C} \in \text{Struct}(\mathscr{K})$ and let f, g, h be $\mathscr{C}$-morphisms such that $fh = gh$. The following two statements are true:*

1. *If h is co-optimal and if $h = \text{coeq}(f, g)$ in $\mathscr{K}$ then $h = \text{coeq}(f, g)$ in $\mathscr{C}$.*
2. *If $h = \text{coeq}(f, g)$ in $\mathscr{C}$ and if h is epi in $\mathscr{K}$ then h is co-optimal.*

Proof. Consider the diagram shown below. For the first statement,

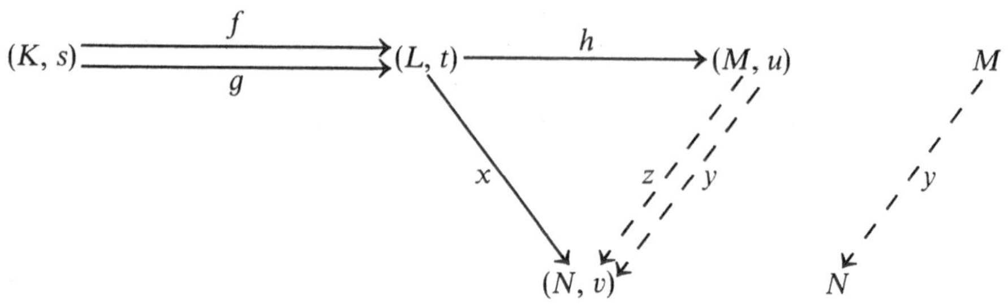

if $fx = gx$ with x in $\mathscr{C}$ as shown then there exists unique y in $\mathscr{K}$ with $hy = x$; y is admissible since h is co-optimal. For the second statement, if y is in $\mathscr{K}$ and $x = hy$ is admissible then there exists unique z in $\mathscr{C}$ with $hz = hy$. As h is epi in $\mathscr{K}$, $z = y$ in $\mathscr{K}$ and y is admissible. □

3.22 Example. In $\mathbf{Set}^{\mathbf{T}}$, optimal subsets are subalgebras and co-optimal quotients are surjective **T**-homomorphisms (see 1.4.30 and 1.56). Notice that rings or semigroups have ideals (one- or two-sided) and groups have normal subgroups, so that there are more specialized subobjects.

3.23 Example. Let $\mathscr{C}$ be topological groups in Struct(**Set**). Optimal subsets are subgroups (which we provide with the subspace topology) whereas co-optimal quotients are quotient groups—i.e., the kernel pair of the surjection in question must be a subgroup of the product—(which is provided with the quotient topology). The situation changes for *separated* groups. Here, optimal subsets are closed subgroups and co-optimal quotients must satisfy the additional condition that the kernel pair is closed in the product topology.

3.24 Example. Let $\mathscr{C}$ be Banach spaces and norm-decreasing linear maps in Struct(**Set**). Optimal subsets are closed linear subspaces. The coequalizer $h = \text{coeq}(f, g)$ in $\mathscr{C}$ is formed by dividing out by the closed subspace $K = \ker(f - g)$ and imposing the quotient norm $\|\alpha\| = \text{Inf}\{\|x\| : x + K = \alpha\}$. From this and Lemma 3.21 it follows that the co-optimal quotients are the coequalizers, i.e., are those surjections h for which the kernel pair of h is a closed linear subspace of the product.

3.25 Example. Let $f: A \longrightarrow A$ be the squaring homomorphism on the nonzero reals in the category of 2-divisible abelian groups, so that f is the noninjective monomorphism of 1.47. Notice that f is *not* optimal since the absolute value function $A \longrightarrow A$ is not a homomorphism even though $|x|^2 = x^2$. The optimal subsets are the subgroups which, as groups, are 2-divisible.

3.26 Example. Let $\mathscr{C}$ be the category of C^∞ manifolds and C^∞ mappings. The optimal subsets are the well-known submanifolds. For a proof see [Lang '72, page 25].

We now introduce an important property in Struct($\mathscr{K}$):

3.27 Definition. *$\mathscr{C} \in$ Struct$(\mathscr{K})$ is fibre-complete if every (not necessarily small!) family $f_i: K \longrightarrow (L_i, t_i)$ has an optimal lift and if every family $f_i:(L_i, t_i) \longrightarrow K$ has a co-optimal lift.* Notice that each fibre $(\mathscr{C}(K), \leqslant)$ is indeed complete since for any subset A of $\mathscr{C}(K)$, the infimum of A is the optimal lift of the A-indexed family $\mathrm{id}_K: K \longrightarrow (K, a)$ and the supremum of A is constructed, dually, as a co-optimal lift.

3.28 Proposition. *If $\mathscr{C} \in$ Struct$(\mathscr{K})$ is fibre complete then $U:\mathscr{C} \longrightarrow \mathscr{K}$ has left and right adjoints.*

Proof. The free object over K is the least element 0 of $(\mathscr{C}(K), \leqslant)$, that is, the co-optimal lift of the empty family out of K; the reasoning is "f:

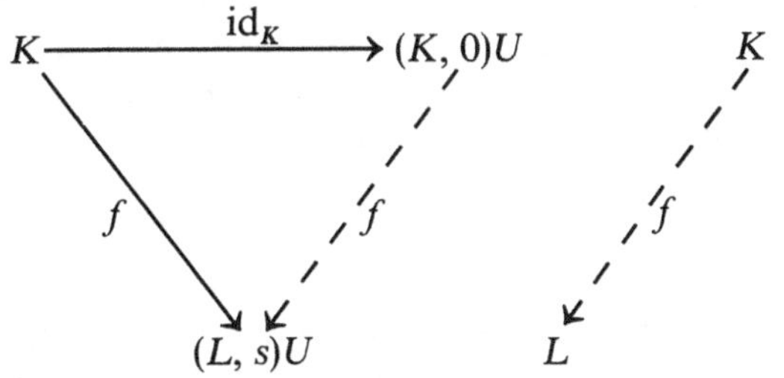

$(K, 0) \longrightarrow (L, s)$ is admissible if and only if it is admissible preceded by every element of the empty family" and this is true for every $f: K \longrightarrow L$. Dually, the cofree object is the greatest element 1 of $\mathscr{C}(K)$. □

3.29 Proposition. *If $\mathscr{C} \in$ Struct$(\mathscr{K})$ is fibre complete then $U:\mathscr{C} \longrightarrow \mathscr{K}$ constructs limits and colimits. A morphism $h:(L, t) \longrightarrow (M, u)$ in $\mathscr{C}$ is a coequalizer in $\mathscr{C}$ if and only if h is co-optimal and is a coequalizer in $\mathscr{K}$.*

Proof. The first statement follows immediately from the definition in 3.17. If $h = \mathrm{coeq}(f, g)$ in $\mathscr{C}$ then $h = \mathrm{coeq}(f, g)$ in $\mathscr{K}$ because U has a right adjoint and, hence, must preserve colimits. By 3.21, h is co-optimal. Conversely, let h be co-optimal and let $h = \mathrm{coeq}(f, g)$ in $\mathscr{K}$. Let s be the optimal lift of $f, g: K \longrightarrow (L, t)$. It follows from 3.21 that $h = \mathrm{coeq}(f, g)$ in $\mathscr{C}$. □

3.30 Example. Topological spaces is fibre complete over sets as discussed in 3.15. Other "topological categories" are fibre complete (see [Wyler '71–A]). For example let $\mathscr{C}$ be the category of uniform spaces and uniformly continuous mappings in Struct(**Set**). Then $\mathscr{C}$ is fibre complete. Given $f_i: K \longrightarrow (L_i, \mathscr{V}_i)$, the optimal lift is the uniformity $\mathscr{U}$ whose entourages are supersets of finite intersections of subsets of $K \times K$ of form $\alpha(f_i \times f_i)^{-1}$ with $\alpha \in \mathscr{V}_i$. Co-optimal lifts do not have an equally direct description but must exist in view of the following useful

3.31 Proposition. *Let $\mathscr{C} \in$ Struct$(\mathscr{K})$. Then if every family $f_i: K \longrightarrow (L_i, t_i)$ has an optimal lift, $\mathscr{C}$ is fibre complete.*

Proof. We must show that $f_i:(K_i, s_i) \longrightarrow L$ has a co-optimal lift. Let B be the set of all $u \in \mathscr{C}(L)$ such that $f_i: s_i \longrightarrow u$ is admissible for all i. B is nonempty since the empty optimal lift is in B. Let t be the optimal lift

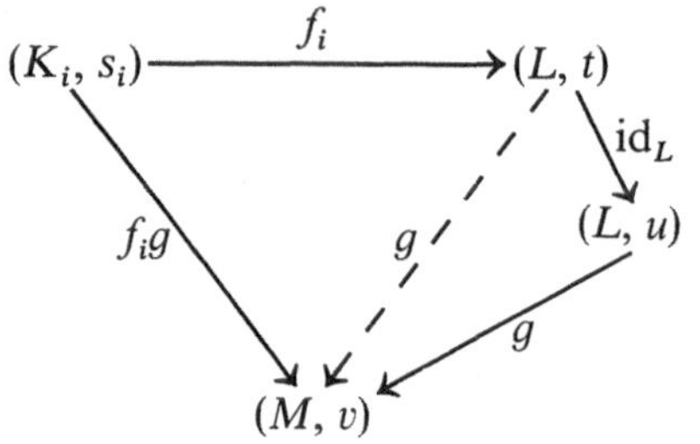

of the B-indexed family $\mathrm{id}_L: L \longrightarrow (L, u)$. Suppose given $g: L \longrightarrow (M, v)$ such that $f_i g: s_i \longrightarrow v$ is admissible for all i. Let u be the optimal lift of $g: L \longrightarrow (M, v)$. As u is optimal, u belongs to B and $g: t \longrightarrow v$ is admissible being the composition of $\mathrm{id}_L: t \longrightarrow u$ and $g: u \longrightarrow v$. □

3.32 Proposition. *Any product in* Struct($\mathscr{K}$) *(as described in 3.7) of fibre-complete categories is fibre complete.* □

3.33 "Type Theory". Let (n, k) be a pair of non-negative integers. An (n, k)-*structure* is a pair (K, s) where K is a set and s is an element of $K^n P^k$ where K^n is the n-fold cartesian power and P^k is the kth iterate of the power set operator, $KP = 2^K$. For example, a topology of open subsets of K is a (1, 2)-structure, a uniform structure of entourages on K is a (2, 2)-structure, a ternary operation $K^3 \longrightarrow K$ is (via its graph) a (4, 1)-structure, a partially ordered set is a (2, 1)-structure. and a set with base point is a (1, 0)-structure. For $k > 0$ define fibre-complete categories in Struct(**Set**), $\mathscr{C}_{(n,k)}$, and $\mathscr{C}^{(n,k)}$, whose objects are the (n, k)-structures. Say that $f:(K, s) \longrightarrow (L, t)$ is admissible in $\mathscr{C}_{(n,k)}$ if the direct image sf^n is a subset of t; whereas f: $(K, s) \longrightarrow (L, t)$ is admissible in $\mathscr{C}^{(n,k)}$ if the inverse image $t(f^n)^{-1}$ is a subset of s. Given $f_i: K \longrightarrow (L, t_i)$, the optimal lift in $\mathscr{C}_{(n,k)}$ is the intersection of all $t_i((f_i)^n)^{-1}$ whereas the optimal lift in $\mathscr{C}^{(n,k)}$ is the union of all $t_i((f_i)^n)^{-1}$. Given $f_i:(K_i, s_i) \longrightarrow L$, the co-optimal lift in $\mathscr{C}_{(n,k)}$ is the union of all $s_i(f_i)^n$ whereas the co-optimal lift in $\mathscr{C}^{(n,k)}$ is the i-indexed intersection of the families $\{A: A((f_i)^n)^{-1} \in s_i\}$. Thus topological spaces, ternary algebras, and partially ordered sets are full replete subcategories, respectively, of $\mathscr{C}^{(1,2)}$, $\mathscr{C}_{(4,1)}$, and $\mathscr{C}_{(2,1)}$.

3.34 Pulling Back Structure. Let $\mathscr{C}$ be in Struct($\mathscr{K}$) and let $\mathscr{L}$ be a full replete subcategory of $\mathscr{K}$ with inclusion functor i. The *pullback of* $\mathscr{C}$ *along* $\mathscr{L}$ is the category $\mathscr{P}$ in Struct($\mathscr{K}$) defined by

$$\mathscr{P}(K) = \begin{cases} \mathscr{C}(K) & \text{(if } K \text{ is in } \mathscr{L}) \\ \varnothing & \text{(if } K \text{ is not in } \mathscr{L}) \end{cases}$$

The $\mathscr{P}$-admissible maps between two $\mathscr{P}$-structures are defined to be all the $\mathscr{C}$-admissible maps. The axioms of 3.1 are clear. As shown in 3.35

$$\begin{array}{ccc} \mathscr{P} & \xrightarrow{\quad j \quad} & \mathscr{C} \\ {\scriptstyle V}\big\downarrow & \overset{W}{\searrow} & \big\downarrow{\scriptstyle U} \\ \mathscr{L} & \xrightarrow[\quad i \quad]{} & \mathscr{K} \end{array} \tag{3.35}$$

(which is a pullback in the category of categories and functors), $\mathscr{P}$ is a full replete subcategory of $\mathscr{C}$ and $\mathscr{P}$ is in Struct($\mathscr{L}$).

For example, if $\mathscr{K}$ is topological spaces, $\mathscr{C}$ is topological groups and $\mathscr{L}$ is Hausdorff (or just T_0) spaces, $\mathscr{P}$ is separated topological groups.

3.36 Proposition. *Let $\mathscr{C}$ be in* Struct($\mathscr{K}$) *and let $\mathscr{L}$ be a full replete subcategory of $\mathscr{K}$. Then if $\mathscr{C}$ is fibre complete over $\mathscr{K}$, its pullback $\mathscr{P}$ along $\mathscr{L}$ is again fibre complete over $\mathscr{L}$.*

Proof. If $f_i:(L, s) \longrightarrow (L_i, t_i)$ is optimal in $\mathscr{C}$ with L and all L_i in $\mathscr{L}$ then this family is *a fortiori* optimal in $\mathscr{P}$. This is all we need check by 3.31. □

3.37 Definition. *$H:\mathscr{C} \longrightarrow \mathscr{D}$ in* Struct($\mathscr{K}$) *is taut if H preserves optimal families, that is $f_i:(K, s)H \longrightarrow (L_i, t_i)H$ is optimal in $\mathscr{D}$ whenever $f_i:(K, s) \longrightarrow (L_i, t_i)$ is optimal in $\mathscr{C}$. Dually, H is cotaut if H preserves co-optimal families.* Examples are developed in the exercises.

In 3.33 we saw how familiar categories of sets with structure admit natural representations as full replete subcategories of fibre-complete categories. Such embeddings are often taut but rarely cotaut. In exercise 12 it is shown that every category of $\mathscr{K}$-objects with structure admits a simultaneously taut and cotaut embedding as a full replete subcategory of a fibre-complete category.

Notes for Section 3

While the definition of "a category of $\mathscr{K}$-objects with structure" is an obvious one, there is very little literature on the subject. The earliest reference (with $\mathscr{K}$ = **Set**) we know of is [Krishnan '51]. Ježek, in his treatise [Ježek '70], refers to a 1958 paper of Mal'cev (cf. [Mal'cev '71, page 52]). See also [Ehresmann '65, page 55] (which is presented in [Bucur and Deleanu '68, page 84]), [Bourbaki '57], [Blanchard '71], and [Guitart '74]. Traditionally, "concrete category" means over **Set**.

Not every locally small category $\mathscr{A}$ admits a $U:\mathscr{A} \longrightarrow$ **Set** such that $(\mathscr{A}, U)$ is concrete over **Set** (see [Isbell '63, Example 2.4] and [Freyd '64, page 108]). For necessary and sufficient conditions for concreteness over **Set** see [Freyd '73].

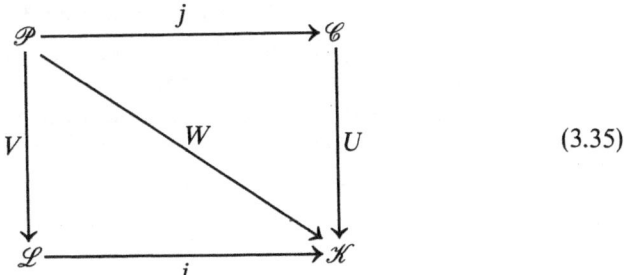

Exercises for Section 3

1. Prove that the category of reflexive and transitively ordered sets and order-preserving maps is isomorphic in Struct(**Set**) to the category of topological spaces in which every intersection of open sets is open and continuous maps. [Hint: $x \leqslant y$ if and only if $x \in \{y\}^-$]. Use this as an aid to constructing the lattice of all topologies on a three-element set; there are 29 such topologies and 9 homeomorphism classes.
2. Exercise 1 raises the question whether a "familiar presentation" of a category $\mathscr{C}$ in Struct(**Set**) can be recovered from the isomorphism class of $\mathscr{C}$ in Struct(**Set**).
   (a) Show that a standard presentation of "abelian group" can be recovered. [Hint: The object **Z** of integers is distinguished by the facts that it admits infinitely many endomorphisms and is such that whenever A admits a monomorphism into **Z** then either A is an initial object or A admits an isomorphism to **Z**; define an element of A to be a map from **Z** to A; for any A the canonical map $A + A \longrightarrow A \times A$ is an isomorphism and the map $A \times A = A + A \longrightarrow A$ which is the identity of A preceded by each coproduct injection defines addition of elements.]

   The hint in (a) achieved much more than was required since it used only the category and not the underlying set functor. Consider:

   (b) Show that a standard presentation of "monoid" can be recovered. [Hint: Since $U:\mathscr{C} \longrightarrow$ **Set** is given, the required n-ary operations will be natural transformations $U^n \longrightarrow U$; there are only two 2-ary operations which are not constant and are associative (because each such operation is a word on two symbols containing say n occurrences of the first and m of the second, and the equations imposed on n and m by associativity are very restrictive—it is perfectly valid to reason first in the "standard" category so long as all transits under isomorphism in Struct(**Set**)) and either one will do (after all, $M \longmapsto M^{\text{op}}$ is an automorphism of the category so there is no categorical way to make this choice).]

   In practice, "recovery" problems are difficult and ad hoc. It is not known at this writing whether there exist "real" categories in Struct(**Set**) which cannot be recovered.
3. Let $\mathscr{C}$ be a category of $\mathscr{K}$-objects with structure. An object S in $\mathscr{C}$ is a *Sierpinski object* if for every X in $\mathscr{C}$ the family of all $\mathscr{C}$-admissible maps $X \longrightarrow S$ is optimal.
   (a) Let $\mathscr{C}$ be topological spaces and continuous maps in Struct(**Set**). Let $S = \{0, 1\}$ be the well-known Sierpinski space (i.e., $\{1\}$ is open, $\{0\}$ is not open). Show that S is a Sierpinski object.
   (b) Let $\mathscr{C}$ be the category of real C^∞ manifolds and differentiable mappings in Struct(**Set**). Show that the real line is a Sierpinski object. [Hint: any chart at p agrees with a globally defined C^∞ map on some neighborhood of p.]

(c) Let $\mathscr{C}$ in Struct($\mathscr{K}$) be fibre complete and let S be any object of $\mathscr{C}$. Let $\mathscr{P}$ be the full replete subcategory of all C for which the family of all admissible maps $C \longrightarrow S$ is optimal. Prove that $\mathscr{P}$ is fibre complete with taut embedding in $\mathscr{C}$ and reflective in such a way that the reflector is again over $\mathscr{K}$. Observe that S is a Sierpinski object in $\mathscr{P}$.

4. Show that Struct($\mathscr{K}$) has equalizers. [Hint: the obvious ones.]
5. Let Cat/$\mathscr{K}$ be the category whose objects are pairs $(\mathscr{A}, U)$ with U: $\mathscr{A} \longrightarrow \mathscr{K}$ an arbitrary functor and whose morphisms are functors over $\mathscr{K}$ just as in 3.3. Show that every such $(\mathscr{A}, U)$ has a reflection in Con($\mathscr{K}$), and hence in Struct($\mathscr{K}$). [Hint: keep the same objects and divide out by the obvious equivalence relation on morphisms.]
6. Show that the reflectivity of partially ordered sets in reflexive and transitively ordered sets (cf. exercise 3 of section 2) is a corollary of 3.4.
7. Let $\mathscr{C}$ be in Struct(**Set**). Prove that $\mathscr{C}$ is isomorphic to topological spaces and continuous maps in Struct(**Set**) if and only if there exists a "Sierpinski space" S with underlying set $\{0, 1\}$ satisfying the following five conditions:
   (a) Every family $f_i: X \longrightarrow S$ has an optimal lift.
   (b) The supremum map $\mathrm{Sup}: S^I \longrightarrow S$ is admissible for every set I.
   (c) The infimum map $\mathrm{Inf}: S^I \longrightarrow S$ is admissible for every finite set I.
   (d) S is a Sierpinski object in $\mathscr{C}$ as defined in exercise 3.

   If (X, s) is in $\mathscr{C}$ say that a subset A of X is *open* in (X, s) if its characteristic function is admissible to S. Say that a family of open subsets is a *subbase* if the corresponding family of characteristic functions to S is an optimal family. The final condition is:

   (e) If $\mathscr{A}$ is a subbase and if A is open then A is a union of finite intersections of elements of $\mathscr{A}$.

   In particular, this solves the recovery problem for topological spaces (in the sense of exercise 2).
8. Let T:**Set** $\longrightarrow$ **Set** be an arbitrary functor. A *T-model* is a pair (X, ξ) where $\xi: XT \longrightarrow X$ is a relation from XT to X, i.e., ξ is a subset of $XT \times X$. Say that $f:(X, \xi) \longrightarrow (Y, \theta)$ is a *T-model map* if $fT \times f$: $XT \times X \longrightarrow YT \times Y$ maps ξ into (but not necessarily onto) θ.
   (a) Show that the resulting category, T-mod, in Struct(**Set**) is fibre-complete. [Hint: the optimal lift ξ of $f_i: X \longrightarrow (Y_i, \theta_i)$ is the intersection of all $\theta_i(f_iT \times f_i)^{-1}$.]
   (b) If (X, ξ) is a T-model and if A is a subset of X define $A^- = \{x \in X$: there exists $\omega \in AT$ with $x = \langle \omega, (\mathrm{inc}_A)T \rangle\}$. A is *closed* if $A = A^-$. Show that $A \subset A^-$ and that $A \subset B$ implies $A^- \subset B^-$. Show that every intersection of closed sets is closed. [Warning: in general, A^- may be smaller than the intersection of all closed sets containing A.]
   (c) If (X, ξ) is a T-model say that a subset A of X is *open* if whenever $(\omega, x) \in \xi$ with $x \in A$ there exists $\omega_0 \in AT$ with $\langle \omega_0, (\mathrm{inc}_A)T \rangle = \omega$. Show that the open sets form a topology on X.

(d) Let $S = (\{0, 1\}, s)$ where $s = (\{0, 1\}T \times \{0\}) \cup (\{1\}T \times \{1\})$. Show that the following three conditions on T are equivalent: (i) the passage from T-models to topological spaces described in (c) is a homomorphism over **Set** from T-mod to topological spaces and continuous maps; (ii) given a pullback diagram as shown below with P

$$\begin{array}{ccc} P & \longrightarrow & A \\ \big\downarrow & & \big\downarrow{\scriptstyle i} \\ X & \longrightarrow & Y \end{array}$$

nonempty and with i injective, T of the diagram is again a pullback; (iii) for every T-model (X, ξ), the open sets of (X, ξ) are the admissible S-valued maps.

(e) Let $T = \beta$ and let ξ be the convergence relation of a topology. Show that "closed" and "open" have their usual meanings. Show that this embedding of topological spaces in T-mod is taut but not cotaut.

(f) Let XT be the set of filters on X with functorial action $\langle \mathscr{F}, fT \rangle = \{B \subset Y : Bf^{-1} \in \mathscr{F}\}$ (cf. exercise 4 of section 1.5). A *convergence structure* [Fischer '74] is a T-model (X, ξ) subject to the three axioms (i) $(\text{prin}(x), x) \in \xi$; (ii) if $(\mathscr{F}, x) \in \xi$ and if $\mathscr{G} \supset \mathscr{F}$ then $(\mathscr{G}, x) \in \xi$; (iii) if $(\mathscr{F}, x) \in \xi$ and if $(\mathscr{G}, x) \in \xi$ then $(\mathscr{F} \cap \mathscr{G}, x) \in \xi$. Show that the category of convergence structures is fibre complete with taut reflective embedding in T-mod; show that this embedding is not cotaut. The closed sets of (b) and the induced topology of (c) coincide with the usual notions in the theory of convergence structures. Verify that T satisfies the pullback condition of (d).

9. Let (T, η, μ) be an algebraic theory in **Set**. If $R : X \longrightarrow Y$ is a relation, $R \subset X \times Y$ with projections $p : R \longrightarrow X$, $q : R \longrightarrow Y$, define $RT : XT \longrightarrow YT$ to be the image of $(pT, qT) : RT \longrightarrow XT \times YT$. [Warning: T does not preserve composition of relations.]

(a) Let (X, ξ) be a T-model (as in exercise 8). Prove that the conditions (i) and (ii) below are equivalent.

$$\begin{array}{ccc} X & \xrightarrow{X\eta} & XT \\ & \searrow^{\text{id}_X} \;\; \leqslant & \big\downarrow{\scriptstyle \xi} \\ & & X \end{array} \qquad \begin{array}{ccc} XTT & \xrightarrow{\xi T} & XT \\ {\scriptstyle X\mu}\big\downarrow & \geqslant & \big\downarrow{\scriptstyle \xi} \\ XT & \xrightarrow[\xi]{} & X \end{array} \qquad \text{(i)}$$

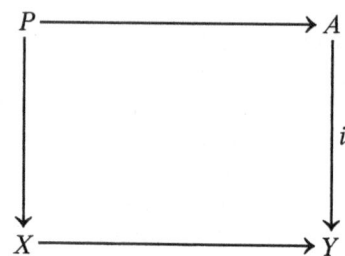

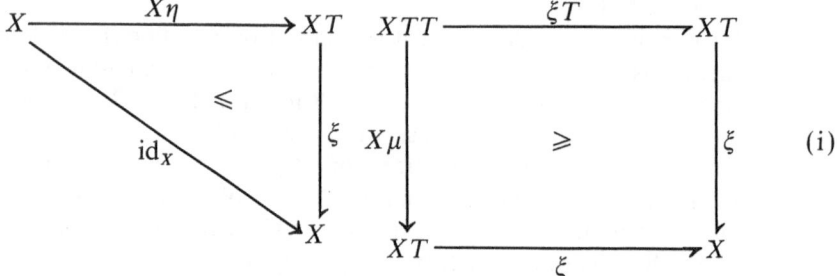

(Here, single-headed arrows are relations, double-headed arrows are functions considered as relations via their graph, and composition is composition of relations; the inequalities specify inclusions of relations—these diagrams do not commute in general). These axioms generalize the **T**-algebra axioms.

(ii) $\xi = (X\eta^{-1})^{-}$ (where $X\eta^{-1}: XT \longrightarrow X$ is the relational inverse of the graph of $X\eta$ and $(-)^{-}$ is the operator of exercise 8(b) of the T-model $(XT, X\mu) \times (X, \xi)$).

(b) A **T**-*model* is a T-model satisfying either of the equivalent conditions of (a). Prove that A^{-} is closed in any **T**-model. Verify that the full subcategory **T**-mod of all **T**-models in T-mod is fibre complete with taut reflective embedding.

(c) Prove *Barr's theorem* [Barr '70]: **β**-mod = topological spaces and continuous maps. (When **T** is the filter theory (see exercise 4 of section 1.5) it is not known at this writing what sort of "convergence structure" characterizes the **T**-models.)

(d) A **T**-model (X, ξ) is *compact* if for every ω in XT there exists x in X with (ω, x) in ξ; *Hausdorff* if whenever both (ω, x) and (ω, y) are in ξ then $x = y$; $T1$ if whenever $(\langle x, X\eta\rangle, y)$ is in ξ then $y = x$. Prove that these adjectives take on their usual meaning when $\mathbf{T} = \boldsymbol{\beta}$. Show that each of these properties is closed under products. Observe that the compact Haudorff **T**-models are precisely the **T**-algebras.

(e) Prove the generalized β-compactification theorem: every **T**-model has a **T**-algebra reflection. [Hint: the proof of 2.8 goes through.]

10. A *coalgebraic theory* in a category $\mathscr{K}$ is an algebraic theory $\mathbf{G} = (G, \varepsilon, \circ, \delta)$ in $\mathscr{K}^{\mathrm{op}}$. A **G**-*coalgebra* (in terms of $\mathscr{K}$) is just a **G**-algebra (in terms of $\mathscr{K}^{\mathrm{op}}$); the coalgebra axioms then look like

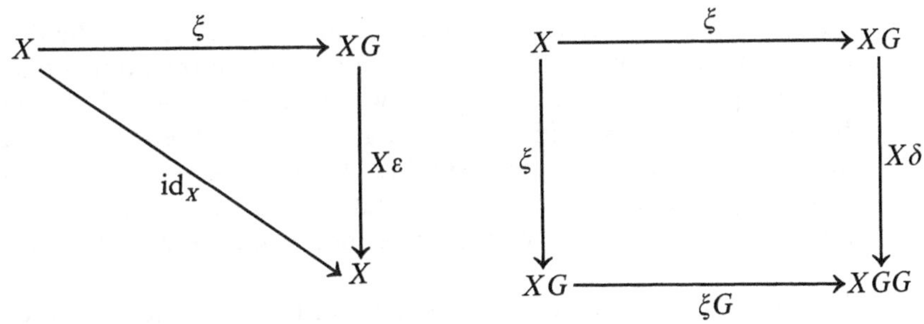

(a) Verify that the coalgebraic theories on a partially ordered category are interior operators, as discussed in 3.13.

(b) Let $\mathbf{T} = (T, \eta, \mu)$ be an algebraic theory in $\mathscr{K}$. Show that (G, ε, δ) is a coalgebraic theory in $\mathscr{K}^{\mathbf{T}}$ where $(K, \xi)G = (KT, K\mu)$, $(K, \xi)\varepsilon$: $KG \longrightarrow K = \xi, (K, \xi)\delta: KG \longrightarrow KGG = K\eta T$. (For many **T** in **Set**, the coalgebras of the induced cotheory in **T**-algebras can be identified with **Set**; see [Barr '69]).

(c) Let $\mathscr{K}$ be a category and let K be an object. The category $\mathscr{K}/K$ of *objects over* K is that category in Struct($\mathscr{K}$) with objects all pairs (A, f) with $f: A \longrightarrow K$ and admissible maps $\psi:(A, f) \longrightarrow (B, g)$ all $\psi: A \longrightarrow B$ satisfying $f = \psi.g$ (cf. Cat/$\mathscr{K}$ in exercise 5). Assume that $K \times A$ exists for all A. Define a coalgebraic theory in $\mathscr{K}$ by $AG = K \times A$, $A\varepsilon: K \times A \longrightarrow A = A$-projection, $A\delta: K \times A \longrightarrow K \times K \times A = \Delta \times \mathrm{id}_A$ where $\Delta: K \longrightarrow K \times K$ is the "diagonal map," i.e., is id_K when followed by either projection. Show that the category of **G**-coalgebras is isomorphic over $\mathscr{K}$ to $\mathscr{K}/K$.

11. (Cf. exercise 1.5.1.) Let $\mathscr{C}$ be in Struct(**Set**), let $s \in \mathscr{C}(2)$ where $2 = \{0, 1\}$ and assume that $(2, s)^X$ exists for all X (via the optimal lift of all projections 2^X —— $(2, s)$). Set XT_s to be the set of all subsets of X whose characteristic function is admissible from $(2, s)^X$ to $(2, s)$. Prove that T_s is a subtheory of the double power-set theory.

12. In this exercise we develop a "fibre-completion" for arbitrary $\mathscr{C}$ in Struct($\mathscr{K}$).

(a) For K in $\mathscr{K}$ consider all (L, t, a) with t in $\mathscr{C}(L)$, $a: L \longrightarrow K$ and say that $(L, t, a) \leqslant (L', t', a')$ if for all $f: K \longrightarrow (M, u)$ we have

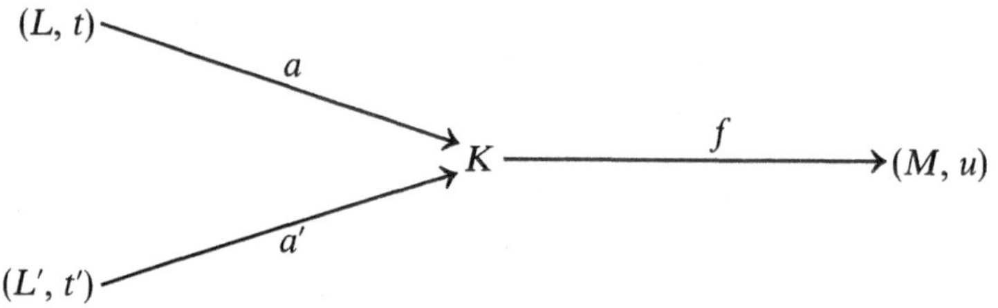

"if $a'f: t' \longrightarrow u$ then $af: t \longrightarrow u$." Define $\mathscr{C}^{\text{co-opt}}(K)$ to be the class of all $\leqslant$-antisymmetry classes $[L, t, a]$ and define the admissibility of f by $[L, t, af] \leqslant [L', t', a']$. Show that $H: \mathscr{C} \longrightarrow \mathscr{C}^{\text{opt}}$ defined by $(K, s)H = (K, [K, s, \mathrm{id}_K])$ is a taut and cotaut full replete subcategory over $\mathscr{K}$, that each morphism $f:(K, [L, t, a]) \longrightarrow K'$ has co-optimal lift $[L, t, af]$ and that H is an isomorphism if $\mathscr{C}$ already had this property that single morphisms admit co-optimal lifts.

(b) For A a subset of $\mathscr{C}(K)$ define $\mathscr{C}\sup(A)$ (if it exists) to be the co-optimal lift of the A-indexed family $\mathrm{id}_K:(K, a) \longrightarrow K$. A is a *$\mathscr{C}$-ideal on* K if (i) given $s \leqslant a$ with s in $\mathscr{C}(K)$ and a in A then s is in A and (ii) given $B \subset A$ such that $s = \mathscr{C}\sup(B)$ exists in $\mathscr{C}(K)$, then s is in A. Let $\mathscr{C}^{\text{id}}(K)$ be the class of $\mathscr{C}$-ideals on K and say that f: $(K, A) \longrightarrow (L, B)$ is admissible in $\mathscr{C}^{\text{id}}$ if for every s in A there exists t in B such that $f: s \longrightarrow t$ is admissible in $\mathscr{C}$. Assuming that $\mathscr{C}$ has the property that single morphisms admit co-optimal lifts (as in (a)), prove that $\mathscr{C}^{\text{id}}$ is fibre-complete, that $H: \mathscr{C} \longrightarrow \mathscr{C}^{\text{id}}$ defined by

$(K, s)H = (K, \{t : t \leqslant s\})$ is a taut and cotaut full replete subcategory over $\mathscr{X}$, and that H is an isomorphism if $\mathscr{C}$ was already fibre complete.

(c) What are the dual constructions to (a) and (b)? Look up the MacNeille "completion by cuts" of a partially ordered set [MacNeille '37] and investigate the possibility of using this construction instead of ideals in (b).

(d) Let H be as in (a) and let $J : \mathscr{C} \longrightarrow \mathscr{D}$ be a functor over $\mathscr{X}$ with $\mathscr{D}$ fibre complete. Define Γ by $\alpha\Gamma = \sigma$ in $\mathscr{D}(K)$ such that the family $a : (L, t)J \longrightarrow (K, \sigma)$ indexed by all (L, t, a) with $[L, t, a] = \alpha$ is co-optimal in $\mathscr{D}$. Show that Γ is a functor over $\mathscr{X}$ satisfying $H\Gamma = J$

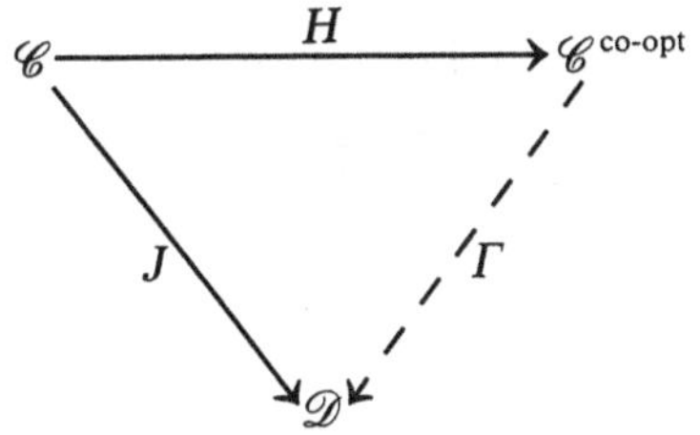

and "if Γ' is another functor over $\mathscr{X}$ with $H\Gamma' = J$ then $\Gamma \leqslant \Gamma'$." It is an open question how to characterize any of the fibre-completion constructions by universal properties involving functors over $\mathscr{X}$.

(e) Show that if $\mathscr{C}$ can be represented as a full replete subcategory of any fibre-complete category which is small over $\mathscr{X}$ then $(\mathscr{C}^{\text{co-opt}})^{\text{id}}$ is small over $\mathscr{X}$. [Hint: use (d).]

(f) Let $\mathscr{C}$ in Struct(**Set**) be the subcategory of **Set** having all sets as objects but with $f : X \longrightarrow Y$ admissible if and only if X and Y have the same cardinality. Show that $\mathscr{C}$ cannot be represented as a full replete subcategory of any fibre-complete category which is small over $\mathscr{X}$.

13. Let Fib(**Set**) be the full replete subcategory of Struct(**Set**) of all fibre complete categories which are small over **Set**. Show that Fib(**Set**) is cartesian closed (as defined in exercise 7 of section 2). [Hint: if $\mathscr{C}$, $\mathscr{D}$ are in Fib(**Set**) define $\mathscr{D}^{\mathscr{C}}(X)$ to be the set of all order-preserving maps ψ from $(\mathscr{C}(X), \leqslant)$ to $(\mathscr{D}(X), \leqslant)$, and say that $f : (X, \psi_1) \longrightarrow (Y, \psi_2)$ is admissible if whenever $f : s \longrightarrow t$ in $\mathscr{C}$, $f : s\psi_1 \longrightarrow t\psi_2$ in $\mathscr{D}$.]

14. (R. Paré.) Show that an arbitrary functor H admits a factorization $H =$

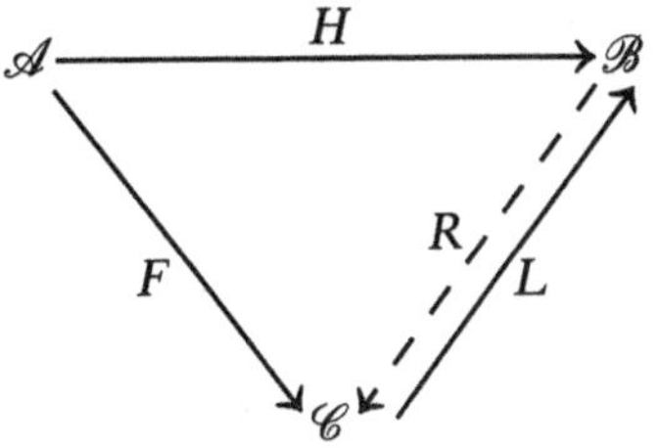

FL where F is a full subcategory and L has a right adjoint. [Hint: to define $\mathscr{C}$ adjoin to the coproduct category of $\mathscr{A}$ and $\mathscr{B}$ all morphisms $AH \longrightarrow B$, considered as $\mathscr{C}$-morphisms from A to B; F and R are injections and L is defined by $AL = AH$, $BL = B$.]

15. Let $\mathbf{T}$ be an algebraic theory in **Set** and fix a set I and an element p of IT. For $f, g: I \longrightarrow X$ say that $f \sim_p g$ if f and g agree on a support of p.
   (a) Prove that $\sim_p$ is an equivalence relation on X^I [Hint: use exercise 1.5.2.]
   (b) Set ${}^pX = X^I/{\sim_p}$. If (X, ξ) is a $\mathbf{T}$-algebra show that $f \sim_p g$ implies that $\langle p, f^\# \rangle = \langle p, g^\# \rangle$, so that $m(x) = \{[f] \in {}^pX : \langle p, f^\# \rangle = x\}$ is well-defined.

When $\mathbf{T} = \boldsymbol{\beta}$ and p has no finite supports, ${}^pX$ is the well-known *ultraproduct* construction. For (X, ξ) a $\boldsymbol{\beta}$-algebra, $m(x)$ is called the *monad of* x in nonstandard topology. The definition of $m(x)$ extends easily to $\mathbf{T}$-models as in exercise 9.

Chapter 3

Algebraic Theories in a Category

This chapter serves as an introduction to categorical universal algebra. Necessary and sufficient conditions for a functor $U:\mathscr{A} \longrightarrow \mathscr{K}$ to be a category of algebras are provided. Theories in $\mathscr{K}$ are interpreted as monoids in the category of endofunctors of $\mathscr{K}$. Epimorphic quotient theories characterize the abstract Birkhoff subcategories; when the base category is regular, the Birkhoff subcategory generated by a class $\mathscr{A}$ of algebras is the class of all quotients of subalgebras of products of elements of $\mathscr{A}$. As a generalization of topological algebra, for each fibre-complete category $\mathscr{C}$ over **Set** the category of all (X, s, ξ) with s a $\mathscr{C}$-structure and (X, ξ) a **T**-algebra in such a way that the **T**-operations are admissible in $\mathscr{C}$ is seen to be algebraic over $\mathscr{C}$. Given two algebraic theories in **Set**, the category of bialgebras—the two sorts of operation commute with each other—is studied; it is often itself algebraic. A general colimit theorem is applied to prove that many categories of algebras have small colimits.

1. Recognition Theorems

This section provides a number of useful theorems which stipulate necessary and sufficient conditions for an arbitrary (not necessarily faithful) functor $U:\mathscr{A} \longrightarrow \mathscr{K}$ to be (in fact in Struct($\mathscr{K}$) and) isomorphic in Struct($\mathscr{K}$) to $U^{\mathbf{T}}:\mathscr{K}^{\mathbf{T}} \longrightarrow \mathscr{K}$ for some algebraic theory **T** in $\mathscr{K}$. Special theorems for the case $\mathscr{K}$ = **Set** are presented and, in particular, we are able to prove two theorems (1.26 and 1.27) which fill in the gaps left in Chapter 1.

Definition 2.3.8 admits a mild generalization:

(1.1) An arbitrary functor $U:\mathscr{A} \longrightarrow \mathscr{K}$ is *algebraic* if $(\mathscr{A}, U)$ is in Struct($\mathscr{K}$) (see 2.3.3) and $(\mathscr{A}, U)$ is algebraic as in 2.3.8; in other words, U is algebraic if and only if there exists an algebraic theory **T** in $\mathscr{K}$ and an isomorphism $\Phi:\mathscr{A} \longrightarrow \mathscr{K}^{\mathbf{T}}$ over $\mathscr{K}$ as shown below:

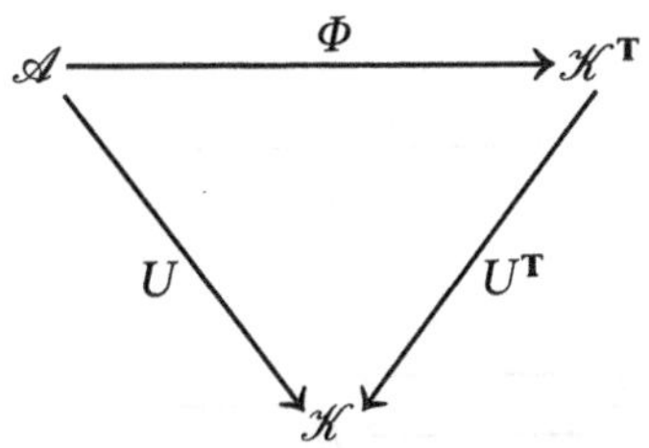

This section is devoted to characterizing algebraic functors. A necessary condition that U be algebraic is that U have a left adjoint (1.4.12). This will always be one of the sufficient conditions. This is not too unsatisfactory in view of the existence theorems of 2.24 and 2.29; since these theorems tell us little about the specific structure of the adjointness, it is important to complete the list of sufficient conditions with properties of U which are independent of adjointness. The first clue along these lines is the following commutative diagram associated with any algebra (X, ξ) over an algebraic theory (T, η, μ) in $\mathscr{K}$.

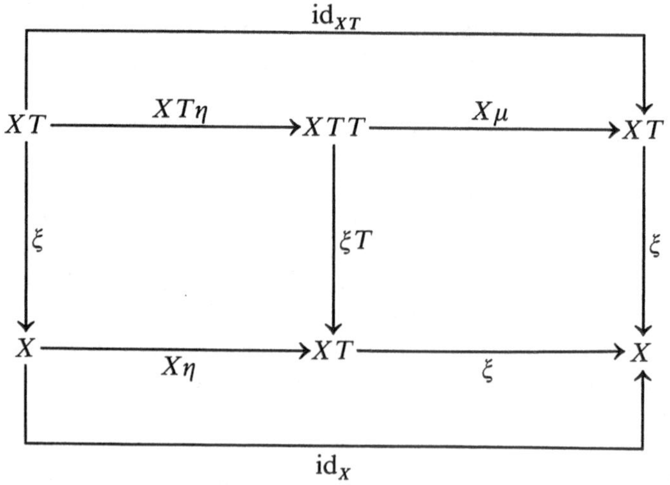

(1.2)

The obvious abstraction is a diagram of form

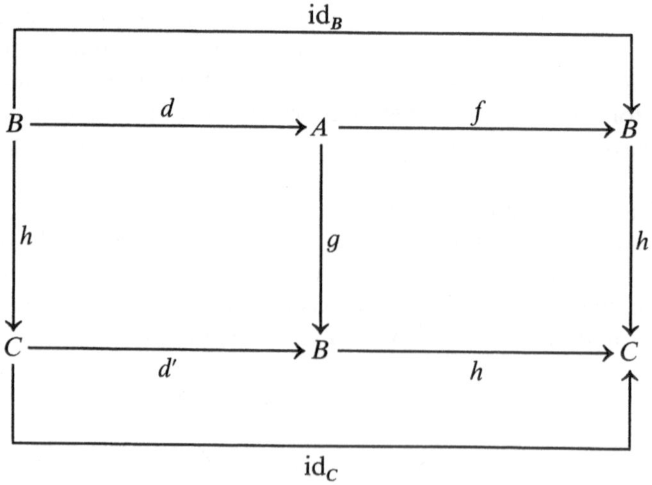

(1.3)

Given

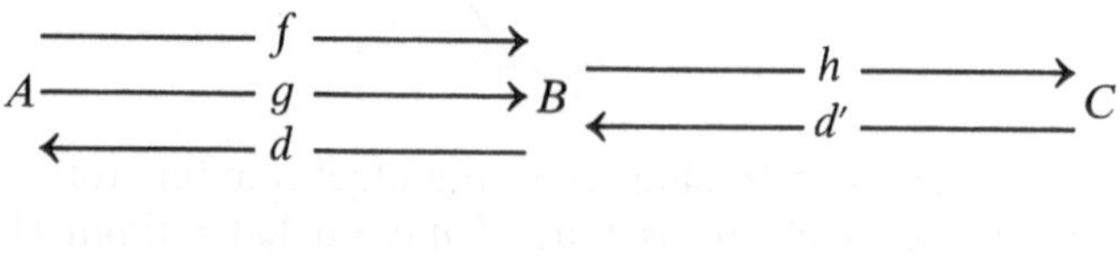

in a category $\mathscr{K}$, $(f, g, h; d, d')$ is a *contractible coequalizer* (also: (f, g, h) is a contractible coequalizer with (d, d') as *contraction*) if the diagram of 1.3 is commutative. A diagram

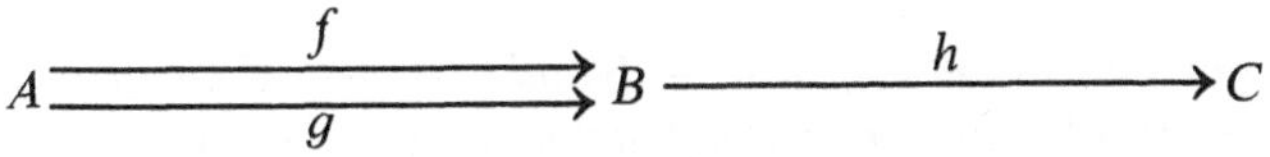

in a category $\mathscr{K}$ is an *absolute coequalizer* if for every functor $H:\mathscr{K} \longrightarrow \mathscr{L}$, $hH = \text{coeq}(fH, gH)$; since H may be the identity functor, every absolute coequalizer is a coequalizer.

1.4 Proposition. *Let $(f, g, h; d, d')$ be a contractible coequalizer in a category $\mathscr{K}$. Then (f, g, h) is an absolute coequalizer.*

Proof. Let $H:\mathscr{K} \longrightarrow \mathscr{L}$ be any functor. It is obvious that $(fH, gH, hH; dH, d'H)$ is again a contractible coequalizer in $\mathscr{L}$, so it suffices to prove that $h = \text{coeq}(f, g)$. Suppose $f.h' = g.h'$ as shown below. Then $h.d'.h' =$

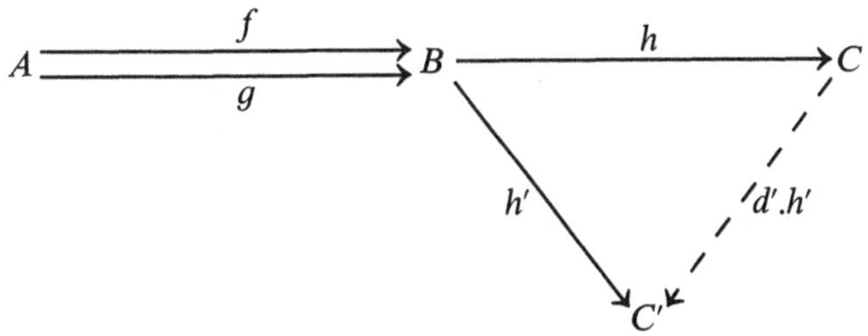

$d.g.h' = d.f.h' = h'$; the uniqueness of $d'.h'$ is secure since h is (split) epi. □

Paré [Paré '71, Proposition 5.3] proves that if (f, g, h) is an absolute coequalizer then there exist $n \geqslant 1$ and $d^0, d^1, \ldots, d^n$ as shown below

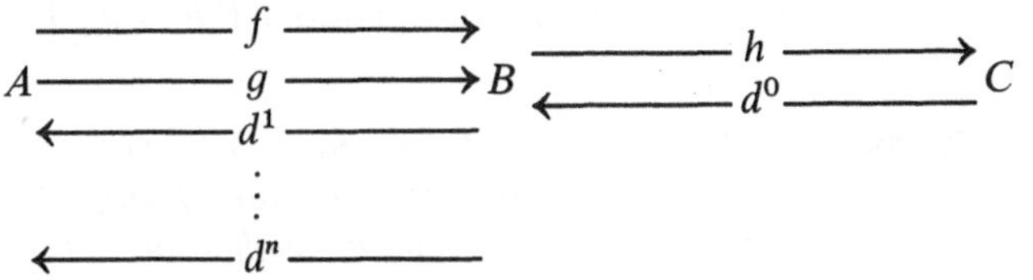

subject to a set of commutativity conditions which make it clear, as in the proof of 1.4, why (f, g, h) is an absolute coequalizer. It is not true that every absolute coequalizer is contractible [Paré '71, page 86].

It is easy to prove that if (f, g, h) is an absolute coequalizer then h is split epi (see exercise 1). The following is a sort of converse:

1.5 Proposition. *Let $h:B \longrightarrow C$ be split epi in $\mathscr{K}$ and let $d':C \longrightarrow B$ be any right inverse for h. Assume that the kernel pair $f, g:A \longrightarrow B$ of h exists. Then there exists $d:B \longrightarrow A$ such that $(f, g, h; d, d')$ is a contractible coequalizer.*

Proof. Consult diagram 1.3. The desired d is uniquely induced by the universal property of a kernel pair. □

As a consequence of 1.5, dividing out by an equivalence relation always gives rise to an absolute coequalizer diagram in **Set**.

1.6 Proposition. *Let $f, g: A \longrightarrow B$ and $h: B \longrightarrow C$ be given in a category $\mathscr{K}$, with $h = \text{coeq}(f, g)$. Then the following two conditions are equivalent:*

1. *There exist $d: B \longrightarrow A$ and $d': C \longrightarrow B$ such that $(f, g, h; d, d')$ is a contractible coequalizer.*
2. *There exists $d: B \longrightarrow A$ such that 1.7 below commutes*

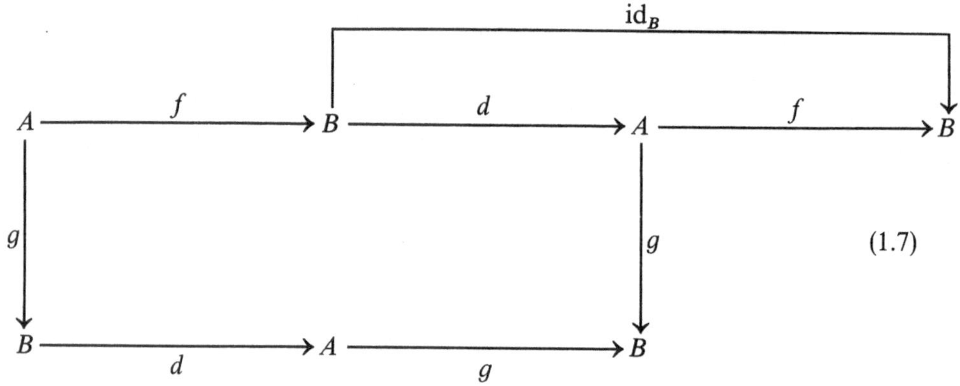

(1.7)

that is, $d.f = \text{id}_B$ and $f.d.g = g.d.g$.

Proof. If $(f, g, h; d, d')$ is a contractible coequalizer as in 1.3 then $f.d.g = f.h.d' = g.h.d' = g.d.g$. Conversely, suppose $d: B \longrightarrow A$ exists subject to 1.7. As $f.d.g = g.d.g$ and $h = \text{coeq}(f, g)$, there exists a unique $d': C \longrightarrow B$ such that $h.d' = d.g$; as h is epi and $h.d'.h = d.g.h = d.f.h = h$, $d'.h = \text{id}_C$. □

1.8 Definitions. *Let $U: \mathscr{A} \longrightarrow \mathscr{K}$ be a functor. U creates coequalizers of U-absolute pairs if whenever we are given a pair of maps $f, g: A \longrightarrow B$ in $\mathscr{A}$ and a morphism $h: BU \longrightarrow K$ in $\mathscr{K}$ such that (fU, gU, h) is an absolute coequalizer in $\mathscr{K}$, we may conclude that h has a unique lift $\bar{h}$ (that is, there exists a morphism $\bar{h}: B \longrightarrow \bar{K}$ in $\mathscr{A}$ such that $\bar{K}U = K$ and $\bar{h}U = h$, and if $h': B \longrightarrow K' \in \mathscr{A}$ is such that $K'U = K$ and $hU = h$ then $K' = \bar{K}$ and $h' = \bar{h}$); and then, moreover, $\bar{h} = \text{coeq}(f, g)$ in $\mathscr{A}$. Similarly, U creates coequalizers of U-contractible pairs if whenever we are given a pair of maps $f, g: A \longrightarrow B$ in $\mathscr{A}$ and a morphism $h: BU \longrightarrow K$ in $\mathscr{K}$ such that there exist d and d' with respect to which $(fU, gU, h; d, d')$ is a contractible coequalizer in $\mathscr{K}$, we may conclude that (just as before) h has a unique lift $\bar{h}$; and, moreover, $\bar{h} = \text{coeq}(f, g)$ in $\mathscr{A}$.*

We are now ready to prove the fundamental recognition theorem:

1.9 Characterization Theorem for Algebraic Functors (J. Beck). Let $U: \mathscr{A} \longrightarrow \mathscr{K}$ *be a functor which has a left adjoint (as defined in 2.2.19+). Then the following three conditions on U are equivalent:*

1. *U is algebraic (1.1);*
2. *U creates coequalizers of U-absolute pairs (1.8); and*
3. *U creates coequalizers of U-contractible pairs (1.8).*

Proof. 1 implies 2. There is no loss in generality in assuming that $U = U^{\mathbf{T}}: \mathscr{K}^{\mathbf{T}} \longrightarrow \mathscr{K}$ for some algebraic theory $\mathbf{T}$ in $\mathscr{K}$. Suppose that $f, g: (L, \theta) \longrightarrow (M, \gamma)$ in $\mathscr{K}^{\mathbf{T}}$ and $h: M \longrightarrow K$ in $\mathscr{K}$ are such that (f, g, h) is an absolute coequalizer in $\mathscr{K}$. In particular, $hT = \text{coeq}(fT, gT)$ in $\mathscr{K}$. Consulting the diagram below,

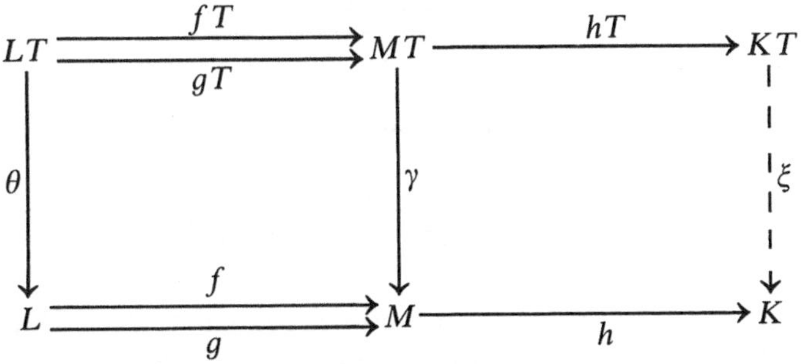

we have that $fT.\gamma.h = gT.\gamma.h$ thereby inducing a unique $\mathscr{K}$-morphism $\xi: KT \longrightarrow K$ such that $\gamma.h = hT.\xi$. It is useful to prove the following lemma, which is also an apprcpriately general statement of some ideas used previously (cf. 1.2.6, 2.1.56).

1.10 Lemma. *Let* $\mathbf{T}$ *be an algebraic theory in* $\mathscr{K}$ *and suppose given a commutative square*

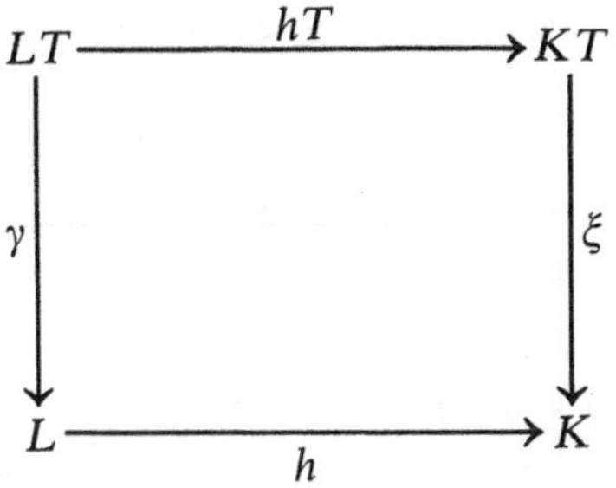

with (L, γ) *a* $\mathbf{T}$*-algebra and with* h, hT *and* hTT *epi in* $\mathscr{K}$. *Then* (K, ξ) *is a* $\mathbf{T}$*-algebra and* $h: (L, \gamma) \longrightarrow (K, \xi)$ *is co-optimal in* $\mathscr{K}^{\mathbf{T}}$.

Proof. We leave this as an exercise in diagram pasting with the following hints: to prove $K\eta.\xi = \text{id}_K$ prove that $h.K\eta.\xi = h$ and use the fact that h

is epi; to prove $K\mu.\xi = \xi T.\xi$, similarly use "hTT is epi"; the argument in 2.1.56 proves that h is co-optimal since hT is epi. □

Returning to the proof of 1.9, h, hT, and hTT are coequalizers and are epimorphisms in particular, so 1.10 completes the proof of "1 implies 2" in view of 2.3.21. "2 implies 3" is obvious from 1.4.

3 implies 1. By 2.2.18 there exists an adjointness of form $(\mathscr{A}, \mathscr{K}, U, F, \eta, \varepsilon)$. Let $\mathbf{T} = (T, \eta, \mu)$ be the induced algebraic theory in $\mathscr{K}$ (2.2.20) and let $\Phi: \mathscr{A} \longrightarrow \mathscr{K}^{\mathbf{T}}$ be the semantics comparison functor as in 2.2.21. We will show Φ is an isomorphism.

(*i*) *The fundamental observations.* Let (K, ξ) be an arbitrary $\mathbf{T}$-algebra. Consider the pair of $\mathscr{A}$-morphisms

$$KFUF \underset{\xi F}{\overset{KF\varepsilon}{\rightrightarrows}} KF$$

Then U of this pair is the pair of $\mathscr{K}$-morphisms

$$KTT \underset{\xi T}{\overset{K\mu}{\rightrightarrows}} KT$$

Since $(X\mu, \xi T, \xi; XT\eta, X\eta)$ is a contractible coequalizer (see 1.2) there exists, by hypothesis, a unique $\mathscr{A}$-morphism $\bar{\xi}: KF \longrightarrow \bar{K}$ such that $\bar{\xi}U = \xi$; and, moreover, $\bar{\xi} = \operatorname{coeq}(KF\varepsilon, \xi F)$ in $\mathscr{A}$. We observe further that because $K\eta.\bar{\xi}U = K\eta.\xi = \mathrm{id}_K$, $\bar{\xi} = (\mathrm{id}_K)^{\#} = \bar{K}\varepsilon$.

(*ii*) *Φ is bijective on objects.* Using the notations of (i), for arbitrary (K, ξ) we have $\bar{K}\Phi = (\bar{K}U, \bar{K}\varepsilon U) = (K, \xi)$ and this proves Φ is surjective on objects. If $A\Phi = (K, \xi) = A'\Phi$ then, by (i), $A\varepsilon: KF \longrightarrow A = \bar{K}\varepsilon = A'\varepsilon: KF \longrightarrow A'$, and $A = A'$ in particular.

(*iii*) *Φ and U are faithful.* Consulting the diagram 2.2.21, since $U^{\mathbf{T}}$ is faithful, Φ is faithful if and only if U is faithful. We choose to prove that U is faithful. Let $f, g: A \longrightarrow B$ in $\mathscr{A}$ be such that $fU = gU$. Because of the naturality squares:

$$\begin{array}{ccc} AUF & \xrightarrow{\;A\varepsilon\;} & A \\ {\scriptstyle fUF}\downdownarrows{\scriptstyle gUF} & & {\scriptstyle f}\downdownarrows{\scriptstyle g} \\ BUF & \xrightarrow[\;B\varepsilon\;]{} & B \end{array}$$

it suffices to observe that $A\varepsilon = \operatorname{coeq}(AUF\varepsilon, A\varepsilon UF)$ (as proved in (i) and (ii)) so that $A\varepsilon$ is epi in $\mathscr{A}$ in particular.

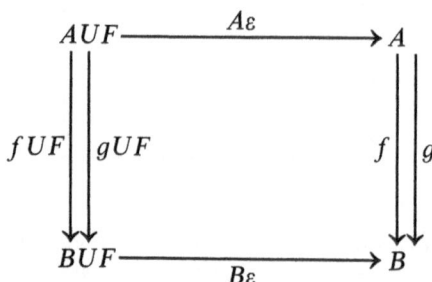

(*iv*) Φ *is full.* Let A, B be objects of $\mathscr{A}$ and let $f:AU \longrightarrow BU$ be a **T**-homomorphism $A\Phi \longrightarrow B\Phi$, that is, in the notation of (i), we have:

$$\begin{array}{ccc} KFU & \xrightarrow{\quad fFU \quad} & LFU \\ {\scriptstyle \bar{\xi}}\downarrow & & \downarrow{\scriptstyle \bar{\theta}} \\ K & \xrightarrow[\quad f \quad]{} & L \end{array}$$

where $A\Phi = (K, \xi)$ and $B\Phi = (L, \theta)$. Since $\bar{\xi}$ is a coequalizer in $\mathscr{A}$ and $\bar{\xi}U = \xi$ is a coequalizer in $\mathscr{K}$, it follows from 2.3.21 that $\bar{\xi}$ is co-optimal in $\mathscr{A}$. Therefore, $f:A \longrightarrow B$ is admissible in $\mathscr{A}$. The proof is complete. □

Let $(\mathscr{A}, U) \in$ Struct(**Set**) be a category of sets with structure. If $A \in \mathscr{A}$ and if R is an equivalence relation on the set AU with coordinate projections $p, q:R \longrightarrow AU$, say that R is a *congruence on* A if R lifts to $\bar{R}$ in $\mathscr{A}$ such that $\bar{R}U = R$ and $p, q:\bar{R} \longrightarrow A$ are admissible in $\mathscr{A}$. If $(\mathscr{A}, U)$ is algebraic, we would expect that the canonical projection $\theta:A \longrightarrow AU/R$ has a unique lift to $\mathscr{A}$ which is, moreover, co-optimal. On the other hand, the important coequalizer in the proof of the Beck theorem of 1.9 replaces (p, q) with the pair

$$ATT \underset{\xi T}{\overset{A\mu}{\rightrightarrows}} AT$$

which does not have the appearence of an equivalence relation (e.g., write down some specific examples for the semigroups theory of 1.4.17). We turn out attention now to a special restatement of 1.9 for **Set** which reconciles the disparity.

1.11 Separators. Consider the diagram scheme (2.1.18) Δ with four nodes and four edges as shown below:

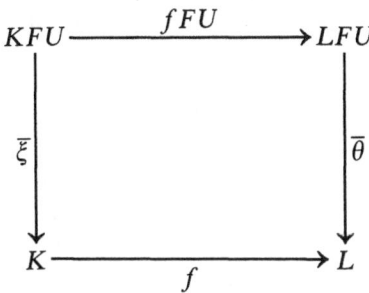

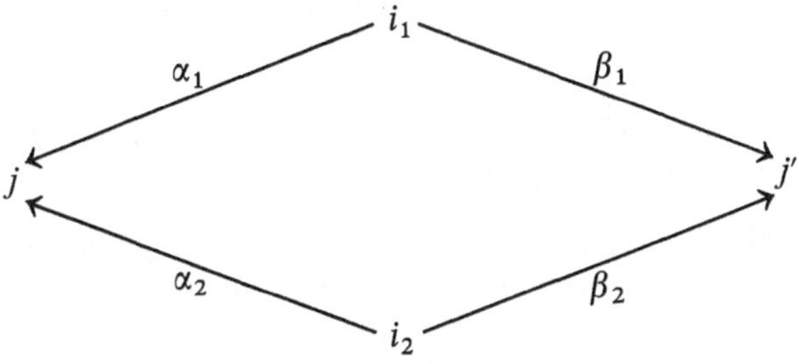

Let $\mathscr{K}$ be any category. A pair $f, g:K \longrightarrow L$ of morphisms in $\mathscr{K}$ induce

the Δ-diagram

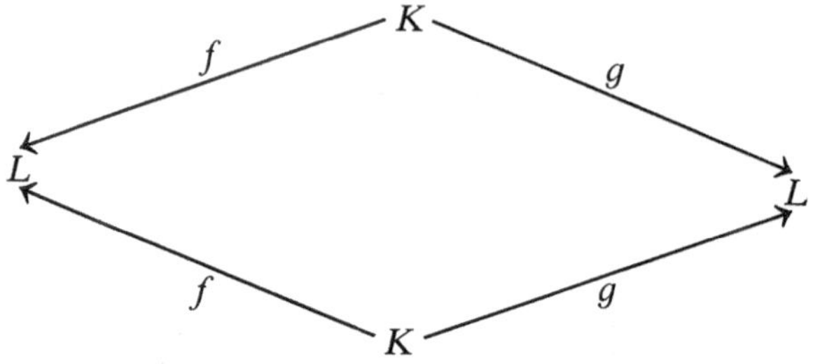

the limit of which, if it exists, is called the *separator of* f and g. Since $\{i_1, i_2\}$ is final, the separator of f and g may be thought of as an object S equipped with two K-valued morphisms $p, q: S \longrightarrow K$ such that $p.f = q.f$ and $p.g = q.g$ and universal with this property, that is, given $p', q': S' \longrightarrow K$ such that $p'.f = q'.f$ and $p'.g = q'.g$ then there exists unique $\psi: S' \longrightarrow S$ such that $\psi.p = p'$ and $\psi.q = q'$.

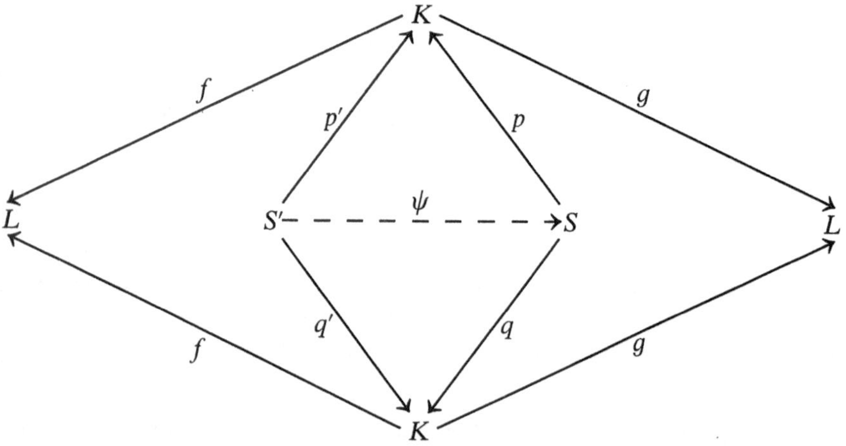

Given a morphism $f: K \longrightarrow L$, the kernel pair of f is the same thing as the separator of f and f. $\mathscr{K}$ *has separators* if every pair f, g have a separator; in particular, if $\mathscr{K}$ has separators then $\mathscr{K}$ has kernel pairs. **Set** has separators. S is the subset of $K \times K$ of all pairs (x, y) such that $xf = yf$ and $xg = yg$, and p, q are the coordinate projections.

1.12 Definition. *Let* $(\mathscr{A}, U)$ *be a category of sets with structure. Then* U *creates quotients of congruences if whenever we are given an object* A *of* $\mathscr{A}$ *and a congruence* R *on* A *(as defined in 1.11–), we may conclude that the canonical projection* $\theta: A \longrightarrow AU/R$ *has a unique lift* $\overline{\theta}: A \longrightarrow A/R$ *(in the sense of 1.8); and then, moreover,* $\overline{\theta}$ *is co-optimal in* $\mathscr{A}$.

1.13 Characterization Theorem for Algebraic Structure on Sets. Let $(\mathscr{A}, U) \in \text{Struct}(\textbf{Set})$ *be a category of sets with structure. Then* $U: \mathscr{A} \longrightarrow \textbf{Set}$ *is algebraic if and only if the following three conditions are satisfied:*

1. *U has a left adjoint;*
2. *$\mathscr{A}$ has separators (1.11); and*
3. *U creates quotients of congruences (1.12).*

Proof of necessity. This follows immediately from 1.4.12, 2.1.11, 2.1.14, 2.1.22, 1.5, and 1.9.

Proof of sufficiency. By 1.9, it is enough to prove that U creates coequalizers of U-contractible pairs. To this end, suppose given a pair $f, g: A \longrightarrow B$ of morphisms in $\mathscr{A}$ and functions $h: BU \longrightarrow K, d: BU \longrightarrow AU$ and $d': K \longrightarrow BU$ such that $(fU, gU, h; d, d')$ is a contractible coequalizer in **Set**.

Claim: It is sufficient to find a congruence R on B, with coordinate projections $p, q: R \longrightarrow BU$, such that $h = \text{coeq}(p, q)$ in **Set**.

For suppose such R exists. Since the canonical projection $\theta: BU \longrightarrow BU/R$ is also a coequalizer of (p, q), there exists an isomorphism Γ: $BU/R \longrightarrow K$ such that $\theta.\Gamma = h$ as shown below:

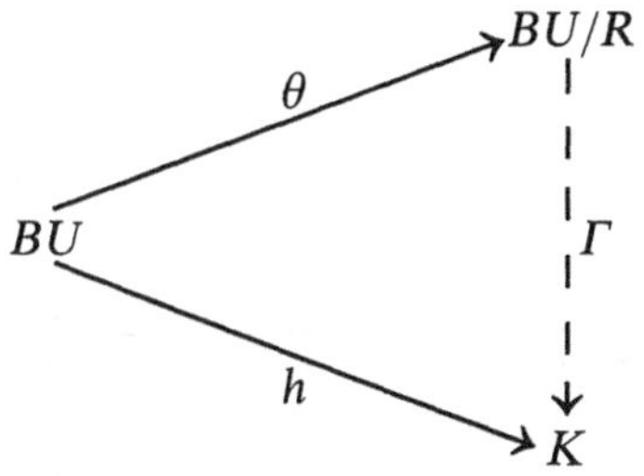

As U creates quotients of congruences, θ lifts uniquely to $\bar{\theta}: B \longrightarrow B/R$ and this lift is co-optimal. As $(\mathscr{A}, U) \in \text{Struct}(\textbf{Set})$, there exists unique $\bar{K}$ such that $\bar{K}U = K$ and $\Gamma: B/R \longrightarrow \bar{K}$ and $\Gamma^{-1}: \bar{K} \longrightarrow B/R$ are admissible. It is clear that $h: B \longrightarrow \bar{K}$ is then the unique admissible lift of h: $BU \longrightarrow K$ and is co-optimal (being the composition of a co-optimal with an isomorphism) so is the coequalizer by 2.3.21. This proves the claim.

Consider the function $d.gU: BU \longrightarrow BU$. Let R be the kernel pair of $d.gU$ in **Set**; that is $R = \{(x, y): x, y \in BU \text{ and } \langle x, d.gU \rangle = \langle y, d.gU \rangle\}$. Then R is an equivalence relation on BU. We will show that R is the desired congruence. Let $p, q: R \longrightarrow BU$ be the coordinate projections. We show first that $h = \text{coeq}(p, q)$. Consult diagram 1.3 and the diagram shown on the following page. We have $p.h = p.d.fU.h = p.d.gU.h = q.d.gU.h = q.h$. As $fU.d.gU = gU.d.gU$ (see 1.6) there exists a unique function $\alpha: AU \longrightarrow R$ such that $\alpha.p = fU$ and $\alpha.q = gU$. Therefore, if $h': BU \longrightarrow K'$ is a function such that $p.h' = q.h'$, then also $fU.h' = gU.h'$ inducing the desired unique $\beta: K \longrightarrow K'$ such that $h.\beta = h'$.

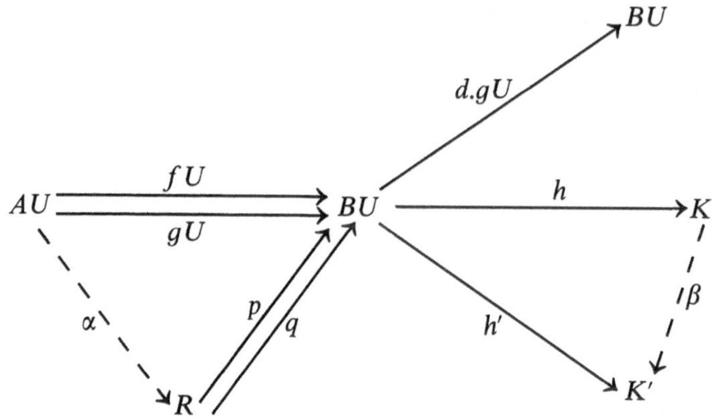

To complete the proof we must demonstrate that R is a congruence. To prove this we will have to appeal once again to the fact that U creates quotients of congruences. We will work around the following diagram:

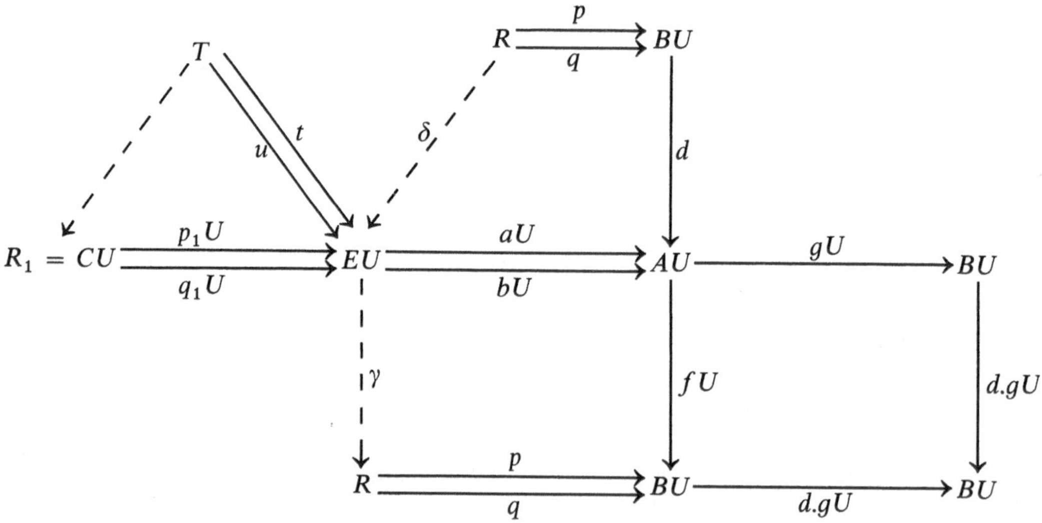

As kernel pairs are separators, the kernel pair (E, a, b) of g exists in $\mathscr{A}$. Because $aU.fU.d.gU = aU.gU.d.gU = bU.gU.d.gU = bU.fU.d.gU$, there exists a unique function $\gamma: EU \longrightarrow R$ such that $\gamma.p = aU.fU$ and $\gamma.q = bU.fU$. Define R_1 to be the separator of $aU.fU$ and $bU.fU$ in **Set**. Clearly, R_1 is an equivalence relation on EU. In fact, R_1 is a congruence (proof: the separator (C, p_1, q_1) of $a.f$ and $b.f$ exists in $\mathscr{A}$ by hypothesis; since U preserves limits (2.2.22) (CU, p_1U, p_2U) is a separator of $aU.fU$ and $bU.fU$, so is isomorphic to R_1 with its coordinate projections; but then, since $(\mathscr{A}, U)$ is in Struct(**Set**), we may transport the isomorphism $CU \cong R_1$ or, even better, assume that (C, p_1, p_2) was over R_1 to begin with).

Claim: It is sufficient to prove that $\gamma = \text{coeq}(p_1 U, q_1 U)$ in **Set**

For then, arguing as in the previous claim, γ admits a co-optimal lift $\gamma: E \longrightarrow \bar{R}$; but then, since $\gamma.p = a.f$ and $\gamma.q = b.f$ are admissible, so are p and q. This supports the second claim.

Our method will be to show that $(p_1 U, q_1 U)$ is the kernel pair of γ and that γ is split epi (and hence a coequalizer) so that, by 2.1.53, $\gamma = \text{coeq}(p_1 U, q_1 U)$.

$(p_1 U, q_1 U)$ is the kernel pair of γ: since $p_1 U$ and $q_1 U$ compose equally with $aU.fU$, $p_1 U.\gamma$ and $q_1 U.\gamma$ compose equally with p. Similarly, $p_1 U.\gamma$ and $q_1 U.\gamma$ compose equally with q. Therefore (cf. 1.20) $p_1 U.\gamma = q_1 U.\gamma$. Suppose that $t.\gamma = u.\gamma$ (see the diagram above). Then, clearly, t and u compose equally with $aU.fU$ and $bU.fU$, inducing the desired unique map $T \longrightarrow CU$.

γ is split epi: as (EU, aU, bU) is the kernel pair of gU (2.2.22) and $p.d.gU = q.d.gU$ there exists a unique function $\delta: R \longrightarrow EU$ such that $\delta.aU = p.d$ and $\delta.bU = q.d$. Since $d.fU = \text{id}_{BU}$ (1.3) $\delta.\gamma.p = p$ and $\delta.\gamma.q = q$ so $\delta.\gamma = \text{id}_R$. The proof is complete. □

1.14 Compact Groups. Let $\mathscr{A}$ be the category of compact (Hausdorff) topological groups and continuous homomorphisms. We illustrate the use of 1.13 by proving that the underlying set functor $U: \mathscr{A} \longrightarrow$ **Set** is algebraic. We use the general adjoint functor theorem of 2.2.24 to see that U has a left adjoint. $\mathscr{A}$ has products (the Tychanoff topology) and equalizers (the subset on which two continuous homomorphisms agree is a closed subgroup) which U preserves. In particular, note that $\mathscr{A}$ has separators. The solution set condition is not hard: given a function $f: n \longrightarrow GU$, f factors

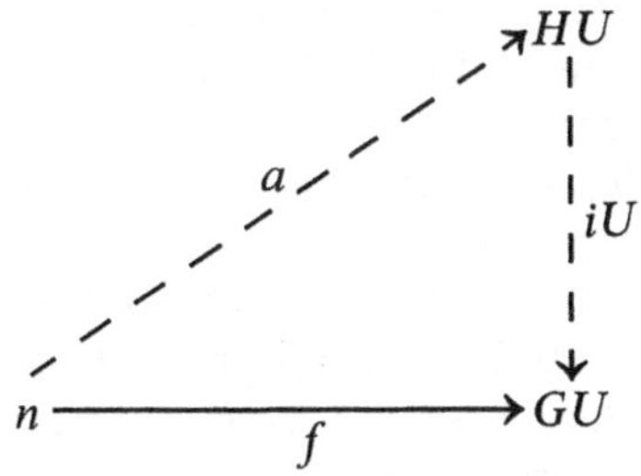

where H is the closure of the subgroup generated by the image of f. It is clear that the cardinal of H is bounded as a function of the cardinal of n (see exercise 2) and so H can be chosen to range over a small set up to isomorphism; the crucial point in this argument is that $H \in \mathscr{A}$, that is "the closure of a subgroup is a subgroup." This completes the proof that every set freely generates a compact group. The structure of such groups is another question which the general adjoint functor theorem does not answer.

That U creates quotients of congruences follows from standard facts about compact groups. A congruence R on G is the same thing as an equivalence relation which is a closed subgroup of $G \times G$ (an injective continuous map from a compact space to a Hausdorff space is a homeomorphism into). The quotient group G/R provided with the quotient topology is compact (a continuous image of G) and Hausdorff (R is closed). Should this argument be unconvincing, notice that $G/R = G/H$ where H is the closed normal subgroup $\{g \in G : (g, e) \in R\}$ and consult [Hewitt and Ross '63, Theorem 5.21].

We will prove later in 6.5 that compact groups are algebraic because compact spaces (1.5.24) and groups (1.4.15) are.

1.15 Definition. *Say that the functor* $U : \mathscr{A} \longrightarrow \mathscr{K}$ *creates limits if whenever we are given a (not necessarily small) diagram* (Δ, D) *in* $\mathscr{A}$ *and a limit* $\psi : L \longrightarrow DU$ *for* (Δ, DU) *in* $\mathscr{K}$ *we may conclude that there exists a unique lift* $(\bar{L}, \bar{\psi})$ *of* (L, ψ) (that is $\bar{L}U = L$, $(\bar{\psi}_i : \bar{L} \longrightarrow D_i)U = \psi_i : L \longrightarrow D_iU$ and given (A, Γ) with $AU = L$ and $\Gamma_i U = \psi_i$ then $A = \bar{L}$ and $\Gamma_i = \bar{\psi}_i$); *and, moreover,* $(\bar{L}, \bar{\psi})$ *is a limit of* D *in* $\mathscr{A}$.

Clearly "creates limits" implies "constructs limits" as in 2.3.17.

1.16 Proposition. *If* $U : \mathscr{A} \longrightarrow \mathscr{K}$ *creates limits then the following statements are true:*

1. $(\mathscr{A}, U) \in \mathrm{Struct}(\mathscr{K})$ *(2.3.3) and is discretely ordered (2.3.9);*
2. *If* $\mathscr{K}$ *is locally small then* $\mathscr{A}$ *is locally small;*
3. *If* $\mathscr{K}$ *is well powered (2.2.29(2)) and has kernel pairs then* $\mathscr{A}$ *is well powered; and*
4. U *reflects isomorphisms, that is if* $f : A \longrightarrow B$ *in* $\mathscr{A}$ *is such that* $fU : AU \longrightarrow BU$ *is an isomorphism in* $\mathscr{K}$*, then* f *was an isomorphism in* $\mathscr{A}$.

Proof. (1). If $f, g : A \longrightarrow B$ with $fU = gU$ then $\mathrm{id}_{AU} = \mathrm{eq}(fU, gU)$ so that id_A, being the unique lift of id_{AU}, is the equalizer of f and g. This proves that U is faithful. If $h : AU \longrightarrow K$ is an isomorphism in $\mathscr{K}$ we may choose to regard it as a unary product so that h^{-1} has a unique lift $h^{-1}: A' \longrightarrow A$ which is again a unary product, that is, an isomorphism. It is clear by now that U reflects isomorphisms and that the concrete category $(\mathscr{A}, U)$ is discretely ordered and so is a full representative subcategory $\Phi : \mathscr{A} \longrightarrow \mathscr{C}$ of some $\mathscr{C}$ in $\mathrm{Struct}(\mathscr{K})$ (2.3.5). But Φ is onto on objects, for if $C \in \mathscr{C}$ there exists $A \in \mathscr{A}$ and an isomorphism $f : A\Phi \longrightarrow C$; there exists (as above) an isomorphism $f : A \longrightarrow A'$ over f in $\mathscr{A}$; since $f^{-1}.f : C \longrightarrow A'\Phi$ and $\mathscr{C}$ is discretely ordered, $A'\Phi = C$.

(2). This is true because U is faithful.

(3). Fix $A \in \mathscr{A}$. If $i : S \longrightarrow A$ is mono so is $iU : SU \longrightarrow AU$ because

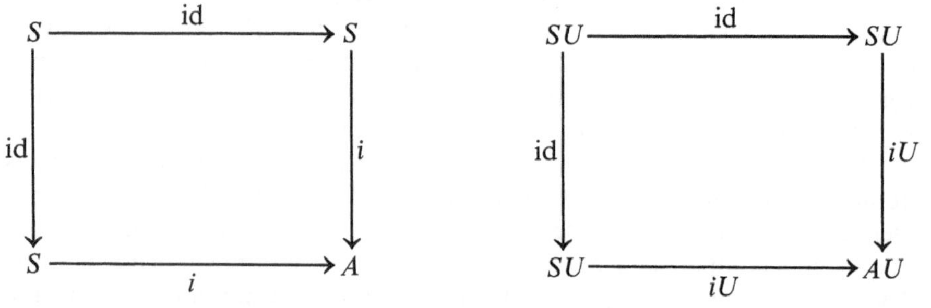

the rightmost square above is a pullback whenever the leftmost square is (consider the kernel pair of iU). It is now clear that the passage from i to iU is a well-defined function from monosubobjects of A into monosubobjects of AU. It suffices to prove that this function is injective. To this end, let $i: S \longrightarrow A$ and $i': S' \longrightarrow A$ be monos and suppose there exists an isomorphism $f: SU \longrightarrow S'U$ of subobjects as

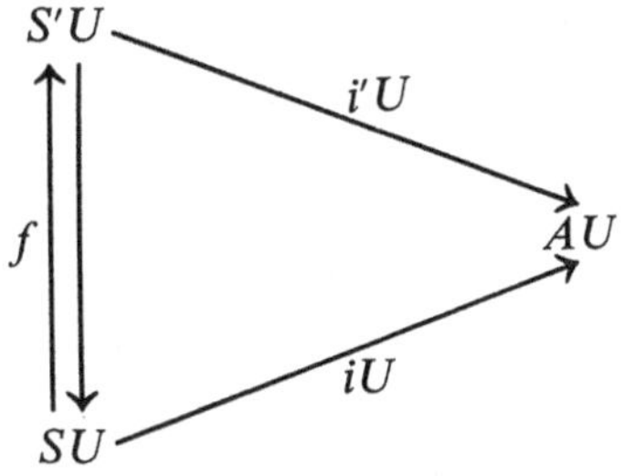

shown above. Then the rightmost square shown below is a pullback diagram,

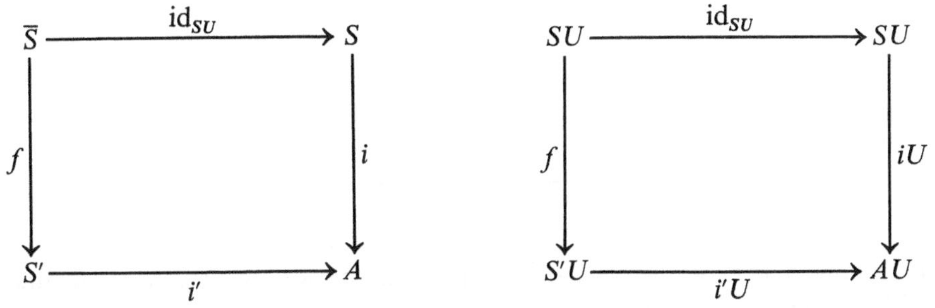

so must lift to a pullback diagram in $\mathscr{A}$ as shown on the left, above. As id_{SU} and f are isomorphisms in $\mathscr{A}$, $(\mathrm{id}_{SU})^{-1}.f$ and $f^{-1}.\mathrm{id}_{SU}$ are the desired $\mathscr{A}$-morphisms showing that (S, i) and (S', i') represent the same subobject of A. □

1.17 Beck Functors. A functor $U: \mathscr{A} \longrightarrow \mathscr{K}$ is a *Beck functor* if U creates limits and if U creates coequalizers of U-absolute pairs.

The proof of the following theorem is an easy exercise:

1.18 Proposition. *The product (2.3.7) of any family of Beck functors in* Struct($\mathscr{K}$) *is again a Beck functor.* □

1.19 Proposition. *Every algebraic functor is Beck.*

Proof. Let $U: \mathscr{A} \longrightarrow \mathscr{K}$ be algebraic. By 1.9, it suffices to show that U creates limits. We may assume without loss of generality that $U = U^{\mathbf{T}}$ for some algebraic theory $\mathbf{T}$ in $\mathscr{K}$. Let (Δ, D) be a diagram in $\mathscr{A} = \mathscr{K}^{\mathbf{T}}$ and write $D_i = (K_i, \xi_i)$. Let $\psi_i: L \longrightarrow K_i$ be a limit of DU in $\mathscr{K}$. For $\alpha \in \Delta(i, j)$

we have

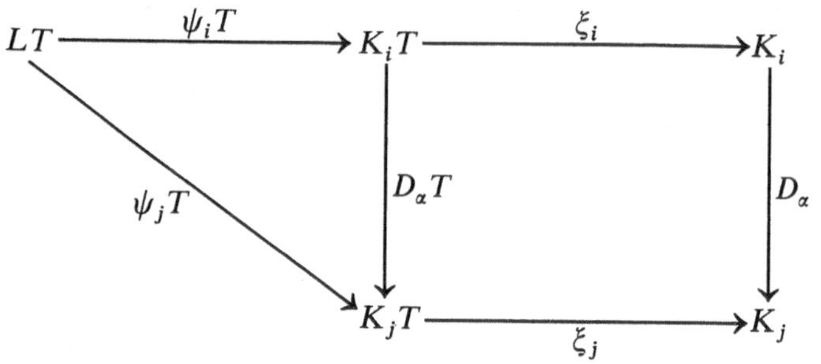

which proves that $(\psi_i)^\# = \psi_i T.\xi_i : LT \longrightarrow K_i$ is a lower bound of (Δ, DU), and there exists a unique $\xi : LT \longrightarrow L$ such that

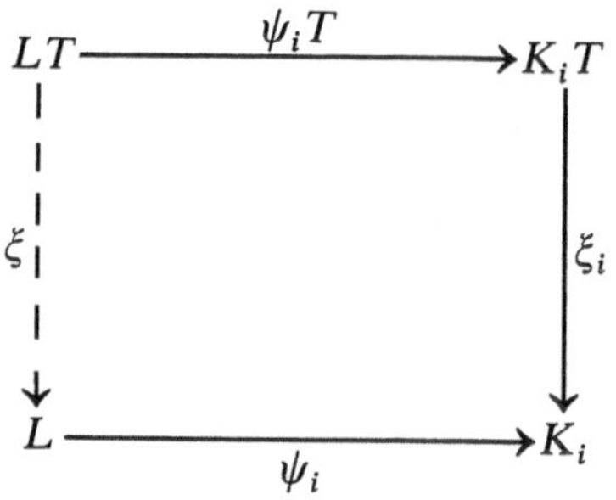

It is now worth our while to make explicit the following general lemma which includes 2.1.15 as a special case:

(1.20) If $\psi : L \longrightarrow D$ is a limit of (Δ, D) in $\mathscr{K}$ then the family $(\psi_i : L \longrightarrow D_i)$ is (collectively) mono in the sense that given any pair of maps $t, u : T \longrightarrow L$ such that $t.\psi_i = u.\psi_i$ for all i, $t = u$. The proof is trivial: the common value $t.\psi_i = u.\psi_i : T \longrightarrow D_i$ is a lower bound of (Δ, D) and t, u are both the unique induced map.

The remaining details of 1.19 are too similar to 1.4.27 and 2.1.11 to bear repeating. □

The following result is easily pieced together from the adjoint functor theorems 2.2.24, 2.2.29, and 1.9, 1.16, and 1.19. Notice, also, that if $U : \mathscr{A} \longrightarrow \mathscr{K}$ creates limits then $\mathscr{A}$ has and U preserves whatever limits $\mathscr{K}$ has.

1.21 Proposition. *Let $U : \mathscr{A} \longrightarrow \mathscr{K}$ be a functor and assume that $\mathscr{K}$ is locally small and has small limits. The following statements are true:*

1. *U is algebraic if and only if U is Beck and U satisfies the solution set condition at K for every object K of $\mathscr{K}$.*

2. *If $\mathscr{A}$ has a cogenerator and if either $\mathscr{A}$ or $\mathscr{K}$ is well powered, U is algebraic if and only if U is Beck.* □

Using 1.13 instead of 1.9, 1.21 takes the following form:

1.22 Proposition. *Let $U:\mathscr{A} \longrightarrow$ **Set** be a set-valued functor. The following statements are true:*

1. *U is algebraic if and only if U creates limits, U creates quotients of congruences, and U satisfies the solution set condition at K for every set K.*
2. *If $\mathscr{A}$ has a cogenerator, U is algebraic if and only if U creates limits and U creates quotients of congruences.* □

1.23 Compact Abelian Groups. Let $\mathscr{A}$ be the category of compact (Hausdorff) abelian groups and continuous homomorphisms and let $U:\mathscr{A} \longrightarrow$ **Set** be the underlying set functor. Then U is algebraic. The circle group S^1 is a cogenerator in $\mathscr{A}$ [Hewitt and Ross '63, Theorem 22.17]. The proof that U creates limits will be left as an easy exercise. The proof that U creates quotients of congruences is essentially the same as in 1.14. While the circle group is, in fact, a cogenerator in the category of locally compact abelian groups the forgetful set-valued functor V from this category is not algebraic. By 1.16 and 1.19 it is enough to observe that V does not reflect isomorphisms. For example, let G be any locally compact abelian group which is not discrete and let H be the same group with the discrete topology; then the identity function $f:H \longrightarrow G$ is a continuous homomorphism of locally compact abelian groups which is not an isomorphism.

The next theorem is just right to patch up the gap we left in 1.5.45.

1.24 Characterization Theorem for Algebraic Structure on Sets. Let $(\mathscr{A}, U) \in$ Struct(**Set**) *be a category of sets with structure. Then* $U:\mathscr{A} \longrightarrow$ **Set** *is algebraic if and only if the following three conditions are satisfied:*

1. *U is tractable (1.5.44);*
2. *U creates limits (1.15); and*
3. *U creates quotients of congruences (1.12).*

Proof of necessity. This follows immediately from 1.5.5 and 1.22.

Proof of sufficiency. Fix a set n. By 1.22, it is enough to show that U satisfies the solution set condition at n. Since U is tractable at n, the class nT of all natural transformations from U^n to U is a small set. Let $n\eta:n \longrightarrow nT$ map i to the ith projection $p_i: U^n \longrightarrow U$. We will show that nT lifts to an $\mathscr{A}$-object $\overline{nT}$ (i.e., $(\overline{nT})U = nT$) in such a way that $\{(\overline{nT}, n\eta)\}$ is a one-element solution set. The crucial observation is:

(1.25) Define a diagram (Δ, D) in $\mathscr{A}$ as follows. The nodes of Δ are pairs (A, f) with A in $\mathscr{A}$ and $f:n \longrightarrow AU$ in **Set** and an edge $\alpha:(A, f) \longrightarrow (A', f')$ is an $\mathscr{A}$-morphism $\alpha: A \longrightarrow A'$ such that $f.\alpha U = f'$; $D_{(A,f)} = A$ and $D_\alpha = \alpha$. Define $\psi_{(A,f)}:nT \longrightarrow AU$ by $\langle \omega, \psi_{(A,f)} \rangle = \langle f, A\omega: AU^n \longrightarrow AU \rangle$. Then (nT, ψ) is the limit of DU in **Set**.

Let us prove 1.25. To see that (nT, ψ) is a lower bound of DU, we use

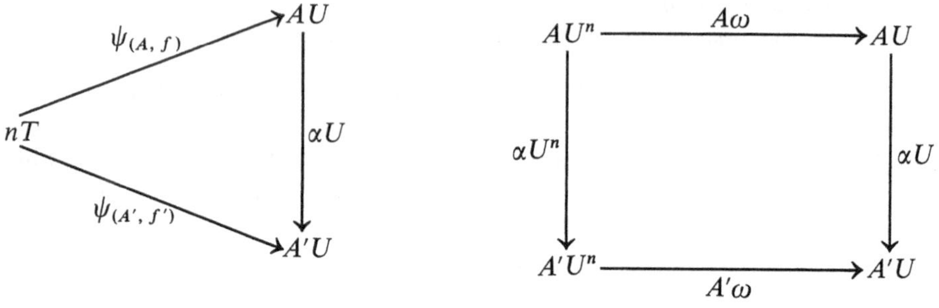

the structure of nT. Let $\omega \in nT$ and suppose that $\alpha: A \longrightarrow A'$ is an edge of D; then $\langle \omega, \psi_{(A,f)}.\alpha U\rangle = \langle f, A\omega.\alpha U\rangle = \langle f, (\alpha U)^n.A'\omega\rangle = \langle f.\alpha U, A'\omega\rangle = \langle f', A'\omega\rangle = \langle \omega, \psi_{(A,f)}\rangle$. Now let (L', ψ') be another lower bound of DU. Define $\Gamma: L' \longrightarrow nT$ by

$$AU^n \xrightarrow{A(x\Gamma)} AU$$
$$f \longmapsto \langle x, \psi'_{(A,f)}\rangle$$

To prove that $x\Gamma \in nT$ for all $x \in L'$, let $\alpha: A \longrightarrow A'$ in $\mathscr{A}$; (it will be helpful to glance at the triangle and square above). For all $f: n \longrightarrow AU$ we have $\langle f, A(x\Gamma).\alpha U\rangle = \langle x, \psi'_{(A,f)}.\alpha U\rangle = \langle x, \psi'_{(A',f.\alpha U)}\rangle = \langle f.\alpha U, A'(x\Gamma)\rangle = \langle f, \alpha U^n.A'(x\Gamma)\rangle$. Since $\langle x, \Gamma.\psi_{(A,f)}\rangle = \langle f, A(x\Gamma)\rangle = \langle x, \psi'_{(A,f)}\rangle$ we have

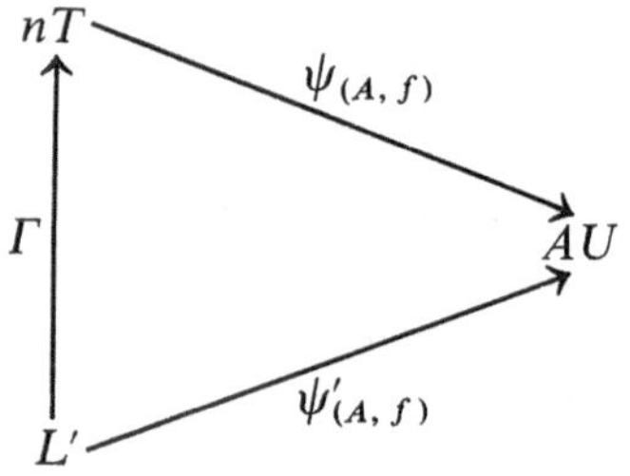

Suppose also that $\Lambda: L' \longrightarrow nT$ satisfies $\Lambda.\psi_{(A,f)} = \psi'_{(A,f)}$. Then for all A in $\mathscr{A}$ and $f: n \longrightarrow AU$ we have $\langle f, A(x\Lambda)\rangle = \langle x\Lambda, \psi_{(A,f)}\rangle = \langle x, \psi'_{(A,f)}\rangle$ which proves $\Lambda = \Gamma$. The proof of 1.25 is complete. To complete the main proof, since U creates limits there exists unique $\overline{nT}$ in $\mathscr{A}$ such that $(\overline{nT})U = nT$ and $\psi_{(A,f)}: \overline{nT} \longrightarrow A$ is admissible in $\mathscr{A}$ for all (A, f). Since

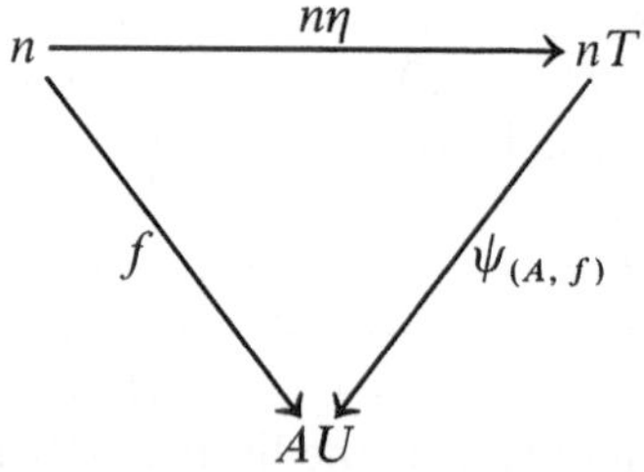

is commutative for every (A, f), the proof is complete. $\square$

1.26 Theorem (Same as 1.5.45). *Let (Ω, E) be an equational presentation as in 1.5.34. If (Ω, E) is tractable (as in 1.5.44), the underlying set functor $U:(\Omega, E)\text{-alg} \longrightarrow$* **Set** *is algebraic.*

Proof. We use 1.24. It suffices to show that U creates limits and that U creates quotients of congruences. Let (Δ, D) be a diagram of (Ω, E)-algebras and write $D_i = (X_i, \delta_i)$. Assume that (L, ψ) is a limit of DU in **Set**. Let $\omega \in \Omega_n$. If $\alpha \in \Delta(i, j)$ we have

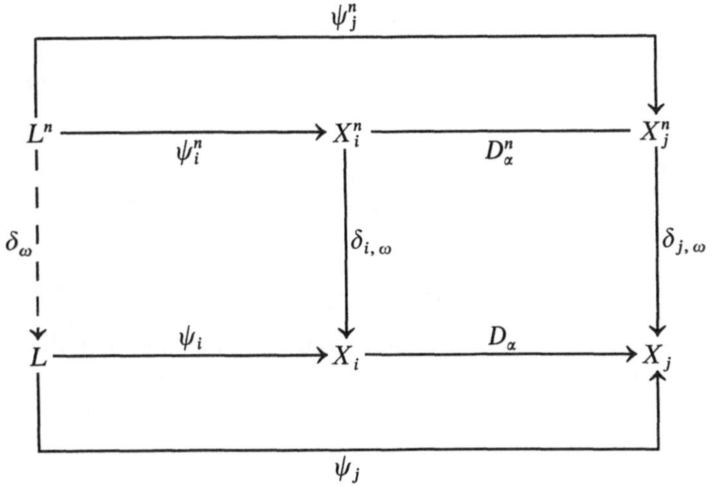

which shows that $(\psi_i^n.\delta_{i,\omega}).D_\alpha = \psi_j^n.\delta_{j,\omega}$, i.e., that $(L^n, \psi^n.\delta_{-,\omega})$ is a lower bound of DU. It follows that there exists unique $\delta_\omega: L^n \longrightarrow L$ such that $\delta_\omega.\psi_i = \psi_i^n.\delta_{i,\omega}$ for all i; or, in other words, that there exists a unique Ω-algebra structure δ on L such that $\psi_i:(L, \delta) \longrightarrow (X_i, \delta_i)$ is an Ω-homomorphism for all i. If $p, q: U^m \longrightarrow U$ with $\{p, q\} \in E$ then for all i we have $(L, \delta)p.\psi_i = \psi_i^n.(X_i, \delta_i)p = \psi_i^n.(X_i, \delta_i)q = (L, \delta)q.\psi_i$. It follows from 1.20 that $(L, \delta)p = (L, \delta)q$, and (L, δ) satisfies E. To prove that $((L, \delta), \psi)$ is a limit for D it suffices to show (2.3.17) that $\psi_i:(L, \delta) \longrightarrow (X_i, \delta_i)$ is an optimal family. Let $f:(X, \gamma) \longrightarrow L$ be such that $f.\psi_i:(X, \gamma) \longrightarrow (X_i, \delta_i)$ is an Ω-homomorphism for all i. Then in the diagram below, all

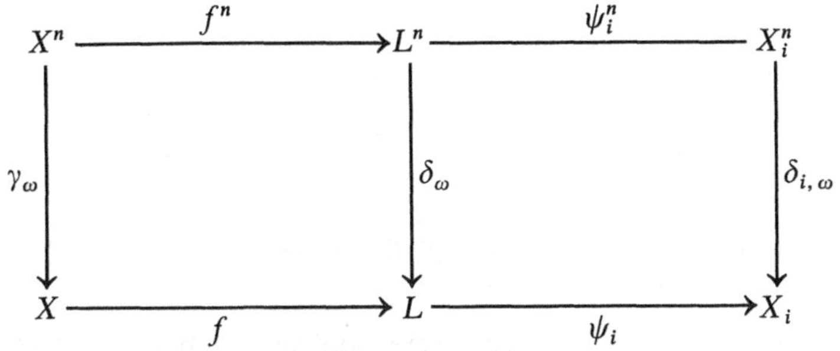

commutes except possibly the leftmost square; now use 1.20.

Let R be a congruence on the (Ω, E)-algebra (X, δ) with coordinate projections $p, q : R \longrightarrow X$. By definition (1.11−) there exists an (Ω, E)-structure (R, γ) such that $p, q : (R, \gamma) \longrightarrow (X, \delta)$ are Ω-homomorphisms. For each $\omega \in \Omega_n$, define $\overline{\delta}_\omega : (X/R)^n \longrightarrow X/R$ by $(x_i\theta)\overline{\delta}_\omega = (x_i)\delta_\omega\theta$. If $(x_i, y_i) \in R^n$ then $(x_i)\delta_\omega = (x_i, y_i)p^n\delta_\omega = (x_i, y_i)\gamma_\omega p$ and $(y_i)\delta_\omega = (x_i, y_i)\gamma_\omega q$

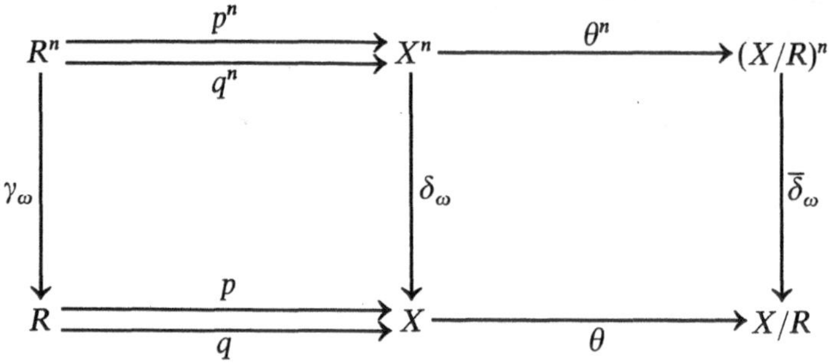

similarly, so that $((x_i)\delta_\omega, (y_i)\delta_\omega) = (x_i, y_i)\gamma_\omega \in R$, which proves that $\overline{\delta}_\omega$ is well defined. As $\theta^n : X^n \longrightarrow (X/R)^n$ is onto, $\overline{\delta}$ is the unique Ω-algebra structure on X/R making θ an Ω-homomorphism. The facts that $(X/R, \overline{\delta})$ satisfies E and that $\theta : (X, \delta) \longrightarrow (X/R, \overline{\delta})$ is co-optimal both follow from the fact that θ^n is onto for every set n, as is clear from the following two diagrams:

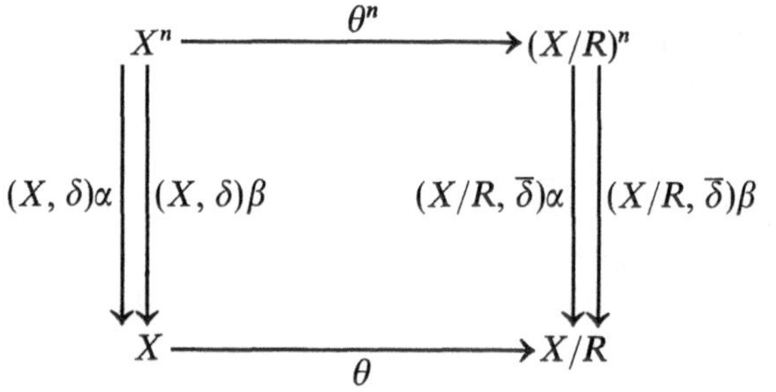

$(\alpha, \beta \in E)$

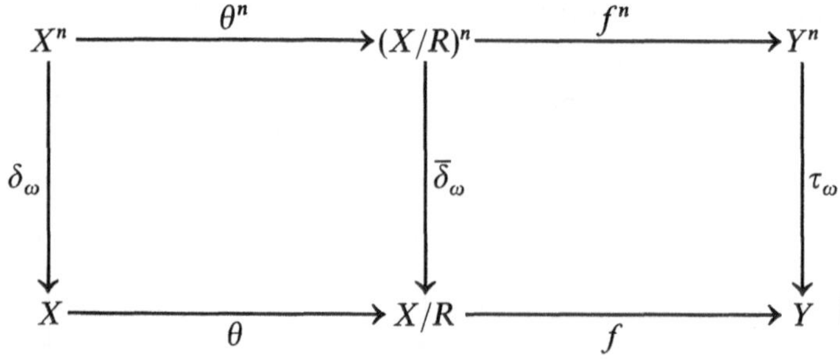

□

Let $\mathscr{A}$ be the category of complete Boolean algebras of 1.5.48 and let $U : \mathscr{A} \longrightarrow$ **Set** be the underlying set functor. We proved in 1.5.48 that U is not algebraic. On the other hand, the proof of 1.26 shows that an equational presentation, tractable or not, always has an underlying set functor (Ω, E)-*alg* $\longrightarrow$ **Set** which creates limits and which creates quotients of congruences. The problem with complete Boolean algebras is tractability.

1.27 Theorem. *Let* (Ω, E) *be an equational presentation as in 1.5.34 and assume that* (Ω, E) *is bounded, i.e., there exists a cardinal* n_0 *such that for all* $n \geqslant n_0$, Ω_n *is empty. Then the underlying set functor* $U:(\Omega, E)$-*alg* $\longrightarrow$ **Set** *is algebraic.*

Proof. As just mentioned above, the proof of 1.26 shows that U creates limits and that U creates quotients of congruences. By 1.22 (1) it suffices to prove that for an arbitrary set S, U satisfies the solution set condition at S. Given an (Ω, E)-algebra (X, δ) and a function $f:S \longrightarrow X$ let A be the intersection of all Ω-subalgebras of (X, δ) containing $f(S)$. It is obvious that A is itself an Ω-subalgebra. Moreover, if $\{p, q\} \in E$ is an n-ary equation then the naturality squares

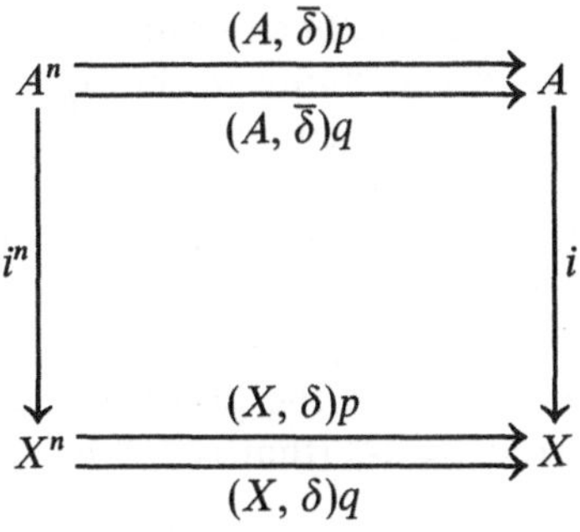

(where $\overline{\delta}$ is the subalgebra structure on A and $i:A \longrightarrow X$ is the inclusion map) prove that $(A, \overline{\delta})$ is an (Ω, E)-algebra, since i is a monomorphism. We clearly have a factorization

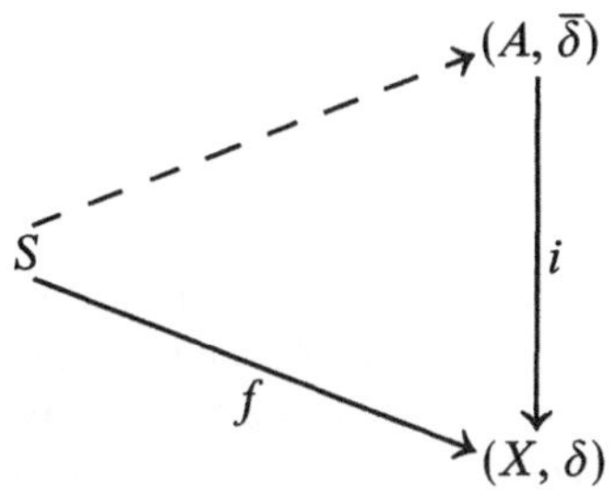

By the condition that (Ω, E) is bounded, $\overline{\delta}$ ranges over a small set once A is fixed. To complete the proof, then, it is enough to show that there exists a cardinal number α, depending only on Ω and S, such that the cardinality of A is at most α.

The remainder of the proof assumes some familiarity with ordinal and cardinal arithmetic (see the primer on set theory at the end of this section), but we have tried to spell things out so that the inexperienced reader can still grasp the flavor of the proof. Let us mention that, for cardinals β and γ, $\beta + \gamma$ is the cardinality of the disjoint union of the sets γ and β and, similarly, $\beta \times \gamma$ and β^γ denote (again somewhat ambiguously) the cardinalities of the cartesian product $\beta \times \gamma$ and the set of all functions from γ to β, respectively. By the axiom of choice, if β is an infinite cardinal then $\beta = \beta \times \beta$. Recall that, if β is an ordinal (in particular, if β is a cardinal) then "$\gamma < \beta$" and "$\gamma \in \beta$" are the same statement and such γ is itself an ordinal. If $(\gamma_i : i \in I)$ is a family in β then $\mathrm{Sup}(\gamma_i)$ is the least element of the set of upper bounds in β of (γ_i), this being the least element 0 of β if I is empty and being the ordinal β in case no such upper bounds exist.

Let us turn to the proof. α is constructed as follows. Let α_1 be any infinite cardinal $>$ card(S) and $>$ card(Ω_n) for all n. This is possible because (Ω, E) is bounded. Let α_2 be any infinite cardinal $>$ every n for which Ω_n is nonempty. This is possible, again because (Ω, E) is bounded. Define $\alpha = \alpha_1^{\alpha_2}$. We will show that card$(A) \leqslant \alpha$. An outline of our approach is as follows:

Step 1. We construct a transfinite tower A_β of subsets of X, starting with $A_0 = f(S)$, defining each succeeding stage by applying the Ω-operations to the preceding ones.

Step 2. We prove that A_α is an Ω-subalgebra containing $f(S)$ and conclude, therefore, that A_α contains A.

Step 3. We prove that card$(A_\alpha) \leqslant \alpha$.

If we carry out this program then the proof is complete. The reader may want to consider the case $n_0 = \aleph_0$ (that is, (Ω, E) is finitary) for additional intuition.

Step 1. Define a subset A_β of X for each $\beta < \alpha$ by

$$A_0 = f(S)$$
$$A_\beta = \bigcup(A_\gamma : \gamma < \beta) \cup \left(\bigcup_{\omega \in \Omega_n} \delta_\omega((\bigcup(A_\gamma : \gamma < \beta))^n)\right)$$

for $\beta > 0$.

Define $A_\alpha = \bigcup_{\beta < \alpha} A_\beta$.

Step 2. A_α is a subalgebra of (X, δ) containing $f(S)$ as follows: $f(S) = A_0 \subset A_\alpha$. To show that A_α is a subalgebra, let $\omega \in \Omega_n$ and let $(a_i : i \in n) \in (A_\alpha)^n$. We must show that $\delta_\omega(a_i) \in A_\alpha$. For each $i \in n$ there exists $\beta(i) < \alpha$ with $a_i \in A_{\beta(i)}$. Define $\bar{\alpha} = \mathrm{Sup}(\beta(i) : i \in n)$. Since $n \leqslant \alpha_2 = \alpha_2 \times \alpha_2$, $\alpha^n = (\alpha_1^{\alpha_2})^n = \alpha_1^{(\alpha_2 \times n)} \leqslant \alpha_1^{(\alpha_2 \times \alpha_2)} = \alpha_1^{\alpha_2} = \alpha$ and there exists a surjection $\psi : \alpha \longrightarrow \alpha^n$. For $i \in n$ let B_i denote $\{\psi_\beta(i) : \beta < \beta(i)\} \subset \alpha$. Since card$(B_i) \leqslant$ card$\{\beta : \beta < \beta(i)\} = \beta(i) < \alpha$, $B_i \neq \alpha$ and there exists $(\gamma_i : i \in n) \in \alpha^n$ with $\gamma_i \notin B_i$ for all i. By the definition of ψ, there exists $\tilde{\beta} < \alpha$ with $(\psi_{\tilde{\beta}}(i)) = (\gamma_i)$. There cannot exist $i \in n$ with $\beta(i) > \tilde{\beta}$ since then $\gamma_i = \psi_{\tilde{\beta}}(i) \in B_i$ which contradicts the definition of

γ_i. Therefore $\beta(i) \leqslant \tilde{\beta}$ for all i and $\bar{\alpha} = \mathrm{Sup}(\beta(i)) \leqslant \tilde{\beta} < \alpha$. Therefore $\delta_\omega(a_i) \in A_{\bar{\alpha}} \subset A_\alpha$ as desired.

Step 3. $\mathrm{card}(A_\alpha) \leqslant \alpha$ as follows. We have $\mathrm{card}(A_0) \leqslant \mathrm{card}(S) < \alpha_1 \leqslant \alpha$. Now let $0 < \beta < \alpha$ and assume that $\mathrm{card}(A_\gamma) \leqslant \alpha$ for all $\gamma < \beta$. Then $\mathrm{card}(\bigcup(A_\gamma : \gamma < \beta)) \leqslant \beta \times \alpha \leqslant \alpha \times \alpha = \alpha$. Therefore, for $\omega \in \Omega_n$,

$$\mathrm{card}(\delta_\omega((\bigcup(A_\gamma : \gamma < \beta))^n)) \leqslant \mathrm{card}((\bigcup(A_\gamma : \gamma < \beta))^n) \leqslant (\beta \times \alpha)^n$$
$$\leqslant (\alpha \times \alpha)^n = \alpha^n \leqslant \alpha$$

(the last statement was observed in step 2). Therefore,

$$\mathrm{card}(A_\beta) \leqslant \alpha + \mathrm{card}(\bigcup(\Omega_n \times \alpha : \Omega_n \neq \varnothing))$$
$$\leqslant \alpha + \mathrm{card}(\bigcup(\alpha_1 \times \alpha : \Omega_n \neq \varnothing))$$
$$\leqslant \alpha + (\alpha_2 \times \alpha_1 \times \alpha) \leqslant \alpha + \alpha = \alpha.$$

By transfinite induction, $\mathrm{card}(A_\beta) < \alpha$ for all $\beta < \alpha$. Then $\mathrm{card}(A_\alpha) = \mathrm{card}(\bigcup(A_\beta : \beta < \alpha)) \leqslant \beta \times \alpha = \alpha$. $\quad\square$

1.28 Proposition. *Given the commutative diagram of functors the follow-*

ing statements are true:

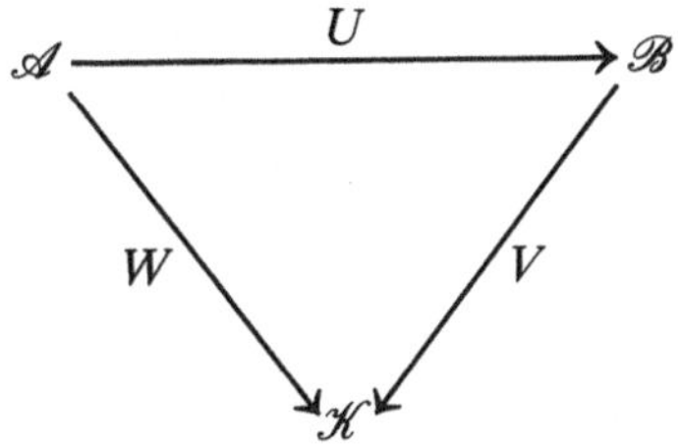

1. *If W creates coequalizers of W-absolute pairs and if V creates coequalizers of V-absolute pairs then U creates coequalizers of U-absolute pairs.*

2. *If W and V create limits and if V preserves limits then U creates limits.*

Proof. 1. Suppose $f, g : A_1 \longrightarrow A_2$ in $\mathscr{A}$ and $h : A_2 U \longrightarrow B \in \mathscr{B}$ are such that (fU, gU, h) is an absolute coequalizer. We must show that there exists a unique lift $\bar{h} : A_2 \longrightarrow A$ in $\mathscr{A}$ with $\bar{h}U = h$ and that moreover $\bar{h} = \mathrm{coeq}(f, g)$ in $\mathscr{A}$. Applying V and observing that (obviously) any functorial image of an absolute coequalizer is an absolute coequalizer, (fW, gW, hV) is an absolute coequalizer in $\mathscr{K}$.

$$\mathscr{A} \qquad A_1 \underset{g}{\overset{f}{\rightrightarrows}} A_2 \;\text{-} \text{-} \text{-} \text{-} \overset{\bar{h}}{\text{-} \text{-} \text{-} \text{-}}\!\!\rightarrow A$$

$$\mathscr{K} \qquad A_1 W \underset{gW}{\overset{fW}{\rightrightarrows}} A_2 W \overset{hV}{\longrightarrow} BV$$

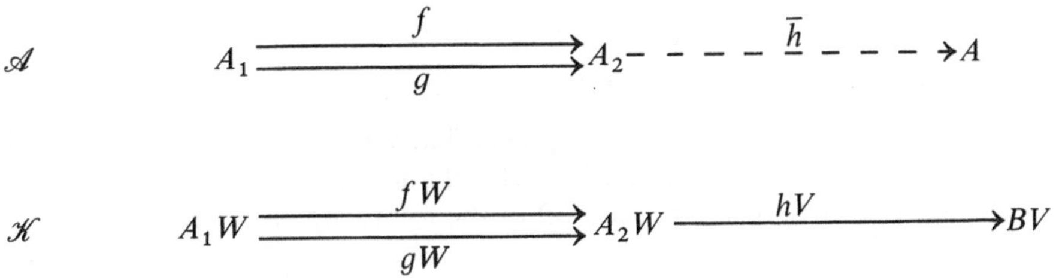

By the hypothesis on W there exists a unique lift $\bar{h}: A_2 \longrightarrow A$ with $\bar{h}W = hV$; and, moreover, $\bar{h} = \text{coeq}(f, g)$ in $\mathscr{A}$. Given an $\mathscr{A}$-morphism $h': A_2 \longrightarrow A'$, $h'U = h$ if and only if $h'UV = hV$ (as V creates coequalizers of V-absolute pairs) if and only if $h'W = hV$. Since $h'W = hV$ has unique solution $h' = \bar{h}$, $h'U = h$ has unique solution $\bar{h}$ as desired.

2. Let (Δ, D) be a diagram in $\mathscr{A}$ and let $\psi_i: B \longrightarrow D_i U$ be a limit of (Δ, DU) in $\mathscr{B}$. We must show that there exists a unique lift $\bar{\psi}_i: A \longrightarrow D_i$ in $\mathscr{A}$ with $\bar{\psi}_i U = \psi_i$ and that moreover $(A, \bar{\psi})$ is a limit of (Δ, D) in $\mathscr{A}$. Applying V, and using the hypothesis that V preserves limits we have that $\psi_i V: BV \longrightarrow D_i W$ is a limit of (Δ, DW) in $\mathscr{K}$. The remaining details are based on the same principles as the corresponding parts of the proof of (1). □

The following result is a major theorem in universal algebra:

1.29 Sandwich Theorem. *Assume given a commutative diagram of*

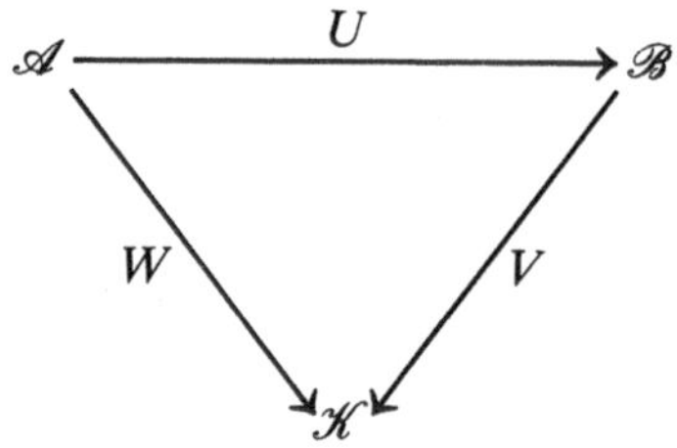

functors and assume that V is algebraic and that $\mathscr{A}$ has coequalizers. Then the following statements are true:

1. *U has a left adjoint if and only if W does.*
2. *If W is algebraic, so is U.*

Proof. Since V has a left adjoint, if U has a left adjoint then W has a left adjoint by 2.2.30. In view of 1.9 and 1.28 it suffices to prove that if W has a left adjoint then so does U. We may, and do, assume without loss of generality that $\mathscr{B} = \mathscr{K}^{\mathbf{T}}$ and $V = U^{\mathbf{T}}$ for an appropriate algebraic theory $\mathbf{T}$ in $\mathscr{K}$. Let $(\mathscr{A}, \mathscr{K}, W, F', \eta', \varepsilon')$ be an adjointness (2.2.18) and fix $(K, \xi) \in \mathscr{B}$. We must find (A, f) free over (K, ξ) with respect to U. Our method of proof will not involve the adjoint functor theorems.

(1.30) The fundamental observations: $KF'U$ has the form $(KF'W, \theta_K: KF'WT \longrightarrow KF'W)$. Define $\mathscr{A}$-morphisms $a, b: KTF' \longrightarrow KF'$ as follows:

$$a = KTF' \xrightarrow{K\eta' TF'} KF'WTF' \xrightarrow{\theta_K F'} KF'WF' \xrightarrow{KF'\varepsilon'} KF'$$
$$b = \xi F': KTF' \longrightarrow KF'$$

Let $p: KF' \longrightarrow A$ be any $\mathscr{A}$-morphism. Then $K\eta'.pW: K \longrightarrow AW$ is a $\mathbf{T}$-homomorphism from (K, ξ) to AU if and only if $a.p = b.p$ in $\mathscr{A}$.

The proof of 1.30 can be read off of the following large diagram:

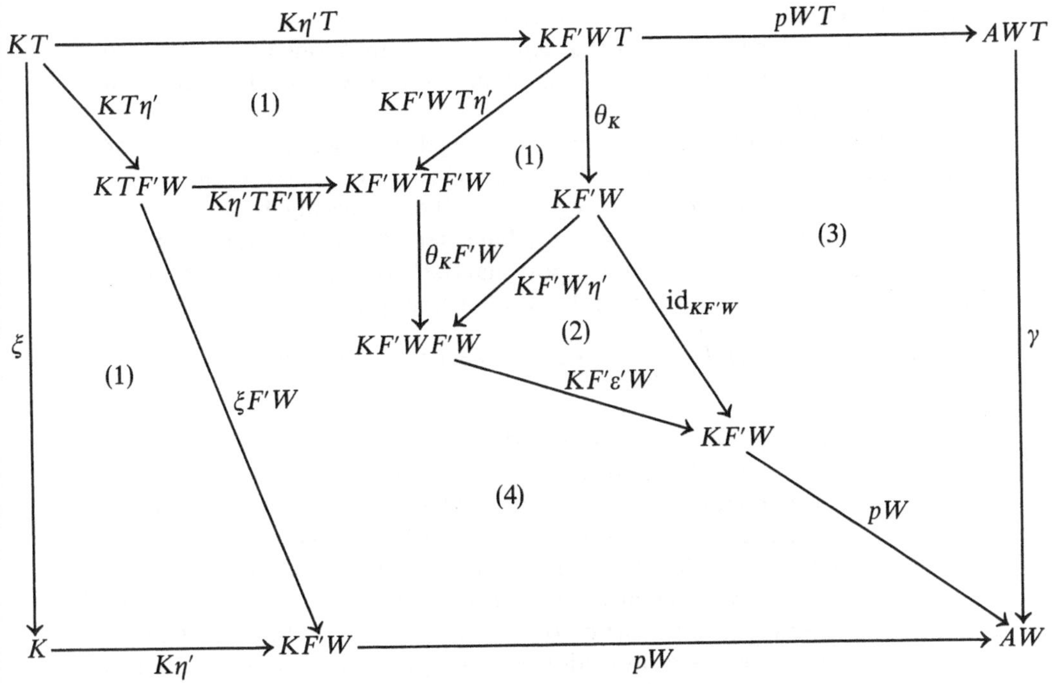

Here, $AU = (AW, \gamma : AWT \longrightarrow AW)$. The regions (1) all commute because η' is a natural transformation; (2) commutes, being one of the triangular identities (2.2.16); (3) commutes because pU is a morphism in $\mathscr{B}$; $K\eta'.pW$: $(K, \xi) \longrightarrow AU$ is a **T**-homomorphism if and only if the perimeter of the diagram commutes; and (4) commutes if and only if $(a.p)W = (b.p)W$. Therefore $a.p = b.p$ implies that $K\eta'.pW$ is a **T**-homomorphism. Conversely, if $K\eta'.pW$ is a **T**-homomorphism then at least $KT\eta'.(a.p)W = KT\eta'.(b.p)W$; but since $(KTF', KT\eta')$ is free over KT with respect to W, $a.p = b.p$.

The rest is easy. Let $p: KF' \longrightarrow A = \text{coeq}(a, b)$ in $\mathscr{A}$. Since $a.p = b.p$, $f = K\eta'.pW : (K, \xi) \longrightarrow AU$ is a **T**-homomorphism. Let A' in $\mathscr{A}$ be arbitrary and let $f' : (K, \xi) \longrightarrow A'U$ be any **T**-homomorphism. As $(KF', K\eta')$ is free over K with respect to W, there exists unique $p' : KF' \longrightarrow$

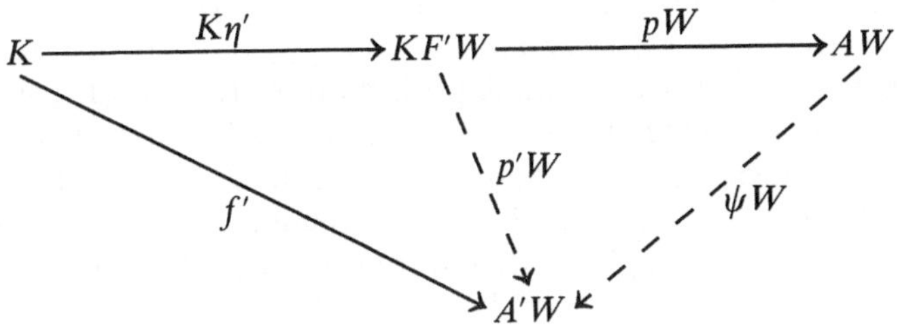

A' in $\mathscr{A}$ such that $K\eta'.p'W = f'$. Since $f'\colon(K, \xi) \longrightarrow A'U$ was assumed a **T**-homomorphism it follows that $a.p' = b.p'$ so that, since $p = \text{coeq}(a, b)$, there exists a unique $\psi\colon A \longrightarrow A'$ in $\mathscr{A}$ such that $p.\psi = p'$. The remaining details are clear. □

In the context of 1.29, if W and V are algebraic then U is just a homomorphism (as in 2.3.3) from $(\mathscr{A}, W)$ to $(\mathscr{B}, V)$; 1.29 then asserts that homomorphisms between algebraic categories are themselves algebraic (assuming the existence of coequalizers; theorems that assert $\mathscr{K}^{\mathbf{T}}$ has colimits will appear later in section 7). It is pleasant to report that $\mathbf{Set}^{\mathbf{T}}$ is always small co-complete (7.10). For the immediate present let us offer a simple proof of

1.31 Lemma. *If* $U\colon\mathscr{A} \longrightarrow \mathbf{Set}$ *is a set-valued algebraic functor then* $\mathscr{A}$ *has coequalizers.*

Proof. Let $f, g\colon A_1 \longrightarrow A_2$ be admissible morphisms in $\mathscr{A}$. Let $S = \{(a_1 f, a_1 g)\colon a_1 \in A_1 U\}$. Let R be the intersection of all congruences (defined in (1.11−) on A_2 containing S. We argue that R is a congruence as follows: Let $(R_i\colon i \in I)$ be the set of all congruences on A_2 which contain S. Define a diagram scheme Δ with $N(\Delta) = I \cup \{t, t'\}$ where $t, t' \notin I$ and such that $\Delta(i, j)$ has exactly one element if $i \in I$ and $j \in \{t, t'\}$ but is empty otherwise. Define the diagram (Δ, D) in **Set** by $D_i = R_i$ for $i \in I$, $D_t = D_{t'} = A_2 U$, $D_\alpha =$ first or second coordinate projection $R_i \longrightarrow A_1$ accordingly as $\alpha \in \Delta(i, t)$ or $\alpha \in \Delta(i, t')$. Then I is final. (R, ψ), where $\psi_i\colon R \longrightarrow R_i$ is the inclusion map,

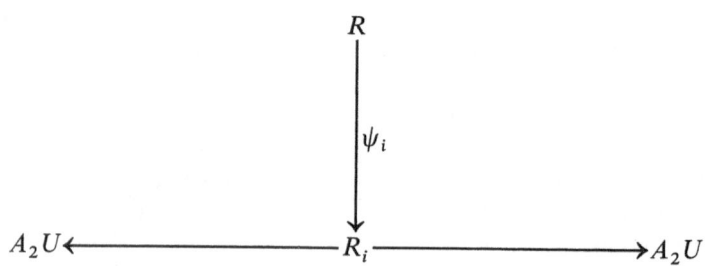

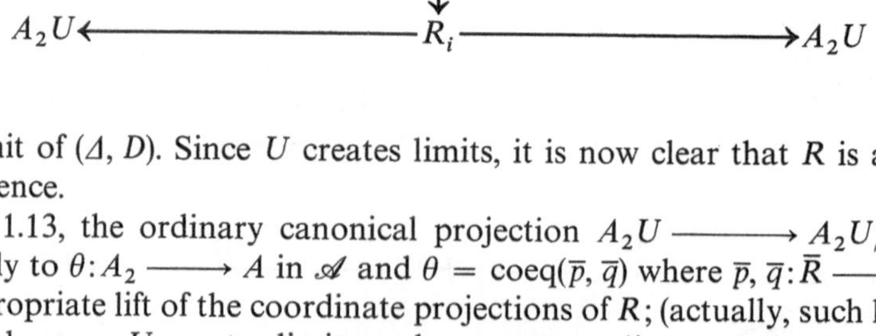

is a limit of (Δ, D). Since U creates limits, it is now clear that R is again a congruence.

By 1.13, the ordinary canonical projection $A_2 U \longrightarrow A_2 U/R$ lifts uniquely to $\theta\colon A_2 \longrightarrow A$ in $\mathscr{A}$ and $\theta = \text{coeq}(\bar{p}, \bar{q})$ where $\bar{p}, \bar{q}\colon \bar{R} \longrightarrow A_2$ is an appropriate lift of the coordinate projections of R; (actually, such lifts are unique because U creates limits and congruences live over the kernel pair of a homomorphism). Now let $h\colon A_2 \longrightarrow A_3$ in $\mathscr{A}$ be such that $f.h = g.h$. Let $E = \{(x, y)\colon xh = yh\}$. Then E is a congruence on A_2 because (E, a, b) is the kernel pair of h in **Set** (if a, b are the coordinate projections) and so admit a unique lift $(\bar{E}, \bar{a}, \bar{b})$ in $\mathscr{A}$ because U creates limits.

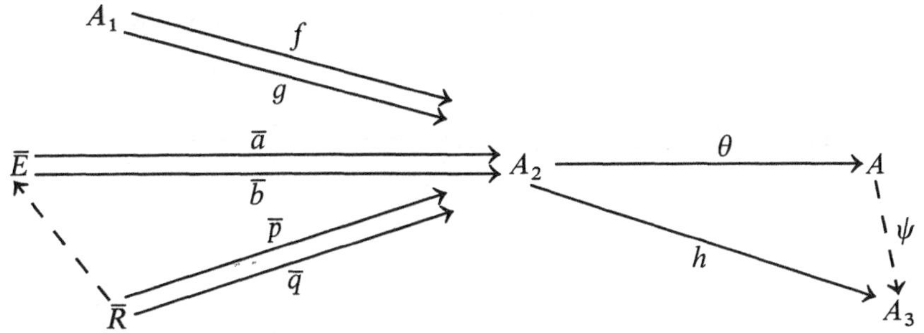

Since $f.h = g.h$, S is a subset of E. Therefore, R is also a subset of E, and $\overline{p}.h = \overline{q}.h$ inducing the desired unique ψ. ψ is admissible by 2.1.56. □

1.32 Examples of Algebraic Homomorphisms over Set. By 1.29 and 1.31, every homomorphism $U:(\mathscr{A}, W) \longrightarrow (\mathscr{B}, V)$ with $W:\mathscr{A} \longrightarrow$ **Set** and $V:\mathscr{B} \longrightarrow$ **Set** algebraic is itself algebraic $U:\mathscr{A} \longrightarrow \mathscr{B}$ and in particular has a left adjoint $F:\mathscr{B} \longrightarrow \mathscr{A}$. We give a few specific examples. Let $\mathscr{A}$ be rings (with unit) and unitary ring homomorphisms, let $\mathscr{B}$ be monoids, and let U be the obvious forgetful functor. If M is a monoid, MF is usually called the *monoid ring over M*; this is best known when M is a group and then MF is called the *integral group ring over M*. [Mac Lane '63, Chapter IV, section 1]. Or, let $\mathscr{A}$ be associative linear algebras over a fixed commutative ring R with unit and let $\mathscr{B}$ be Lie algebras over R. There is a well-known homomorphism U which transforms the associative algebra A into a Lie algebra on the same underlying set with Lie bracket given by $[x, y] = xy - yx$. For a Lie algebra L, LF is called the *universal enveloping algebra over L* [Jacobson '62, Chapter V]. Our previous Example 2.2.7 arises from $\mathscr{A}$ = abelian groups and $\mathscr{B}$ = groups. Compact abelian groups (1.23) admits obvious homomorphisms to, say, compact spaces, abelian groups, and monoids; all of these functors must have left adjoints.

1.33 Proposition. *Let $U:\mathscr{A} \longrightarrow \mathscr{K}$ be the inclusion of a full replete reflective subcategory. Then U is algebraic.*

Proof. There exists a left adjoint $F:\mathscr{K} \longrightarrow \mathscr{A}$ to U such that UF is the identity functor of $\mathscr{A}$. We will use 1.9. Let

$$AU \underset{gU}{\overset{fU}{\rightrightarrows}} BU \xrightarrow{\quad h \quad} K$$

be an absolute coequalizer in $\mathscr{K}$. By applying F, we have

$$A \underset{g}{\overset{f}{\rightrightarrows}} B \xrightarrow{\quad hF \quad} KF$$

which is again a coequalizer in $\mathscr{A}$. Similarly applying FU, we have

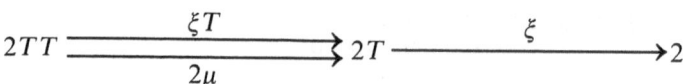

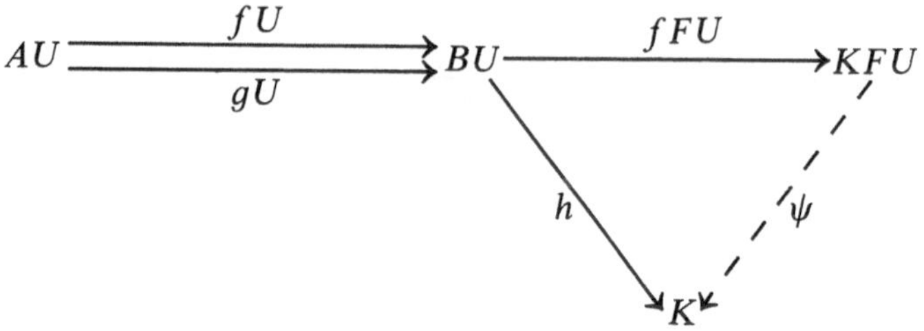

that $fFU = \text{coeq}(fU, gU)$ and there exists a unique isomorphism ψ: $KFU \longrightarrow K$ in $\mathscr{K}$ such that $fFU.\psi = h$. By the definition of "replete" in 2.3.2, $\psi: KFU \longrightarrow K$ has a unique lift $\bar{\psi}: KF \longrightarrow A$. The remaining details are clear. □

1.34 Corollary. *A full replete reflective subcategory is closed under limits.*

Proof. 1.33 and 1.19. □

1.35 The Composition of Algebraic Functors Need Not Be Algebraic. Let $\mathscr{A}$ be the category of abelian groups with algebraic forgetful functor V: $\mathscr{A} \longrightarrow$ **Set**. Let $\mathscr{P}$ be the full replete subcategory of torsion-free abelian groups (a group is *torsion-free* if it has no elements of finite order). Then $\mathscr{P}$ is reflective (the reflection of A is $\theta: A \longrightarrow A/C$ where C is the torsion subgroup of A—the "C" of 2.2.13; since A is abelian, C is simply the subset of elements of finite order). By 1.33, the inclusion functor $U: \mathscr{P} \longrightarrow \mathscr{A}$ is algebraic. Although $UV: \mathscr{P} \longrightarrow$ **Set** has a left adjoint (2.2.30), UV is not algebraic. Let 2 be a two-element set and let (T, η, μ) be the algebraic theory in **Set** whose algebras are the abelian groups. Then $(2TT, 2T\mu)$ and $(2T, 2\mu)$ are in $\mathscr{P}$ since free abelian groups are torsion free. Let $(2, \xi)$ be the two-element group. Then

$$2TT \underset{2\mu}{\overset{\xi T}{\rightrightarrows}} 2T \xrightarrow{\quad \xi \quad} 2$$

is a contractible coequalizer in **Set** which has no lift to $\mathscr{P}$ since the two-element group is not torsion free. □

We conclude the section with a pair of results about "pulling back structure" as in 2.3.34.

1.36 Proposition. *Let* $(\mathscr{P}, V) \in \text{Struct}(\mathscr{L})$ *be the pullback of* $(\mathscr{A}, U) \in \text{Struct}(\mathscr{K})$ *along the full replete subcategory* $\mathscr{L}$. *Then the following three statements are valid*:

$$\begin{array}{ccc} \mathscr{P} & \xrightarrow{\ j\ } & \mathscr{A} \\ {\scriptstyle V}\downarrow & & \downarrow{\scriptstyle U} \\ \mathscr{L} & \xrightarrow[\ i\ ]{} & \mathscr{K} \end{array}$$

1. *If U creates coequalizers of U-absolute pairs then V creates coequalizers of V-absolute pairs.*

2. *If U creates limits and if i preserves limits then V creates limits.*

3. *If U has a left adjoint $F:\mathscr{K} \longrightarrow \mathscr{A}$ and if either $FU:\mathscr{K} \longrightarrow \mathscr{K}$ maps $\mathscr{L}$ into $\mathscr{L}$ or $\mathscr{P}$ is reflective in $\mathscr{A}$ then V has a left adjoint.*

Proof. 1. Let

$$P_1V \underset{gV}{\overset{fV}{\rightrightarrows}} P_2V \xrightarrow{\quad h \quad} L$$

be an absolute coequalizer in $\mathscr{L}$. Then

$$P_1jU \underset{gjU}{\overset{fjU}{\rightrightarrows}} P_2jU \xrightarrow{\quad hi \quad} Li$$

is an absolute coequalizer in $\mathscr{K}$ so that there exists a unique lift $\bar{h}:P_2j \longrightarrow A$ with $\bar{h}U = hi$; and, moreover, $\bar{h} = \text{coeq}(fj, gj)$ in $\mathscr{A}$. As $AU = Li$, A is in $\mathscr{P}$. The rest is clear.

2. The proof is similar to that of (1). Let D be a diagram in $\mathscr{P}$ and let $\psi:L \longrightarrow DV$ be a limit of DV. As i preserves limits, $\psi i:Li \longrightarrow DjU$ is a limit of DjU in $\mathscr{K}$ and there exist unique lifts $\bar{\psi}:\bar{L} \longrightarrow Dj$ with $\bar{\psi}U = \psi i$; and, moreover, $(\bar{L}, \bar{\psi})$ is a limit of Dj. As $\bar{L}U = Li$, $\bar{L}$ is in $\mathscr{P}$. The rest is clear.

3. If FU maps $\mathscr{L}$ into $\mathscr{L}$ then for all $L \in \mathscr{L}$, $(LiF)U \in \mathscr{L}$, that is, $LF \in \mathscr{P}$; but then LF is *a fortiori* free over L with respect to V. The second statement follows immediately from 2.2.30. □

1.37 Countable Algebras. Let (Ω, E) be an equational presentation with each Ω_n countable and with Ω_n empty when n is infinite (i.e., (Ω, E) is a familiar everyday finitary equational presentation). Then the forgetful functor from countable (Ω, E)-algebras to countable sets is algebraic. This is an easy application of 1.36 with $i:\mathscr{L} \longrightarrow$ **Set** being "countable sets." The detail to check is that LT is countable when L is. But this is clear from 1.1.7.

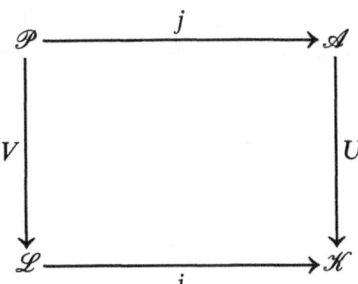

1.38 Proposition. *Let* $(\mathscr{P}, V) \in \text{Struct}(\mathscr{L})$ *be the pullback of* $(\mathscr{A}, U) \in$ $\text{Struct}(\mathscr{K})$ *along the full replete subcategory* $\mathscr{L}$ *and let W be the common value*

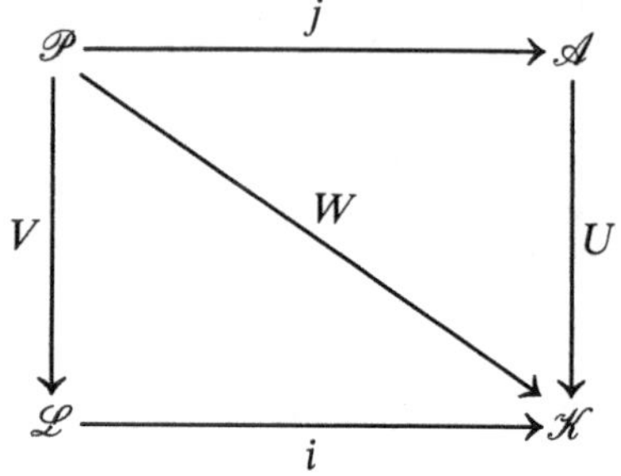

of Vi and jU as shown above. Assume further that $\mathscr{P}$ *is reflective in* $\mathscr{A}$ *and that there exists a functor* $s: \mathscr{K} \longrightarrow \mathscr{L}$ *with* $i.s = \text{id}_{\mathscr{L}}$ *(e.g. if* $\mathscr{L}$ *is either reflective or coreflective in* $\mathscr{K}$*). Then if U is algebraic, W is algebraic.*

Proof. W has a left adjoint by 2.2.30. Suppose that

$$P_1 W \underset{gW}{\overset{fW}{\rightrightarrows}} P_2 W \overset{h}{\longrightarrow} K$$

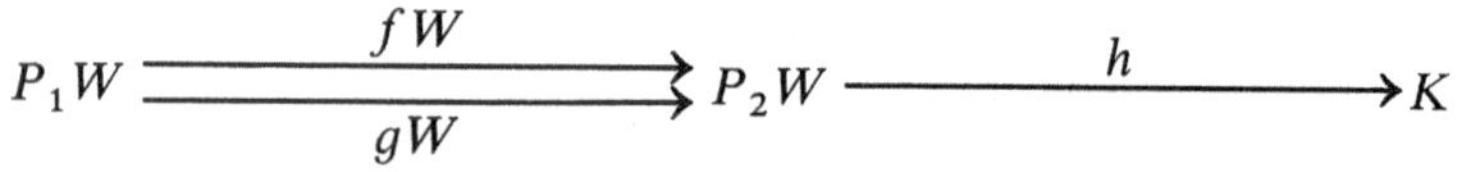

is an absolute coequalizer in $\mathscr{K}$. We will show that $K \in \mathscr{L}$. As $P_1 W si = P_1 V isi = P_1 V i = P_1 W$ and, similarly, $P_2 W si = P_2 W$, (fW, gW, hsi) is a

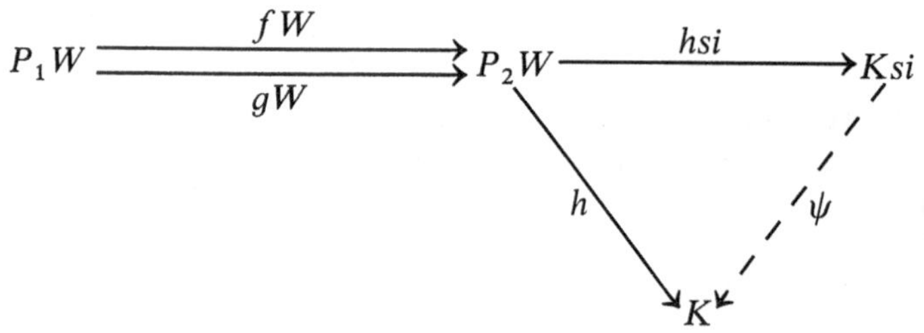

coequalizer in $\mathscr{K}$ and there exists an isomorphism $\psi: K si \longrightarrow K$ such that $hsi.\psi = h$. Since $\mathscr{L}$ is replete, K is in $\mathscr{L}$. $\square$

Primer on Set Theory

We extend the primer on set theory at the end of section 1.5 with a few facts on cardinal arithmetic which will help the uninitiated in reading the proof of 1.27. Proofs can be found in [Monk '69, Chapter 4]. For cardinals α, β their *sum* $\alpha + \beta$, *product* $\alpha \times \beta$, and *exponential* β^{α} are defined as in the proof of 1.27. We have

$$\begin{aligned}(\alpha + \beta) + \gamma &= \alpha + (\beta + \gamma)\\ \alpha + \beta &= \beta + \alpha\\ (\alpha \times \beta) \times \gamma &= \alpha \times (\beta \times \gamma)\\ \alpha \times \beta &= \beta \times \alpha\\ (\gamma^{\beta})^{\alpha} &= \gamma^{(\beta \times \alpha)}\end{aligned}$$

Moreover, if $\alpha_1 \leqslant \alpha_2$ then for all β

$$\begin{aligned}\beta + \alpha_1 &\leqslant \beta + \alpha_2\\ \beta \times \alpha_1 &\leqslant \beta \times \alpha_2\\ \beta^{\alpha_1} &\leqslant \beta^{\alpha_2}\\ \alpha_1^{\beta} &\leqslant \alpha_2^{\beta}\end{aligned}$$

A proof that $\alpha \times \alpha = \alpha$ when α is infinite (which is one of the equivalent forms of the axiom of choice!) appears as [Monk, Theorem 21.10].

Notes for Section 1

The characterization theorem of 1.9 was proved by J. Beck circa 1965 and remained an "untitled manuscript" until it appeared in Beck's thesis [Beck '67, Theorem 1]. 1.4–1.6 is due to Beck. The emphasis on U-absolute coequalizers follows [Paré '71]. The proof of 1.13 is an adaptation of [Duskin '69] (see exercise 7) and [Linton '66]. While a direct proof of 1.13 (bypassing 1.9) is neither difficult nor longer, the proof offered clarifies what is needed to relate contractible coequalizers to the more intuitive coequalizers of congruences. The grandfather of all of these theorems is [Lawvere '63, theorem 1, page 79]. See also [Felscher '68–A]. 1.25 is based on the "codensity triple" construction known to the Zürich school at least as early as 1965 and described in [Appelgate and Tierney '69, section 3] (who discuss the dual notion of "model-induced cotriple") and [Linton '69]; see also exercise 3.2.12. 1.27 is adapted from [Pierce '68, Proposition 4.1.3]. 1.29 (in the case of finitary theories of sets) was emphasized by Lawvere (see [Lawvere '63, theorem, page 94]) and is well known to universal algebraists (see e.g. [Cohn '65, III.4.2]). For generalizations of 1.29 see [Dubuc '68, Theorem 1], [Huq '70, Main Theorem] and [Tholen '76, Proposition 6].

Exercises for Section 1

1. Given $h: A \longrightarrow B$, prove that h is an absolute epimorphism (i.e., for every functor H, hH is an epimorphism) if and only if h is a split epimorphism. [Hint: let H be the appropriate representable functor.]
2. In the context of 1.14, show that the cardinality of H is at most the cardinality of the set of all ultrafilters on the set $n \times \aleph_0$.
3. Let $f, g: A \longrightarrow B$, $d: B \longrightarrow A$ satisfy $fdg = gdg$ and $df = \mathrm{id}_B$. Prove that if $\mathrm{eq}(dg, \mathrm{id}_B)$ exists then (f, g, d) extends to a contractible coequalizer $(f, g, h; d, d')$.
4. For any category $\mathscr{K}$ let $\mathscr{K}^{\rightarrow}$ be the category whose objects are morphisms

in $\mathscr{K}$ and whose morphisms are commutative squares:

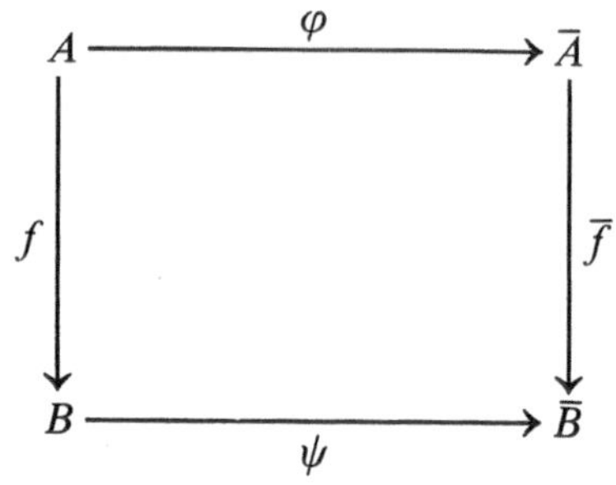

(Here, (φ, ψ) is a morphism from f to $\bar{f}$). Given (f, g, h) with $fh = gh$, show that (f, g, h) extends to a contractible coequalizer if and only if $(f, h): g \longrightarrow h$ is a split epimorphism in $\mathscr{K}^{\rightarrow}$.

5. Show that the intersection (infimum) of subobjects represented by monos $f_1: S_1 \longrightarrow K$, $f_2: S_2 \longrightarrow K$ coincides with the subobject represented by $j: P \longrightarrow K$ in the pullback

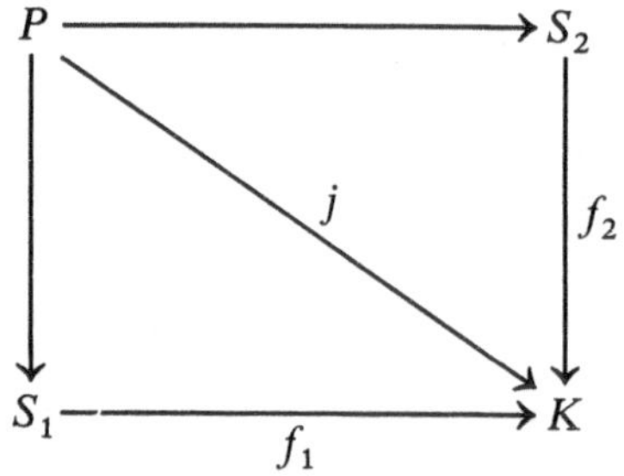

(in the sense that if the pullback exists they are equal). Show that if $f, g: K \longrightarrow L$ and if $K \times K$, ker pair (f) and ker pair (g) exist, then the separator of f and g is just the intersection of ker pair (f) and ker pair (g).

6. (M. Barr, personal communication.) Let $U: \mathscr{C} \longrightarrow$ **Set** in Struct(**Set**) construct finite limits and have the property that for all admissible $f:(X, s) \longrightarrow (Y, t)$ the image Xf is an optimal subset of (Y, t). Many algebraic categories in mathematics are over such a base category. This exercise (the crux of the construction of 1.13; also, see the next exercise) provides insight as to how to interpret ξ as in 1.2 as the coequalizer of a congruence.

   Specializing the definition of the text, an *optimal congruence on* (X, s) is an optimal subset R of $(X, s) \times (X, s)$ which is an equivalence relation.

   (a) Given optimal subsets R, S on (X, s) show that $R^{-1} = \{(y, x):(x, y) \in R\}$ and $RS = \{(x, z)$: for some $y, (x, y) \in R$ and $(y, z) \in S\}$ are optimal subsets of (X, s).

   (b) Let $(f, g, h; d, d')$ be a contractible coequalizer in $\mathscr{C}$. Set $R = \{(af, ag): a \in A\}$. Show that RR^{-1} is the smallest optimal congruence containing R and that h is "dividing out by RR^{-1}." [Hint: to prove that

R is a subset of RR^{-1}, observe that $af = (afd)f$, $ag = (agd)f$, $(afd)g = (agd)g$.]

7. [Duskin '69, Theorem 3.2.] Let $\mathscr{K}$ have finite limits and let $U:\mathscr{A} \longrightarrow \mathscr{K}$. Prove that U is algebraic if and only if (a) U has a left adjoint; (b) $\mathscr{A}$ has separators; (c) given an equivalence relation $p, q: A \longrightarrow B$ in $\mathscr{A}$ (see section 2.1 exercise 15) such that (pU, qU) extends to a contractible coequalizer $(pU, qU, h; d, d')$ in $\mathscr{K}$, h admits a unique lift which is, moreover, coeq(p, q) in $\mathscr{A}$. [Hint: a proof of the same assertion, substituting "kernel pairs" for "equivalence relations" in (c) follows the proof of 1.13 and is itself an interesting result; to prove the assertion as stated requires more work.]
8. A pair of maps $f, g: A \longrightarrow B$ is *reflexive* if there exists $d: B \longrightarrow A$ with $df = \mathrm{id}_B = dg$. (In **Set**, (f, g) is reflexive if and only if $\{(af, bf): a \in A\}$ is a reflexive relation.) In the context of 1.29, show that it is enough to postulate that every reflexive pair of $\mathscr{A}$-morphisms has a coequalizer.
9. Where in the proof of 1.27 did we use the fact that $\bigcup(A_\gamma : \gamma < \beta)$ is a subset of A_β?
10. (M. Barr.) A Beck functor $U:\mathscr{A} \longrightarrow \mathscr{B}$ is *crude* if whenever $h = \mathrm{coeq}(f, g)$ then $hU = \mathrm{coeq}(fU, gU)$. 1 is an *isolated terminal object* of $\mathscr{A}$ if 1 is a terminal object of $\mathscr{A}$ and if, for each A, every morphism $1 \longrightarrow A$ is an isomorphism (e.g. the empty set is an isolated initial object of **Set**). A Beck functor $U:\mathscr{A} \longrightarrow \mathscr{B}$ is *vulgar* if given $f, g: A_1 \longrightarrow A_2$, $h: A_2 \longrightarrow A_3$ in $\mathscr{A}$ such that A_3 is not an isolated terminal object and such that $(fU, gU, hU; d, d')$ is a contractible coequalizer in $\mathscr{B}$ for some d, d', then (f, g, h) is an absolute coequalizer in $\mathscr{A}$. Prove that, given Beck functors $U_1:\mathscr{A} \longrightarrow \mathscr{B}$, $U_2:\mathscr{B} \longrightarrow \mathscr{C}$, $U_3:\mathscr{C} \longrightarrow \mathscr{D}$, if U_1 is crude and if U_3 is vulgar then $U_1U_2U_3$ is Beck. Also, if U_1 and U_2 are vulgar prove that U_1U_2 is vulgar.
11. Given a diagram

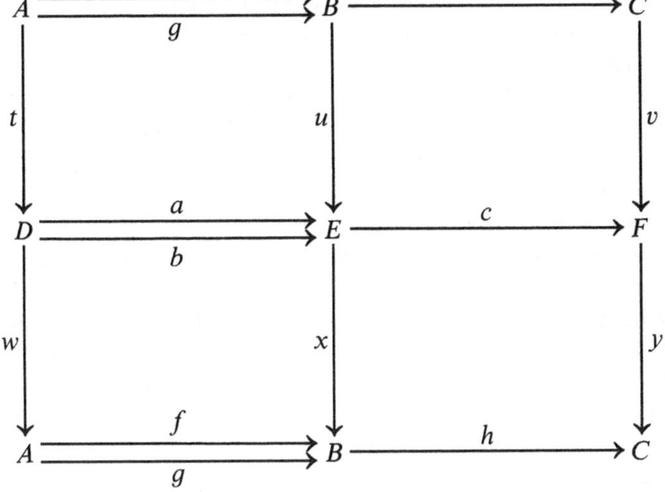

with $fu = ta$, $gu = tb$, $ax = wf$, $bx = wg$, $hv = uc$, $cy = xh$, $ux = \mathrm{id}_B$, and $vy = \mathrm{id}_C$ (it is not necessary that $tw = \mathrm{id}_A$), prove that if $c = \mathrm{coeq}(a, b)$ then $h = \mathrm{coeq}(f, g)$.

12. Let $\mathbf{T}$ be an algebraic theory in **Set**. A $\mathbf{T}$-algebra (X, ξ) is an *algebraic generator* (cf. [Lawvere '63, section 2 of Chapter III]) if the representable functor $\mathbf{Set}^{\mathbf{T}}((X, \xi), -): \mathbf{Set}^{\mathbf{T}} \longrightarrow \mathbf{Set}$ is weakly algebraic. Show that for nonempty n, $(nT, n\mu)$ is an algebraic generator. [Hint: use exercises 10, 11 with respect to the functor $U_3 = (-)^n: \mathbf{Set} \longrightarrow \mathbf{Set}$.]

13. A *finitary phylum* (cf. [Birkhoff and Lipson '70], [Grätzer '69] and [Higgins '63]) is a pair (P, Ω) where P is a nonempty set and $\Omega = (\Omega_w)$ is a disjoint family of (possibly empty) sets indexed by words $w = p_1 \cdots p_n$ on the alphabet P of length $n > 0$. A (P, Ω)-*algebra* is a pair (X, δ) where $X = (X(p))$ is a P-indexed family of sets and δ assigns to each $\omega \in \Omega_{p_1 \ldots p_{n+1}}$ a function $\delta_\omega: X(p_1) \times \cdots \times X(p_n) \longrightarrow X(p_{n+1})$. A (P, Ω)-*homomorphism* $f:(X, \delta) \longrightarrow (X', \delta')$ is a family of functions of form $f(p): X(p) \longrightarrow X'(p)$ satisfying

$$\begin{array}{ccc} X(p_1) \times \cdots \times X(p_n) & \xrightarrow{\delta_\omega} & X(p_{n+1}) \\ {\scriptstyle f(p_1) \times \cdots \times f(p_n)} \big\downarrow & & \big\downarrow {\scriptstyle f(p_{n+1})} \\ X'(p_1) \times \cdots \times X'(p_n) & \xrightarrow[\delta'_\omega]{} & X'(p_{n+1}) \end{array}$$

(For examples refer to the papers cited above.)

Let $\mathbf{Set}^{\mathsf{P}}$ denote the obvious (functor) category whose objects are families $(X(p): p \in \mathsf{P})$ and whose morphisms are families $(f(p))$. Show that the category of (P, Ω)-algebras is algebraic over $\mathbf{Set}^{\mathsf{P}}$. [Hint: free algebras are constructed in [Birkhoff and Lipson '70, Proposition 15].]

For further generalizations see [Bénabou '66], [Davis '67], [Walters '69], and [Hoehnke '74].

14. Construct interesting examples of algebraic homomorphisms $U: \mathbf{Set}^{\mathbf{S}} \longrightarrow \mathbf{Set}^{\mathbf{T}}$ over **Set** for which the "inclusion of the generators" morphism is (i) always injective, (ii) always surjective, (iii) neither of these.

15. Let M be a monoid. Define $\mathscr{K}^M$ in Struct($\mathscr{K}$) to be the category whose objects are pairs (A, s) with $s: M \longrightarrow \mathscr{K}(A, A)$ a monoid homomorphism, and whose morphisms $f:(A, s) \longrightarrow (B, t)$ satisfy the following diagram (for all $m \in M$).

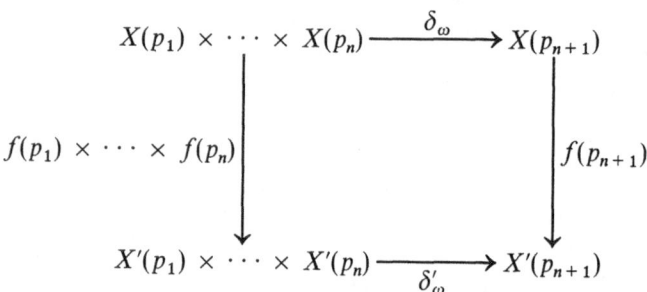

$$\begin{array}{ccc} A & \xrightarrow{f} & B \\ {\scriptstyle s_m}\downarrow & & \downarrow{\scriptstyle t_m} \\ A & \xrightarrow[f]{} & B \end{array}$$

(a) Show that $U:\mathscr{K}^M \longrightarrow \mathscr{K}$ is both Beck and co-Beck. [Hint: simplify $(\mathscr{K}^M)^{\text{op}}$.]

(b) Show that $U:\mathscr{K}^M \longrightarrow \mathscr{K}$ is algebraic providing each object of $\mathscr{K}$ has an M-fold copower and, dually, show that U is coalgebraic providing each object of $\mathscr{K}$ has an M-fold power. [Hint: the cofree object over A is A^M with action $L_m.-:A^M \longrightarrow A^M$ defined by $(L_m.-)\text{pr}_{m'} = \text{pr}_{mm'}$.] This generalizes exercise 2.2.4.

The right adjoint of (b) is an example of "right Kan extension along the functor $e:1 \longrightarrow M$" [Mac Lane '71, Chapter X, section 3].

(c) Let $M = G$ be a compact (separated) topological group and let $\mathscr{K}$ be the category of Banach spaces and norm-decreasing linear maps. Objects of $\mathscr{K}^G$ are "representations of G by isometries in a Banach space." Let $\mathscr{A}$ be the full subcategory of continuous representations (A, s), i.e., the map $G \longrightarrow A, g \longmapsto as_g$ is continuous for each a in A, and let $V:\mathscr{A} \longrightarrow \mathscr{K}$ be the restriction of U. Show that V is coalgebraic. [Hint: The co-Beck condition is easy. The cofree object over A is the subspace $[G, A]$ of continuous functions in A^G; the only hard part is showing that $G \longrightarrow ([G, A], [G, A])$ is continuous and for this [Hewitt and Ross '63, Theorems 4.9, 4.15] are helpful.]

In the context of (c), if $\mathbf{C}$ is the complex ground field, a scalar multiple of a morphism $[G, \mathbf{C}] \longrightarrow \mathbf{C}$ is an "integral" and the co-Kleisli composition of two such integrals is their convolution [Hewitt and Ross '63, sections 14, 19].

(d) (open question) It would be interesting to generalize (c) when G is locally compact. One is tempted to set $[G, A]$ to be the continuous functions which vanish at infinity and to set $\mathscr{A}$ to be those representations for which the maps $g \longmapsto as_g$ are continuous and vanish at infinity; discover why this doesn't work.

16. In this exercise we indicate how commutative C^*-algebras arise as the algebras over a theory in **Set** with rank $\aleph_1$ (where $\aleph_1 = (\aleph_0)^+$). By a *commutative C^*-algebra* we mean a commutative complex Banach algebra with unit and involution $x \longmapsto x^*$ satisfying $\|xx^*\| = \|x\|^2$. Let $\mathscr{A}$ be the category of commutative C^*-algebras and algebra homomorphisms which preserve the involution and the unit. As is well known, such

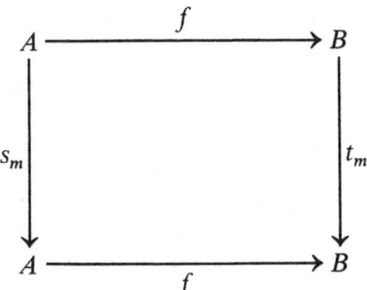

homomorphisms are necessarily bounded of norm 1. The underlying set functor will not do since

(a) Prove that $\mathscr{A}$ has products but that the underlying set functor does not preserve products.

(b) Let $U:\mathscr{A} \longrightarrow \mathbf{Set}$ be the unit disc functor. Let $\mathbf{D}$ denote the unit disc of the complex plane and, for any compact Hausdorff space Y, let $C(Y)$ denote the C^*-algebra of continuous complex-valued functions on Y in the sup norm. Show that $C(\mathbf{D}^X)$ is the free C^*-algebra over X with respect to U. [Hint: $X\eta$ assigns to x its evaluation map $f \longmapsto xf$; if A is a C^*-algebra there is no problem extending $g:X \longrightarrow A$ to the subalgebra-with-involution generated by the evaluations; prove that this partial extension is uniformly continuous and use the Stone-Weierstrass theorem to extend the rest of the way.]

(c) Prove that if M is any metric space and if $f:M^I \longrightarrow M$ is uniformly continuous then f has a countable support. [Hint: for each integer $n > 0$ there exists finite $F_n \subset I$ such that $d(af, bf) < 1/n$ if a and b agree on F_n; consider the union of the sets F_n]. Use this to prove that the algebraic theory $\mathbf{T}$ induced in $\mathbf{Set}$ by the adjointness of (b) is of rank $\aleph_1$.

(d) Prove that the semantics comparison functor $\Phi:\mathscr{A} \longrightarrow \mathbf{Set}^{\mathbf{T}}$ is a full representative subcategory (so that $(\mathscr{A}, U)$ is weakly algebraic). [Hint (suggested by J. R. Isbell): let (X, ξ) be a $\mathbf{T}$-algebra (i.e., it suffices to establish the Beck condition for U-contractible coequalizers as in 1.2); the set of scalar multiples of $\{0\}\xi^{-1} \subset XT$ is a closed ideal of $C(\mathbf{D}^X)$ and so has form $\{f:Hf = 0\}$ for some closed subset H of $\mathbf{D}^X$; show that (X, ξ) is $\mathbf{T}$-isomorphic to $C(H)$.]

For similar results see [Negrepontis '71], [Isbell '74], and [Semadeni '74]. Isbell has in fact shown that the (unit disc of) real Banach algebras isomorphic to $C(X, \mathbf{R})$ have an equational presentation in terms of the five operations:

$$\begin{aligned} &\text{0-ary:} && 1 \\ &\text{1-ary:} && x \longmapsto -x \\ & && x \longmapsto \operatorname{Max}(\operatorname{Min}(2x, 1), -1) \\ &\text{2-ary:} && x, y \longmapsto xy \\ &\aleph_0\text{-ary:} && (x_i) \longmapsto \sum_{i=1}^{\infty} 2^{-i}x_i \end{aligned}$$

It is not known whether a nice (e.g. finite) set of equations can be provided.

17. In this exercise we establish the (unpublished) theorem of Linton which asserts that "a category algebraic over complete atomic Boolean algebras is also algebraic over **Set**." The suggested method of proof is due to Manes and Paré.

(a) Given $f, g:A \longrightarrow B$ and $h:B \longrightarrow C$ in $\mathscr{K}$ such that $hH = \operatorname{coeq}(fH, gH)$ for H any of the representable functors $\mathscr{K}(A, -)$,

$\mathscr{K}(B, -)$ or $\mathscr{K}(C, -)$, prove that (f, g, h) is an absolute coequalizer. [Hint: prove directly that $h = \text{coeq}(f, g)$; conclude that $hH = \text{coeq}(fH, gH)$ whenever H is full and faithful; now use 2.2.22 and exercise 2.3.14.]

(b) Establish the 3×3 lemma ([Duskin '69, 0.4]): Given

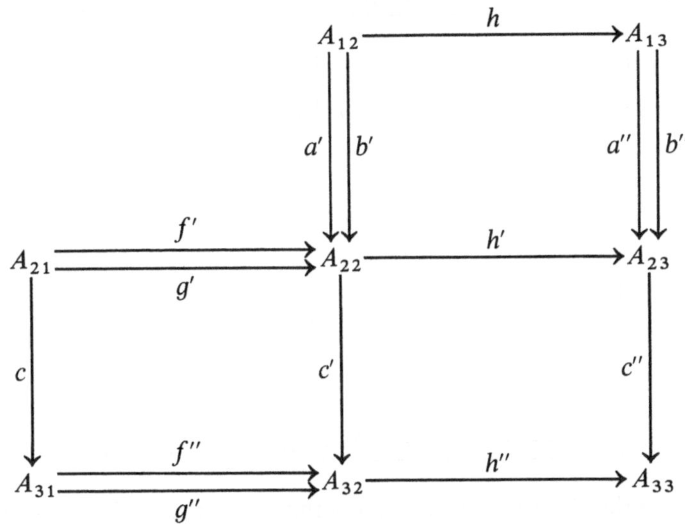

such that $f'c' = cf''$, $g'c' = cg''$, $ha'' = a'h'$, $hb'' = b'h'$, $h'c'' = c'h''$, $a'c' = b'c'$; c, c', and h are epi; $h' = \text{coeq}(f', g')$, $c'' = \text{coeq}(a'', b'')$; prove that $h'' = \text{coeq}(f'', g'')$.

(c) See exercise 10 for terminology. Let $U: \mathscr{A} \longrightarrow \mathscr{B}$ be an algebraic functor and let $\mathscr{A}$ have the property that given any object X which is not an isolated terminal object, X is projective with respect to coequalizers, that is, whenever $h: Y \longrightarrow Z$ is a coequalizer then for

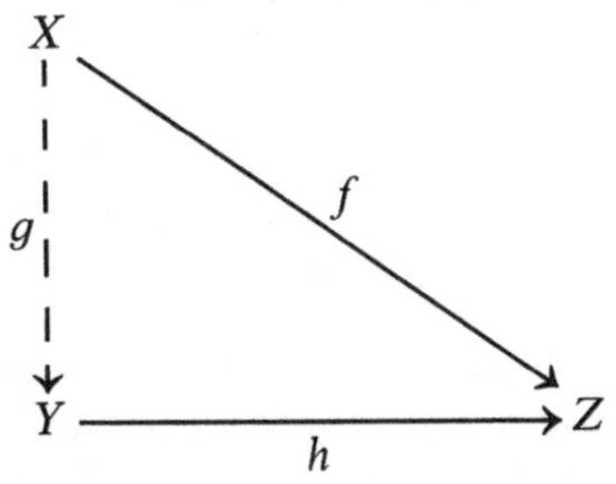

all $f: X \longrightarrow Z$ there exists $g: X \longrightarrow Y$ with $gh = f$. Show that U is vulgar. [Hint: given $f, g: A \longrightarrow B$, $h: B \longrightarrow C$ with C (and hence A, B) not isolated terminal and (fU, gU, hU) an absolute coequalizer, for X any of A, B, C, $X\varepsilon: XUF \longrightarrow X$ is split epi—$d_X.X\varepsilon = \text{id}$—giving rise to the absolute equalizer diagram

$$X \xrightarrow{\;d_X\;} XUF \underset{\text{id}}{\overset{X\varepsilon . d_X}{\rightrightarrows}} XUF$$

Apply (a) and (b) to the diagram

$$\begin{array}{ccccc}
(XUF, A) & \rightrightarrows & (XUF, B) & \longrightarrow & (XUF, C) \\
\downdownarrows & & \downdownarrows & & \downdownarrows \\
(XUF, A) & \rightrightarrows & (XUF, B) & \longrightarrow & (XUF, C) \\
\downarrow & & \downarrow & & \downarrow \\
(X, A) & \rightrightarrows & (X, B) & \longrightarrow & (X, C)
\end{array} \quad].$$

(d) Use the Beck theorem to show that the structural reflection U of the functor **Set** $(-, 2)$: $\mathbf{Set}^{\mathrm{op}} \longrightarrow$ **Set** corepresented by the 2-element set is algebraic. Then check that U is isomorphic to the category of complete atomic Boolean algebras of 1.5 [Hint: show that the algebraic theory induced by the obvious adjoint is the double power-set theory.]

This result is generalized in [Paré '74] where it is shown that $\Omega^{(-)}: \mathscr{E}^{\mathrm{op}} \longrightarrow \mathscr{E}$ is weakly algebraic for any topos (exercise 2.2.16) $\mathscr{E}$. Since it is easy to prove that any topos has finite limits, this provides an elegant way to see that a topos must also have finite colimits.

(e) Use exercise 10 to conclude that if $(\mathscr{A}, U)$ in Struct($\mathbf{Set}^{\mathrm{op}}$) is algebraic then $(\mathscr{A}, \mathbf{Set}((-)U, 2))$ is weakly algebraic over **Set**.

Paré and Linton have shown [private communication] that (e) does not generalize to an arbitrary topos.

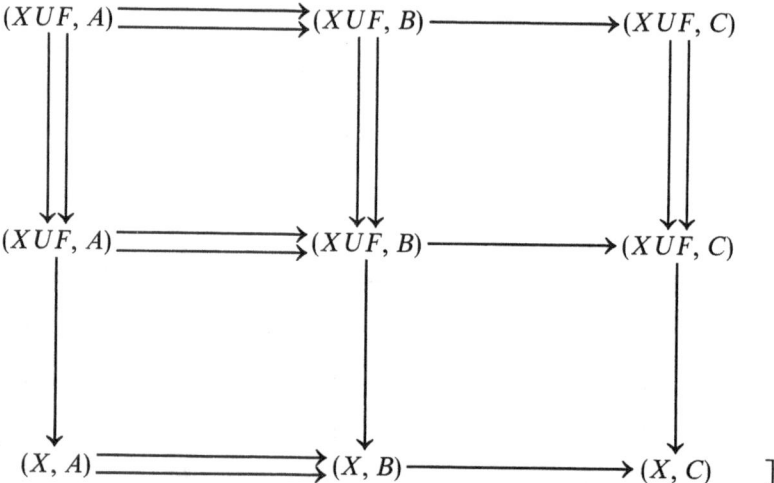

The cohomology and homology of groups, rings, associative, or Lie algebras over a commutative ring and other (Ω, E)-algebras has been of interest since the 1940's; see [Cartan and Eilenberg '56], [Mac Lane '63] and the references cited there. More recently, André, Appelgate, Barr, Beck, and Rinehart followed by Duskin, Van Osdol, and others (see [Duskin '74, '75], [Van Osdol '73] and their bibliographies) have used the algebraic theory **T** to define the cohomology of a **T**-algebra, thereby unifying and clarifying older cohomology theories as well as contributing new ones. This work was anticipated by [Godement '58, appendix] who defined our "algebraic theories in monoid form" but not their algebras.

The subject of "triple cohomology" is much vaster than exercises 18–24 below indicate.

In its widest sense, a *cohomology theory on* $\mathscr{K}$ is a functor $H: \mathscr{K}^{\mathrm{op}} \longrightarrow \mathscr{A}$ where $\mathscr{A}$ is an abelian category ([Freyd '64], [Mitchell '65]). While such H may "lose information," that is we may have KH and LH isomorphic without K and L being isomorphic, H may conveniently classify important properties of objects in $\mathscr{K}$. For our purposes, we fix $\mathscr{A}$ to be the category of abelian groups. A *cochain complex* is a diagram

$$0 \xrightarrow{d^0} A_0 \xrightarrow{d^1} A_1 \xrightarrow{d^2} \cdots \longrightarrow A_n \xrightarrow{d^{n+1}} A_{n+1} \longrightarrow \cdots$$

in $\mathscr{A}$ satisfying $d^n.d^{n+1} = 0$; equivalently, $\mathrm{Im}(d^n) \subset \mathrm{Ker}(d^{n+1})$. An element of A_n, $\mathrm{Ker}(d^{n+1})$, $\mathrm{Im}(d^n)$ is called, respectively, an *n-cochain, n-cocycle, n-coboundary*. The quotient group $\mathrm{Ker}(d^{n+1})/\mathrm{Im}(d^n)$ is the *cohomology group in dimension n* of the cochain complex; this construction is functorial (where morphisms of cochain complexes are diagram morphisms). In practice, a cohomology theory on $\mathscr{K}$ usually arises by defining a cochain complex valued functor on $\mathscr{K}^{\mathrm{op}}$, so that there is a sequence of cohomology theories, one in each dimension.

18. Let $\mathscr{K}$ have finite products. An *abelian group in* $\mathscr{K}$ (cf. 3.2.3 and [Eckmann and Hilton '62]) is a quadruple $(Y, +, i, 0)$ where the $\mathscr{K}$-morphisms $+: Y \times Y \longrightarrow Y$, $i: Y \longrightarrow Y$, $0: 1 \longrightarrow Y$ satisfy the usual axioms:

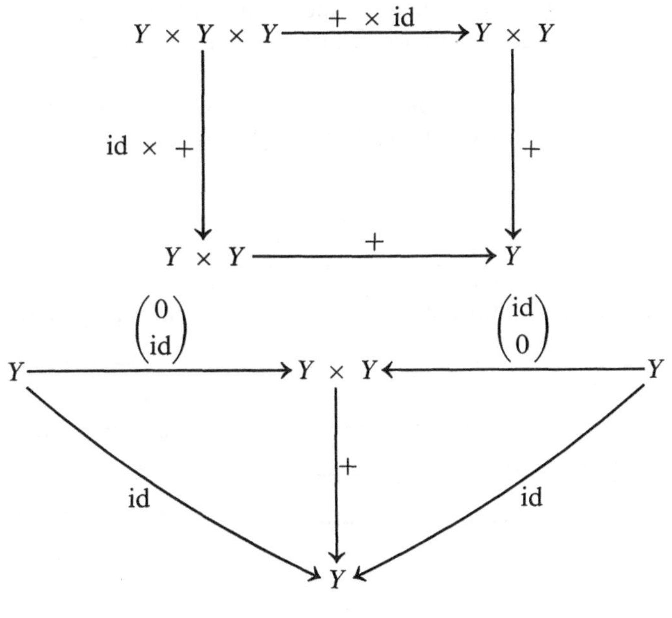

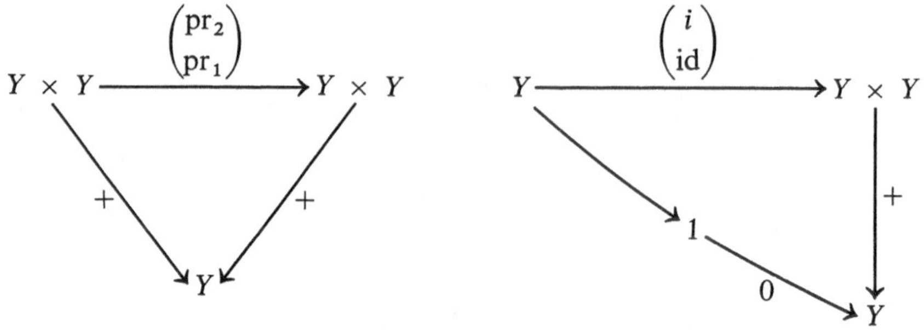

(a) Show that an abelian group in the category of topological spaces is the same thing as a topological abelian group.

(b) Show that for each object Y in $\mathscr{K}$ there is a bijective correspondence between abelian group structures $(Y, +, i, 0)$ and functorial liftings F

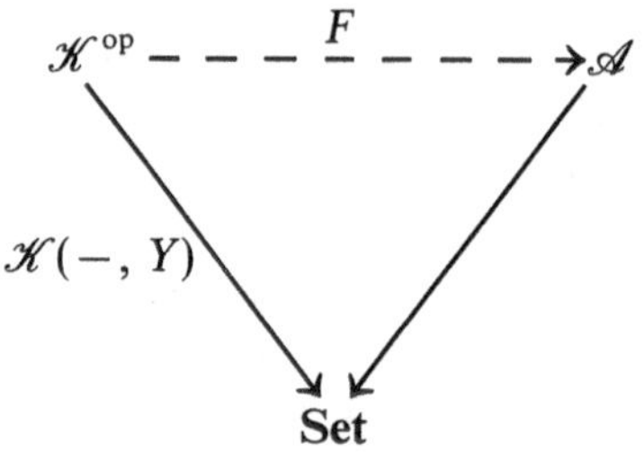

as shown above, where the unnamed functor is the forgetful functor from abelian groups. [Hint: $A \xrightarrow{f+g} Y =$

$A \xrightarrow{\binom{f}{g}} Y \times Y \xrightarrow{+} Y$.]

Thus it is possible to define abelian group objects in any (locally small) category; but we continue to use the original definition for clarity.

19. Let $\mathbf{T}$ be an algebraic theory in $\mathscr{K}$ and let F be an arbitrary functor $\mathscr{K}^{\text{op}} \longrightarrow \mathscr{A}$. For each $\mathbf{T}$-algebra (X, ξ), consider the diagram of abelian groups

$$0 \xrightarrow{d^0} (XT, X\mu)F \xrightarrow{d^1} \cdots \xrightarrow{d^n} (XT^{n+1}, XT^n\mu)F \xrightarrow{d^{n+1}} (XT^{n+2}, XT^{n+1}\mu)F \to \cdots$$

induced by the "canonical resolution" of $\mathbf{T}$-homomorphisms

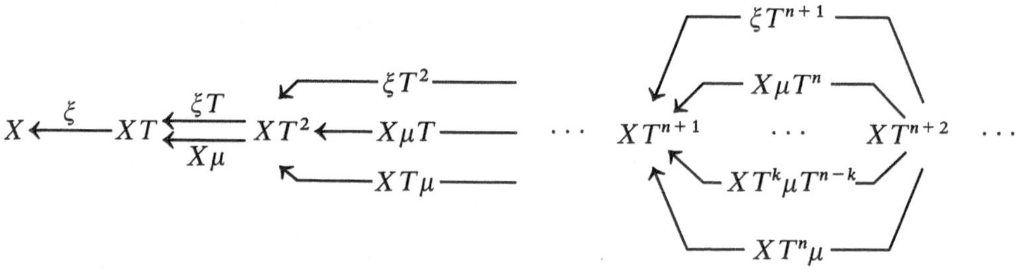

according to the alternating sum formula

$$d^{n+1} = \xi T^{n+1}F + \sum_{k=0}^{n} (-1)^{k+1}(XT^k\mu T^{n-k})F$$

Verify that this construction defines a functor from $(\mathscr{K}^{\mathbf{T}})^{\mathrm{op}}$ to the category of cochain complexes.

20. Let $\mathbf{T}$ be an algebraic theory in $\mathscr{K}$ and let $(Y, \theta; +, i, 0)$ be an abelian group in $\mathscr{K}^{\mathbf{T}}$, inducing a functor as in (18b) and hence, for each $\mathbf{T}$-algebra (X, ξ), the cochain complex of (19):

$$0 \xrightarrow{d^0} [XT, Y] \xrightarrow{d^1} \cdots \xrightarrow{d^n} [XT^{n+1}, Y] \xrightarrow{d^{n+1}} [XT^{n+2}, Y] - \cdots$$

(where $[XT^{n+1}, Y]$ denotes the abelian group $\mathscr{K}^{\mathbf{T}}((XT^{n+1}, XT^n\mu), (Y, \theta))$), with d^{n+1} given by

$$(XT^{n+1} \xrightarrow{a} Y)d^{n+1} = \xi T^{n+1}.a + \sum_{k=0}^{n} (-1)^{k+1}(XT^k\mu T^{n-k}.a)$$

The nth ("*triple*") *cohomology group*, $H^n(X, \xi)$, *of* (X, ξ) *with coefficients in* $(Y, \theta; +, i, 0)$ is the n-dimensional cohomology group, $\mathrm{Ker}(d^{n+1})/\mathrm{Im}(d^n)$, of this complex.

Show that $H^0(X, \xi)$ may be identified with the abelian group $\mathscr{K}^{\mathbf{T}}((X, \xi), (Y, \theta))$. [Hint: $\xi = \mathrm{coeq}(X\mu, \xi T)$ in $\mathscr{K}^{\mathbf{T}}$.]

The next exercise summarizes [Beck '67, Theorem 5] and deals with an interpretation of $H^1(X, \xi)$. The reader will have to work hard to expand the hints into proofs!

21. (a) Let the notation be as in 20. We introduce the notation $Z^1(X, \xi)$ for the subgroup $\ker(d^2)$ of 1-cocycles in $[XTT, Y]$. Show that the passages

$$\gamma \mapsto a = (X\mu.sT.\gamma.\mathrm{pr}_Y - \xi T.sT.\gamma.\mathrm{pr}_Y) \in [XTT, Y]$$

(where $s = X \xrightarrow{\binom{0}{\mathrm{id}_X}} Y \times X$)

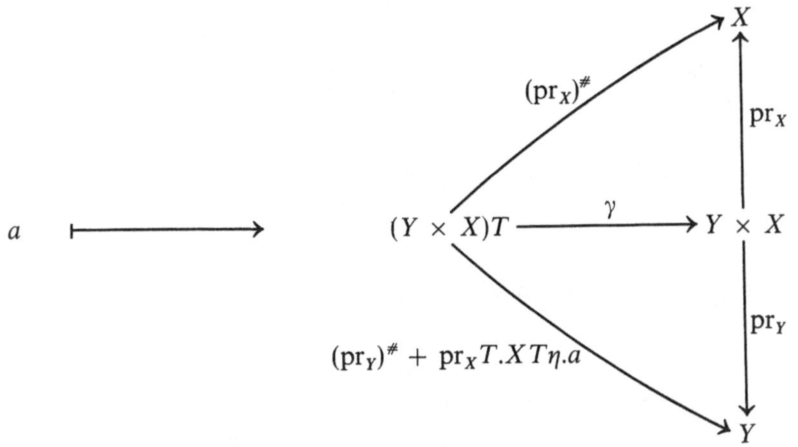

establish mutually inverse bijections between $Z^1(X, \xi)$ and the set of all **T**-algebra structures γ on $Y \times X$ which satisfy

(i) $\mathrm{pr}_X:(Y \times X, \gamma) \longrightarrow (X, \xi)$ is a **T**-homomorphism, and

(ii) $Y \times (Y \times X) \cong (Y \times Y) \times X \xrightarrow{+\times \mathrm{id}} Y \times X$ is a **T**-homomorphism from $(Y, \theta) \times (Y \times X, \gamma)$ to $(Y \times X, \gamma)$.

(b) Via (a), identify $H^1(X, \xi)$ with the group of all equivalence classes of γ under the equivalence relation generated by the following relation: $\gamma \sim \gamma'$ if there exists $q: Y \times X \longrightarrow Y$ such that

$$\begin{pmatrix} q \\ \mathrm{pr}_X \end{pmatrix}:(Y \times X, \gamma) \longrightarrow (Y \times X, \gamma')$$

is a **T**-homomorphism and such that

$$\begin{array}{ccc} Y \times Y \times X & \xrightarrow{+\times \mathrm{id}} & Y \times X \\ {\scriptstyle \mathrm{id}\times q}\big\downarrow & & \big\downarrow{\scriptstyle q} \\ Y \times Y & \xrightarrow[+]{} & Y \end{array}$$

(c) A *principal object over* (X, ξ) is (E, γ, p, m, s) where $p:(E, \xi) \longrightarrow (X, \xi)$ and $m:(Y, \theta) \times (E, \gamma) \longrightarrow (Y, \theta)$ are **T**-homomorphisms and $s: X \longrightarrow E$ is a $\mathscr{K}$-morphism subject to the following four conditions:

$$\begin{array}{ccc} Y \times Y \times E & \xrightarrow{\mathrm{id}\times m} & Y \times E \\ {\scriptstyle +\times \mathrm{id}}\big\downarrow & & \big\downarrow{\scriptstyle m} \\ Y \times E & \xrightarrow[m]{} & E \end{array} \qquad \begin{array}{ccc} E & \xrightarrow{\begin{pmatrix} 0 \\ \mathrm{id} \end{pmatrix}} & Y \times E \\ & {\scriptstyle \mathrm{id}}\searrow \quad \swarrow{\scriptstyle m} & \\ & E & \end{array}$$

$$\begin{array}{ccc} (Y, \theta) \times (E, \gamma) & \xrightarrow{m} & (E, \gamma) \\ {\scriptstyle \mathrm{pr}_E}\big\downarrow & \text{pullback in } \mathscr{K}^{\mathbf{T}} & \big\downarrow{\scriptstyle p} \\ (E, \gamma) & \xrightarrow[p]{} & (X, \xi) \end{array} \qquad \begin{array}{ccc} X & \xrightarrow{s} & E \\ & {\scriptstyle \mathrm{id}}\searrow & \big\downarrow{\scriptstyle p} \\ & & X \end{array}$$

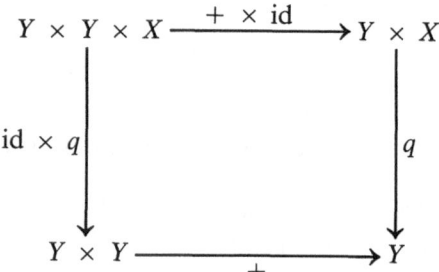

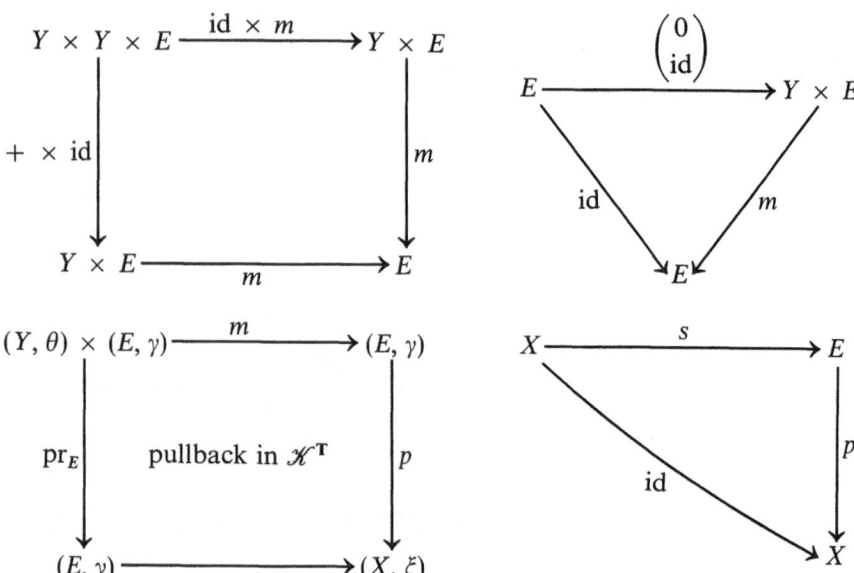

A *morphism* of principal objects is a **T**-homomorphism which preserves m and p (but ignores the choice of "section" s). Show that isomorphism classes of principal objects correspond to elements of $H^1(X, \xi)$. [Hint: (m, pr_E, p) is an absolute coequalizer using 1.5; use the pullback property to define the $\mathscr{K}$-morphism σ as shown below:

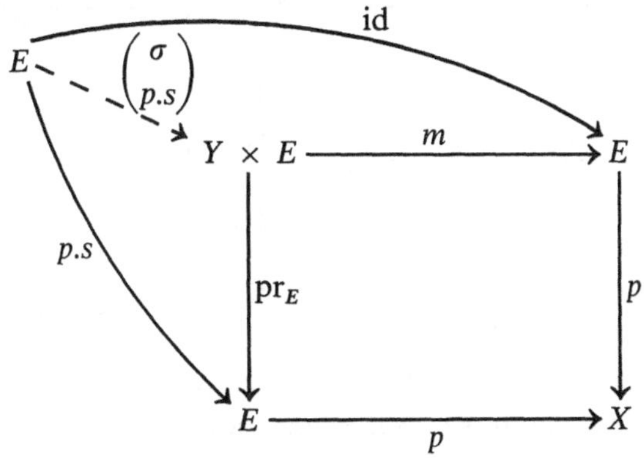

and show that

$$Y \xleftarrow{\sigma} E \xrightarrow{p} X$$

is a product diagram in $\mathscr{K}$.

(d) $m: Y \times E \longrightarrow E$ is an *action of* Y *on* E if the first two diagrams of (c) hold (see exercise 3.2.4). Let $\mathscr{K} = \mathbf{Set}$. Say that the action (Y, E, m) is *fixpoint-free* if $(q, -)m: Y \longrightarrow E$ is injective for each q in E. In the context of (c), show that elements of $H^1(X, \xi)$ correspond to (suitable isomorphism classes of) fixpoint-free actions whose orbit-space is a quotient **T**-algebra isomorphic to (X, ξ).

(e) [Duskin.] Let $\mathscr{K} = \mathbf{Set}$, let $\mathbf{T} = \boldsymbol{\beta}$ and let $Y = \{0, 1\}$ be the two-element compact group. Let E be the circle of radius 2, with arc length parameter q, $0 \leqslant q < 4\pi$. Then E is a $\boldsymbol{\beta}$-algebra and Y acts freely via

$$q0 = q; q1 = q + 2\pi \quad (\mathrm{mod}\ 4\pi)$$

Show that the orbitspace is the circle S and conclude that $H^1(\mathsf{S}) \neq 0$. [Hint: E is connected whereas the trivial principal bundle $\mathsf{S} \times Y$ is not.]

[Van Osdol '75] has shown how to identify 1-cocycles with certain functors $(\mathscr{K}^{\mathbf{T}})^{\mathrm{op}} \longrightarrow \mathbf{Set}$.

For results in higher dimensions see [Duskin '74, '75].

Essentially nothing is known about the relationship between the "triple" cohomology of $\boldsymbol{\beta}$-algebras and the various cohomologies of algebraic topology. We record a few elementary comments concerning "Eilenberg-Steenrod" axioms ([Eilenberg and Steenrod '52]).

22. Let **T** be an algebraic theory in **Set** and let $\mathscr{K}$ be the category whose objects are $(\bar{X}, \bar{\xi}, \bar{A})$ with A a subalgebra of (X, ξ) and whose morphisms

$f:(X, \xi, A) \longrightarrow (\bar{X}, \bar{\xi}, \bar{A})$ are **T**-homomorphisms $f:(X, \xi) \longrightarrow (\bar{X}, \bar{\xi})$ such that f maps A into $\bar{A}$. Fix an abelian group $(Y, \theta; +, i, 0)$ in $\mathbf{Set}^{\mathbf{T}}$. We use the notations of 20.

(a) If (A, ξ_0) is a subalgebra of (X, ξ) with inclusion j, show that

$$\begin{array}{ccccccccc}
0 \xrightarrow{d^0} & [XT, Y]_A & \xrightarrow{d^1} & \cdots & \to [XT^{n+1}, Y]_A & \xrightarrow{d^{n+1}} & [XT^{n+2}, Y]_A & \to & \cdots \\
& \downarrow & & & \downarrow & & \downarrow & & \\
0 \xrightarrow{d^0} & [XT, Y] & \xrightarrow{d^1} & \cdots & \to [XT^{n+1}, Y] & \xrightarrow{d^{n+1}} & [XT^{n+2}, Y] & \to & \cdots \\
& \downarrow{\scriptstyle jT.(-)} & & & \downarrow{\scriptstyle jT^{n+1}.(-)} & & \downarrow{\scriptstyle jT^{n+2}.(-)} & & \\
0 \xrightarrow{d^0} & [AT, Y] & \xrightarrow{d^1} & \cdots & \to [AT^{n+1}, Y] & \xrightarrow{d^{n+1}} & [AT^{n+2}, Y] & \to & \cdots
\end{array}$$

$jT^{n+1}.(-):[XT^{n+1}, Y] \longrightarrow [AT^{n+1}, Y]$ is a cochain complex morphism whose pointwise kernel $[XT^{n+1}, Y]_A = \mathrm{Ker}(jT^{n+1}.(-))$ is a subcomplex of $[XT^{n+1}, Y]$. Verify that the passage $(X, \xi, A) \longmapsto [XT^{n+1}, Y]_A$ is functorial, thereby defining cohomology groups $H^n(X, \xi, A)$. It follows from a standard theorem ([Mac Lane '63, II.4.1]) that each (X, ξ, A) induces a "long exact sequence"

$$\begin{array}{l} 0 \longrightarrow H^0(X, \xi) \longrightarrow H^0(A, \xi_0) \longrightarrow H^0(X, \xi, A) \\ \qquad\qquad \longrightarrow H^1(X, \xi) \longrightarrow H^1(A, \xi_0) - \cdots \end{array}$$

Let 0 denote the subalgebra of (X, ξ) generated by the empty set. Show that $[0, Y] = 0$ and conclude that $H^n(X, \xi, 0) = H^n(X, \xi)$. Thus the long exact sequence is also induced by the $\mathscr{K}$-diagram

$$(A, \xi_0, 0) \longrightarrow (X, \xi, 0) \longrightarrow (X, \xi, A).$$

(b) In the general context of (20), show that $H^n(XT, X\mu) = 0$ if $n > 0$. [Hint: construct a "contracting homotopy" ([Mac Lane '63, page 41])

$$\begin{array}{ccccccccc}
0 & \longrightarrow & [XT, Y] & \longrightarrow & (XT^2, Y] & \longrightarrow & [XT^3, Y] & \longrightarrow & \cdots \\
& \swarrow{\scriptstyle s_0} & & \swarrow{\scriptstyle s_1} & & \swarrow{\scriptstyle s_2} & & \swarrow & \\
0 & \longleftarrow & [XT, Y] & \longrightarrow & [XT^2, Y] & \longrightarrow & [XT^3, Y] & \longrightarrow & \cdots
\end{array}$$

(s_n) by setting $s_n = X\eta T^n.(-)$ for $n \geqslant 1$.] This establishes the "dimension axiom": $H^n(1T, 1\mu) = 0$.

(c) Show that $\boldsymbol{\beta}: \mathbf{Set} \longrightarrow \mathbf{Set}$ preserves binary coproducts. [Hint: use 1.5.25 (3).]

(d) Assume that $T: \mathbf{Set} \longrightarrow \mathbf{Set}$ preserves binary coproducts. Prove the "excision axiom": if $f:(X, \xi, A) \longrightarrow (\bar{X}, \bar{\xi}, \bar{A})$ induces a bijection from $X - A$ to $\bar{X} - \bar{A}$ then $H^n f: H^n(X, \xi, A) \longrightarrow H^n(\bar{X}, \bar{\xi}, \bar{A})$ is an isomorphism for all $n \geqslant 0$. [Hint: to prove $H^n f$ is injective, let $\bar{a}$, $\bar{b}$ in $[\bar{X}T^{n+1}, Y]_{\bar{A}}$ be n-cocycles with $fT^{n+1}.\bar{a} - fT^{n+1}.\bar{b} = d^n(t)$ for some t in $[XT^n, Y]_A$ then $\bar{a} - \bar{b} = \bar{d}^n(\bar{c})$ where $\bar{c}$ in $[\bar{X}T^n, Y]_{\bar{A}}$ is induced by the coproduct property:

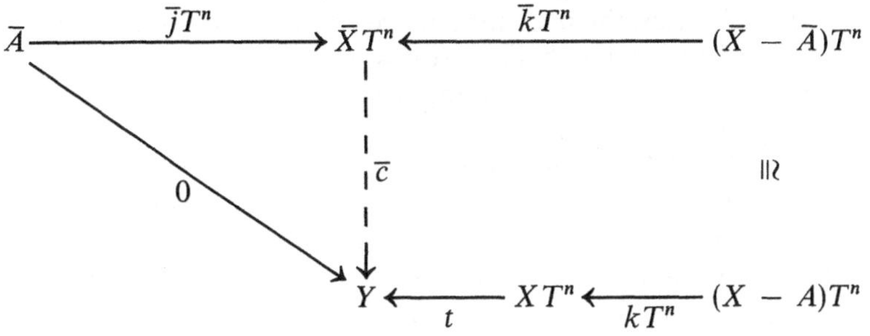

To show $H^n f$ is onto, let a in $[XT^{n+1}, Y]_A$ be an n-cocycle and show that $fT^{n+1}.\bar{a} = a$, where $\bar{a}$ in $[\bar{X}T^{n+1}, Y]_{\bar{A}}$ is defined by

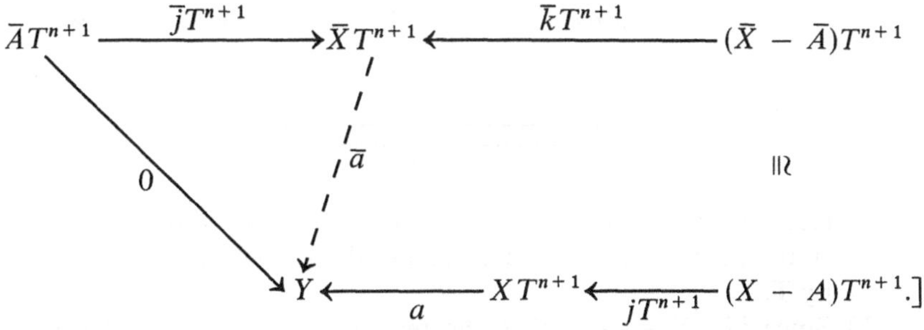

We do not know the extent to which the homotopy axiom holds.

23. For K in $\mathscr{K}$, the category $(\mathscr{K}, K)$ of *objects over* K has as objects all (A, a) with $a: A \longrightarrow K$, and has as morphisms $f:(A, a) \longrightarrow (B, b)$ all $f: A \longrightarrow B$ with $f.b = a$.
    (a) Show that $(\mathscr{K}, K)$ has finite products when $\mathscr{K}$ has pullbacks.
    (b) Let **T** be an algebraic theory in $\mathscr{K}$ and let (Z, γ) be a **T**-algebra. Show that the obvious forgetful functor $(\mathscr{K}^{\mathbf{T}}, (Z, \gamma)) \longrightarrow (\mathscr{K}, Z)$ is algebraic with algebraic theory $(A, a)\tilde{T} = (AT, a^{\#})$, $(A, a)\tilde{\eta} = A\eta$: $(A, a) \longrightarrow (AT, a^{\#})$, $(A, a)\tilde{\mu} = A\mu:(ATT, (a^{\#})^{\#}) \longrightarrow (AT, a^{\#})$.
    (c) [Beck '67.] Let $\mathscr{A}$ be the category of abelian groups in $(\mathscr{K}, R)$, where $\mathscr{K}$ is the category of commutative rings with unit and unitary ring homomorphisms. An object of $\mathscr{A}$, then, is a ring homomorphism $f: Y \longrightarrow R$ with addition defined on each rf^{-1} and whose zero is a group homomorphism $z: R \longrightarrow Y$ satisfying $z.f = \text{id}$. Show that the structural reflection of $U: \mathscr{A} \longrightarrow$ **Set** defined by $(Y, f, +, z)U = \text{Ker}(f)$ is isomorphic over **Set** to the category of right R-modules.

[Hint: for a in $\mathrm{Ker}(f)$ and r in R, ar is the Y-product $(0, a)(r, 0)$, identifying Y with $R \times \mathrm{Ker}(f)$ as abelian groups; conversely, invent the "semi-direct product" of R with an R-module.]

Many algebraic cohomology theories coincide with triple cohomology at the level $(\mathscr{K}^{\mathbf{T}}, (Z, \gamma))$ with coefficients in a "(Z, γ)-module," that is, an abelian group in $(\mathscr{K}^{\mathbf{T}}, (Z, \gamma))$ (see [Beck '67] for examples). In such cases, the need for a more complex theory is explained by the lack of nontrivial abelian groups in $\mathscr{K}^{\mathbf{T}}$.

(d) In the category of commutative rings, prove that the zero ring is the only abelian group.

24. [Duskin '75.] This exercise outlines how the Čech cohomology group of a topological space may be defined using triple cohomology; the reader familiar with the standard definition should verify that the following agrees with the usual construction. Let X be a topological space, let $\mathscr{U}$ be an open cover of X and let Y be a topological abelian group. Let $\mathscr{K}$ be the category of topological spaces and continuous maps. Let Z be the coproduct in $\mathscr{K}$ of the elements of $\mathscr{U}$ (each having the subspace topology) and let $\gamma: Z \longrightarrow X$ be the codiagonal map, i.e., $in_U.\gamma$ is the inclusion for each U in $\mathscr{U}$.

(a) Using the notations of exercise 23, show that U = "pullback along γ"

$$(\mathscr{K}, X) \underset{F}{\overset{U}{\rightleftarrows}} (\mathscr{K}, Z)$$

(see the diagram above) is algebraic with left adjoint F = "composition with γ." Denote the corresponding algebraic theory in $(\mathscr{K}, Z)$ by $\mathbf{T}$.

(b) Since $\mathrm{id}_X: X \longrightarrow X$ is the terminal object in $(\mathscr{K}, X)$ it is a $\mathbf{T}$-algebra. Interpret the "canonical resolution" (exercise 19) of this algebra. [Hint: in degree n consider n-fold intersections of elements of $\mathscr{U}$.]

(c) Show that Y, via $\mathrm{pr}_Y: Y \times X \longrightarrow X$, is an abelian group in $(\mathscr{K}, X)$ in a natural way.

The resulting first triple cohomology group is naturally denoted as $H^1(X, \mathscr{U})$. The Čech group is then the colimit

$$H(X) = \operatorname*{colim}_{\mathscr{U}} H^1(X, \mathscr{U})$$

2. Theories as Monoids

In this section we explicate the analogy between monoids and algebraic theories in monoid form (1.3.17) by demonstrating that both are examples of the more general notion of "monoid in a monoidal category." The obvious monoid homomorphisms $\lambda: \mathbf{T} \longrightarrow \mathbf{S}$ are identified with the homomorphisms $V: (\mathscr{K}^{\mathbf{S}}, U^{\mathbf{S}}) \longrightarrow (\mathscr{K}^{\mathbf{T}}, U^{\mathbf{T}})$ over $\mathscr{K}$.

2.1 Monoids in a Monoidal Category. An (ordinary) monoid is a triple (X, m, e) where X is a set and $m: X \times X \longrightarrow X$ is a binary associative operation having $e \in X$ as a two-sided unit. A "monoidal" category should be a category provided with whatever additional structure seems appropriate to allow the context for the definition of an ordinary monoid.

Define the *product* $\mathscr{K}_1 \times \cdots \times \mathscr{K}_n$ of the categories $\mathscr{K}_i$ $(n > 0)$ to have as objects all ordered n-tuples $(A_1, \ldots, A_n)$ with A_i an object of $\mathscr{K}_i$ and to have as morphisms $(A_1, \ldots, A_n) \longrightarrow (B_1, \ldots, B_n)$ all n-tuples $(f_1, \ldots, f_n)$ with $f_i : A_i \longrightarrow B_i$ in $\mathscr{K}_i$. Composition is given by $(f_1, \ldots, f_n)(g_1, \ldots, g_n) = (f_1 g_1, \ldots, f_n g_n)$ and $\mathrm{id}_{(A_1, \ldots, A_n)} = (\mathrm{id}_{A_1}, \ldots, \mathrm{id}_{A_n})$. Then if $\mathscr{K}$ has binary products $A \times B$, $\times : \mathscr{K} \times \mathscr{K} \longrightarrow \mathscr{K}$ is a functor, the action on morphisms being given by

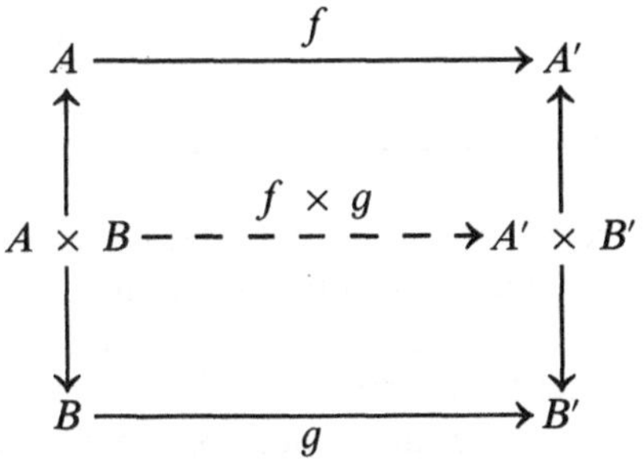

(Notice that the definition requires us to choose a fixed product

$$A \longleftarrow A \times B \longrightarrow B$$

for each pair of objects (A, B).) Each three objects A, B, C induce the "associativity isomorphism" $(A, B, C)a : A \times (B \times C) \longrightarrow (A \times B) \times C$ by

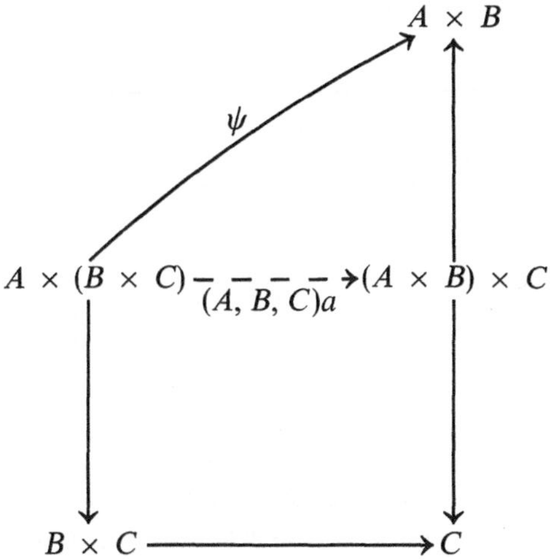

where $\psi: A \times (B \times C) \longrightarrow A \times B$ is defined by

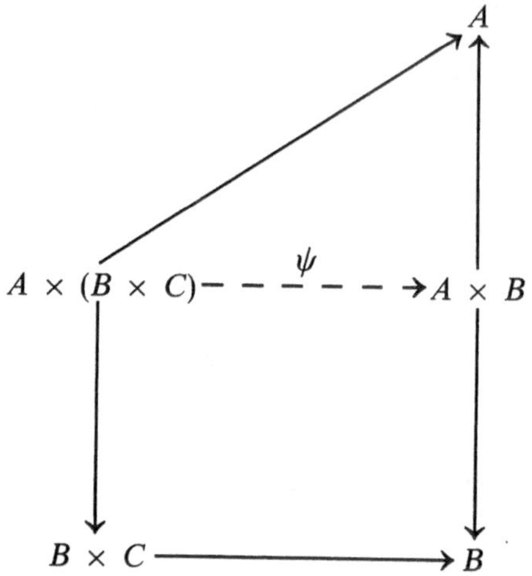

We leave it as an exercise for the reader that $(A, B, C)a$ is an isomorphism and that both a and a^{-1} are natural transformations of functors from $\mathscr{K} \times \mathscr{K} \times \mathscr{K}$ to $\mathscr{K}$; that is, a is a *natural equivalence* (as defined in 2.5 below). Similarly, if $\mathscr{K}$ has a terminal object 1, the projections $Ab: 1 \times A \longrightarrow A$, $Ac: A \times 1 \longrightarrow A$ are natural equivalences of functors from $\mathscr{K}$ to itself.

We may then set $\mathscr{K}$ = **Set** and observe that an ordinary monoid is (X, m, e) where $m: X \times X \longrightarrow X$ and $e: 1 \longrightarrow X$ are morphisms satisfying

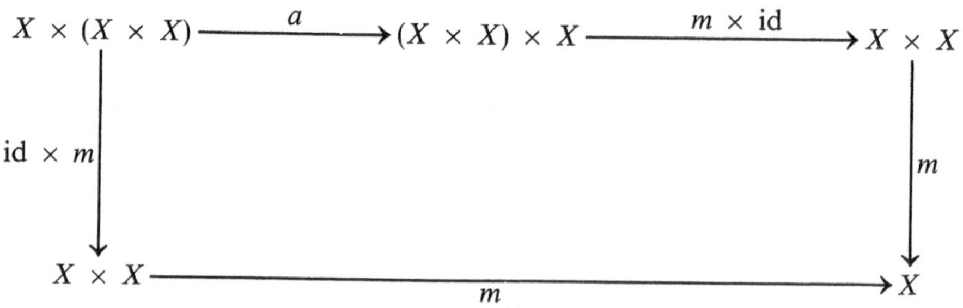

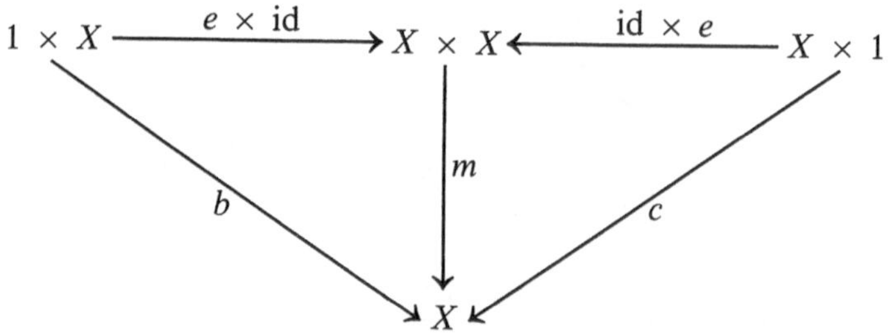

We have motivated the following definition:

(2.2) A *monoidal category* is a 6-tuple $(\mathscr{K}, \otimes, I, a, b, c)$ where $\mathscr{K}$ is a category, $\otimes: \mathscr{K} \times \mathscr{K} \longrightarrow \mathscr{K}$ is a functor, I is an object of $\mathscr{K}$ (the "unit for

$\otimes$"), and $(A, B, C)a: A \otimes (B \otimes C) \longrightarrow (A \otimes B) \otimes C$, $Ab: I \otimes A \longrightarrow A$, $Ac: A \otimes I \longrightarrow A$ are natural equivalences subject to the commutativity of the following three diagrams:

$$\begin{array}{ccccc}
A \otimes (B \otimes (C \otimes D)) & \xrightarrow{a} & (A \otimes B) \otimes (C \otimes D) & \xrightarrow{a} & ((A \otimes B) \otimes C) \otimes D \\
\Big\downarrow{\scriptstyle \mathrm{id} \otimes a} & & & & \Big\downarrow{\scriptstyle a \otimes \mathrm{id}} \\
A \otimes ((B \otimes C) \otimes D) & & \xrightarrow[a]{\qquad\qquad} & & (A \otimes (B \otimes C)) \otimes D
\end{array}$$

$$\begin{array}{ccc}
A \otimes (I \otimes C) & \xrightarrow{a} & (A \otimes I) \otimes C \\
{\scriptstyle \mathrm{id} \otimes b} \searrow & & \swarrow {\scriptstyle c \otimes \mathrm{id}} \\
& A \otimes C &
\end{array}
\qquad\qquad
I \otimes I \underset{c}{\overset{b}{\rightrightarrows}} I$$

The reader should check that these axioms hold for the motivating example $(\mathscr{K}, \times, 1, a, b, c)$.

Let $(\mathscr{K}, \otimes, I, a, b, c)$ be a monoidal category. A *monoid in* $(\mathscr{K}, \otimes, I, a, b, c)$ is a triple (K, m, e) where $m: K \times K \longrightarrow K$ and $e: I \longrightarrow K$ are $\mathscr{K}$-morphisms subject to the laws

$$\begin{array}{ccccc}
K \otimes (K \otimes K) & \xrightarrow{a} & (K \otimes K) \otimes K & \xrightarrow{\mathrm{m} \otimes \mathrm{id}} & K \otimes K \\
\Big\downarrow{\scriptstyle \mathrm{id} \otimes m} & & & & \Big\downarrow{\scriptstyle m} \\
K \otimes K & & \xrightarrow[m]{\qquad\qquad} & & K
\end{array} \tag{2.3}$$

$$\begin{array}{ccccc}
I \otimes K & \xrightarrow{e \otimes \mathrm{id}} & K \otimes K & \xleftarrow{\mathrm{id} \otimes e} & K \otimes I \\
& {\scriptstyle b} \searrow & \Big\downarrow{\scriptstyle m} & \swarrow {\scriptstyle c} & \\
& & K & &
\end{array}$$

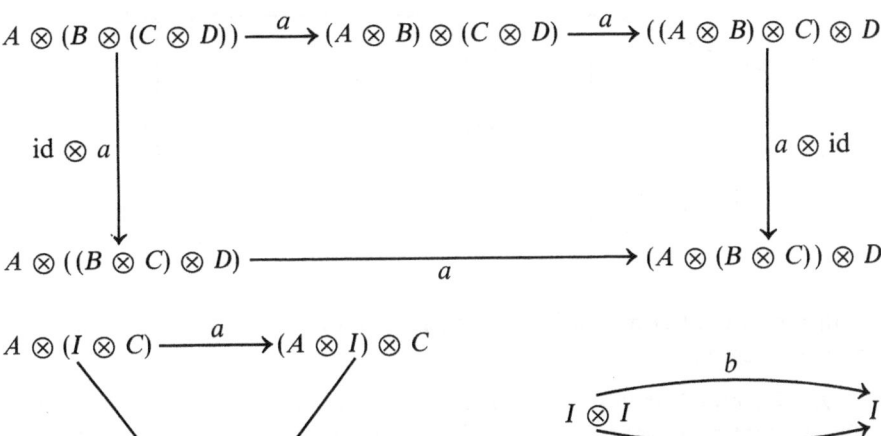

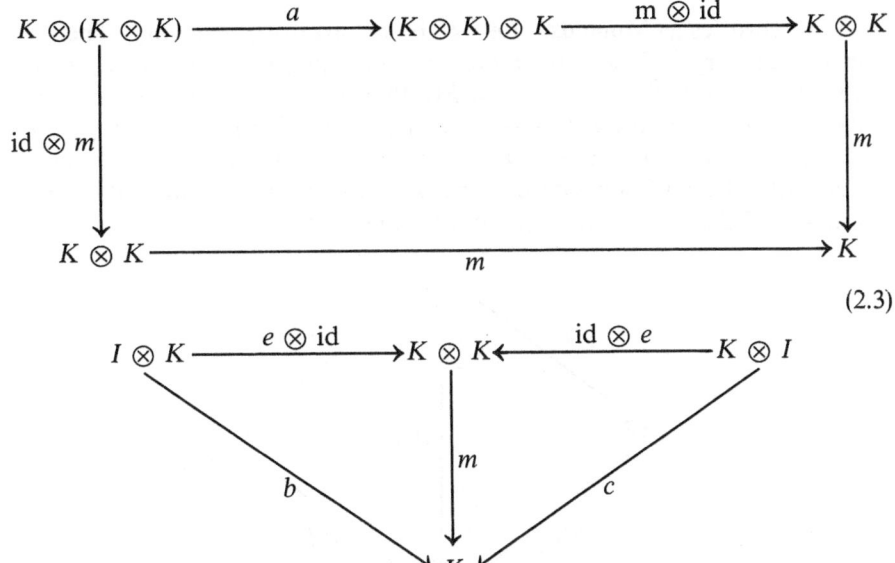

A *monoid homomorphism* $f:(K, m, e) \longrightarrow (K', m', e')$ is a $\mathcal{K}$-morphism $f: K \longrightarrow K'$ such that

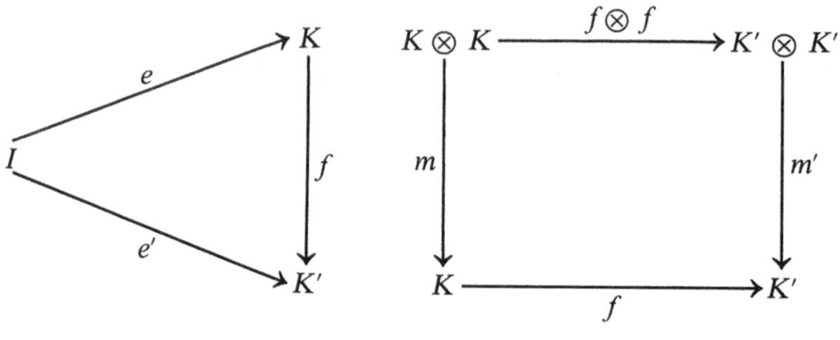

(2.4)

With the evident composition and identities, the $(\mathcal{K}, \otimes, I, a, b, c)$-monoids form a category.

2.5 Functor Categories. It is an important observation in many areas of mathematics that the homomorphisms between two objects of interest can be structured itself so as to become such an object. Witness "function space" theory in analysis or vector spaces of linear maps; also, see exercise 5. In category theory, it is the case that the functors between two categories $\mathcal{A}$ and $\mathcal{K}$ form a category $\mathcal{K}^{\mathcal{A}}$. Such categories are called *functor categories* (cf. section 2.2, exercise 9). Specifically, the objects of $\mathcal{K}^{\mathcal{A}}$ are functors $H: \mathcal{A} \longrightarrow \mathcal{K}$ and a morphism $\alpha: H_1 \longrightarrow H_2$ is a natural transformation. Composition is horizontal composition (2.2.31). The identities are provided by $A(\mathrm{id}_H) = \mathrm{id}_{AH}$. As mentioned in 2.1, the isomorphisms in $\mathcal{K}^{\mathcal{A}}$ are called *natural equivalences* and are the same thing as natural transformations each of whose components is an isomorphism in $\mathcal{K}$.

2.6 Theories as Monoids. Let $\mathcal{K}$ be an arbitrary category. Then the functor category $\mathcal{K}^{\mathcal{K}}$ becomes a monoidal category in a natural way. Given $F, G: \mathcal{K} \longrightarrow \mathcal{K}$ define $F \otimes G = FG$, their usual composition; explicitly, $K(FG) = (KF)G$ and for $f: K \longrightarrow K'$, $f(FG) = (fF)G$. Given $\alpha: F \longrightarrow F'$ and $\beta: G \longrightarrow G'$ define $\alpha \otimes \beta: FG \longrightarrow F'G'$—which we will write simply as $\alpha\beta$ (not to be confused with the functor category composition $\alpha.\beta$)— as the path from FG to $F'G'$ in the square shown below:

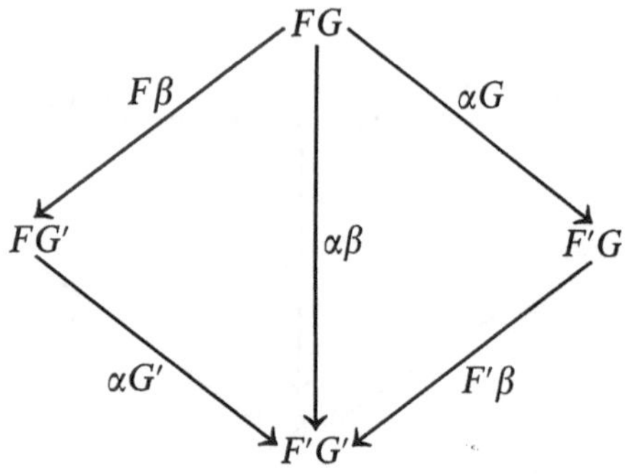

This is called the *vertical composition* of α and β. The two paths are the same because β is a natural transformation. $F\beta$, $\alpha G'$, αG, $F'\beta$ are all natural transformations and we have $F\beta.\alpha G' = \alpha\beta = \alpha G.F'\beta$. $\otimes:\mathscr{K} \times \mathscr{K} \longrightarrow \mathscr{K}$ is a well-defined functor which is strictly associative on objects and morphisms. The following diagrams will aid the reader in the verification.

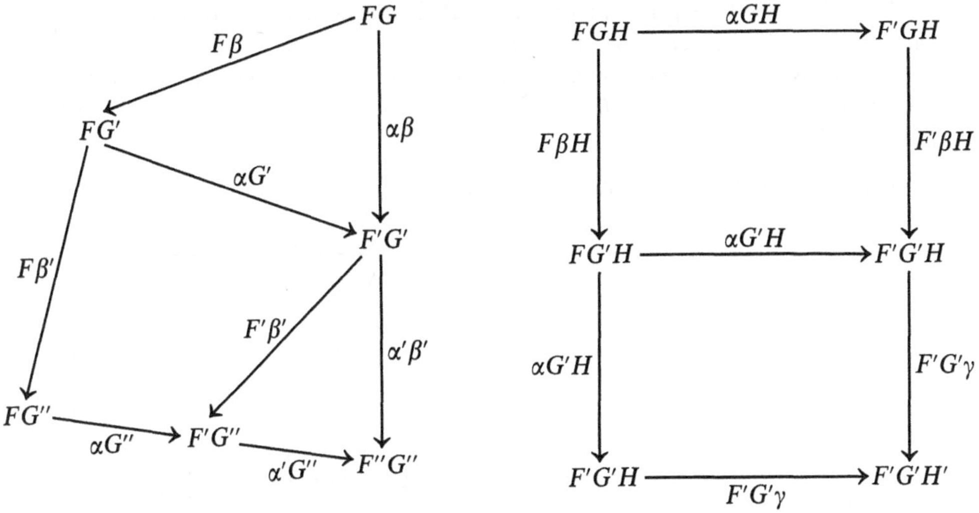

The unit object I is just the identity functor $\mathrm{id}_{\mathscr{K}}:\mathscr{K} \longrightarrow \mathscr{K}$. The remaining verification that $(\mathscr{K}^{\mathscr{K}}$, comp, $\mathrm{id}_{\mathscr{K}}$, id, id, id) is a monoidal category will be left to the reader; this statement effectively sums up the rules of the Godement calculus (2.2.31; see exercise 2). We have written comp instead of $\otimes$ since this seems more natural. $(\mathscr{K}^{\mathscr{K}}$, comp, id) is a *strict* monoidal category in the sense that a, b, c are all identity transformations, a pleasant property.

According to 2.3, a monoid of $(\mathscr{K}^{\mathscr{K}}$, comp, id) is the same thing as an algebraic theory (T, η, μ) in $\mathscr{K}$ in monoid form (1.3.17). Since the monoids of a monoidal category form a category we have learned, now, that the algebraic theories in a category form a category. By 2.4 a *theory map* $\lambda:\mathbf{T} \longrightarrow \mathbf{T}'$ is a natural transformation $\lambda: T \longrightarrow T'$ such that the following two diagrams commute:

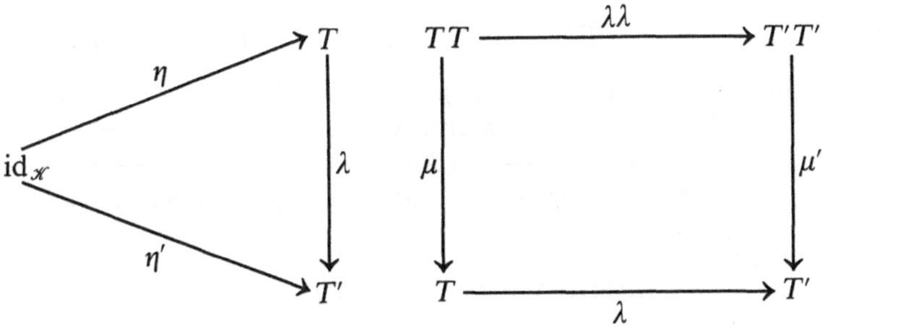

(2.7)

The composition of the theory map $\lambda:\mathbf{T} \longrightarrow \mathbf{T}'$ and $\lambda':\mathbf{T}' \longrightarrow \mathbf{T}''$ is just $\lambda.\lambda': T \longrightarrow T''$.

The following proposition shows how to translate the definition of "theory map" for theories in clone form (1.3.2).

2.8 Proposition. *Let* $\mathbf{T} = (T, \eta, \circ, \mu)$ *and* $\mathbf{T}' = (T', \eta', \circ', \mu')$ *be algebraic theories in* $\mathcal{K}$ *and let* $K\lambda : KT \longrightarrow KT'$ *be given for each* $K \in \mathcal{K}$. *The following two conditions are equivalent:*

1. $\lambda : \mathbf{T} \longrightarrow \mathbf{T}'$ *is a theory map as in 2.6, that is,* λ *is a natural transformation* $T \longrightarrow T'$ *subject to the diagrams 2.7.*

2. $K\eta . K\lambda = K\eta'$ *for all* $K \in \mathcal{K}$; *also, "composition with* λ*" preserves Kleisli composition:* $(\alpha \circ \beta).C\lambda = (\alpha.B\lambda) \circ' (\beta.C\lambda)$ for all $\alpha : A \longrightarrow BT$, $\beta : B \longrightarrow CT$.

Proof. 1 implies 2. Recall 1.3.14 and use

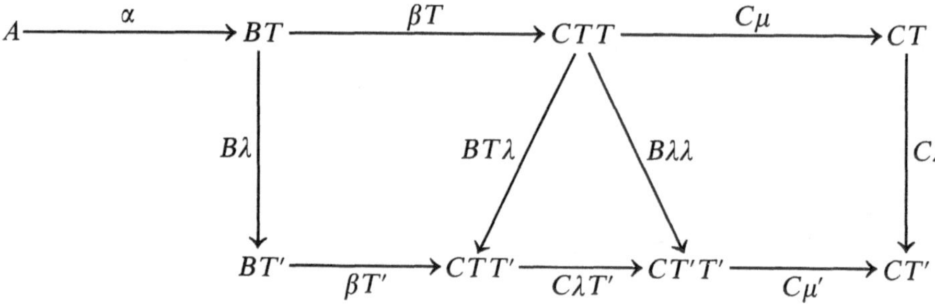

2 implies 1. To prove that $\lambda : T \longrightarrow T'$ is natural let $f : A \longrightarrow B$ and recall 1.3.10. We have $fT.B\lambda = (\mathrm{id}_{AT} \circ f^{\Delta}).B\lambda = A\lambda \circ' (f^{\Delta}.B\lambda) = A\lambda \circ' (f.B\eta') = A\lambda.fT'$ (the last step by 1.3.12). Also, $A\mu.A\lambda = (\mathrm{id}_{ATT} \circ \mathrm{id}_{AT}).A\lambda =$

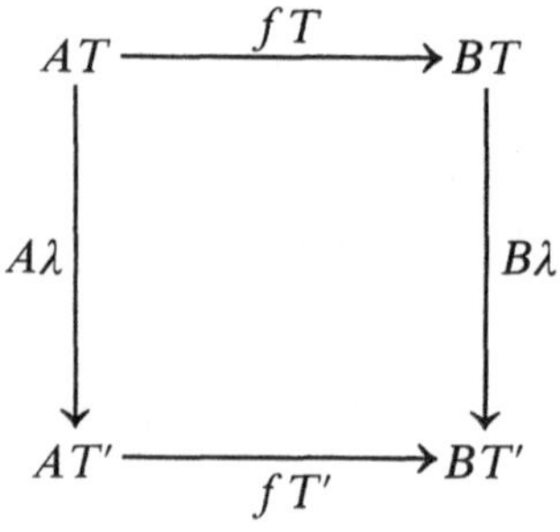

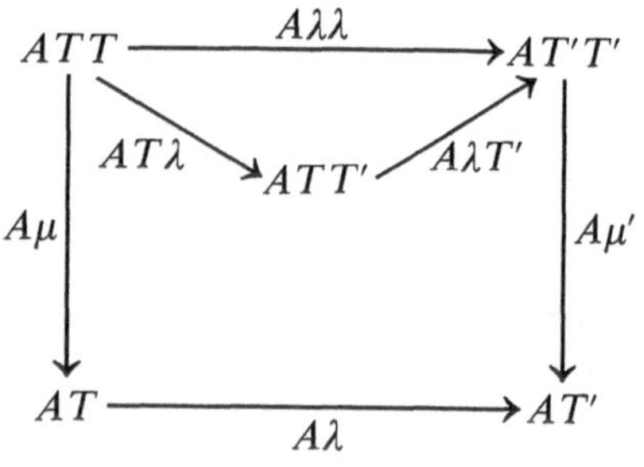

$(\mathrm{id}_{ATT}.AT\lambda) \circ' (\mathrm{id}_{AT}.A\lambda) = AT\lambda \circ' A\lambda = AT\lambda.A\lambda T'.A\mu' = A\lambda\lambda.A\mu'$. $\square$

2.9 Proposition. *Let* $\mathbf{T}$ *and* $\mathbf{T}'$ *be algebraic theories in* $\mathcal{K}$. *Then each theory map* $\lambda : \mathbf{T} \longrightarrow \mathbf{T}'$ *induces a homomorphism* $V : (\mathcal{K}^{\mathbf{T}'}, U^{\mathbf{T}'}) \longrightarrow (\mathcal{K}^{\mathbf{T}}, U^{\mathbf{T}})$ *defined by* $(X, \xi' : XT' \longrightarrow X)V = (X, X\lambda.\xi' : XT \longrightarrow X)$. *Conversely, each such homomorphism* V *induces a theory map* λ *defined by* $X\lambda : XT \longrightarrow XT' =$

$$XT \xrightarrow{X\eta' T} XT'T \xrightarrow{\xi_X} XT'$$

where (XT', ξ_X) *is defined to be* $(XT', X\mu')V$. *Moreover, the two passages are mutually inverse bijections.*

Proof. *λ to V is well defined.*

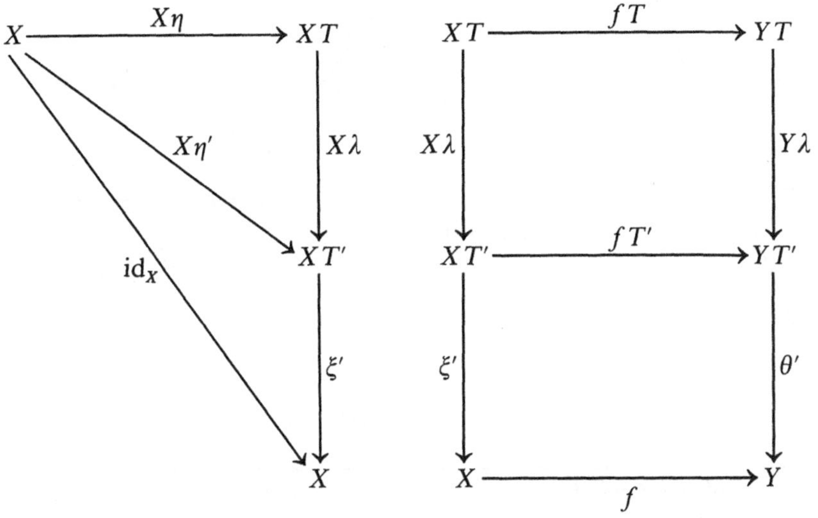

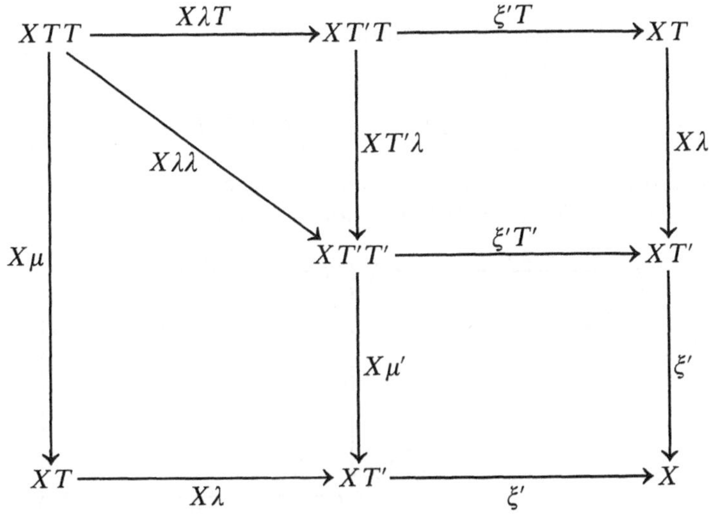

V to λ is well defined.

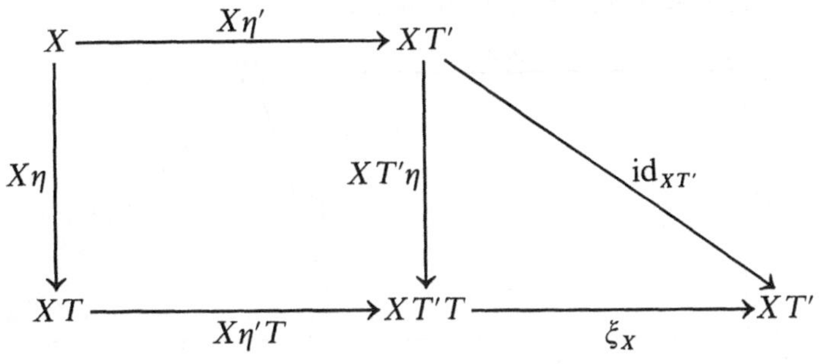

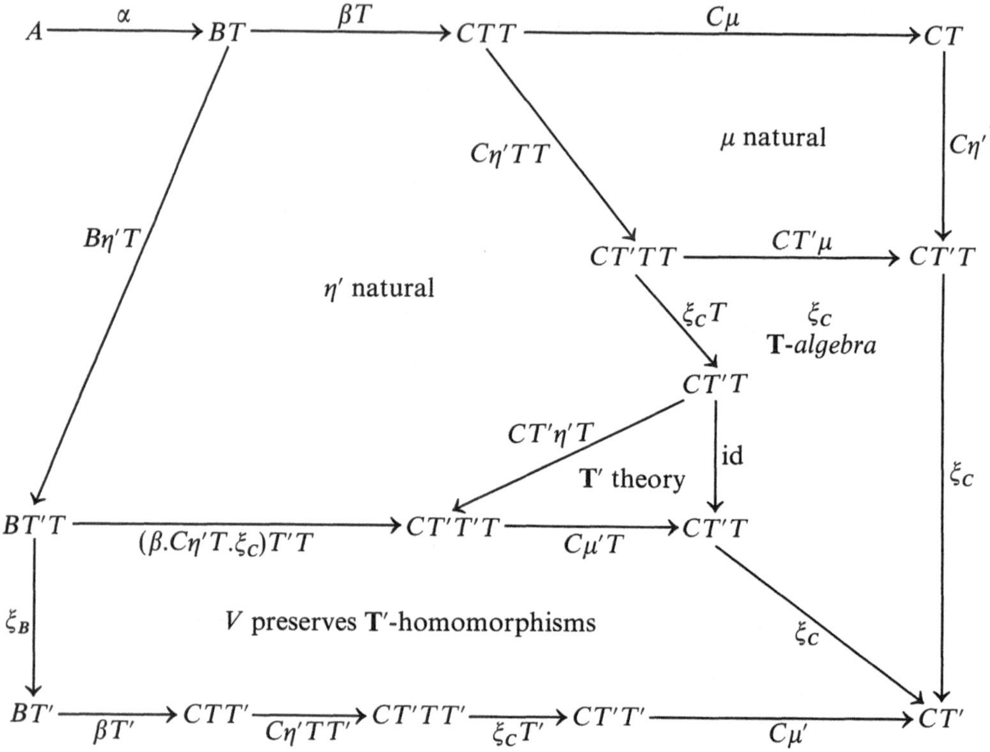

$$A \xrightarrow{\alpha} BT \xrightarrow{\beta T} CTT \xrightarrow{C\mu} CT$$

$C\eta'TT$

μ natural

$C\eta'T$

$Bn'T$

η' natural

$$CT'TT \xrightarrow{CT'\mu} CT'T$$

$\xi_C T$ ξ_C
                   T-*algebra*

$CT'T$

$CT'\eta'T$ id

                T' theory

$$BT'T \xrightarrow{(\beta . C\eta'T . \xi_C)T'T} CT'T'T \xrightarrow{C\mu'T} CT'T$$

ξ_B *V* preserves **T'**-homomorphisms

ξ_C

ξ_C

$$BT' \xrightarrow{\beta T'} CTT' \xrightarrow{C\eta'TT'} CT'TT' \xrightarrow{\xi_C T'} CT'T' \xrightarrow{C\mu'} CT'$$

V to λ to $\bar{V}$, $\bar{V} = V$.   Let (X, ξ') be a **T'**-algebra and let $(X, \theta) = (X, \xi')V$.
$(X, \xi')\bar{V} = (X, X\eta'T . \xi_X . \xi')$. As the **T'**-homomorphism $\xi' : (XT', X\mu') \longrightarrow$

$$XT \xrightarrow{X\eta'T} XT'T \xrightarrow{\xi_X} XT'$$

id_X $\xi'T$ ξ'

$$XT \xrightarrow{\theta} X$$

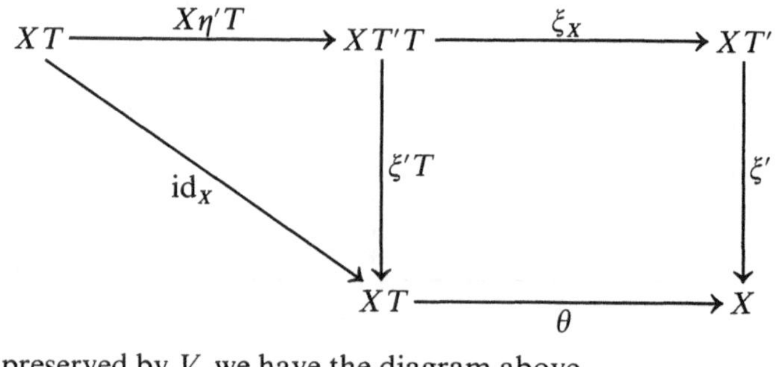

(X, ξ') is preserved by *V*, we have the diagram above.
    λ to V to $\bar{\lambda}$, $\bar{\lambda} = \lambda$. The result follows from

$$XT \xrightarrow{X\lambda} XT'$$

$X\eta'T$ $X\eta'T'$ $\mathrm{id}_{XT'}$

$$XT'T \xrightarrow{XT'\lambda} XT'T' \xrightarrow{X\mu'} XT'$$

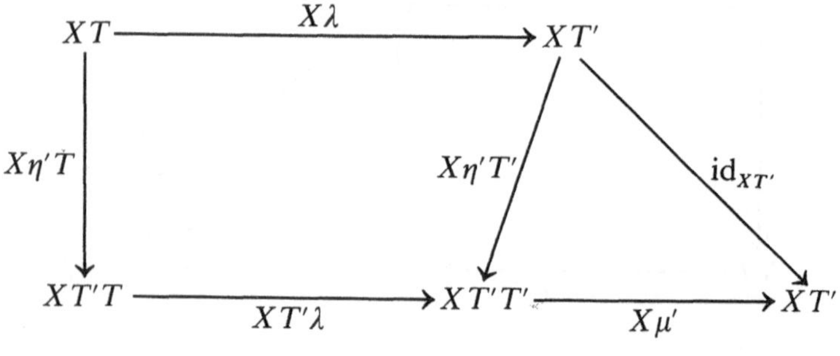

since $(XT', X\mu')V = (XT', XT'\lambda.X\mu')$. □

2.10 Theorem. *Let* Th($\mathscr{K}$) *be the category of algebraic theories* **T** *in* $\mathscr{K}$ *and theory maps* λ:**T** ⟶ **T**′ *as in 2.7 and let* Alg($\mathscr{K}$) *be the full subcategory of* Struct($\mathscr{K}$) *of all algebraic functors. Then the passages of 2.9 establish a full representative subcategory* (2.3.2) $(\mathrm{Th}(\mathscr{K}))^{\mathrm{op}} \longrightarrow \mathrm{Alg}(\mathscr{K})$.

Proof. The only details left to check are the preservation of identities and composition, and these are immediate:

$$X\lambda.\xi = \xi \text{ if } X\lambda = \mathrm{id}_{XT}$$
$$(X\lambda.X\lambda').\xi'' = X\lambda.(X\lambda'.\xi'') \quad \square$$

2.11 Corollary. *Let* $(\mathscr{A}, U)$ *and* $(\mathscr{A}', U') \in$ Struct($\mathscr{K}$) *be algebraic so that there exist algebraic theories* **T**, **T**′ *in* $\mathscr{K}$ *with* $(\mathscr{A}, U)$ *isomorphic to* $(\mathscr{K}^{\mathbf{T}}, U^{\mathbf{T}})$ *and* $(\mathscr{A}', U')$ *isomorphic to* $(\mathscr{K}^{\mathbf{T}'}, U^{\mathbf{T}'})$. *Then* $(\mathscr{A}, U)$ *and* $(\mathscr{A}', U')$ *are isomorphic if and only if* **T** *and* **T**′ *are isomorphic.* □

2.12 η as a theory map. If **T** is a theory in $\mathscr{K}$ then $\eta:\mathrm{id} \longrightarrow T$ is a theory map $\eta:(\mathrm{id}, \mathrm{id}, \mathrm{id}) \longrightarrow$ **T** and the corresponding homomorphism is just $U^{\mathbf{T}}: \mathscr{K}^{\mathbf{T}} \longrightarrow \mathscr{K}$.

2.13 Abelianization of a Free Group. Let $\mathscr{K}^{\mathbf{T}'}$ be abelian groups and let $\mathscr{K}^{\mathbf{T}}$ be groups. Then the abelianization (2.2.7) of the free group $(XT, X\mu)$ is the free abelian group $(XT', X\mu')$ and the canonical projection $\lambda: T \longrightarrow T'$ is the theory map which corresponds to the inclusion from abelian groups into groups.

2.14 Complete Atomic Boolean Algebras as Compact Spaces. Let **T** be the double power-set theory of 1.3.19 whose algebras are complete atomic Boolean algebras (1.5.17). Let **β** be the ultrafilter theory of 1.3.21 whose algebras are the compact Hausdorff spaces (1.5.24). Then inclusion is a theory map **β** ⟶ **T**, inducing a homomorphism V from complete atomic Boolean algebras to compact Hausdorff spaces. Another description of V is as follows: Let X be a complete atomic Boolean algebra with set of atoms A. Then the function

$$X \longrightarrow 2^A$$
$$x \longmapsto \{a \in A : a \leqslant x\}$$

is an isomorphism of complete atomic Boolean algebras. Thus X is isomorphic to the A-fold power of two-element Boolean algebras. Since V must preserve products (1.28) $(2^A)V$ must be the Tychonoff product of two-element discrete spaces.

Notes for Section 2

Mac Lane has referred to "monoid" as a candidate for the fundamental notion of category theory [Mac Lane '71, preface]. Monoidal categories are a subject of much current research interest in category theory; see [Dubuc '70], [Kelly, Laplaza, Lewis, and Mac Lane '72], and the bibliographies there, as well as the exercises. Theorem 2.10 is implicit in [Lawvere '63, section 1 of Chapter III] and was proved in [Applegate '65].

Exercises for Section 2

1. Let $H: \mathscr{K} \times \mathscr{L} \longrightarrow \mathscr{M}$ be a "prefunctor" assigning to each object (K, L) an object $(K, L)H$ of $\mathscr{M}$ and to each morphism $(f, g):(K, L) \longrightarrow (K', L')$ a morphism $(f, g)H:(K, L)H \longrightarrow (K', L')H$. Show that H is a functor if and only if for every (K, L), $H(K, -):\mathscr{L} \longrightarrow \mathscr{M}$ is a functor and $H(-, L):\mathscr{K} \longrightarrow \mathscr{M}$ is a functor.
2. Prove that $(\mathscr{K}^{\mathscr{K}}$, comp, id, id, id, id) is a monoidal category from the Godement rules of 2.2.31, and conversely.
3. A *symmetric monoidal category* is a monoidal category equipped with a *symmetry* natural equivalence $(A, B)d: A \otimes B \longrightarrow B \otimes A$ satisfying

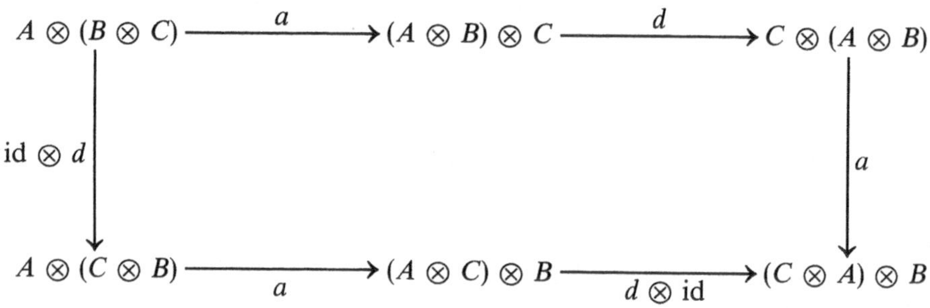

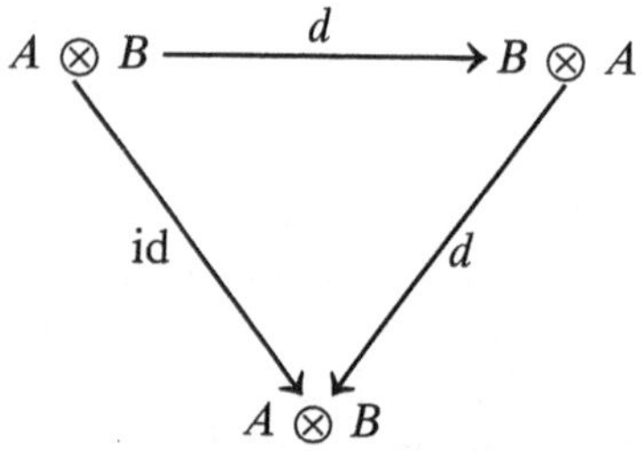

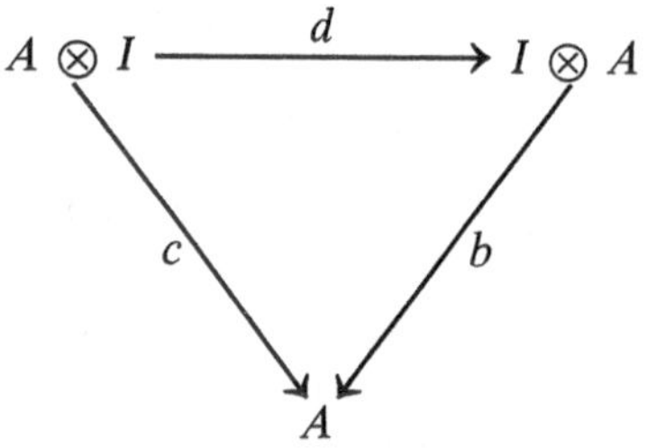

Verify the following examples of symmetric monoidal categories and their monoids.

(a) $(\mathscr{K}, \times\ 1, a, b, c, d)$ for any category with finite products (this is the *cartesian* monoidal structure). $\mathscr{K}$ = topological spaces produces topological monoids (the multiplication is jointly continuous). When $\mathscr{K}$ is the category of categories and functors, the monoids are strict monoidal categories.

(b) Let $\mathscr{C}$ in Struct(**Set**) be fibre complete. Define $(X, s) \otimes (Y, t) = (X \times Y, s \otimes t)$ where $s \otimes t$ is the co-optimal lift of the maps $(X, s) \longmapsto X \times Y$, $x \longmapsto (x, y)$ $(y \in Y)$, $(Y, t) \longrightarrow X \times Y$, $y \longmapsto (x, y)$ $(x \in X)$. Define I to be the empty co-optimal lift in $\mathscr{C}(1)$. The monoids are characterized by having separately admissible multiplications.

(c) Let $\mathscr{K}$ be the category of abelian groups, let $\otimes$ be the usual tensor products and let $I = \mathbf{Z}$. The monoids are rings with unit.

4. Let (X, m, e) be a monoid in the monoidal category $\mathscr{K}$. Define $AT =$

$A \otimes X, A\eta = c^{-1}.(\mathrm{id} \otimes e), \alpha \circ \beta = \alpha.(\beta \otimes \mathrm{id}).(a^{-1}).(\mathrm{id} \otimes m)$. Show that $T = (T, \eta, \circ)$ is an algebraic theory in $\mathscr{K}$. $\mathscr{K}^{\mathbf{T}}$ is the category of (X, m, e)-*actions*. Examine the previous exercise for well-known examples of actions. Show that each monoid homomorphism induces a theory map.

5. A symmetric monoidal category is *closed* if $- \otimes A: \mathscr{K} \longrightarrow \mathscr{K}$ has a right adjoint for every object A (cf. the cartesian closed categories of section 2.2, exercise 7). The cofree object over B with respect to $- \otimes A$ is denoted B^A, so that

$$\frac{C \otimes A \longrightarrow B}{C \longrightarrow B^A}$$

and, setting $C = I$, "elements" $I \longrightarrow B^A$ are in bijective correspondence with morphisms from A to B. Using this as an aid, show that all of the specific examples of exercise 3 are closed, except "topological spaces."
6. Show that the forgetful functor from the category of $(\mathscr{K}, \otimes, I, a, b, c)$-monoids to $\mathscr{K}$ is Beck.
7. In this exercise we present an account of Lawvere's original theories as elevated to arbitrary categories by Linton. A *Lawvere theory* in $\mathscr{K}$ is a category $\mathscr{T}$ with the same objects as $\mathscr{K}$ together with a functor R: $\mathscr{K}^{\mathrm{op}} \longrightarrow \mathscr{T}$ such that R is the identity function on objects and such that R has a left adjoint. A *morphism* $\psi:(\mathscr{T}, R) \longrightarrow (\mathscr{T}', R')$ is a functor ψ with $R\psi = R'$ as shown below:

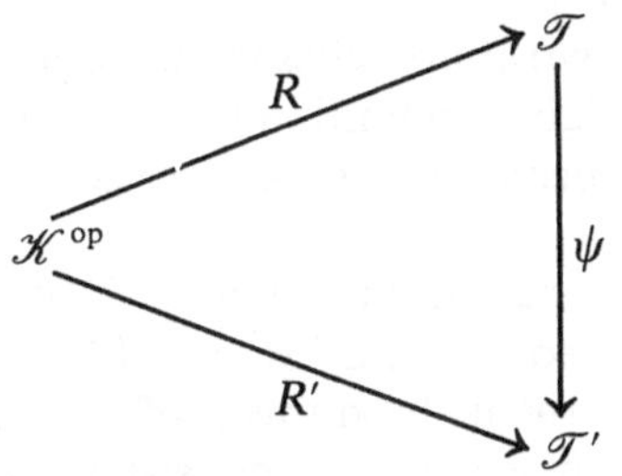

The category of $(R, \mathscr{T})$-*algebras* is that $\mathscr{C}$ in Struct($\mathscr{K}$) with $\mathscr{C}(A)$ the class of all functors $H: \mathscr{T} \longrightarrow$ **Set** such that

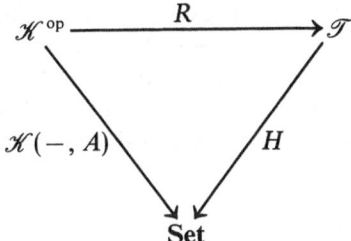

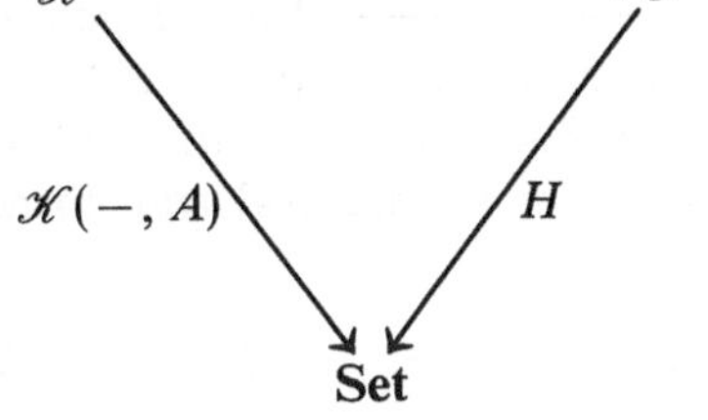

commutes. $f:(A, H) \longrightarrow (A', H')$ is $\mathscr{C}$-admissible just in case $-.f: \mathscr{K}(B, A) \longrightarrow \mathscr{K}(B, A')$ describes a natural transformation from H to H'.

(a) *If* $\mathbf{T} = (T, \eta, \circ)$ is an algebraic theory in clone form in $\mathscr{K}$ show that $(\mathscr{T}, R)$ is a Lawvere theory in $\mathscr{K}$ where $\mathscr{T} = (\mathscr{K}_{\mathbf{T}})^{\mathrm{op}}$, $AR = A$, $fR = f^{\Delta}$. [Hint: a left adjointness for R is $AL = AT$, $(\alpha: A \rightharpoondown B)L = \alpha^{\#}: AT \mathrel{-\!\!\!-\!\!<} BT$ (see 1.5.35 for the $\rightharpoondown$ notation), $A\iota$: $A \rightharpoondown ALR = \mathrm{id}_{AT}: A \rightharpoondown AT$, $A\eta: ARL \mathrel{-\!\!\!-\!\!<} A = A\eta$: $AT \mathrel{-\!\!\!-\!\!<} A$; here the "$\eta$" and "$\varepsilon$" of 2.2.15 are being written "ι" and "η."]

(b) If $\lambda: \mathbf{T} \longrightarrow \mathbf{S}$ is a theory map as in 2.7, show that $\alpha: BT \mathrel{-\!\!\!-\!\!<} A \longmapsto \alpha.B\lambda: BS \mathrel{-\!\!\!-\!\!<} A$ describes a morphism between the associated Lawvere theories.

(c) Show that (a) and (b) establish the category of algebraic theories (as in 2.10) as a full representative subcategory of Lawvere theories. [Hint: given $(\mathscr{T}, R)$ with associated adjointness (L, ι, η) (using the same notation as in (a)) set $AT = ARL$, $A\eta: A \longrightarrow ARL = A\eta$: $ARL \mathrel{-\!\!\!-\!\!<} A$, $(\alpha: A \longrightarrow BT) \circ (\beta: B \longrightarrow CT) = \alpha.\beta^{\#}$ where $\beta^{\#}$: $BRL \longrightarrow CRL = \beta RL.CR\iota L$].

(d) If (A, ξ) is a $\mathbf{T}$-algebra as in 1.4.8 show that (A, H) is an algebra over the associated Lawvere theory where $\alpha: B \rightharpoondown C$ induces $\mathscr{K}(B, A) \longrightarrow \mathscr{K}(C, A)$ by $f: B \longrightarrow A \longmapsto \alpha.f^{\#}: C \longrightarrow A$.

(e) Using the construction of (d), show that the two sorts of algebra categories are isomorphic in Struct($\mathscr{K}$). [Hint: given H, define ξ to be the value of id_A of the map $\mathscr{K}(A, A) \longrightarrow \mathscr{K}(AT, A)$ induced by $\alpha: A \rightharpoondown AT$.]

In short, the two approaches to algebraic structure are equivalent.

Lawvere's original definition (only slightly generalized in 1.5.35) is not, strictly speaking, a special case of the above. See exercise 2.1.26.

In general, one could replace $R: \mathscr{K}^{\mathrm{op}} \longrightarrow \mathscr{T}$ with R^{op}: $\mathscr{K} \longrightarrow \mathscr{T}^{\mathrm{op}}$—so that algebras are functors $\mathscr{T}^{\mathrm{op}} \longrightarrow$ **Set**—without changing the theory, and this is often done in the literature. Our choice of notation seems most consistent with 1.5.35–1.5.40. However, Linton has pointed out to us that the opposite notation is essential for universal algebra relative to closed or not necessarily symmetric monoidal categories.

8. (a) [Kock '69.] Generalizing exercise 1.17d, show that for any set X, $\mathbf{Set}(-, X): \mathbf{Set}^{\mathrm{op}} \longrightarrow \mathbf{Set}$ is algebraic. If $\mathbf{T}_X = (T_X, \eta_X, \mu_X)$ is the induced algebraic theory, observe that $AT_X = A^{(A^X)}$.

(b) Let $T: \mathbf{Set} \longrightarrow \mathbf{Set}$ be any functor and let X be any set. Show that the passage from the natural transformation $\gamma: T \longrightarrow T_X$ to the function

$$\xi = XT \xrightarrow{X\lambda} X^{(X^X)} \xrightarrow{\mathrm{pr}_{\mathrm{id}_X}} X$$

is bijective. [Hint: Yoneda lemma!]

(c) Let $\mathbf{T}$ be an algebraic theory in **Set**. Show that the passage of (b) establishes a bijection between $\mathbf{T}$-algebras (X, ξ) and theory maps

$\lambda:\mathbf{T} \longrightarrow \mathbf{T}_X$. Interpretation: $n\lambda$ interprets elements of nT as n-ary operations $X^n \longrightarrow X$ in the expected way.

9. (a) [Manes '67, 1.4.2.] Let $U:\mathscr{A} \longrightarrow \mathscr{K}$ have a left adjoint and let $\mathscr{F}$ be the full subcategory generated by objects of the form KF where F is left adjoint to U. Let $\mathbf{S}$ be an algebraic theory in $\mathscr{L}$ and let $H:\mathscr{K} \longrightarrow \mathscr{L}$ be any functor. Show that for any functor $J:\mathscr{F} \longrightarrow \mathscr{L}^{\mathbf{S}}$ with $J.U^{\mathbf{S}} = i.U.H$ there exists a unique functor $\Gamma:\mathscr{A} \longrightarrow \mathscr{L}^{\mathbf{S}}$

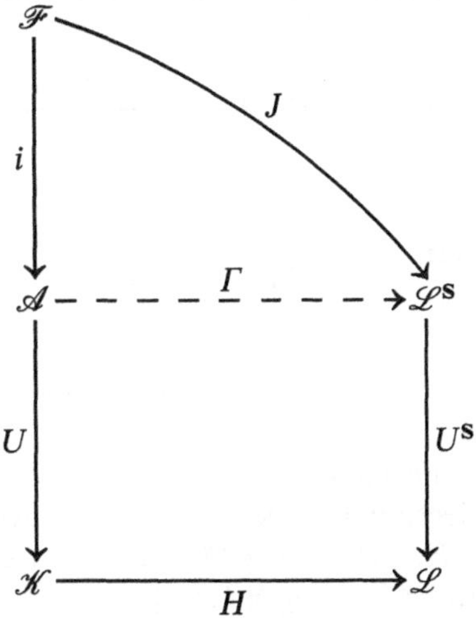

with $i.\Gamma = J$ and $\Gamma.U^{\mathbf{S}} = UH$. [Hint: if $\mathbf{T}$ is the algebraic theory in $\mathscr{K}$ induced by U and F, each $\mathscr{A}$-object A induces, via the semantic comparison functor, a $\mathbf{T}$-algebra structure on AU, H of whose corresponding contractible coequalizer diagram lifts to $\mathscr{L}^{\mathbf{S}}$ to define $A\Gamma$.]

(b) Consider the category whose objects are functors $U:\mathscr{A} \longrightarrow \mathscr{K}$ (for fixed $\mathscr{K}$) with a left adjoint and whose morphisms $H:(\mathscr{A}, U) \longrightarrow (\mathscr{A}', U')$ are functors $H:\mathscr{A} \longrightarrow \mathscr{A}'$ over $\mathscr{K}$ (i.e., $HU' = U$). Show that the semantics comparison functor of $(\mathscr{A}, U)$ is a reflection of $(\mathscr{A}, U)$ in the full subcategory of algebraic functors over $\mathscr{K}$.

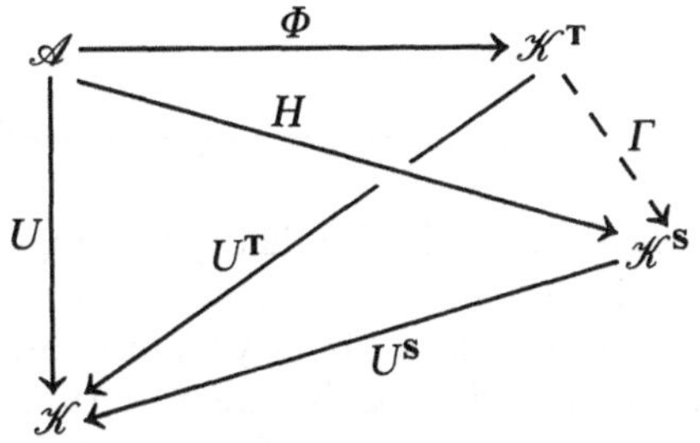

[Hint: let U have left adjoint F and let Φ be the semantics comparison functor as shown above; each **T**-algebra (A, ξ) induces the pair $AF\varepsilon H$, $\xi FH: AFUFH \longrightarrow AFH$, $U^{\mathbf{S}}$ of which extends via ξ to a contractible coequalizer whose lift defines $(A, \xi)\Gamma$; for uniqueness, use (a).] The earliest result of this type is [Maranda '66].

10. ([Appelgate '65], [Manes '67, 1.4.5].) Let **T** be a theory in $\mathscr{L}$, let **S** be a theory in $\mathscr{K}$, and let $H: \mathscr{K} \longrightarrow \mathscr{L}$ be a functor. A functor $\Gamma: \mathscr{K}^{\mathbf{S}} \longrightarrow \mathscr{L}^{\mathbf{T}}$ is said to be *over* H if $\Gamma . U^{\mathbf{T}} = U^{\mathbf{S}}. H$, generalizing the case $H = \text{id}$. A *theory map relative to* H is a natural transformation $\lambda: HT \longrightarrow SH$ satisfying

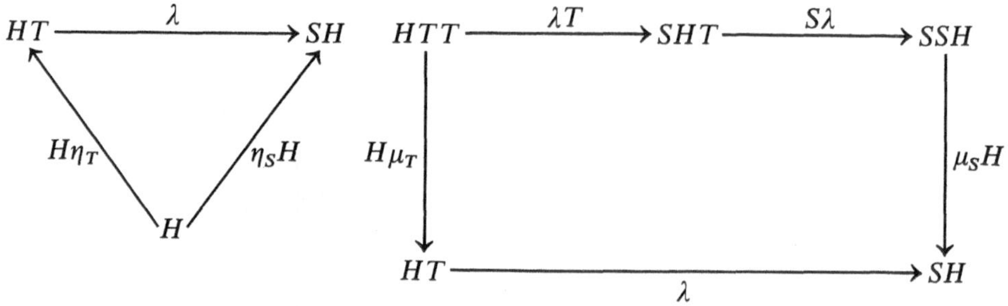

(a) Show that the passage $(A, \xi: AS \longrightarrow A)\Gamma = (AH, A\lambda.\xi H)$ establishes a bijection between functors Γ over H and theory maps λ relative to H. [Hint: if $(AS, A\mu_{\mathbf{S}})\Gamma = (ASH, \xi_A: ASHT \longrightarrow ASH)$, define $A\lambda = A\eta_{\mathbf{S}}HT.\xi_A$.]

(b) Let $H: \mathscr{K} \longrightarrow \mathscr{L}$ be the underlying set functor from topological spaces. Interpret the unique continuous (surjective) map induced by the identity function from discrete X to the space (X, τ) between their β-compactifications as a theory map relative to H; show that the corresponding Γ is the identity functor on compact Hausdorff spaces.

(c) Compute the theory map relative to H in the context shown below:

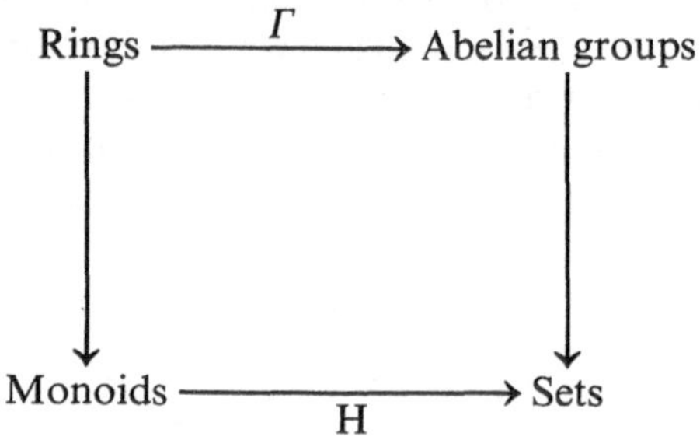

[Hint: recall from 1.32 the theory for rings over monoids.]

(d) Let **T** be the algebraic theory in **Set** corresponding to lattices [i.e., (Ω, E)-algebras with two binary commutative associative idempotent operations $\wedge$ and $\vee$ satisfying $x \wedge (x \vee y) = x = x \vee (x \wedge y)$]. Let **Poset** be the category of partially-ordered sets and order-preserving maps. Prove that the forgetful functor $\mathbf{Set}^{\mathbf{T}} \longrightarrow \mathbf{Poset}$ (via $x \leqslant y$ if $x \wedge y = x$) is algebraic by using the general adjoint functor theorem and the Beck theorem. Show that the theory map $\lambda\colon HT \longrightarrow SH$ induced by

$$\begin{array}{ccc} \mathbf{Poset}^{\mathbf{S}} = \mathbf{Set}^{\mathbf{T}} & \xrightarrow{\ \mathrm{id}\ } & \mathbf{Set}^{\mathbf{T}} \\ \Big\downarrow & & \Big\downarrow \\ \mathbf{Poset} & \xrightarrow[\ H\ ]{} & \mathbf{Set} \end{array}$$

is componentwise surjective.

11. A topological space is a *Cantor space* if it is homeomorphic to a cartesian power of copies of the discrete two-element space. It is well known that every compact Hausdorff space is a continuous image of a closed subset of a Cantor space. Give a more conceptual proof than [Kelley '55, page 166, exercise 0] by proving that βX is closed in $2^{(2^X)}$.
12. ([Lawvere '63, Chapter III, Theorem 2], [Linton '66, section 2].) Let $U\colon \mathscr{A} \longrightarrow \mathbf{Set}$ be a tractable functor (1.5.44). For each set n, let nT be the set of natural transformations from U^n to U. Define $n\eta\colon n \longrightarrow nT$, $i \longmapsto \mathrm{pr}_i\colon U^n \longrightarrow U$. Given $\alpha\colon n_1 \longrightarrow n_2T$, $\beta\colon n_2 \longrightarrow n_3T$ define $\langle i, \alpha \circ \beta\rangle = (\beta_j\colon j \in n_2).\alpha_i\colon U^{n_3} \longrightarrow U\ (i \in n_1)$.
    (a) Prove that $\mathbf{T} = (T, \eta, \circ)$ is an algebraic theory in **Set**.
    (b) Define $\Phi\colon \mathscr{A} \longrightarrow \mathbf{Set}^{\mathbf{T}}$ by $A\Phi = (AU, \xi_A)$ where, for $\omega \in AUT$, $\omega\xi_A = \langle \mathrm{id}_{AU}, A\omega\rangle$. Show that Φ is a well-defined functor over **Set** and that Φ is an algebraic reflection in that for any $H\colon \mathscr{A} \longrightarrow \mathbf{Set}^{\mathbf{S}}$ with $H.U^{\mathbf{S}} = U$ there exists unique $\Gamma\colon \mathbf{Set}^{\mathbf{T}} \longrightarrow \mathbf{Set}^{\mathbf{S}}$ in Struct(**Set**) with $\Phi.\Gamma = H$. [Hint: using the notation of 1.25, $\psi_{(A,f)}\colon (nT, n\mu) \longrightarrow A\Phi$ is a (created) limit which must be preserved by Γ; now use exercise 9(a).]
    (c) Prove that every set-valued functor with a left adjoint is tractable and that the Φ of (b) is isomorphic to the semantic comparison functor in this case.

        The induced $\mathbf{T}$ we call the *algebraic completion of* U. The Φ of (b) is, of course, called the *semantic comparison functor of* U.
    (d) Show that if $\mathscr{A}$ is *skeletally small* (i.e., has a small full representative subcategory) then every functor $U\colon \mathscr{A} \longrightarrow \mathbf{Set}$ is tractable.
    (e) The category $\mathscr{A}$ of finite sets is skeletally small. Show that the algebraic completion of the inclusion functor into **Set** is the ultrafilter theory, with semantic comparison functor "finite discrete space." [Hint: each ultrafilter on n induces an element of nT since finite sets are compact spaces; conversely, given $\omega \in nT$, show that $2\omega\colon 2^n \longrightarrow 2$ is a Boolean homomorphism.]

    For more about algebraic completion over **Set** see section 6, exercise 8.

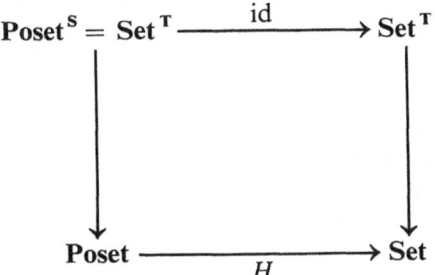

13. (Zürich school; cf. [Linton '69, page 29].) Given $U:\mathscr{A} \longrightarrow \mathscr{K}$, let K be an object of $\mathscr{K}$. Let (K, U) be the diagram scheme with nodes (A, f), $f:K \longrightarrow AU$, and with edges $e:(A, f) \longrightarrow (A', f')$ all $e:A \longrightarrow A'$ such that $f.eU = f'$. Consider the obvious diagram $D:(K, U) \longrightarrow \mathscr{K}$, $D_{(A,f)} = AU$, $D_e = eU$. Say that U is *tractable* if for every K, D has a limit.
   (a) Show that $U:\mathscr{A} \longrightarrow$ **Set** is tractable in the sense above if and only if U is tractable in the sense of 1.5.44.
   (b) Let $U:\mathscr{A} \longrightarrow \mathscr{K}$ be tractable and let $\psi_{K;(A,f)}:KT \longrightarrow AU$ be the corresponding limit. Define η, $\circ$ according to the following hints:

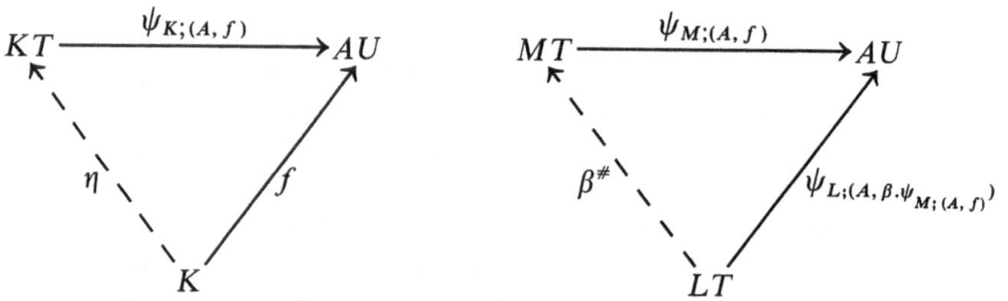

   Verify that $\mathbf{T} = (T, \eta, \circ)$ is an algebraic theory in $\mathscr{K}$ and that gT and μ are given by

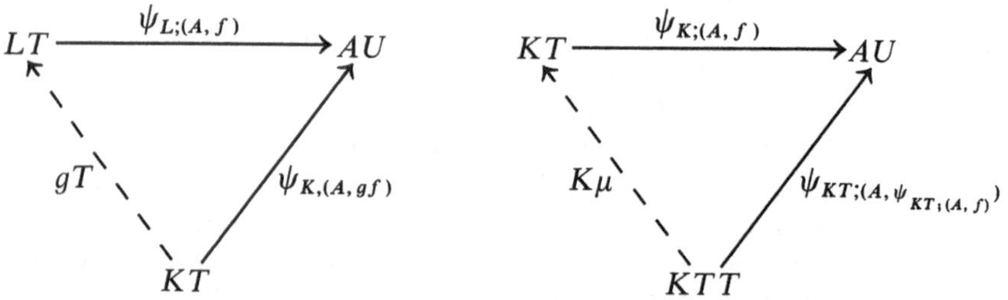

   (c) Show that $\Phi:\mathscr{A} \longrightarrow \mathscr{K}^{\mathbf{T}}$, $A\Phi = (AU, \psi_{AU;(AU, \mathrm{id}_{AU})})$ is an algebraic reflection of U. [Hint: use the construction of part (b) of the previous exercise.]
14 Let $\lambda:\mathbf{T} \longrightarrow \mathbf{T}'$ be a theory map with corresponding $V:\mathscr{K}^{\mathbf{T}'} \longrightarrow \mathscr{K}^{\mathbf{T}}$. Show that $K\lambda:(KT, K\mu) \longrightarrow (KT', K\mu')V$ is a **T**-homomorphism.
15. (See exercise 12.) Let $U:\mathscr{A} \longrightarrow$ **Set** be faithful and tractable such that U is a cogenerator in the functor category $\mathbf{Set}^{\mathscr{A}}$. Prove that the semantic comparison functor of U is full. [Hint: let $f:A\Phi \longrightarrow B\Phi$; let $a, b: U^{AU} \longrightarrow Q$ be the cokernel pair of the inclusion $j:\mathscr{A}(A, -) \longrightarrow U^{AU}$ so that $j = \mathrm{eq}(a, b)$; there exists pointwise mono $\psi:Q \longrightarrow U^n$ for n sufficiently large; for each i in n, f commutes with the U-operations $a.\psi_i, b.\psi_i: U^{AU} \longrightarrow U$; evaluate at id_{AU} to deduce that $f \in \mathscr{A}(A, B)$.]

16. (*The contravariant representation theorem* [Linton '70].) Let $\mathscr{A}$ be locally small and let J in $\mathscr{A}$ be such that the power J^n exists for every set n. Then $\mathscr{A}(-, J): \mathscr{A}^{\mathrm{op}} \longrightarrow \mathbf{Set}$ has $n \longmapsto J^n$ as left adjoint, giving rise to the *double dualization theory* $\mathbf{J}$.
   (a) Let $\mathbf{T}$ be a theory in **Set** such that every monomorphism in $\mathbf{Set}^{\mathbf{T}}$ is an equalizer. Write $|A|$ for $AU^{\mathbf{T}}$ and $[A, B]$ for $\mathbf{Set}^{\mathbf{T}}(A, B)$. Let J be an injective cogenerator (Exercises 2.1.20 and 2.1.59) of $\mathbf{Set}^{\mathbf{T}}$. Prove that the structural reflection of $[-, J]: (\mathbf{Set}^{\mathbf{T}})^{\mathrm{op}} \longrightarrow \mathbf{Set}$ is obtained, via the semantic comparison functor $\Phi: (\mathbf{Set}^{\mathbf{T}})^{\mathrm{op}} \longrightarrow \mathbf{Set}^{\mathbf{J}}$, as the full subcategory over **Set** of $\mathbf{Set}^{\mathbf{J}}$ of all A for which the evaluation map
$$A \xrightarrow{\ ev\ } \bar{J}^{\langle A, \bar{J}\rangle}$$
is injective, where $\bar{J}$ is the **J**-algebra $(|J|, pr_{\mathrm{id}}: J^{(J^{|J|})} \longrightarrow J)$ and $\langle -, - \rangle$ means **J**-homomorphisms. For each set A, show that the **T**-homomorphic extension of
$$A \xrightarrow{\ ev\ } J^{(|J|^A)}$$
is injective and establishes a theory isomorphism of **T** with $\bar{\mathbf{J}}$; thus **T** is a double-dualization theory.
   (b) Let $\mathbf{Set}^{\mathbf{T}}$ be Boolean rings with unit and let $J = 2$. Show that $\mathbf{Set}^{\mathbf{J}}$ is compact Hausdorff spaces and that Φ identifies $(\mathbf{Set}^{\mathbf{T}})^{\mathrm{op}}$ with the full subcategory of totally disconnected spaces. This captures part of Stone duality.

   See [Kock '69] for more on double dualization theories, [Reynolds '74] for an extension of the contravariant representation theorem, and [Negrepontis '71] for a discussion of Gelfand duality (cf. exercise 1.16) and the compact-discrete Pontrjagin duality (cf. exercise 1.5.20.)

17. Let $\mathscr{K}$ have pullbacks and let **T** be a theory in $\mathscr{K}$.
   (a) Prove that $i_A: E_A \longrightarrow AT = \mathrm{eq}(AT\eta, A\eta T)$ exists for all A.
   (b) Prove that $A\eta = \mathrm{eq}(AT\eta, A\eta T)$ for all A if and only if T reflects isomorphisms and $i_A T$ is mono for all A. [Hint: if $A\eta = \mathrm{eq}(AT\eta, A\eta T)$ and fT is an isomorphism, f^{-1} is induced by the equalizer property; conversely, if $f: A \longrightarrow E_A$ with $f.i_A = A\eta$, use a pullback of form
$$\begin{array}{ccc} E_AT & \xrightarrow{\ \mathrm{id}\ } & E_AT \\ \big\downarrow & & \big\downarrow{\scriptstyle i_AT} \\ AT & \xrightarrow[\ A\eta T\ ]{} & E_AT \end{array}$$
to show that fT is an isomorphism.]

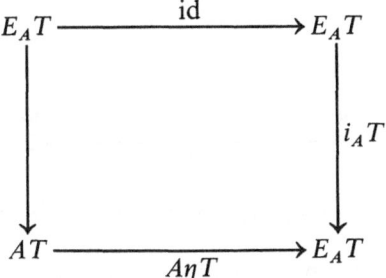

(c) Given $f:A \longrightarrow B$ with $B\eta$ split mono and fT an isomorphism, prove that f is an isomorphism. [Hint: use (b) and exercise 2.1.21.] Conclude that T reflects isomorphisms for every nontrivial **T** in **Set**.

18. Let **T** be an arbitrary theory in **Set**, let **S** be the subsets theory and let **P** be the double power-set theory. Define $A\sigma:AT \longrightarrow AP$, $p \longmapsto$ set of supports of p (as in 1.5.10.)

(a) Prove that $\sigma:\mathbf{T} \longrightarrow \mathbf{P}$ is a theory map if and only if for every $f:A \longrightarrow B$ and mono $i:I \longrightarrow B$ the squares

$$\begin{array}{ccc} (If^{-1})T & \longrightarrow & IT \\ \downarrow & & \downarrow{\scriptstyle iT} \\ AT & \xrightarrow[fT]{} & BT \end{array} \qquad \begin{array}{ccc} I & \xrightarrow{I\eta} & IT \\ {\scriptstyle i}\downarrow & & \downarrow{\scriptstyle iT} \\ B & \xrightarrow[B\eta]{} & BT \end{array}$$

$$\begin{array}{ccc} ITT & \xrightarrow{I\mu} & IT \\ {\scriptstyle iTT}\downarrow & & \downarrow{\scriptstyle iT} \\ BTT & \xrightarrow[B\mu]{} & BT \end{array}$$

are pullbacks.

(b) Prove that $\sigma:\mathbf{T} \longrightarrow \mathbf{P}$ is a theory map if and only if T preserves the pullback

$$\begin{array}{ccc} I & \longrightarrow & 1 \\ \downarrow & & \downarrow \\ B & \xrightarrow[\chi_I]{} & 2 \end{array}$$

for each $I \subset B$ and $\chi_{IT}:2T \longrightarrow 2$ is a **T**-algebra. [Hint: use exercise 8.]

(c) Show that **S** is a subtheory of **P** in two ways:

$$\begin{array}{ccc} AS & \xrightarrow{A(\text{prin})} & AP \\ B & \longmapsto & \{S:S \supset B\} \end{array} \qquad \begin{array}{ccc} AS & \longrightarrow & AP \\ B & \longmapsto & \{S:S \cap B \neq \varnothing\} \end{array}$$

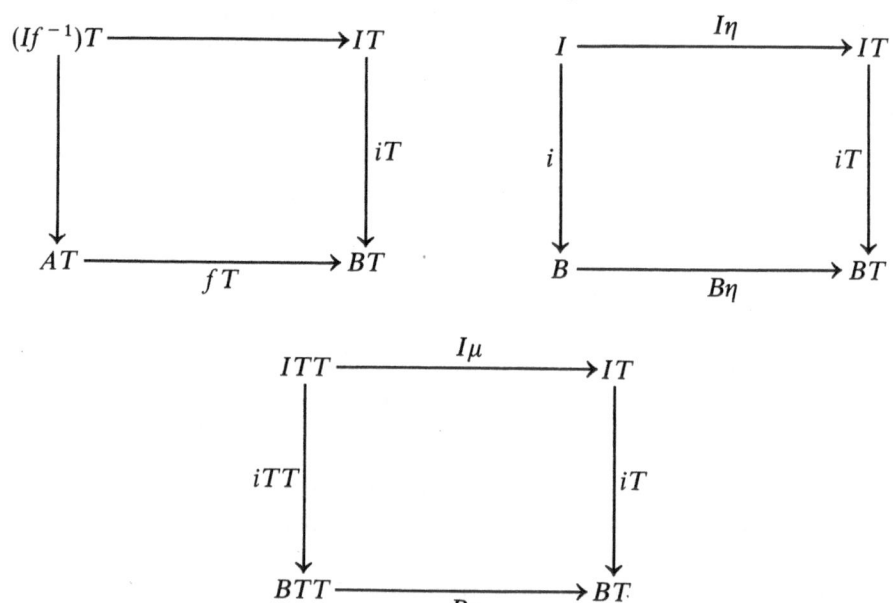

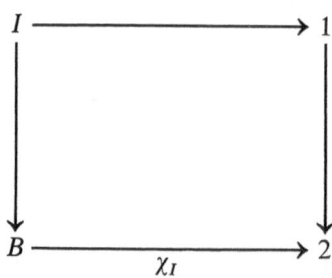

[Hint: a complete atomic Boolean algebra is a complete semilattice in two ways.]

(d) Say that **T** is *variabled* if $\sigma:\mathbf{T} \longrightarrow \mathbf{P}$ is a theory map which factors through prin:

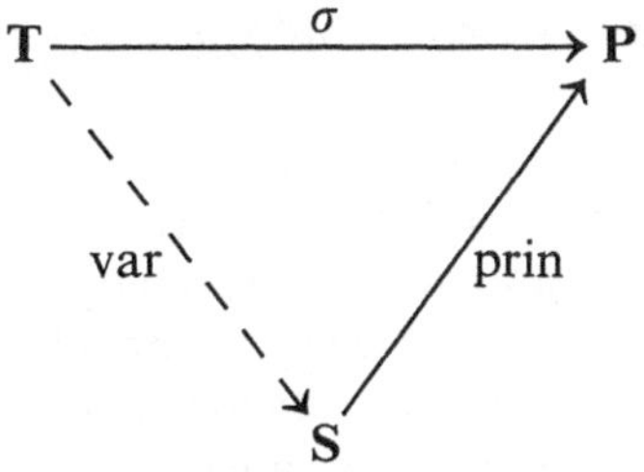

Show that var: $\mathbf{T} \longrightarrow \mathbf{S}$ is a theory map if **T** is variabled. For finitary **T**, prove that **T** is variabled if and only if **T** admits an equational presentation such that, in each equation, the same set of variables appear in each term (cf. exercise 1.3.8.)

(e) If **T** is variabled, show that (A, ξ) is a compact **T**-model [exercise 2.3.9(d)] if $\xi = \{(p, x): x \in \mathrm{var}(p)\}$ for all sets A.

19. [Bergman '75.] Let $\mathscr{A}$ be the category of groups but let $U:\mathscr{A} \longrightarrow \mathbf{Set}$ be the subfunctor of the forgetful functor defined by $GU = \{x \in G: x^2 = e\}$. An *involution algebra* is a triple (X, m, e) with $m: X^2 \longrightarrow X$, $e \in X$ subject to

$$\begin{aligned} exm &= e & xymym &= x \\ xem &= x & xyzmm &= xzmymzm \\ xxm &= x \end{aligned}$$

Prove that the algebraic completion (as in exercise 12) of U is the equationally-definable class of involution algebras.

20. [Guitart '75.] A *theory with involution* in $\mathscr{K}$ is a pair $(\mathbf{T}, I)$ where **T** is an algebraic theory in $\mathscr{K}$ and $I:\mathscr{K}_{\mathbf{T}} \longrightarrow \mathscr{K}_{\mathbf{T}}$ is an involution on the Kleisli category of **T**, that is, I is the identity on objects and $I^2 = \mathrm{id}$ on morphisms. A *contravariant monad* in $\mathscr{K}$ is a triple (F, i, ψ) where $F:\mathscr{K} \longrightarrow \mathscr{K}^{\mathrm{op}}$, $Ai: A \longrightarrow AF$, $A\psi: AF \longrightarrow AF^2$ are subject to the four equations

(i) for all $f: A \longrightarrow B$, $f.Bi.B\psi = Ai.A\psi.fF^2$,

(ii) for all $f: A \longrightarrow B$, $A\psi.fF^2.BiF.B\psi = A\psi.fF^2$,

(iii) for all A, $AFi.AF\psi.A\psi F.AiF = \mathrm{id}_{AF}$,

(iv) for all A, $AF\psi.A\psi F.AiF.A\psi = AF\psi.A\psi F$.

(a) Prove that the passage from $(\mathbf{T}, I)$ to (F, i, ψ) defined by $KF = KT$, $Ai = A\eta$, $fF = \alpha^{\#}$ where $\alpha: \longrightarrow BT = (f.A\eta)I$ and $A\psi = (\mathrm{id}_{AT}I)^{\#}$ establishes a bijection between theories with involution and contravariant monads. [Hint: given (F, i, ψ) set $AT = AF$, $A\eta = Ai$, $\alpha^{\#} = A\psi.\alpha F^2.B\psi F.BiF: AT \longrightarrow BT$; for $\alpha: A \longrightarrow BT$, $\alpha I = Bi.B\psi.\alpha F$.]

(b) Show that $(\mathbf{T}, I)$ is a theory with involution in **Set** if **T** is the power-set theory and I provides a relation with its inverse. Compute the corresponding contravariant monad.

(c) Show that $(F^2, i.\psi:\mathrm{id} \longrightarrow F^2,\ FtF:F^4 \longrightarrow F^2)$ is an algebraic theory for each contravariant monad (F, i, ψ). Observe that in (b) this constructs the double power-set theory.

3. Abstract Birkhoff Subcategories

We define when the full replete subcategory $\mathscr{A}$ of $\mathscr{K}^{\mathbf{T}}$ is an abstract Birkhoff subcategory. When $\mathscr{K} = \mathbf{Set}$, $\mathscr{A}$ is Birkhoff if and only if $\mathscr{A}$ is closed under products, subalgebras, and homomorphic images. In general, Birkhoff subcategories are classified as pointwise epimorphic quotient theories $\lambda: \mathbf{T} \longrightarrow \mathbf{T}'$; thus if $\mathscr{A}$ is Birkhoff then $\mathscr{A}$ is algebraic over $\mathscr{K}$.

For the purposes of this section fix a category $\mathscr{K}$ and an algebraic theory $\mathbf{T} = (T, \eta, \circ, \mu)$ in $\mathscr{K}$.

3.1 Abstract Birkhoff Subcategories of $\mathscr{K}^{\mathbf{T}}$. Let $\mathscr{A}$ be a full replete subcategory of $\mathscr{K}^{\mathbf{T}}$. $\mathscr{A}$ is an *abstract Birkhoff subcategory of* $\mathscr{K}^{\mathbf{T}}$ providing

1. For every X in $\mathscr{K}$, the $\mathbf{T}$-algebra $(XT, X\mu)$ has a reflection $X\lambda: (XT, X\mu) \longrightarrow A$ in $\mathscr{A}$ such that $(X\lambda)U^{\mathbf{T}}$ is epi in $\mathscr{K}$, and
2. $\mathscr{A}$ *is closed under $U^{\mathbf{T}}$-split epimorphisms*, that is, given A in $\mathscr{A}$ and $f:A \longrightarrow (X, \xi)$ in $\mathscr{K}^{\mathbf{T}}$ such that $fU^{\mathbf{T}}$ is a split epimorphism in $\mathscr{K}$ then (X, ξ) is in $\mathscr{A}$.

3.2 Lemma. *Let $\mathscr{A}$ be a full replete subcategory of $\mathscr{K}^{\mathbf{T}}$. Assume that for each X in $\mathscr{K}$, $(XT, X\mu)$ has a reflection in $\mathscr{A}$ which we will (choose and) denote by $X\lambda:(XT, X\mu) \longrightarrow (XT', \xi_X:XT'T \longrightarrow XT')$. Then there exists unique $(\eta', \circ')$ such that "$\mathbf{T}' = (T', \eta', \circ')$ is an algebraic theory admitting $\lambda:\mathbf{T} \longrightarrow \mathbf{T}'$ as a theory map and $(XT', X\mu')V = (XT', \xi_X)$" holds, where $V:\mathscr{K}^{\mathbf{T}'} \longrightarrow \mathscr{K}^{\mathbf{T}}$ corresponds to λ as in 2.9.*

Proof. Given $\beta:B \longrightarrow CT'$ let $\beta^{\#}:(BT, B\mu) \longrightarrow (CT', \xi_C)$ be the

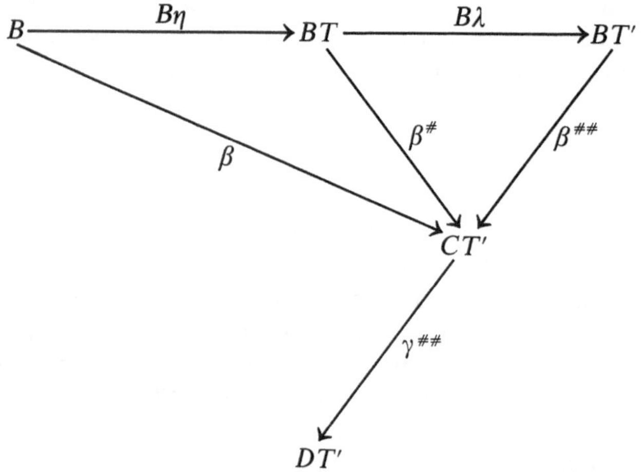

unique $\mathbf{T}$-homomorphism with $B\eta.\beta^{\#} = \beta$, and let $\beta^{\#\#}:(BT', \xi_B) \longrightarrow (CT', \xi_C)$ be the unique $\mathbf{T}$-homomorphism such that $B\lambda.\beta^{\#\#} = \beta^{\#}$. For

$\gamma: C \longrightarrow DT'$ we clearly have $\beta^{\#\#}.\gamma^{\#\#} = (\beta.\gamma^{\#\#})^{\#\#}$. For $\alpha: A \longrightarrow BT'$, define $\alpha \circ \beta = \alpha.\beta^{\#\#}$. Then $(\alpha \circ' \beta) \circ' \gamma = (\alpha.\beta^{\#\#}).\gamma^{\#\#} = \alpha.(\beta.\gamma^{\#\#})^{\#\#} = \alpha \circ' (\beta \circ' \gamma)$. Define $X\eta': X \longrightarrow XT'$ by $X\eta' = X\eta.X\lambda$. Then for $f: A \longrightarrow B$, $(f.B\eta') \circ' \beta = f.B\eta.B\lambda.\beta^{\#\#} = f.\beta$. Also the diagram

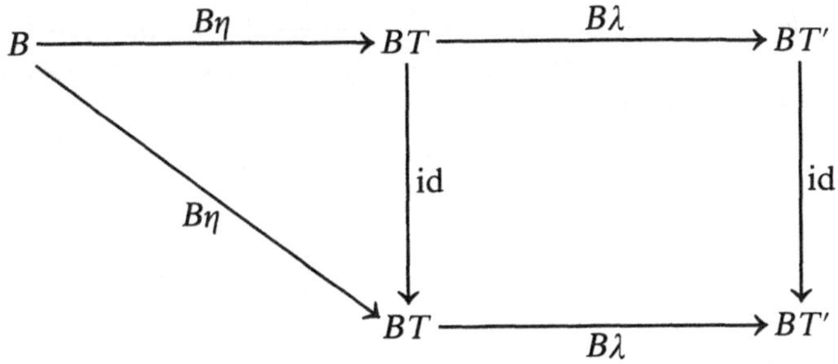

shows that $(B\eta.B\lambda)^{\#\#} = \mathrm{id}_{BT'}$, so $\alpha \circ' \mathrm{B}\eta' = \alpha.(\mathrm{B}\eta.\mathrm{B}\lambda)^{\#\#} = \alpha$. Given $a: A \longrightarrow BT$ and $b: B \longrightarrow CT$ we have

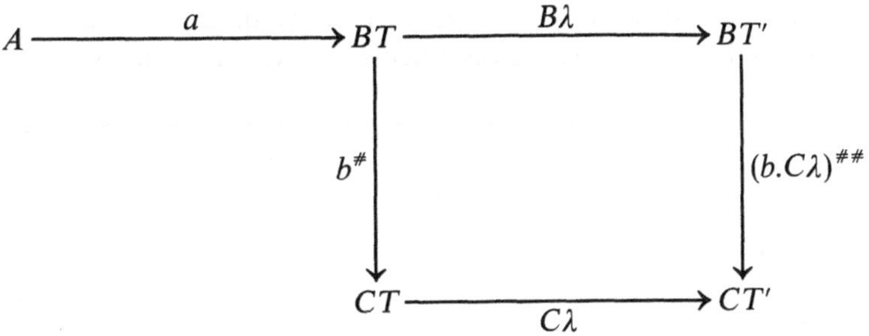

(where $B\lambda.(b.C\lambda)^{\#\#} = (b.C\lambda)^{\#} = b^{\#}.C\lambda$, as $C\lambda$ is a **T**-homomorphism). This shows that $\mathbf{T}' = (T', \eta', \mu')$ is an algebraic theory and that $\lambda: \mathbf{T} \longrightarrow \mathbf{T}'$ is a theory map. To see that $(XT', X\mu')V = (XT', \xi_X)$, that is, that $\xi_X = XT'\lambda.X\mu'$, observe that $\xi_X:(XT'T, XT'\mu) \longrightarrow (XT', \xi_X)$ and $XT'\lambda: (XT'T, XT'\mu) \longrightarrow (XT'T', \xi_{XT'})$ are **T**-homomorphisms by definition and that $X\mu':(XT'T', \xi_{XT'}) \longrightarrow (XT', \xi_X)$ is a

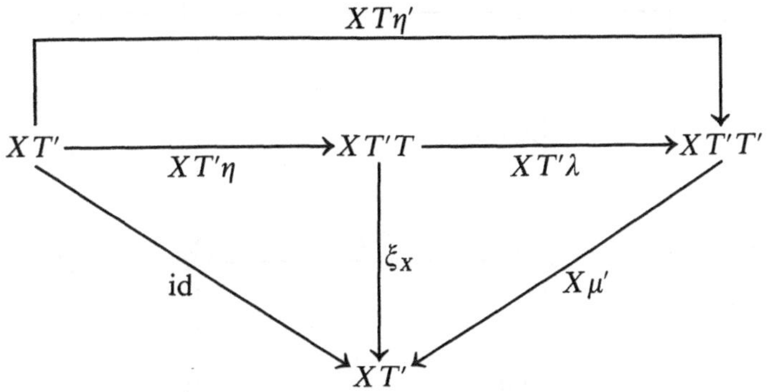

T-homomorphism since, by 1.3.13, $X\mu' = \mathrm{id}_{XT'T'} \circ' \mathrm{id}_{XT'} = (\mathrm{id}_{XT'})^{\#\#}$. Now consult the diagram above. Finally, we check that $(\eta', \circ')$ is unique with these properties. Clearly $X\eta' = X\eta.X\lambda$ is forced. By 1.3.14, we must have $\alpha \circ' \beta = \alpha.\beta^{\#\#\#}$ where $\beta^{\#\#\#}:(BT', B\mu') \longrightarrow (CT', C\mu')$ is the unique

T′-homomorphism such that $B\eta'.\beta^{\#\#\#} = \beta$. Applying V, $\beta^{\#\#\#}: (BT', \xi_B) \longrightarrow (CT', \xi_C)$ is a **T**-homomorphism such that $B\eta . B\lambda.\beta^{\#\#\#} = \beta$. Therefore, $\beta^{\#\#\#} = \beta^{\#\#}$. □

3.3 Theorem. *Let* $W: \mathscr{A} \longrightarrow \mathscr{K}^{\mathbf{T}}$ *be an abstract Birkhoff subcategory of* $\mathscr{K}^{\mathbf{T}}$. *Let* $\mathbf{T}'$, $\lambda: \mathbf{T} \longrightarrow \mathbf{T}'$ *and* $V: \mathscr{K}^{\mathbf{T}'} \longrightarrow \mathscr{K}^{\mathbf{T}}$ *be induced by* $\mathscr{A}$ *as in 3.2. Then* V *and* W *are isomorphic, that is, there exists an isomorphism of categories as shown below:*

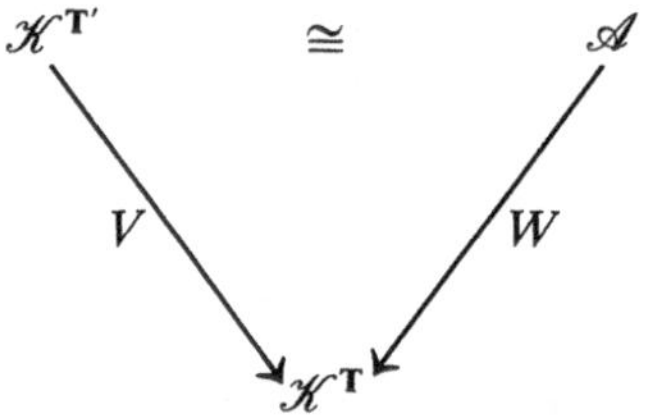

In particular, $(\mathscr{A}, U)$ *is algebraic in* Struct($\mathscr{K}$), *where* $U = WU^{\mathbf{T}}$.

Proof. The passage from ξ to $X\lambda.\xi$ is injective because $X\lambda$ is an epimorphism in $\mathscr{K}$, so that V is injective on objects. The diagram below

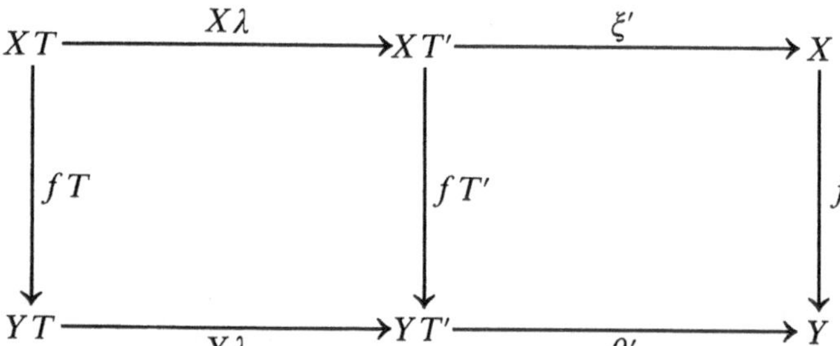

shows that $f:(X, \xi') \longrightarrow (Y, \theta')$ is a **T**′-homomorphism if and only if $f:(X, \xi')V \longrightarrow (X, \theta')V$ is a **T**-homomorphism ($X\lambda$ is epi!). Therefore, V is a full subcategory. Let $(X, \xi) \in \mathscr{A}$. There exists a unique **T**-homomor-

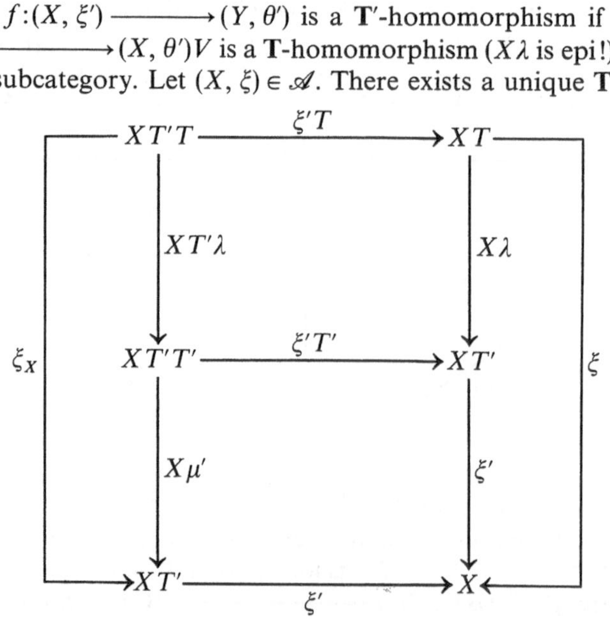

phism $\xi':(XT', \xi_X) \longrightarrow (X, \xi)$ (for notation, see 3.2) such that $X\lambda.\xi' = \xi$. As ξ' is a $\mathbf{T}$-homomorphism we have the perimeter of the diagram above. Since $XT'\lambda$ is epi, $X\mu'.\xi' = \xi'T'.\xi'$. That $X\eta'.\xi' = \mathrm{id}_X$ follows immediately from

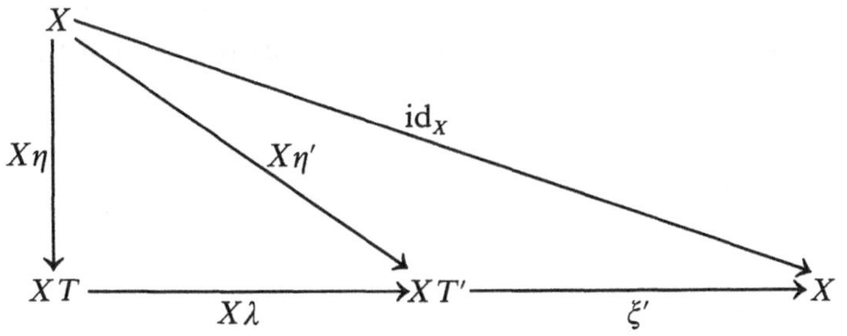

This proves that $\mathscr{A}$ is contained in V. Conversely, if (X, ξ') is in $\mathscr{K}^{\mathbf{T}'}$ then $\xi':(XT', \xi_X) \longrightarrow (X, X\lambda.\xi')$ is a $U^{\mathbf{T}}$-split epi, which proves that V is contained in $\mathscr{A}$. □

The next theorem is a categorical generalization of 1.4.22.

3.4 Birkhoff Variety Theorem. *Let $V:(\mathscr{A}, U) \longrightarrow (\mathscr{K}^{\mathbf{T}}, U^{\mathbf{T}})$ be a homomorphism in* Struct($\mathscr{K}$). *Then the following two statements are equivalent.*

1. *"$\mathscr{A}$ is a full subcategory of $\mathscr{K}^{\mathbf{T}}$ closed under the formation of products, subalgebras, and quotients," that is, V is an abstract Birkhoff subcategory of $\mathscr{K}^{\mathbf{T}}$ as defined in 3.1.*
2. *"$\mathscr{A}$ is a full subcategory of $\mathscr{K}^{\mathbf{T}}$ obtained by imposing additional equations to the operations and equations that present* $\mathbf{T}$*-algebras," that is, $(\mathscr{A}, U)$ is algebraic and the theory map $\lambda:\mathbf{T} \longrightarrow \mathbf{T}'$ corresponding to V is such that $X\lambda$ is epi in $\mathscr{K}$ for all X (where $\mathbf{T}'$ such that $(\mathscr{A}, U)$ and $(\mathscr{K}^{\mathbf{T}'}, U^{\mathbf{T}'})$ are isomorphic is unique up to isomorphism by 2.11).*

Proof. *1 implies 2.* This has already been done in the proofs of 3.2 and 3.3.

2 implies 1. The proof that V is a full subcategory follows from the fact that $X\lambda$ is epi just as in the proof of 3.3. V is replete on general principles because $(\mathscr{A}, U)$ and $(\mathscr{K}^{\mathbf{T}}, U^{\mathbf{T}})$ are both in Struct($\mathscr{K}$). $(XT', X\mu')V = (XT', \xi_X)$ where $\xi_X = XT'\lambda.X\mu'$. Therefore the diagram

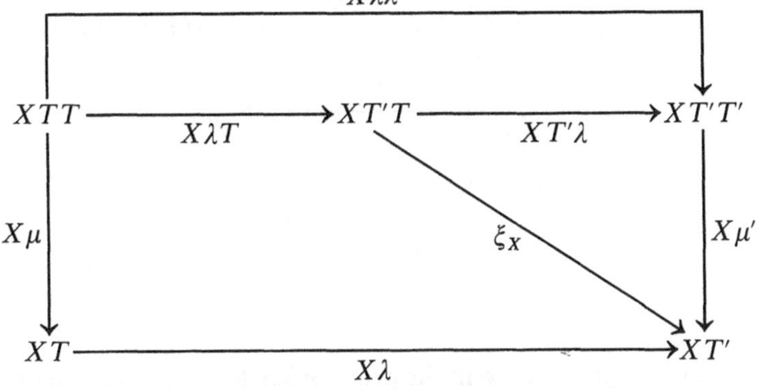

proves that $X\lambda:(XT, X\mu) \longrightarrow (XT', \xi_X)$ is a **T**-homomorphism. Let (Y, θ') be a **T**′-algebra and let $f:(XT, X\mu) \longrightarrow (Y, \theta')V$ be a **T**-homomorphism. As $\bar{f} = (X\eta.f)T'.\theta':(XT', X\mu') \longrightarrow (Y, \theta')$ is a **T**′-homomorphism, $\bar{f}:(XT', \xi_X) \longrightarrow (Y, \theta')V$ is a **T**-homomorphism. The diagram below proves that $X\lambda.\bar{f} = f$ and $\bar{f}$ is unique with this property because $X\lambda$ is epi in $\mathscr{K}$. This proves that $(XT', \xi_X; X\lambda)$ is a reflection of $(XT, X\mu)$ in V.

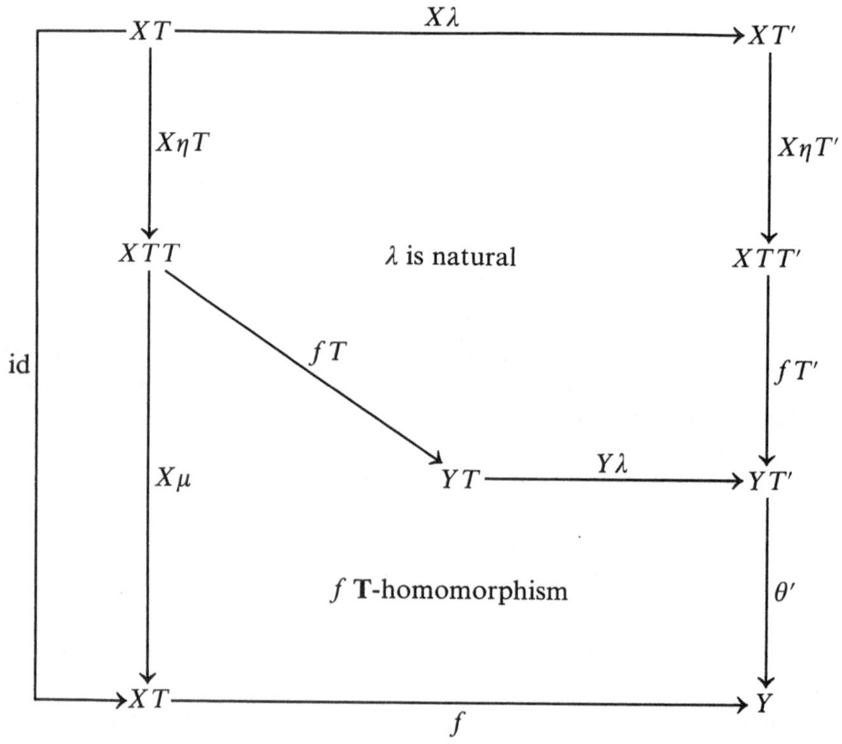

To see that V is closed under $U^{\mathbf{T}}$-split epimorphisms let (X, ξ') be a **T**′-algebra, let $f:(X, X\lambda.\xi') \longrightarrow (Y, \theta)$ be a **T**-homomorphism, and let $d: Y \longrightarrow X \in \mathscr{K}$ be such that $d.f = \mathrm{id}_Y$. It is sufficient to prove that there exists a **T**-homomorphism $\theta':(YT', \xi_Y) \longrightarrow (Y, \theta)$ such that

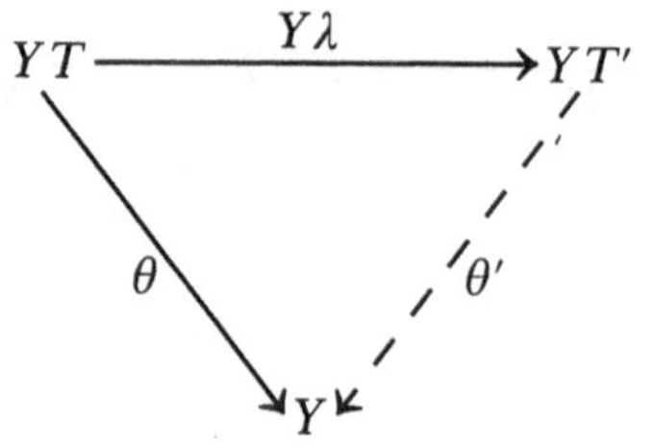

since the same diagrams used in the proof of 3.3 (for ξ') prove that (Y, θ') must be a **T**′-algebra. As is seen from the diagram below, the desired θ' is $dT'.\xi'.f$.

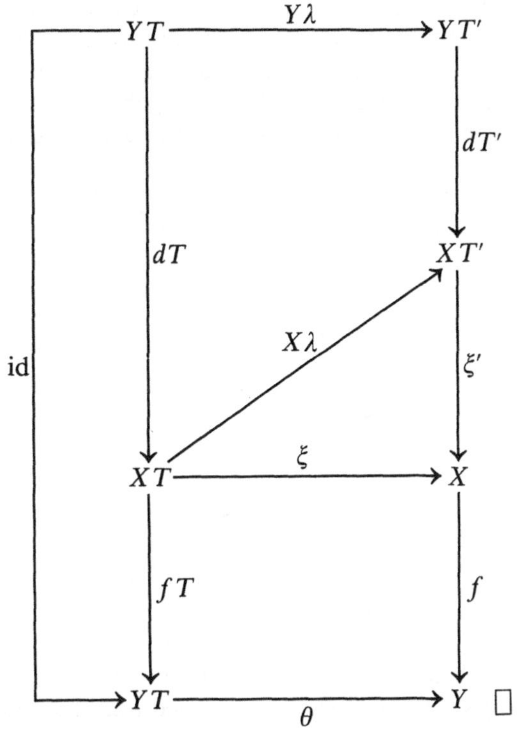

The next result follows immediately from 3.2, 3.3, and 3.4.

3.5 Corollary. *Let $\mathscr{A}$ and $\mathscr{B}$ be two abstract Birkhoff subcategories of $\mathscr{K}^{\mathbf{T}}$ and assume that for all X in $\mathscr{K}$ there exists (XT', ξ_X) in $\mathscr{A} \cap \mathscr{B}$ and a* **T**-*homomorphism* $X\lambda:(XT, X\mu) \longrightarrow (XT', \xi_X)$ *which is simultaneously a reflection of* $(XT, X\mu)$ *in $\mathscr{A}$ and in $\mathscr{B}$. Then* $\mathscr{A} = \mathscr{B}$. □

We now explicate the situation in the category of sets.

3.6 Theorem. *Let* **T** *be an algebraic theory in* **Set** *and let $\mathscr{A}$ be a full subcategory of* $\mathbf{Set}^{\mathbf{T}}$ *with inclusion functor* $V:\mathscr{A} \longrightarrow \mathbf{Set}^{\mathbf{T}}$. *Then the following conditions are equivalent:*

1. *V is closed under the formation of products, subalgebras, and quotient algebras in the sense of 1.4.22.*
2. *V is an abstract Birkhoff subcategory as in 3.1.*

Proof. 1 implies 2. $\mathscr{A}$ has and V preserves small limits. If f: $(X, \xi) \longrightarrow (Y, \theta)$ is a **T**-homomorphism with (Y, θ) in $\mathscr{A}$ then the **T**-subalgebra Xf of (Y, θ) (1.4.32) is again in $\mathscr{A}$ and there is a commutative diagram

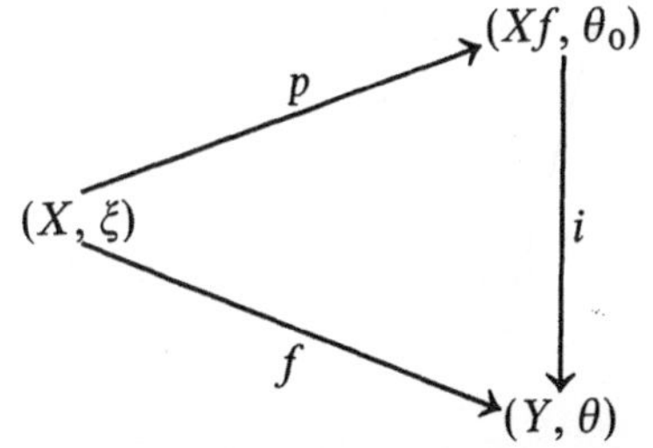

As the cardinal of Xf is at most as large as the cardinal of X, it follows from the general adjoint functor theorem of 2.2.24 that every **T**-algebra has a reflection in $\mathscr{A}$. Moreover, all such reflections are surjective. To prove it, let $\Gamma : X \longrightarrow A$ denote the reflection of the **T**-algebra X in $\mathscr{A}$. Let (p, i) be the image factorization of Γ. Since $X\Gamma$ is in $\mathscr{A}$ there exists unique $\bar{p} : A \longrightarrow X\Gamma$

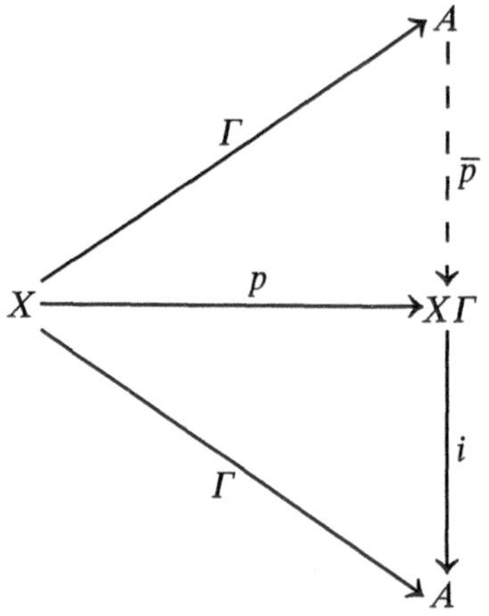

with $\Gamma.\bar{p} = p$. As $\Gamma.\bar{p}.i = p.i = \Gamma$, $\bar{p}.i = \mathrm{id}_A$ and i is surjective. Therefore, Γ is surjective. It is clear that V is closed under $U^{\mathbf{T}}$-split epimorphisms since, with $\mathscr{K} = \mathbf{Set}$, this is the same as "$\mathscr{A}$ is closed under quotients."

2 implies 1. Define $\mathscr{B}$ to be the full subcategory of all **T**-algebras which are a quotient of a subalgebra of an element of $\mathscr{A}$. Clearly, $\mathscr{B}$ is closed under quotients or, equivalently, $\mathscr{B}$ is closed under $U^{\mathbf{T}}$-split epimorphisms. Now let $\Gamma : (XT, X\mu) \longrightarrow A$ be a reflection of $(XT, X\mu)$ in $\mathscr{A}$ with Γ surjective. Let $Q \in \mathscr{B}$ so that there exists A_1 in $\mathscr{A}$, a subalgebra S of A_1 with inclusion $i : S \longrightarrow A_1$ and a surjective **T**-homomorphism $q : S \longrightarrow Q$. Let $f : (XT, X\mu) \longrightarrow Q$ be an arbitrary **T**-homomorphism. Using the

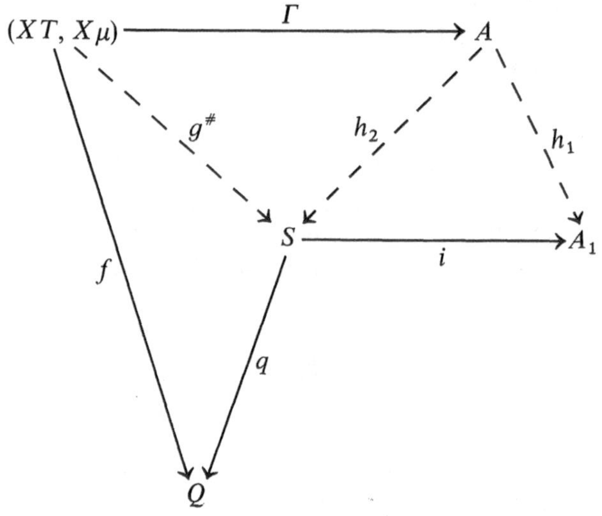

axiom of choice there exists a function $g: X \longrightarrow S$ such that $g.q = X\eta.f$. Therefore, $g^{\#}.q = f$. There exists a unique **T**-homomorphism $h_1: A \longrightarrow A_1$ with $\Gamma.h_1 = g^{\#}.i$. As Γ is surjective, $\mathrm{Im}(h_1) = \mathrm{Im}(\Gamma.h_1) = \mathrm{Im}(g^{\#}.i)$ and there exists unique **T**-homomorphism $h_2: A \longrightarrow S$ with $h_2.i = h_1$. As i is mono, $\Gamma.h_2 = g^{\#}$. We have $\Gamma.(h_2.q) = g^{\#}.q = f$ and $h_2.q$ is unique with this property because Γ is epi. By 3.5, $\mathscr{A} = \mathscr{B}$ and $\mathscr{A}$ is closed under subalgebras and quotients. Since V creates limits (1.28) V is closed under products. □

3.7 Abelianization of a Group. Abelian groups is a Birkhoff subcategory of groups. This is clear both from 1.4.22 and from 2.13 via 3.4.

3.8 Simple Theories. Quotient groups of a group are the same thing as normal subgroups; quotient rings of a ring are the same thing as ideals. Theorem 3.4 asserts that (componentwise epimorphic) quotient theories of a theory may be identified with abstract Birkhoff subcategories. Let **T** be an algebraic theory in **Set**. $\mathbf{Set}^{\mathbf{T}}$ and $\{1\}$ are always Birkhoff subcategories of $\mathbf{Set}^{\mathbf{T}}$. If $\varnothing$ is a **T**-algebra, then $\{\varnothing, 1\}$ is also a Birkhoff subcategory. Any other Birkhoff subcategory of $\mathbf{Set}^{\mathbf{T}}$ is called *proper*. **T** is a *simple theory* if $\mathbf{Set}^{\mathbf{T}}$ possesses no proper Birkhoff subcategories. For example, let **T** be the double power-set theory of 1.3.19 whose algebras are complete atomic Boolean algebras (1.5.17). Then **T** is simple. For let $\mathscr{A}$ be a Birkhoff subcategory containing an algebra A which has at least two elements. Then the two-element Boolean algebra 2 is a subalgebra of A, so belongs to $\mathscr{A}$. Let (X, ξ) be an arbitrary **T**-algebra. Then the cartesian power $2^{(2^X)}$ is isomorphic to $(XT, X\mu)$, so (X, ξ) belongs to $\mathscr{A}$.

Notes for Section 3

Some form of 3.4 was known to the Zürich school (see [Manes '67, section 1.6]). 3.1–3.4 we believe to be new.

Exercises for Section 3

1. Show that the ultrafilter theory and the power-set theory are simple.
2. Let **T** be a bounded theory in **Set** of rank α and let $\beta = \alpha$ (if α is infinite), $\beta = 2\alpha$ (if α is finite).
   (a) Show that every **T**-algebra is a quotient of a subalgebra of a product of subalgebras of $(\beta T, \beta\mu)$ so that "every bounded theory is singly generated as a Birkhoff subcategory of itself." [Hint: this is much like 1.4.23; if $p \neq q$ in XT observe that p and q are in AT where $A \subset X$ has cardinal at most α.]
   (b) Show that $\mathbf{Set}^{\mathbf{T}}$ has at most $2^{\beta T \times \beta T}$ distinct Birkhoff subcategories. [Hint: if $\lambda: \mathbf{T} \longrightarrow \mathbf{S}$ is a pointwise onto theory map, $(\beta S, \beta\mu_{\mathbf{S}})$ generates $\mathbf{Set}^{\mathbf{S}}$ as in (a).]
3. Let Ω have a single unary operation and let **T** be the corresponding algebraic theory in **Set**. Show that **T** admits but countably many Birkhoff subcategories. Give equational presentations. [Hint: quotient theories correspond to monoid quotients of the natural numbers.]
4. Let **T** be an algebraic theory in **Set**. A congruence R on a **T**-algebra

(X, ξ) is *fully invariant* if every **T**-endomorphism $f:(X, \xi) \longrightarrow (X, \xi)$ is such that $f \times f$ maps R into R.

(a) If $\lambda:\mathbf{T} \longrightarrow \mathbf{S}$ is a theory map and if X is a set, show that the kernel pair of $X\lambda$ is a fully invariant congruence on $(XT, X\mu)$.

(b) For any set X, show that the passage from R to the class $\mathscr{B}$ of all **T**-algebras with the property that each **T**-homomorphism g from $(XT, X\mu)$ is such that R is contained in the kernel pair of g is a well-defined function from fully invariant congruences on $(XT, X\mu)$ to Birkhoff subcategories of $\mathbf{Set}^{\mathbf{T}}$.

(c) Starting with R, passing to $\mathscr{B}$ as in (b) and then to $\bar{R}$ as in (a), show that $R = \bar{R}$. [Hint: $R \subset \bar{R}$ is trivial; conversely, let λ correspond to $\mathscr{B}$ and show that the canonical projection $p:(XT, X\mu) \longrightarrow (XT/R, \theta)$ factors through $X\lambda$ by choosing, for each $f:X \longrightarrow Q$, a function $g:X \longrightarrow XT$ such that $g.p = f$ and applying full invariance to $g^{\#}$.]

(d) Starting with $\mathscr{B}$, passing to R as in (a) and then to $\bar{\mathscr{B}}$ as in (b) show that $\mathscr{B} \subset \bar{\mathscr{B}}$.

(e) Assume that **T** is bounded of rank α and that X is the disjoint union of two copies of α. Show that (a) and (b) establish a bijective correspondence between fully invariant congruences on $(XT, X\mu)$ and Birkhoff subcategories of $\mathbf{Set}^{\mathbf{T}}$. [Hint: in the context of (d) we must show that $\bar{\mathscr{B}} \subset \mathscr{B}$; for (Y, θ) in $\bar{\mathscr{B}}$ it suffices to show, given p, q in YT with $\langle p, Y\lambda\rangle = \langle q, Y\lambda\rangle$, that $p\theta = q\theta$; there exists a function f: $X \longrightarrow Y$ and a subset $i:A \longrightarrow X$ such that $i.f$ is injective and with both p and q in the image of $iT.fT$; now consult the diagram

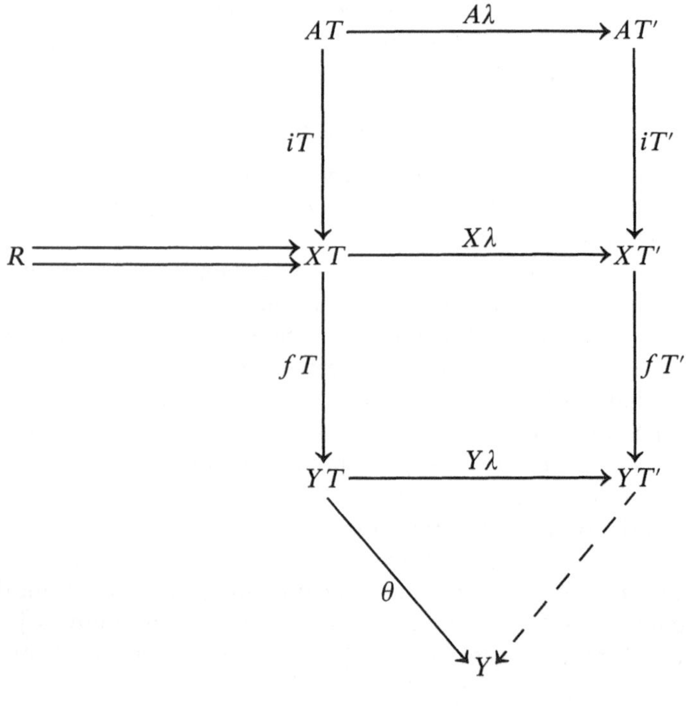

and use the fact that $iT'.fT'$ is injective.] This generalizes a standard theorem (see e.g. [Cohn '65, IV.1.2]).

5. (a) ([Scott '56].) Let **T** be any algebraic theory in **Set**. Show that **T** has at most $2^{2T \ltimes 2T}$ simple Birkhoff subcategories (and hence only finitely many if $2T$ is finite).
   (b) Show that every nontrivial finitary **T** in **Set** possesses a simple Birkhoff subcategory. [Hint: construct a maximal proper fully invariant congruence using Zorn's lemma; because operations are finitary, each nested union of subalgebras is a subalgebra.]
   (c) (Open question.) Does there exist an algebraic theory **T** in **Set** possessing no simple Birkhoff subcategories? Any nontrivial **T** corresponding to a bounded equational presentation will contain simple Birkhoff subcategories because, by imposing equations, there is a nontrivial finitary Birkhoff subcategory.
6. (a) In **Set**, show that $\lambda:\mathbf{T} \longrightarrow \mathbf{S}$ is an epimorphism in the category of theories and theory maps if and only if the corresponding V: $\mathbf{Set}^{\mathbf{S}} \longrightarrow \mathbf{Set}^{\mathbf{T}}$ is injective on objects. [Hint: use exercise 8 of section 2.]
   (b) Show that the λ corresponding to V:monoids $\longrightarrow$ semigroups is an epimorphism in the category of theories in **Set** but that each $X\lambda$ is injective and not onto. Observe that V is not full.
   (c) Show that V:groups $\longrightarrow$ monoids over **Set** is a full non-Birkhoff subcategory such that $X\lambda:(XT, X\mu_T) \longrightarrow (XS, X\mu_S)V$ is an epimorphism in the category of monoids.
   (d) (Open questions.) Is there an interesting example of an epimorphism of theories whose V is not injective on objects? How does one characterize those λ whose V is a full subcategory? A sufficient condition for V to be full is that $X\lambda$ is an epimorphism in $\mathscr{K}^{\mathbf{T}}$. Isbell has proved a "Beth definability theorem" for full V over **Set**; see [Isbell '73].
7. Let **T** be an algebraic theory in **Set**. The **T**-algebra (X, ξ) is *simple* if X has at least two elements and if (X, ξ) has no quotients other than itself and 1. Show that if **T** is finitary and nontrivial then there exists a simple **T**-algebra. [Hint: choose any **T**-algebra with at least two elements and use Zorn's lemma to construct a maximal proper congruence.] This result does not generalize to infinitary theories (see [Nelson '74].)
8. For any set X, define $C_X:X^4 \longrightarrow X$ by

$$(a, b, x, y)C_X = \begin{cases} x & \text{if } a = b \\ y & \text{if } a \neq b \end{cases}$$

Show that the simple comparison algebras [see the previous exercise and exercise 1.5.21(b)] are precisely those isomorphic to an (X, C_X). [Hint: if R is a congruence on (X, C_X) and if $(a, b) \in R$ then $((a, b, x, y)C_X, (a, a, x, y)C_X) \in R$; if (X, C) is simple and a, b are such that $(a, b, x, y)C \neq (a, b, x, y)C_X$, consider the congruence $\{(x, y): C(a, b, x, t) = C(a, b, y, t)$ for all $t\}$.]

4. Regular Categories

In 2.1.48 we defined two types of image factorization for morphisms in a category. In a given category, there may be a multiplicity of other "nice" ways to construct epi-mono factorizations of morphisms. In this section we axiomatize rather than specify such image factorization systems. A regular category will be defined to be a category together with an image factorization system satisfying certain completeness and smallness conditions. Regular categories provide a proper setting in which to characterize abstract Birkhoff subcategories as subcategories closed under products, "subalgebras," and "quotient algebras."

4.1 Image Factorization Systems. An image factorization system in a category $\mathscr{K}$ is a pair (E, M) where E and M are subclasses of the class of morphisms of $\mathscr{K}$ satisfying the following four axioms:

(4.2) E and M are subcategories of $\mathscr{K}$.

(4.3) Every element of E is an epimorphism and every element of M is a monomorphism.

(4.4) Every isomorphism is both in E and in M.

(4.5) Every $f: A \longrightarrow B$ in $\mathscr{K}$ has a unique E-M factorization. More precisely, there exists (e, m) with $e \in \mathsf{E}$ and $m \in \mathsf{M}$ such that $f = e.m$ (so that the codomain of e is the domain of m—we denote it as $\mathrm{Im}(f)$) and whenever

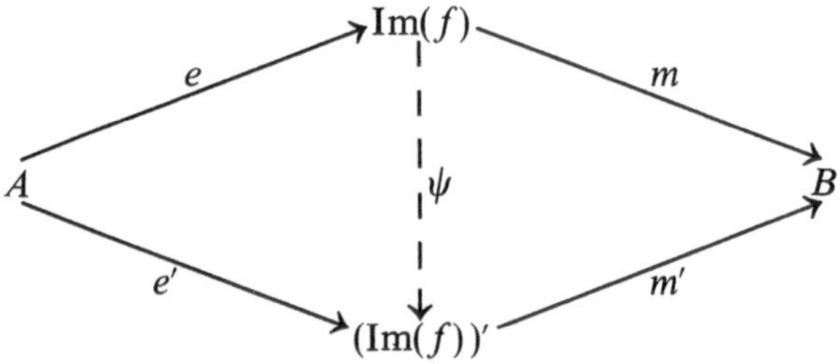

(e', m') satisfies $e' \in \mathsf{E}$, $m' \in \mathsf{M}$, $f = e'm'$ there exists a (necessarily unique by 4.3) isomorphism ψ with $e.\psi = e'$ and $\psi.m' = m$.

4.6 Duality Principle. If (E, M) is an image factorization system in $\mathscr{K}$ then (M, E) is an image factorization system in $\mathscr{K}^{\mathrm{op}}$. □

4.7 Coequalizer-Mono and Epi-Equalizer Factorizations. Assume that $\mathscr{K}$ has coequalizer-mono factorizations. Then (E, M) is an image factorization system if E = all coequalizers and M = all monomorphisms. This follows at once from 2.1.57 (1), 2.1.44, 2.1.36, and 2.1.49. Dually, if $\mathscr{K}$ has epi-equalizer factorizations the E = epimorphisms and M = equalizers forms an image factorization system in $\mathscr{K}$.

4.8 Hausdorff Spaces. Let $\mathscr{K}$ be the category of Hausdorff topological spaces and continuous maps. Then $\mathscr{K}$ has three reasonable image factorization systems. $\mathscr{K}$ has epi-equalizer factorizations; here E = all maps with dense image and M = all homeomorphisms onto closed subspaces. There is no problem in verifying the image factorization axioms; hints to see that these are epi-equalizer factorizations can be found in 2.1.38. $\mathscr{K}$ has coequalizer-mono factorizations and they are constructed at the level of all topological spaces. To see this, let $f:A \longrightarrow B \in \mathscr{K}$ and construct the coequalizer-mono factorization of f in all topological spaces as in the diagram of 2.1.54; here, E is Hausdorff because E is a subspace of $A \times A$ and C is Hausdorff because C admits a continuous injection into B. Thus E = all surjections which induce the quotient topology and M = all injections. Finally, E = all continuous surjections and M = all homeomorphisms onto a subspace is a third image factorization system, as is easy to check.

We now derive some formal consequences of the image factorization axioms. For the next six propositions we fix a category $\mathscr{K}$ provided with a specific image factorization system (E, M).

4.9 Proposition. *If $f:A \longrightarrow B$ is both in E and in M then f is an isomorphism.*

Proof. This is a formal consequence of 4.5. Use the same argument as in 2.1.50. □

4.10 Diagonal Fill-In Proposition. *Given the commutative square*

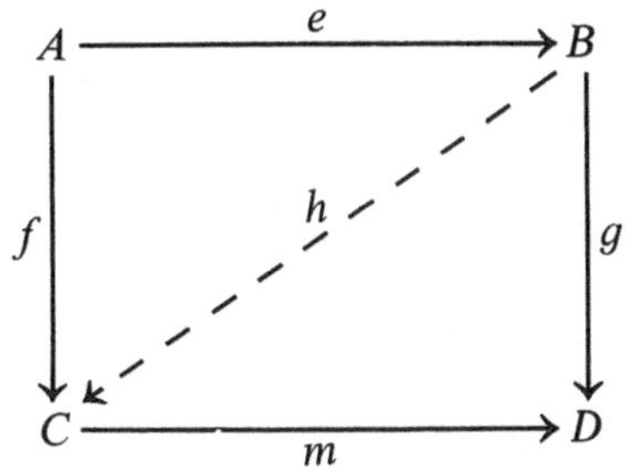

with $e \in \mathsf{E}$ and $m \in \mathsf{M}$ there exists unique h with $e.h = f$ and $h.m = g$.

Proof. The uniqueness assertion is clear since e is epi (4.3) or since m is mono and, in fact, either triangle implies the other. To establish existence let (e_1, m_1) be the E-M factorization of g and let (e_2, m_2) be the E-M factorization of f as shown below. Then (e_2, m_2m) and (ee_1, m_1) are both E-M

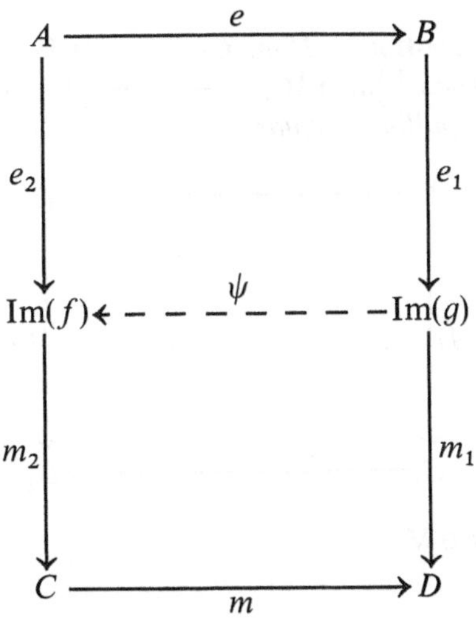

factorizations of $fm = eg$ giving rise to ψ as shown by 4.5. Now define $h:B \longrightarrow C = e_1.\psi.m_2$. $\square$

4.11 Proposition (E Determines M). *Let $t:C \longrightarrow D$ have the property that for every commutative square of form*

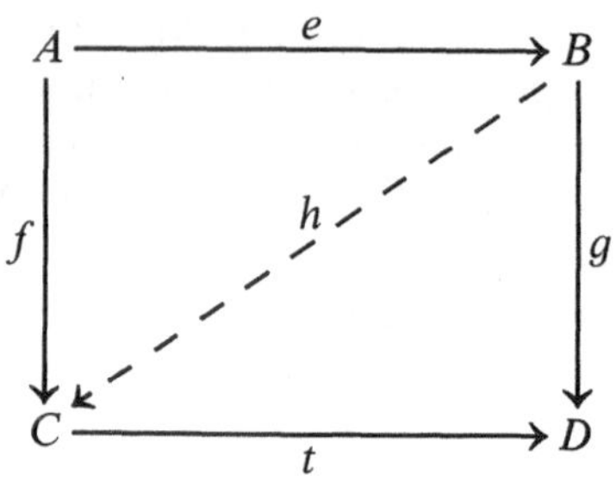

with $e \in$ E there exists $h:B \longrightarrow C$ with $e.h = f$. Then $t \in$ M.

Proof. Let (e, m) be the E-M factorization of t. Then there exists h:

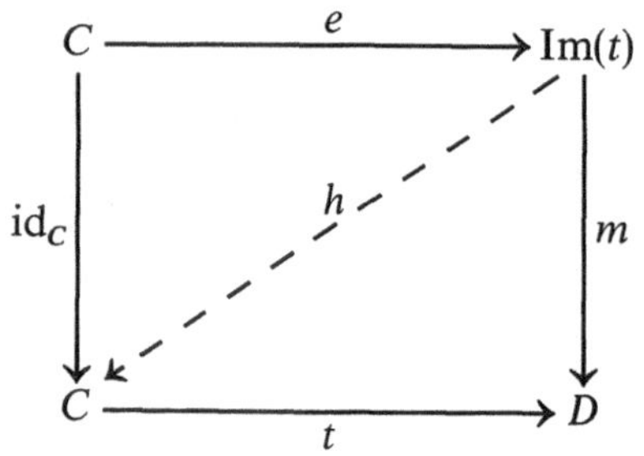

$\mathrm{Im}(t) \longrightarrow C$ such that $e.h = \mathrm{id}_C$. As e is epi, $h.t = m$. Since m is mono and $h.e.m = h.t = m$, $h.e = \mathrm{id}_{\mathrm{Im}(t)}$. Therefore h is an isomorphism. Using 4.4 and 4.2 we have $t = h^{-1}m \in$ M. $\square$

Propositions 4.10 and 4.11 show that M is determined by E in a straightforward way. This is useful in a context where there is a natural candidate for E since there is only one M to try! Dually, E is determined by M and similar comments apply.

4.12 Stability Proposition. *If $m_i:C_i \longrightarrow D_i$ is a family in* M *and if $\prod C_i$ and $\prod D_i$ exist in $\mathscr{K}$ then $\prod m_i:\prod C_i \longrightarrow \prod D_i$ (defined by $(\prod m_i).p_j = p_j.m_j$) is in* M. *Given a pullback square*

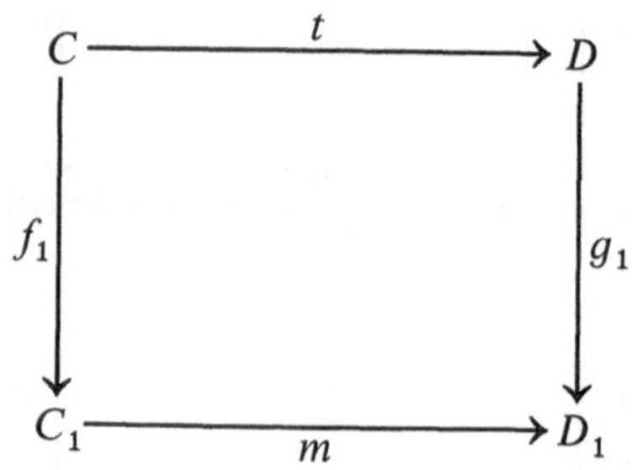

with $m \in$ M then also $t \in$ M.

Proof. We use 4.11 to prove both statements. Consider a commutative square $f.\prod m_i = e.g$ with $e \in \mathsf{E}$ as shown below. By 4.10, there exists h_j:

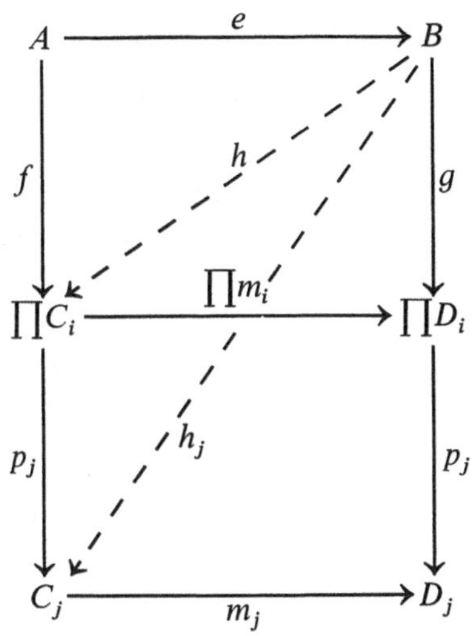

$B \longrightarrow C_j$ with $e.h_j = f.p_j$ for all j. Let h be the unique morphism such that $h.p_j = h_j$. Then $e.h.p_j = f.p_j$ for all j, so $e.h = f$. Turning to the second statement, consider a commutative square $f.t = e.g$ with $e \in \mathsf{E}$.

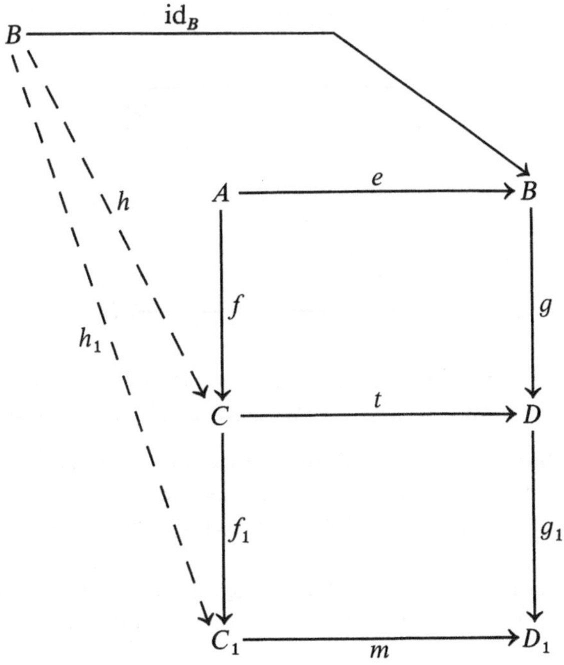

By 4.10 there exists $h_1 : B \longrightarrow C_1$ with $e.h_1 = f.f_1$ and $h_1.m = g.g_1$. By the pullback property, there exists unique $h : B \longrightarrow C$ such that $h.f_1 = h_1$ and $h.t = g$. We have $e.h.t = e.g = f.t$ and $e.h.f_1 = e.h_1 = f.f_1$. Therefore (1.20) $e.h = f$. □

The next proposition generalizes 2.1.57 (2).

4.13 Proposition. *Let $f : A \longrightarrow B$ and $g : B \longrightarrow C$ in $\mathscr{K}$. Then if $f.g \in \mathsf{M}$, $f \in \mathsf{M}$. If $f.g \in \mathsf{E}$ then $g \in \mathsf{E}$.*

Proof. By duality, it suffices to prove the second statement. Let (e, m) be an E-M factorization of g and let (e_1, m_1) be an E-M factorization of $f.e$. By 4.10 there exists h such that $(h.m_1).m = \mathrm{id}_C$ as shown below.

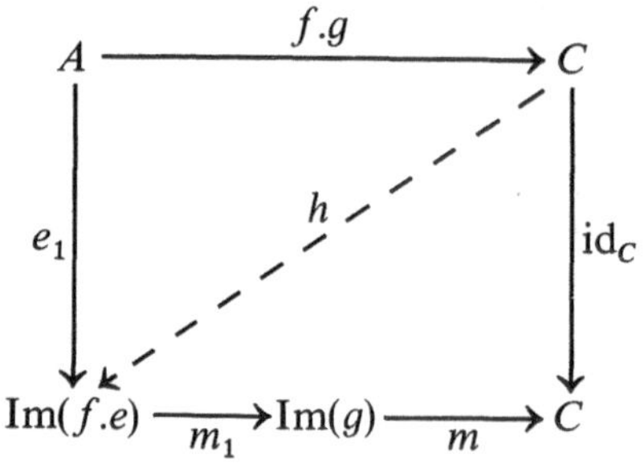

Therefore m is mono and split epi, so is an isomorphism, and $g = e.m$ is in E. □

4.14 Proposition. *Every split mono is in* M. *Every split epi is in* E.

Proof. We need only prove the first statement. We use 4.11. Let

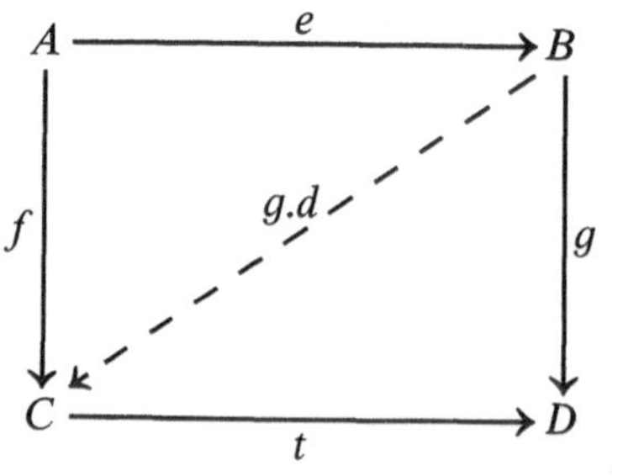

$t : C \longrightarrow D$ be split mono so that there exists $d : D \longrightarrow C$ with $t.d = \mathrm{id}_C$. Let $e.g = f.t$ be a commutative square with $e \in \mathsf{E}$. Then $e.(g.d) = f.t.d = f$. □

4.15 Regular Categories. A *regular category* is a triple $(\mathscr{K}, \mathsf{E}, \mathsf{M})$ where $\mathscr{K}$ is a locally small category with small limits, (E, M) is an image factorization system in $\mathscr{K}$ and $\mathscr{K}$ is E co-well-powered (cf. 2.2.29 (2)) as is explained immediately below.

If A is an object in $\mathscr{K}$ and $e : A \longrightarrow Q$, $e' : A \longrightarrow Q' \in \mathsf{E}$, define $e \sim e'$ if there exists f, g with $e.f = e'$ and $e'.g = e$. As e and e' are epi by 4.3, f and g

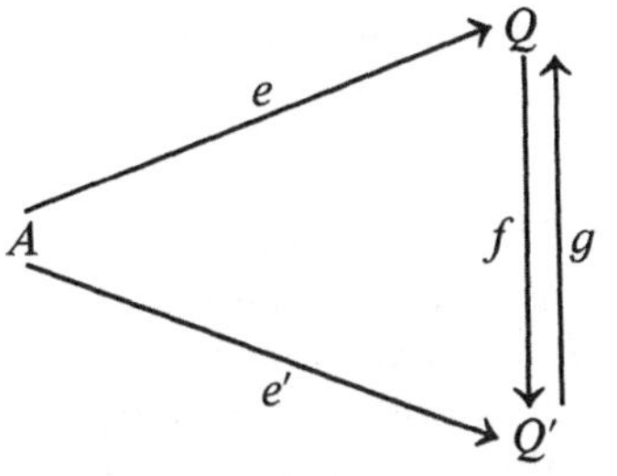

are mutually inverse isomorphisms. Hence it is natural to call a $\sim$-equivalence class $[e]$ of E-morphisms with domain A an E-*quotient object of A*. Compare with the dual of 2.1.65. $\mathscr{K}$ is E *co-well-powered* providing the class E(A) of E-quotient objects of A is a small set for every object A in $\mathscr{K}$.

The following proposition is easily proved by the reader.

4.16 Proposition. *Let* $(\mathscr{K}, \mathsf{E}, \mathsf{M})$ *be a regular category and let* $(\mathscr{A}, U) \in$ Struct$(\mathscr{K})$ *be fibre-complete. Define*

$$\begin{aligned}
\mathsf{E}_1 &= \{e \in \mathscr{A} : e \text{ is co-optimal and } eU \in \mathsf{E}\}\\
\mathsf{M}_1 &= \{m \in \mathscr{A} : mU \in \mathsf{M}\}\\
\mathsf{E}_2 &= \{e \in \mathscr{A} : eU \in \mathsf{E}\}\\
\mathsf{M}_2 &= \{m \in \mathscr{A} : m \text{ is optimal and } mU \in \mathsf{M}\}
\end{aligned}$$

Then $(\mathscr{A}, E_1, M_1)$ *and* $(\mathscr{A}, E_2, M_2)$ *are regular categories.* □

4.17 Proposition. *Let* $(\mathscr{K}, \mathsf{E}, \mathsf{M})$ *be a regular category and let* $\mathbf{T} = (T, \eta, \circ, \mu)$ *be an algebraic theory in* $\mathscr{K}$ *such that* $eT \in \mathsf{E}$ *whenever* $e \in \mathsf{E}$. *Define*

$$\begin{aligned}
\mathsf{E}^{\mathbf{T}} &= \{e \in \mathscr{K}^{\mathbf{T}} : eU^{\mathbf{T}} \in \mathsf{E}\}\\
\mathsf{M}^{\mathbf{T}} &= \{m \in \mathscr{K}^{\mathbf{T}} : mU^{\mathbf{T}} \in \mathsf{M}\}
\end{aligned}$$

Then $(\mathscr{K}^{\mathbf{T}}, \mathsf{E}^{\mathbf{T}}, \mathsf{M}^{\mathbf{T}})$ *is a regular category.*

Proof. All is obvious except 4.5. To this end, let $f:(X, \xi) \longrightarrow (Y, \theta)$ be a $\mathbf{T}$-homomorphism and let (e, m) be the E-M factorization of $f: X \longrightarrow Y$ in $\mathscr{K}$. Let I denote Im(f). By diagonal fill-in, there exists unique $\theta_0 : IT \longrightarrow$

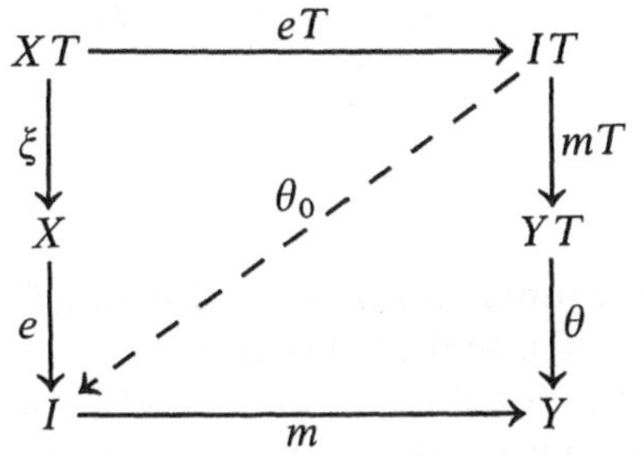

I such that $eT.\theta_0 = \xi.e$ and $\theta_0.m = mT.\theta$. It is crucial, however, that we know eT is in E. Since m is mono (4.3), (I, θ_0) is a $\mathbf{T}$-algebra (use the reasoning of 1.4.28). The remaining details are clear. □

The previous proposition is limited by the rather unnatural requirement that T preserve E. But this is no problem if $\mathscr{K} = \mathbf{Set}$ (1.4.29) and we have at once the

4.18 Corollary. *Let* **T** *be an algebraic theory in* **Set**. *Let* E *be the class of surjective* **T**-*homomorphisms and let* M *be the class of injective* **T**-*homomorphisms. Then* ($\mathbf{Set}^{\mathbf{T}}$, E, M) *is a regular category.* □

Using 1.13 and 2.1.46 it is not hard to show that, in the context of the previous proposition, E is the class of coequalizers in $\mathbf{Set}^{\mathbf{T}}$ and M is the class of monomorphisms in $\mathbf{Set}^{\mathbf{T}}$.

4.19 Hausdorff Spaces. All three image factorization systems on $\mathscr{K}$ = Hausdorff spaces as in 4.8 render $\mathscr{K}$ a regular category. The following hint is useful to prove E co-well-powered for E = maps with dense image: if X is Hausdorff and A is a dense subset of X then each element of X is the limit of an ultrafilter on A and such ultrafilters converge uniquely; therefore the cardinality of X is at most the cardinality of the set of all ultrafilters on A.

4.20 Quasivarieties. Let ($\mathscr{K}$, E, M) be a regular category. A *quasivariety in* ($\mathscr{K}$, E, M) is a full replete subcategory $\mathscr{B}$ of $\mathscr{K}$ such that every object A of $\mathscr{K}$ admits a reflection $e: A \longrightarrow B$ in $\mathscr{B}$ such that $e \in$ E.

The following theorem serves to motivate the definition of a regular category.

4.21 Quasivariety Theorem. *Let* ($\mathscr{K}$, E, M) *be a regular category and let* $\mathscr{B}$ *be a full replete subcategory of* $\mathscr{K}$. *Then* $\mathscr{B}$ *is a quasivariety in* ($\mathscr{K}$, E, M) *if and only if* $\mathscr{B}$ *is closed under limits and closed under* M *in the sense that if* $m: A \longrightarrow B \in$ M *and* $B \in \mathscr{B}$ *then* $A \in \mathscr{B}$.

Proof. First assume that $\mathscr{B}$ is a quasivariety. $\mathscr{B}$ is closed under limits by 1.34. Let $m: A \longrightarrow B \in$ M with $B \in \mathscr{B}$. Let $e: A \longrightarrow B'$ be the reflection of A

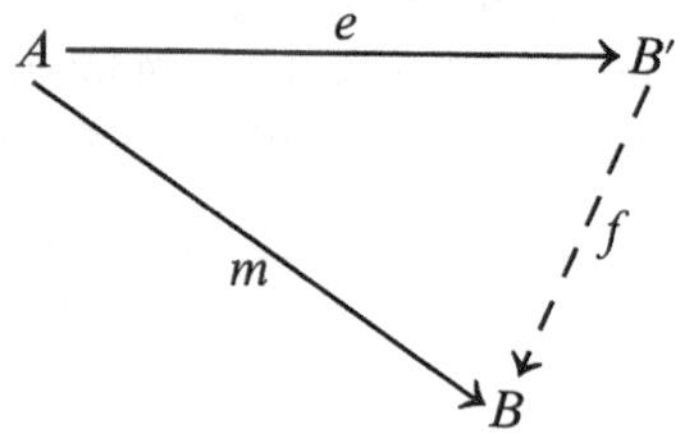

in $\mathscr{B}$ so that there exists (unique) $f: B' \longrightarrow B$ with $e.f = m$. By 4.13 and 4.9, e is an isomorphism. As $\mathscr{B}$ is replete, A is in $\mathscr{B}$.

Conversely, let $\mathscr{B}$ be closed under limits and closed under M. To prove that $\mathscr{B}$ is reflective we make use of the general adjoint functor theorem 2.2.24. The inclusion functor $U: \mathscr{B} \longrightarrow \mathscr{K}$ preserves products and equalizers by hypothesis, and we have only to check the solution set condition. Let $f: A \longrightarrow B \in \mathscr{K}$ with $B \in \mathscr{B}$. Let (e, m) be the E-M factorization of f.

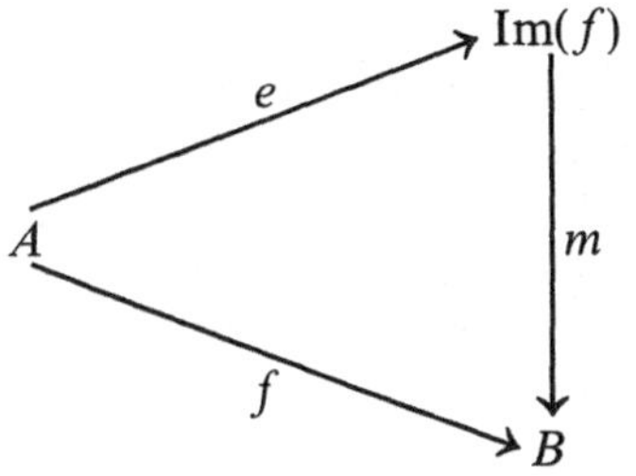

Then $\mathrm{Im}(f) \in \mathscr{B}$ since $m \in \mathsf{M}$ and $\mathrm{Im}(f)$ can be made to range over a small set of objects since $e \in \mathsf{E}$.

This completes the argument that each A in $\mathscr{K}$ has a reflection $r: A \longrightarrow B$ in $\mathscr{B}$. We must show that r is in E. Let (e, m) be the E-M factorization of r. As we know $\mathrm{Im}(r) \in \mathscr{B}$ there exists unique f with $r.f = e$. As $r(f.m) = r, f.m =$

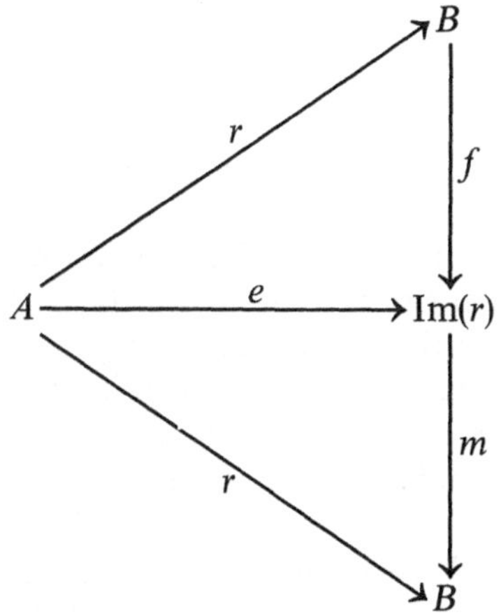

id_B and f is split mono and epi (f is epi because e is) so f is an isomorphism and $r = e.f^{-1}$ is in E. □

It is often true that if (E, M) is an image factorization system in $\mathscr{K}$ then every equalizer is in M. For example, this occurs for all three image factorizations systems of 4.8. This allows the simplification expressed in the following corollary which recaptures the most prevalent definition of quasivariety in the literature, namely a class closed under products and subobjects.

4.22 Corollary. *Let $(\mathscr{K}, \mathsf{E}, \mathsf{M})$ be a regular category such that every equalizer is in* M. *Then a full subcategory $\mathscr{B}$ of $\mathscr{K}$ is a quasivariety if and only if $\mathscr{B}$ is closed under products and closed under* M.

Proof. This follows at once from 2.1.22. □

4.23 Theorem. Let $(\mathscr{K}, \mathsf{E}, \mathsf{M})$ be a regular category and let $\mathbf{T}$ be an algebraic theory in $\mathscr{K}$ such that $eT \in \mathsf{E}$ whenever $e \in \mathsf{E}$. A full replete subcategory

$\mathscr{B}$ of $\mathscr{K}^{\mathbf{T}}$ is an E-*Birkhoff subcategory of* $\mathscr{K}^{\mathbf{T}}$ if $\mathscr{B}$ is an abstract Birkhoff subcategory of $\mathscr{K}^{\mathbf{T}}$ (3.1) such that the reflection $X\lambda:(XT, X\mu) \longrightarrow (X\bar{T}, \xi_X)$ of each free **T**-algebra is such that $X\lambda: XT \longrightarrow X\bar{T}$ is (not only an epimorphism, but) in E. The following two statements are true:

1. *$\mathscr{B}$ is an* E-*Birkhoff subcategory of $\mathscr{K}^{\mathbf{T}}$ if and only if $\mathscr{B}$ is closed under limits, closed under* M *(i.e., closed under $M^{\mathbf{T}}$ in the sense of 4.17), and closed under $U^{\mathbf{T}}$-split epimorphisms.*
2. *If every equalizer is in* M *then $\mathscr{B}$ is an* E-*Birkhoff subcategory of $\mathscr{K}^{\mathbf{T}}$ if and only if $\mathscr{B}$ is closed under products, closed under* M, *and closed under $U^{\mathbf{T}}$-split epimorphisms.*

Proof. Let $\mathscr{B}$ be an E-Birkhoff subcategory. Then $\mathscr{B}$ is closed under $U^{\mathbf{T}}$-split epimorphisms by definition. In the context of the diagram

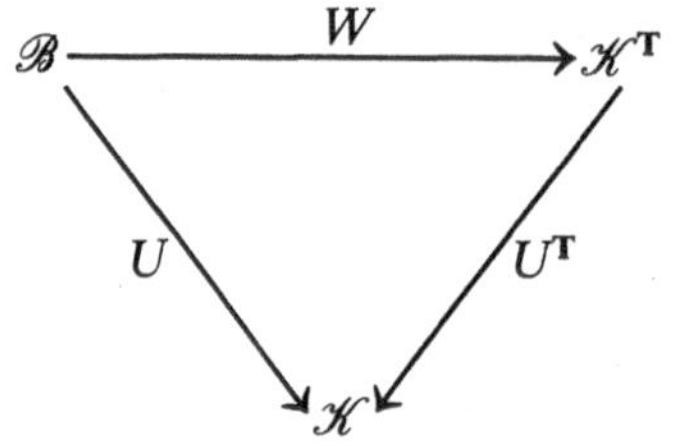

U is algebraic by 3.3 and, by 1.28 (2), $\mathscr{B}$ is closed under limits. To prove that $\mathscr{B}$ is closed under M, let $m:(X, \xi) \longrightarrow (B, \theta)$ be a **T**-homomorphism with (B, θ) in $\mathscr{B}$ and $m \in$ M. In the context of the regular category $(\mathscr{K}, \mathrm{E}^{\mathbf{T}}, \mathrm{M}^{\mathbf{T}})$ of 4.17 there exists a diagonal fill-in h, as shown in the diagram below, where f is induced by the reflection property and we are using the hypothesis that $X\lambda: XT \longrightarrow X\bar{T} \in$ E. Since h is a $U^{\mathbf{T}}$-split epimorphism $(X\eta.X\lambda.h = \mathrm{id}_X)$

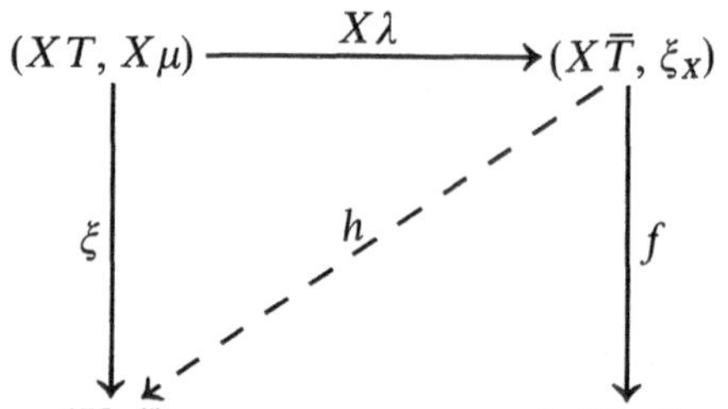

and $(X\bar{T}, \xi_X)$ is in $\mathscr{B}$, (X, ξ) is in $\mathscr{B}$ as desired. The remaining details follow at once from 4.21 and 4.22. □

It is a consequence of 4.23 that E-Birkhoff subcategories are quasivarieties and that all **T**-algebras, not just the free ones, have E-reflections in the subcategory. Theorem 4.23 also provides an alternate proof of 3.6.

4.24 Noniteration Lemma. *Let* $(\mathscr{K}, \mathrm{E}, \mathrm{M})$ *be a regular category in which every equalizer is in* M. *Define operators on full replete subcategories $\mathscr{A}$ of $\mathscr{K}$ by* $\mathrm{P}(\mathscr{A})$ = *the class of all objects isomorphic to a product of elements of $\mathscr{A}$,*

$\mathsf{M}(\mathscr{A})$ = *the class of all objects admitting an* M*-morphism into some element of* $\mathscr{A}$, *and* $\mathsf{S}(\mathscr{A})$ = *the class of all objects admitting a split epimorphism from some element of* $\mathscr{A}$. *Then* $\mathsf{MP}(\mathscr{A})$ *is the quasivariety generated by* $\mathscr{A}$ (*i.e.*, $\mathscr{A} \subset \mathsf{MP}(\mathscr{A})$, $\mathsf{MP}(\mathscr{A})$ *is a quasivariety, and if* $\mathscr{A} \subset \mathscr{B}$ *and* $\mathscr{B}$ *is a quasivariety then* $\mathsf{MP}(\mathscr{A}) \subset \mathscr{B}$). *Similarly,* $\mathsf{SMP}(\mathscr{A})$ *is the smallest quasivariety* $\mathscr{B}$ *containing* $\mathscr{A}$ *and closed under* S *in the sense that* $\mathsf{S}(\mathscr{B}) \subset \mathscr{B}$.

Proof. Clearly P, M, and S are all operators O which are closure operators, that is, which satisfy $\mathscr{A} \subset \mathsf{O}(\mathscr{A})$, $\mathsf{O}(\mathscr{A}) \subset \mathsf{O}(\mathscr{B})$ whenever $\mathscr{A} \subset \mathscr{B}$ and $\mathsf{OO}(\mathscr{A}) \subset \mathsf{O}(\mathscr{A})$. For any such closure operator, $\mathsf{O}(\mathscr{A})$ is the smallest O-closed $\mathscr{B}$ containing $\mathscr{A}$ (where, of course, $\mathscr{B}$ is O-closed just in case $\mathsf{O}(\mathscr{B}) \subset \mathscr{B}$) since $\mathscr{A} \subset \mathsf{O}(\mathscr{A})$, $\mathsf{OO}(\mathscr{A}) \subset \mathsf{O}(\mathscr{A})$ and if $\mathscr{A} \subset \mathscr{B}$ with $\mathsf{O}(\mathscr{B}) \subset \mathscr{B}$ then $\mathsf{O}(\mathscr{A}) \subset \mathsf{O}(\mathscr{B}) \subset \mathscr{B}$. For another general observation, if O_1 and O_2 are closure operators and if $\mathsf{O}_2\mathsf{O}_1$ is again a closure operator (note that $\mathscr{A} \subset \mathsf{O}_2\mathsf{O}_1(\mathscr{A})$ and $\mathsf{O}_2\mathsf{O}_1(\mathscr{A}) \subset \mathsf{O}_2\mathsf{O}_1(\mathscr{B})$ if $\mathscr{A} \subset \mathscr{B}$ are always true) then $\mathsf{O}_2\mathsf{O}_1(\mathscr{A})$ is the smallest $\mathscr{B}$ containing $\mathscr{A}$ and closed under O_1 and O_2. To see this, observe that $\mathscr{A} \subset \mathsf{O}_2\mathsf{O}_1(\mathscr{A}) \subset \mathsf{O}_1\mathsf{O}_2\mathsf{O}_1(\mathscr{A}) \subset \mathsf{O}_2\mathsf{O}_1\mathsf{O}_2\mathsf{O}_1(\mathscr{A}) \subset \mathsf{O}_2\mathsf{O}_1(\mathscr{A})$, and if $\mathscr{A} \subset \mathscr{B}$ with $\mathsf{O}_1(\mathscr{B}) \subset \mathscr{B}$ and $\mathsf{O}_2(\mathscr{B}) \subset \mathscr{B}$ then $\mathsf{O}_2\mathsf{O}_1(\mathscr{A}) \subset \mathsf{O}_2\mathsf{O}_1(\mathscr{B}) \subset \mathsf{O}_2(\mathscr{B}) \subset \mathscr{B}$.

These generalities aside, let us establish the first statement of the lemma. It follows immediately from 4.12 that $\mathsf{PM}(\mathscr{A}) \subset \mathsf{MP}(\mathscr{A})$. Therefore $\mathsf{MPMP}(\mathscr{A}) \subset \mathsf{MMPP}(\mathscr{A}) \subset \mathsf{MP}(\mathscr{A})$ and MP is a closure operator. Similarly, we prove the second statement of the lemma by proving that S(MP) is a closure operator. The diagrams below make it clear that $\mathsf{PS}(\mathscr{A}) \subset \mathsf{SP}(\mathscr{A})$. Now

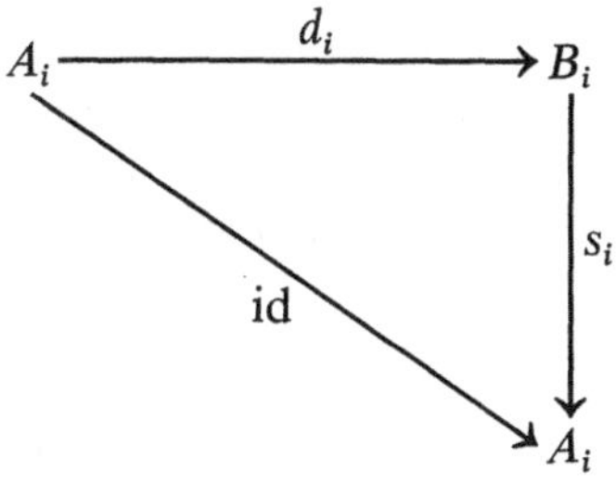

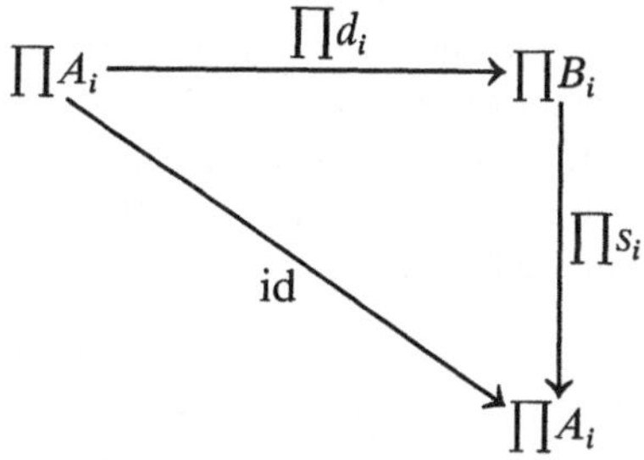

consider a pullback square as shown below with $m{:}B \longrightarrow Q$ in M and $s{:}A \longrightarrow Q$ a split epimorphism. It follows from 4.12 that $m'{:}P \longrightarrow A$ is in M whereas if $d.s = \mathrm{id}_Q$ then, applying the pullback property to id_B and $m.d{:}B \longrightarrow A$, we see that s' is a split epimorphism. Therefore $\mathsf{MS}(\mathscr{A}) \subset$

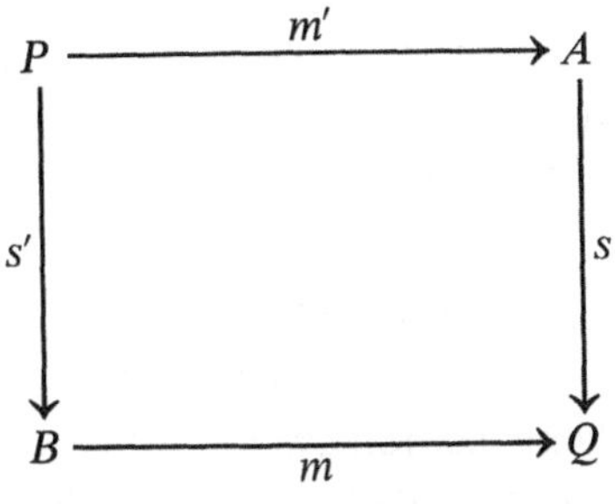

$\mathsf{SM}(\mathscr{A})$. We have $\mathsf{SMPSMP}(\mathscr{A}) \subset \mathsf{SMSPMP}(\mathscr{A}) \subset \mathsf{SSMPMP}(\mathscr{A}) \subset \mathsf{SMP}(\mathscr{A})$. We use 4.22 to prove that $\mathsf{MP}(\mathscr{A})$ and $\mathsf{SMP}(\mathscr{A})$ are quasivarieties. □

A trivial modification of the proofs of 4.24 and 4.23 (2) produces the

4.25 Theorem. *Let* $(\mathscr{K}, \mathsf{E}, \mathsf{M})$ *be a regular category in which every equalizer is in* M *and let* $\mathbf{T}$ *be an algebraic theory in* $\mathscr{K}$ *such that* eT *is in* E *whenever* e *is in* E. *Let* $(\mathscr{K}^{\mathbf{T}}, \mathsf{E}^{\mathbf{T}}, \mathsf{M}^{\mathbf{T}})$ *be the regular category of 4.17, and let* $\mathscr{B}$ *be a full replete subcategory of* $\mathscr{K}^{\mathbf{T}}$. *Then* $\mathsf{S}^{\mathbf{T}}\mathsf{M}^{\mathbf{T}}\mathsf{P}\,(\mathscr{A})$ *is the* E*-Birkhoff subcategory generated by* $\mathscr{A}$, *where* $\mathsf{S}^{\mathbf{T}}(\mathscr{A})$ *is the class of all* $\mathbf{T}$*-algebras admitting a* $U^{\mathbf{T}}$*-split epimorphism from an algebra in* $\mathscr{A}$. □

In light of the previous result, we suggest that the reader look anew at the proof of 1.4.22 where (X, δ) was shown to be a quotient of a subalgebra of a product of elements of $\mathscr{A}$.

4.26 Hausdorff Transformation Groups. Let $\mathscr{K}$ be the category of topological spaces and continuous maps. Let E be the class-of continuous surjections (= epimorphisms) and let M be the class of optimal injections (= equalizers). Then $(\mathscr{K}, \mathsf{E}, \mathsf{M})$ is a regular category in which every equalizer is in M. Let G be a topological group and let $\mathbf{T}$ be the algebraic theory of 1.3.7 whose algebras (1.4.16) are topological transformation groups. If $f: X \longrightarrow Y$ is a continuous surjection then $f \times \mathrm{id}: X \times G \longrightarrow Y \times G$ is again a continuous surjection. Therefore the context of 4.23 (2) and 4.25 is available. Let $\mathscr{B}$ be the Hausdorff transformation groups. $\mathscr{B}$ is closed under products (a product of Hausdorff spaces is Hausdorff). It is also clear that if Y is Hausdorff and if X admits a continuous injection into Y then X is also Hausdorff. This implies that $\mathscr{B}$ is closed both under $\mathsf{M}^{\mathbf{T}}$ and $\mathsf{S}^{\mathbf{T}}$. Therefore Hausdorff transformation groups is a Birkhoff subcategory of $\mathscr{K}^{\mathbf{T}}$ and, in particular, Hausdorff transformation groups is algebraic over topological spaces. The specific structure of the algebraic theory in $\mathscr{K}$ that does the job is by no means clear, but we know that it exists.

Notes for Section 4

The concept of "image factorization system" has been studied by many and dates back at least as far as [Mac Lane '48]. The simplicity and elegance of the development of these axioms makes it surprising that they do not appear in most of the expository literature; indeed, the only books mentioned in the "reader's guide" following section 2.1 which mention image factorization systems are [Arbib and Manes '74] and [Herrlich and Strecker '74]. Our treatment was influenced by [Barr '71].

In proving a special case of the quasivariety theorem ([Schmidt '66, Theorem 2]), Schmidt states [page 74] that a categorical generalization "... ought to be contained in any future text book or monograph on General Algebra." We have complied.

Exercises for Section 4

1. (E, M) is a *factorization system* in $\mathscr{K}$ if 4.2, 4.4, and 4.5 are satisfied.
   (a) Prove that every factorization system satisfies 4.10. [Hint: to prove that h is unique, consider its E-M factorization.]

(b) Prove that 4.11—modified so that the condition on h reads "$e.h = f$ and $h.t = m$—holds for any factorization system. [Hint: to prove $h.e = \mathrm{id}_{\mathrm{Im}(t)}$, use the uniqueness of (e, m).]

(c) Show that 4.12 and 4.14 hold for any factorization system.

2. Let $\mathscr{K}$ be the category of metric spaces with base point as in 2.1.12. Verify that $(\mathscr{K}, \mathsf{E}, \mathsf{M})$ is a regular category with respect to the following three choices of (E, M):
   (a) E = coequalizers, M = injective maps;
   (b) E = surjective maps, M = isometries onto a subspace;
   (c) E = maps with dense image, M = isometries onto a closed subspace.
   (d) Show that the forgetful functor (not the unit disc functor!) from the category of normed linear spaces and norm-decreasing linear mappings to $\mathscr{K}$ is algebraic. [Hint: special adjoint functor theorem.]
   (e) With E as in (c) and $\mathbf{T}$ as in (d), show that "Banach spaces" is an E-Birkhoff subcategory of normed linear spaces. [Hint: use the Hahn-Banach theorem to prove that T preserves E.]

3. Let (E, M) be an image factorization system in $\mathscr{K}$.
   (a) If $F, G: \mathscr{A} \longrightarrow \mathscr{K}$ are functors and $\alpha: F \longrightarrow G$ is a natural transformation, let

$$AF \xrightarrow{Ae} AI \xrightarrow{Am} AG$$

   be an E-M factorization of $A\alpha$ for all A. Show that I is a functor in a unique way so as to render $e: F \longrightarrow I$ and $m: I \longrightarrow G$ natural transformations. Use this construction to show that $(\bar{\mathsf{E}}, \bar{\mathsf{M}})$ is an image factorization system for the functor category $\mathscr{K}^{\mathscr{A}}$ where $\bar{\mathsf{E}}$ is the class of all α with each $A\alpha$ in E, and $\bar{\mathsf{M}}$ is all α with $A\alpha$ in M.
   (b) If $\lambda: \mathbf{T} \longrightarrow \mathbf{S}$ is a map of theories in $\mathscr{K}$ and if

$$KT \xrightarrow{Ke} KI \xrightarrow{Km} KS$$

   is an E-M factorization of $K\lambda$, show that I has unique theory structure such that $e: \mathbf{T} \longrightarrow \mathbf{I}$ and $m: \mathbf{I} \longrightarrow \mathbf{S}$ are theory maps.
   (c) When $(\mathscr{K}, \mathsf{E}, \mathsf{M})$ is regular, show that—in the context of (b)—the E-Birkhoff subcategory corresponding to $e: \mathbf{T} \longrightarrow \mathbf{I}$ is the smallest E-Birkhoff subcategory of $\mathscr{K}^{\mathbf{T}}$ containing all $\mathbf{T}$-algebras of form $(X, \xi)V$, where $V: \mathscr{K}^{\mathbf{S}} \longrightarrow \mathscr{K}^{\mathbf{T}}$ corresponds to λ.
   (d) Let $\mathbf{T}$ be a theory in **Set** and let (X, ξ) be a $\mathbf{T}$-algebra with corresponding theory map $\lambda: \mathbf{T}$ —— $\mathbf{T}_X$ as in exercise 8 of section 2. Prove that the image $e: \mathbf{T} \longrightarrow \mathbf{I}$ as in (b) is the smallest Birkhoff subcategory of $\mathbf{Set}^{\mathbf{T}}$ containing (X, ξ). Conclude that the free $\mathbf{I}$-algebra on n generators is the $\mathbf{T}$-subalgebra of $(X, \xi)^{(X^n)}$ generated by the n projection functions.

4. Let $(\mathscr{K}, \mathsf{E}, \mathsf{M})$ be the regular category of topological linear spaces and continuous linear maps, where M is the subcategory of homeomorphisms into. Show that the quasivariety generated by the scalar field is locally convex spaces.

5. (a) Let $(\mathscr{K}, \mathsf{E}, \mathsf{M})$ be a regular category and let J be an object in $\mathscr{K}$. Show that the quasivariety $\mathsf{MP}(J)$ generated by J consists of all objects K

whose evaluation map (2.1.60)

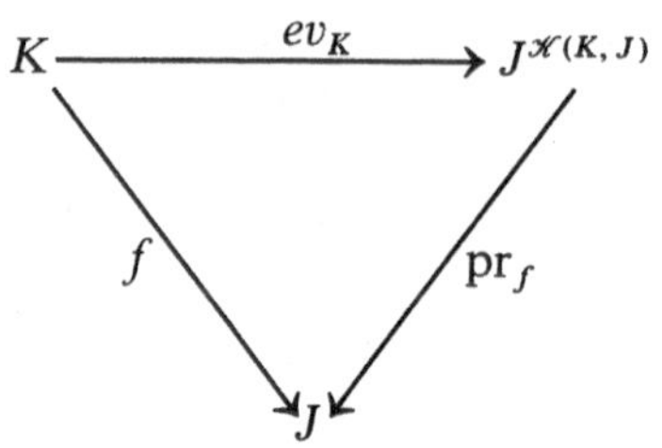

is in M. [Hint: use 4.13.] In particular, J is a cogenerator in the quasivariety it generates.

(b) In contradistinction to exercise 2a of section 3, show that if there is at least one morphism between every pair of $\mathscr{K}$-objects and if $\mathscr{K}$ has no cogenerator (e.g. the category of groups, 2.1.64) then no small set of objects generates $\mathscr{K}$ as a quasivariety.

6. Let $\mathbf{T}$ be an algebraic theory in an arbitrary category $\mathscr{K}$. A $\mathbf{T}$-*equation* is an arbitrary pair of morphisms of form $e, e' : K \longrightarrow AT$. A $\mathbf{T}$-algebra (X, ξ) *satisfies* $e = e'$ if for every $f : A \longrightarrow X$ we have $e.f^{\#} = e'.f^{\#}$. A full replete subcategory of $\mathscr{K}^{\mathbf{T}}$ is *equational* if it consists of all $\mathbf{T}$-algebras satisfying some class of equations. Prove the following version of Birkhoff's theorem (cf. [Hatcher '70], [Herrlich and Ringel '72]): If $(\mathscr{K}, \mathsf{E}, \mathsf{M})$ is a regular category with E the class of all coequalizers, then the E-Birkhoff subcategories coincide with the equational classes in $\mathscr{K}^{\mathbf{T}}$ for any $\mathbf{T}$ such that T preserves E. [Hint: if $\mathscr{B}$ is E-Birkhoff, the appropriate equations are the kernel pairs of the $A\lambda$'s.]

7. Let (E, M) be an image factorization system on $\mathscr{K}$ and let $\mathbf{T}$ be an algebraic theory in $\mathscr{K}$ such that T preserves E. Let (X, ξ) be a $\mathbf{T}$-algebra and let $m : A \longrightarrow X$ be in M. Show that $\langle A \rangle = \mathrm{Im}(mT.\xi)$ is the subalgebra generated by A in the sense of 1.4.31.

8. Let $\mathbf{Set}^{\mathbf{T}}$ be the equationally-definable class corresponding to one binary operation and no equations. Let $\mathscr{B}$ be the full subcategory of all (X, m) with $m : X^2 \longrightarrow X$ bijective. Then $\mathscr{B}$ is equationally-definable (see exercise 1.1.4.) Show that $\mathscr{B}$ is closed neither under subalgebras nor under quotient algebras. As an interesting aside, we note that R. Diaconescu has proved that $\mathscr{B}$ is a topos (exercise 2.2.16); it would be nice to have a direct construction of exponential objects, Ω and $1 + 1$.

5. Fibre-Complete Algebra

The concept of a topological algebra is well known. One provides an Ω-algebra (X, δ) with a topology in such a way that each operation δ_ω: $X^n \longrightarrow X$ is continuous from the product topology on X^n to the topology on X. In this section we define this concept replacing topological spaces with an arbitrary fibre-complete category in Struct(**Set**) and replacing Ω-algebras with **T**-algebras for **T** an arbitrary algebraic theory in **Set**. The main result is the obvious generalization of "topological algebras is algebraic over topological spaces." A Birkhoff subcategory argument is used.

Let $(\mathscr{A}, U) \in \text{Struct}(\mathscr{K})$ be fibre complete and let $(\mathscr{K}^{\mathbf{T}}, U^{\mathbf{T}}) \in \text{Struct}(\mathscr{K})$ be algebraic. Consider $(\mathscr{P}, W) = (\mathscr{A}, U) \times (\mathscr{K}^{\mathbf{T}}, U^{\mathbf{T}})$ in Struct($\mathscr{K}$) (2.3.7) and the associated pullback diagram 5.1 in the category of categories and functors

$$\begin{array}{ccc} \mathscr{P} & \xrightarrow{\quad U_2 \quad} & \mathscr{K}^{\mathbf{T}} \\ {\scriptstyle U_1}\big\downarrow & \searrow^{W} & \big\downarrow{\scriptstyle U^{\mathbf{T}}} \\ \mathscr{A} & \xrightarrow[\quad U \quad]{} & \mathscr{K} \end{array} \tag{5.1}$$

(cf. 2.3.35). Clearly $(\mathscr{P}, U_1)$ is in Struct($\mathscr{A}$) via $\mathscr{P}(K, s) = \{\xi:(K, \xi) \text{ is a } \mathbf{T}\text{-algebra}\}$ and, similarly, $(\mathscr{P}, U_2)$ is in Struct($\mathscr{K}^{\mathbf{T}}$). It is easily seen that $(\mathscr{P}, U_2)$ is fibre complete over $\mathscr{K}^{\mathbf{T}}$ (cf. 2.3.36). We now prove

5.2 Theorem. *In the context of 5.1 above, $(\mathscr{P}, U_1)$ is algebraic.*

Proof. For (K, s) in $\mathscr{A}$, let $\mathscr{F}(K, s)$ denote the class of all (f, L, ξ, t) such that $f:K \longrightarrow L \in \mathscr{K}$, $f:(K, s) \longrightarrow (L, t)$ is admissible in $\mathscr{A}$, and (L, ξ) is a **T**-algebra. Then

$$(KT \xrightarrow{f^{\#} = fT.\xi} (L, t) : (f, L, \xi, t) \in \mathscr{F}(K, s))$$

has an optimal lift $\bar{s} \in (KT)$. Define $(K, s)\bar{T} = (KT, \bar{s})$. Then $K\eta:(K, s) \longrightarrow (K, s)\bar{T}$ is admissible, since if $(f, L, \xi, t) \in \mathscr{F}(K, s)$ then $K\eta.f^{\#} = f:(K, s) \longrightarrow (L, t)$ is admissible. Moreover, if $\beta:(K_2, s_2) \longrightarrow (K_3, s_3)\bar{T}$ is admissible, so is $\beta T.K_3\mu:(K_2, s_2)\bar{T} \longrightarrow (K_3, s_3)\bar{T} \in \mathscr{F}(K_2, s_2)$ so that if also $\alpha:(K_1, s_1) \longrightarrow (K_2, s_2)\bar{T}$ is admissible, $\alpha \circ \beta = \alpha.\beta T.K_3\mu$ is admissible $(K_1, s_1) \longrightarrow (K_3, s_3)\bar{T}$. This defines an algebraic theory $\bar{\mathbf{T}} = (\bar{T}, \bar{\eta}, \bar{\mu})$ in $\mathscr{A}$. We note at once that for $f:(K, s) \longrightarrow (L, t)$, $f\bar{T} = \text{id}_{(K, s)\bar{T}} \circ f^{\Delta} = fT$ qua $\mathscr{K}$-morphism, and $X\bar{\mu} = \text{id}_{(K, s)\bar{T}\bar{T}} \circ \text{id}_{(K, s)\bar{T}} = X\mu$ qua $\mathscr{K}$-morphism. We will show that $(\mathscr{A}^{\bar{\mathbf{T}}}, U^{\bar{\mathbf{T}}})$ and $(\mathscr{P}, U_1)$ are isomorphic in Struct($\mathscr{A}$) by the straightforward passage $(K, s; \xi) \longmapsto (K, s, \xi)$. For let $\xi:(K, s)\bar{T} \longrightarrow (K, s)$ be a $\bar{\mathbf{T}}$-algebra. This is clearly asserting only that $\xi:KT \longrightarrow K$ is a **T**-algebra such that $\xi:(K, s)\bar{T} \longrightarrow (K, s)$ is admissible; but the latter statement is automatically true since $\xi:(K, s)\bar{T} \longrightarrow (K, s) \in \mathscr{F}(X, s)$ (take $f = \text{id}_K$). Therefore, $(K, s; \xi) \longmapsto (K, s, \xi)$ is a bijection. Moreover, a $\bar{\mathbf{T}}$-homomorphism $f:(X, s; \xi) \longrightarrow (Y, t; \xi)$ is, by definition, just an admissible map $f:(X, s) \longrightarrow (Y, t)$ which is also a **T**-homomorphism $(X, \xi) \longrightarrow (Y, \theta)$; that is, a morphism $f:(X, s, \xi) \longrightarrow (Y, t, \theta)$ in $\mathscr{P}$. $\square$

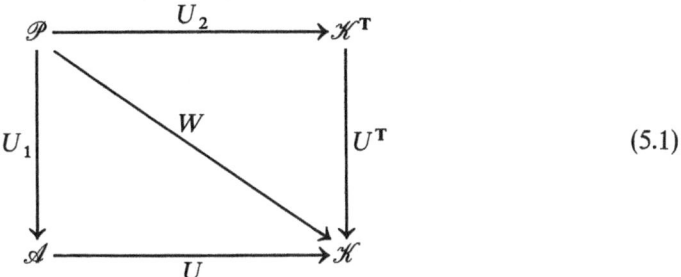

(5.3) $\bar{\mathbf{T}}$ of the proof of 5.2 is called the *canonical lift of* $\mathbf{T}$ *to* $\mathscr{A}$.

5.4 Proposition. *Let* $\lambda:\mathbf{T}_1 \longrightarrow \mathbf{T}_2$ *be a theory map of algebraic theories in* $\mathscr{K}$. *Let* $\bar{\mathbf{T}}_1, \bar{\mathbf{T}}_2$ *be the canonical lifts to* $\mathscr{A}$ *as in 5.3. Then* $\lambda:\bar{\mathbf{T}}_1 \longrightarrow \bar{\mathbf{T}}_2$ *is a theory map.*

Proof. It suffices to show that $K\lambda:(K, s)\bar{T}_1 \longrightarrow (K, s)\bar{T}_2$ is admissible in $\mathscr{A}$ for all (K, s) in $\mathscr{A}$. Using the notations of the proof of 5.2, let $(f, L, \xi_2, t) \in \mathscr{F}_2(K, s)$. Define $\xi_1 = L\lambda.\xi_2$. Then (K, ξ_1) is a $\mathbf{T}_1$-algebra (see

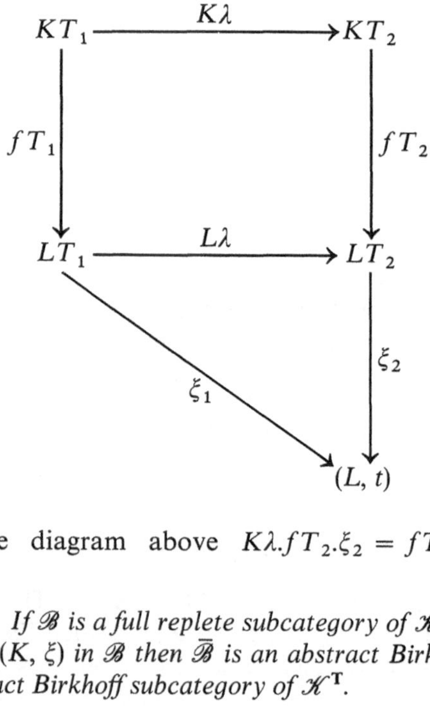

2.9) so, from the diagram above $K\lambda.fT_2.\xi_2 = fT_1.\xi_1 \in \mathscr{F}_1(K, s)$ is admissible. □

5.5 Corollary. *If* $\mathscr{B}$ *is a full replete subcategory of* $\mathscr{K}^{\mathbf{T}}$ *and if* $\bar{\mathscr{B}}$ *consists of all* (K, s, ξ) *with* (K, ξ) *in* $\mathscr{B}$ *then* $\bar{\mathscr{B}}$ *is an abstract Birkhoff subcategory of* $\mathscr{A}^{\bar{\mathbf{T}}}$ *if* $\mathscr{B}$ *is an abstract Birkhoff subcategory of* $\mathscr{K}^{\mathbf{T}}$.

Proof. Use the co-optimal lift of $X\lambda:X\bar{T} \longrightarrow XS$. □

The following theorem shows that not all interesting Birkhoff subcategories of $\mathscr{A}^{\bar{\mathbf{T}}}$ are of the form $\bar{\mathscr{B}}$.

5.6 Theorem. *Let* $\mathbf{T}$ *be an algebraic theory in* **Set**, *let* $(\mathscr{A}, U)$ *be fibre complete in* Struct(**Set**), *and let* $\mathscr{C}$ *be the category with.*

Objects: triples (X, s, ξ) *such that* $(X, s) \in \mathscr{A}$, (X, ξ) *is a* $\mathbf{T}$*-algebra, and for all semantic operations (1.5.3)* $\alpha:(U^{\mathbf{T}})^n \longrightarrow U^{\mathbf{T}}$ *the induced operation* $(X, \xi)\alpha:X^n \longrightarrow X$ *is admissible from* (X^n, s^n) *to* (X, s) *(where* s^n *is the optimal lift of the projections* $p_i:X^n \longrightarrow (X, s)$*, i.e., the product (2.3.29) in* $\mathscr{A}$*).*

Morphisms: $f:(X, s, \xi) \longrightarrow (Y, t, \theta)$ *such that* $f:(X, s) \longrightarrow (Y, t)$ *is admissible in* $\mathscr{A}$ *and* $f:(X, \xi) \longrightarrow (Y, \theta)$ *is a* $\mathbf{T}$*-homomorphism.*

Then the underlying $\mathscr{A}$*-object functor* $V:\mathscr{C} \longrightarrow \mathscr{A}$ *is algebraic.* (The theorem is sharpened in 5.11 below.)

Proof. Let $\bar{\mathbf{T}}$ be the canonical lift of $\mathbf{T}$ to $\mathscr{A}$ (5.3). (**Set**, surjections, injections) is a regular category. It follows from 4.16 that ($\mathscr{A}$, E, M) is a regular category with E = admissible surjections and M = optimal injections. Since T preserves surjections (1.4.29), $\bar{T}$ preserves E (as remarked in the proof of 5.2, $f\bar{T}$ is the function fT). Since $\mathscr{C}$ is a full replete subcategory of $\mathscr{A}^{\bar{\mathbf{T}}}$ it suffices to show that $\mathscr{C}$ is closed under products, "optimal injective **T**-homomorphisms into" and "**T**-homomorphisms which are split epimorphisms in $\mathscr{A}$ out of," since then $\mathscr{C}$ is an E-Birkhoff subcategory of $\mathscr{A}^{\bar{\mathbf{T}}}$ (use 4.23(2) noting that, by 2.3.29, the optimal injective maps are exactly the equalizers in $\mathscr{A}$) and algebraic over $\mathscr{A}$ in particular (3.3).

(5.7) $\mathscr{C}$ is closed under products. Given (X_i, s_i, ξ_i) in $\mathscr{C}$ with product $p_i:(X, s, \xi) \longrightarrow (X_i, s_i, \xi_i)$ in $\mathscr{A}^{\bar{\mathbf{T}}}$ we have, for each $\alpha:(U^{\mathbf{T}})^n \longrightarrow U^{\mathbf{T}}$ the commutative diagram

$$\begin{array}{ccc} X^n & \xrightarrow{(p_i)^n} & (X_i)^n \\ {\scriptstyle (X,\xi)\alpha}\downarrow & & \downarrow{\scriptstyle (X_i,\xi_i)\alpha} \\ X & \xrightarrow[p_i]{} & X_i \end{array}$$

Since $f^n:(K^n, s^n) \longrightarrow (L^n, t^n)$ is admissible whenever $f:(K, s) \longrightarrow (L, t)$ is (it is categorically induced by the universal property of the second product), $(p_i)^n:(X^n, s^n) \longrightarrow (Y^n, t^n)$ is admissible. $(X_i, \xi_i)\alpha$ is admissible from (X_i^n, s_i^n) to (X_i, s_i) by hypothesis. Since $p_i:(X, s) \longrightarrow (X_i, s_i)$ is an optimal family, $(X, \xi)\alpha:(X^n, s^n) \longrightarrow (X, s)$ is admissible, that is, (X, s, ξ) is in $\mathscr{C}$.

(5.8) $\mathscr{C}$ is closed, in fact, under all optimal **T**-homomorphisms into. The argument is essentially the same as 5.7. If $f:(X, s, \xi) \longrightarrow (Y, t, \theta)$ is an optimal **T**-homomorphism with (Y, t, θ) in $\mathscr{C}$ then for each semantic operation $\alpha:(U^{\mathbf{T}})^n \longrightarrow U^{\mathbf{T}}$ we have the commutative square

$$\begin{array}{ccc} X^n & \xrightarrow{f^n} & Y^n \\ {\scriptstyle (X,\xi)\alpha}\downarrow & & \downarrow{\scriptstyle (Y,\theta)\alpha} \\ X & \xrightarrow[f]{} & Y \end{array}$$

to prove that $(X, \xi)\alpha.f$—and so $(X, \xi)\alpha$— is admissible.

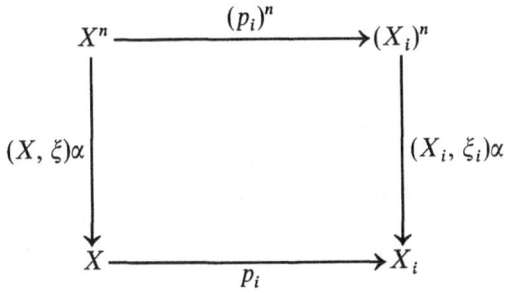

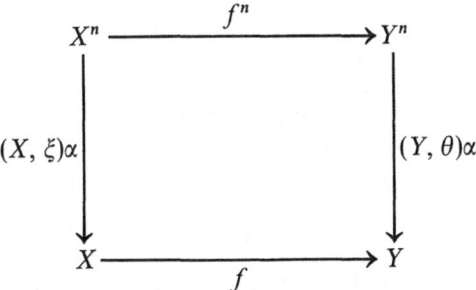

(5.9) $\mathscr{C}$ is closed under $U^{\mathbf{T}}$-split epimorphisms. Let $(X, s, \xi) \in \mathscr{C}$ and let $f:(X, s, \xi) \longrightarrow (Y, t, \theta) \in \mathscr{A}^{\mathbf{T}}$ and $d:(Y, t) \longrightarrow (X, s) \in \mathscr{A}$ satisfy $d.f = \mathrm{id}_Y$. Consider the diagram

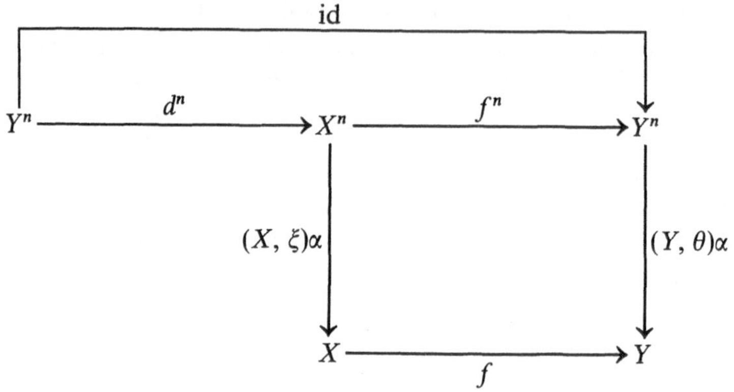

induced by the semantic operation $\alpha:(U^{\mathbf{T}})^n \longrightarrow U^{\mathbf{T}}$. Then $(Y, \theta)\alpha = d^n.(X, \xi)\alpha.f:(Y^n, t^n) \longrightarrow (Y, t)$ is admissible. □

The proof of 5.6 does not provide much information about the nature of the algebraic theory in $\mathscr{A}$ which gives rise to V. The following elementary observation greatly simplifies this problem.

5.10 Taut Birkhoff Subcategories. In the context of 5.1, let $\mathscr{C}$ be an abstract Birkhoff subcategory of $\mathscr{P}$ (of course, U_1 is known to be algebraic

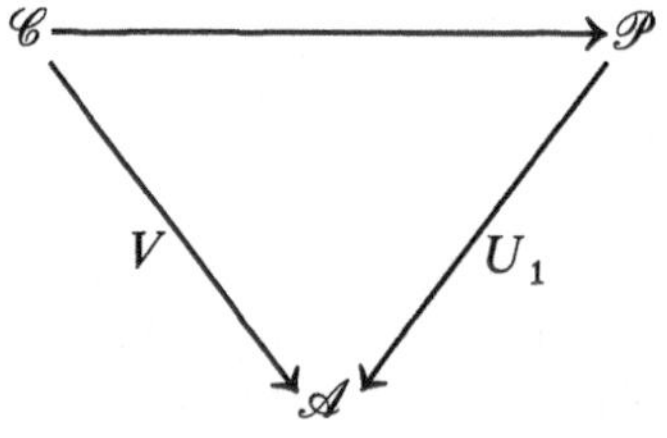

by 5.2) with the additional property that $\mathscr{C}$ is *taut* as in 2.3.37, that is, $\mathscr{C}$ has the property that whenever $f_i:P \longrightarrow C_i$ is a family of morphisms in $\mathscr{P}$ with each C_i in $\mathscr{C}$ such that f_iU_1 constitutes an optimal family in $\mathscr{A}$, then P is also in $\mathscr{C}$. Under these conditions, the algebraic theory in $\mathscr{A}$ giving rise to V is very simple to describe in the style of the $\bar{\mathbf{T}}$ of 5.2. For each (K, s) in $\mathscr{A}$ let $\mathscr{C}(K, s)$ denote the subclass of $\mathscr{F}(K, s)$ of all (f, L, ξ, t) such that $(L, t, \xi) \in \mathscr{C}$. Define $(K, s)\hat{T} = (KT, \hat{s})$ where $\hat{s}$ is the optimal lift of the family

$$(KT \xrightarrow{f^{\#} = fT.\xi} (L, t):(f, L, \xi, t) \in \mathscr{C}(K, s))$$

Then $(KT, \hat{s}, K\mu) \in \mathscr{C}$ by the hypothesis that $\mathscr{C}$ is taut. If $(KT, \bar{s})$ is as in the proof of 5.2, then $\mathrm{id}_{KT}:(KT, \bar{s}) \longrightarrow (KT, \hat{s})$ is admissible because $\mathscr{C}(K, s) \subset \mathscr{F}(K, s)$. Moreover, if $g:(KT, \bar{s}, K\mu) \longrightarrow (L, t, \xi) \in \mathscr{P}$ with $(L, t, \xi) \in \mathscr{C}$ then $g:(KT, \hat{s}, K\mu) \longrightarrow (L, t, \xi) \in \mathscr{C}$ because,

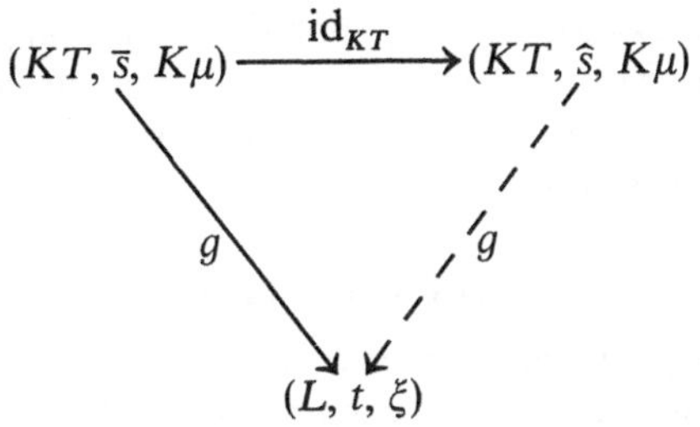

setting $f = K\eta.g$, $g = f^{\#} = fT.\xi$ with $(f, L, \xi, t) \in \mathscr{C}(K, s)$. Therefore id_{KT}: $(KT, \bar{s}, K\mu) \longrightarrow (KT, \hat{s}, K\mu)$ is the reflection $(K, s)\lambda$ of $((K, s)\bar{T}, (K, s)\bar{\mu})$ in $\mathscr{C}$. It follows from the constructions in 3.2 and 3.3 that $\hat{\mathbf{T}}$ is the algebraic theory for V with η, $\circ$, T as a functor and μ all at the level of $\mathbf{T}$ (just as for $\bar{\mathbf{T}}$ in 5.2).

Returning to the context of 5.6, we have:

(5.11) The algebraic theory in $\mathscr{A}$ giving rise to $V: \mathscr{C} \longrightarrow \mathscr{A}$ is given by $(X, s)\hat{T} = (XT, \hat{s})$ where $\hat{s}$ is the cartesian lift of the family $(KT \xrightarrow{g} (L, t)$: there exists ξ with $(L, t, \xi) \in \mathscr{C}$ and $g:(KT, K\mu) \longrightarrow (L, \xi)$ a $\mathbf{T}$-homomorphism). $K\eta:(K, s) \longrightarrow (K, s)\hat{T}$ is admissible and if $\alpha:(K_1, s_1) \longrightarrow (K_2, s_2)\hat{T}$ and $\beta:(K_2, s_2) \longrightarrow (K_3, s_3)\hat{T}$ are admissible then so is $\alpha \circ \beta:(K_1, s_1) \longrightarrow (K_3, s_3)\hat{T}$ thereby providing the η and $\circ$ for the algebraic theory $\hat{\mathbf{T}} = (\hat{T}, \eta, \circ)$. To prove this, we must only be sure that 5.10 applies. But this is clear by the proof of 5.6, 5.7, and 5.8.

5.12 Topological Algebras. By 5.6, if $\mathscr{C}$ is any category of topological algebras and $\mathscr{A}$ is the category of topological spaces then the forgetful functor $V: \mathscr{C} \longrightarrow \mathscr{A}$ is algebraic and has a left adjoint in particular. We deduce that there exists a free topological group, -ring, -lattice, -complete atomic Boolean algebra, and even a free topological compact space over an arbitrary topological space. Moreover, 5.11 tells us that to construct these objects we start with the free group, -ring, -lattice, -complete atomic Boolean algebra, -compact space over the underlying set of the topological space and provide this with the appropriate topology.

Notes for Section 5

The theory of this section is from [Manes '67, Chapter 3]. Similar things have been done by Wyler ([Wyler '71]), Wischnewsky ([Wischnewsky '73]), and others (consult the bibliographies of the papers cited above).

It is interesting to remark that A. A. Markov's 1945 monograph [Markov '45, 201–246] devotes 45 pages to proving, among other things, that there exists a free topological group (his definition is the same as ours—a universal mapping property) over a completely regular space. Not surprisingly, Markov found an appropriate topology on the free group over the underlying set of the space. A more modern proof of Markov's theorem appears as [Hewitt and Ross '63, Theorem 8.8] where the reader will recognize, in context, a version of the adjoint functor theorem. The reader may wish to attempt to generalize theorems about free topological groups as found in [Thomas '74], or the papers of Morris (see [Morris '73] and the bibliography there).

It is known ([Świerczkowski '64]) that if $\mathbf{T}$ is finitary and X is a completely regular Hausdorff space, then the inclusion of the generators, η_X, into the free topological $\mathbf{T}$-algebra is a closed subspace; it is interesting to ask how far this can be generalized.

Exercises for Section 5

1. A *partially ordered group* is $(X, s, \leqslant)$ where (X, s) is a group, $(X, \leqslant)$ is a partially ordered set and the following laws hold: if $x \leqslant y$ and $a \leqslant b$ then $xa \leqslant yb$; *if* $x \leqslant y$ then $y^{-1} \leqslant x^{-1}$. Show that partially ordered groups is algebraic over partially ordered sets. [Hint: start with the canonical lift of the theory for groups to the fibre-complete category of preordered sets, pull back along the property of partially ordered sets using 1.36, and then use an E-Birkhoff subcategory.]
2. [Manes '67, 3.4.9.] Let Λ be the (real or complex) scalar field. A *topological linear space* is a Λ-vector space X which is topologized in such a way that addition $X \times X \longrightarrow X$ and scalar multiplication $\Lambda \times X \longrightarrow X$ are continuous. Show that topological linear spaces and continuous linear maps is algebraic over topological spaces. [Hint: the topology on Λ requires a Birkhoff subcategory of the "topological Λ-vector spaces" of 5.12; observe that split epimorphisms are co-optimal.]
3. (cf. [Morris '70, Theorem 1.13].) Let X be a nonempty topological space and let $\mathbf{T}$ be a nontrivial finitary theory in **Set** with $\varnothing T \neq \varnothing$. Prove that the free topological $\mathbf{T}$-algebra F over X is not connected. [Hint: let F' be the $\mathbf{T}$-algebra F with the discrete topology; consider the continuous $\mathbf{T}$-homomorphic extension of any function from X to F' which is constantly some element not in $\varnothing T$.]
4. (a) Let $\mathbf{T}$ be an algebraic theory in $\mathscr{K}$ and suppose that (E, M), $(\bar{\mathsf{E}}, \bar{\mathsf{M}})$ are image factorization systems on $\mathscr{K}$, $\mathscr{K}^{\mathbf{T}}$ such that $\bar{\mathsf{E}}U^{\mathbf{T}} \subset \mathsf{E}$ and $\bar{\mathsf{M}}U^{\mathbf{T}} \subset \mathsf{M}$. Prove that T preserves E. [Hint: use 4.11 and the diagram

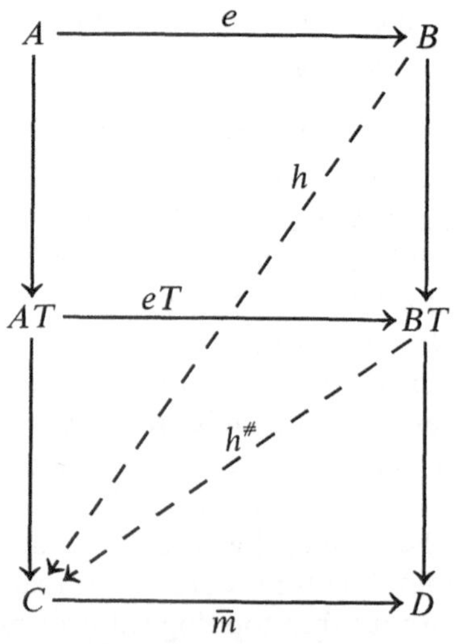

to prove that if e is in E then eT is in $\bar{\mathsf{E}}$.]

(b) In the context of 5.3 with $\mathcal{K} = \mathbf{Set}$ prove that $\bar{T}$ preserves co-optimal surjections. [Hint: use (a).] Thus, for E = co-optimal surjections, Theorem 4.23 applies; note, however, that the proof of 5.6 breaks down.

5. [Manes '72.] Let $(\mathcal{A}, U)$ in Struct(**Set**) be fibre complete, let $\mathbf{T}$ be an algebraic theory in **Set** and let $\mathscr{C}$ be a full replete subcategory of $(\mathcal{A}, U) \times (\mathbf{Set}^{\mathbf{T}}, U^{\mathbf{T}})$. Let $\mathscr{L}$ be a full replete subcatgeory of $\mathcal{A}$ such that if $f: (X, s) \longrightarrow (Y, t)$ is an admissible surjection in $\mathcal{A}$ with (X, s) in $\mathscr{L}$ then also (Y, t) is in $\mathscr{L}$. Consider the pullbacks

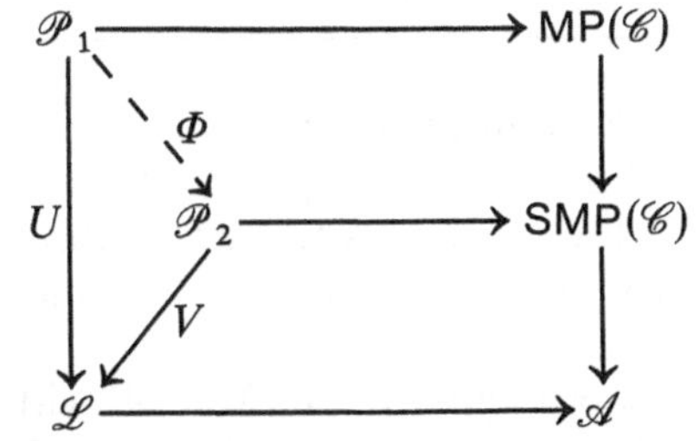

Assume further that whenever (Y, t, θ) is in $\mathsf{MP}(\mathscr{C})$ and $f:(X, s) \longrightarrow (Y, t)$ is admissible in $\mathcal{A}$ with (X, s) in $\mathscr{L}$ there exists an $\mathscr{L}$-structure $\bar{s}$ on XT such that $f^{\#}:(XT, \bar{s}) \longrightarrow (Y, t)$ is admissible in $\mathcal{A}$. Prove that U has a left adjoint, V is algebraic, and Φ is the semantic comparison of of U; moreover, Φ is a full reflective subcategory. [Hint: use 1.36; the assumptions guarantee that the $\bar{T}$ of 5.3 maps $\mathscr{L}$ into $\mathscr{L}$; the left adjoint to Φ works by restricting the left adjoint $\mathsf{SMP}(\mathscr{C}) \longrightarrow \mathsf{MP}(\mathscr{C})$; left adjoints to U and V can be chosen so that

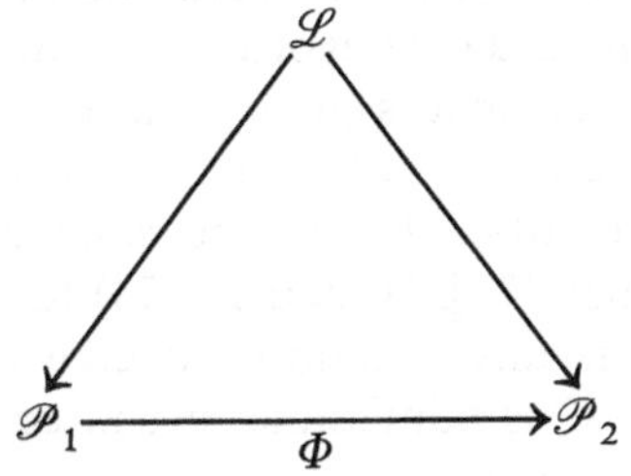

which forces Φ to be the semantics comparison functor by exercise 9(a) of section 2.]

6. In this exercise we illustrate a technique to prove algebraicity without direct verification of the Beck condition. We outline a proof that Banach spaces are algebraic over metric spaces with base point (see exercise 2(d) of section 4).

(a) Define $\mathscr{C}$ in Struct(**Set**) by setting $\mathscr{C}(X)$ to be the set of all pairs (d, A) where $d:X \times X \longrightarrow \mathbf{R} \cup \{\infty\}$ is a "premetric" and A is any subset of X and by defining $f:(X, d, A) \longrightarrow (Y, e, B)$ to be admissible just in case f is "decreasing" (i.e. $f^2 d \leqslant e$ pointwise) and f maps A into B. Prove that $\mathscr{C}$ is fibre complete.

Let $\mathbf{T}$ be the algebraic theory in **Set** whose algebras are vector spaces (scalars may be either real or complex) with canonical lift $\bar{T}$

to $\mathscr{C}$ as in 5.3. The plan of attack is summarized by the following diagram:

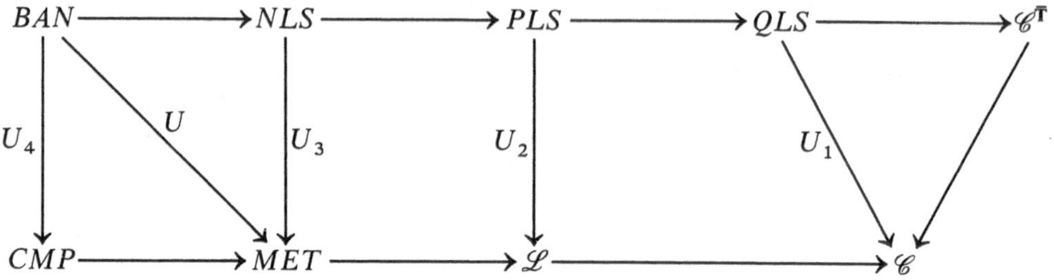

(b) Let QLS ("quasinormed linear spaces") be the full subcategory of $\mathscr{C}^{\mathbf{T}}$ of all (X, d, A, ξ) whose "norm" $\|x\| = d(x, 0)$ satisfies $\|\lambda x\| \leqslant |\lambda|\,\|x\|$ and $\|x + y\| \leqslant \|x\| + \|y\|$ and which are such that $A = \{0\}$. Show that QLS is a Birkhoff subcategory of $\mathscr{C}^{\mathbf{T}}$, and hence that U_1 is algebraic.

(c) Let $\mathscr{L}$ be the full subcategory of all (X, d, A) with $d < \infty$ and $A \neq \varnothing$, and let PLS ("pseudonormed linear spaces") be the pullback along $\mathscr{L}$. Use the previous exercise to prove that U_2 is algebraic. [Hint: given $f:(X, d, A) \longrightarrow (Y, e, \{0\}, \xi)$ in $\mathscr{C}$ with $d < \infty$ and $A \neq \varnothing$, a suitable $\mathscr{L}$-structure $(\bar{d}, \bar{A})$ on the vector space XT is $\bar{d}((\lambda_x), (\mu_x)) = \sum |\lambda_x - \mu_x|\, d(a, x)$ where a is any element of A, and $\bar{A} = \{0\}$.]

(d) Let MET be the full subcategory of $\mathscr{L}$ of metric spaces with base point and let CMP be the full subcategory of MET of complete metric spaces with base point. Verify that the pullbacks along MET and CMP are the usual categories, NLS and BAN or normed linear spaces and Banach spaces (with contractive linear maps). Use 1.36 and 1.38 to prove that U_3, U_4, and U are algebraic.

7. Let L be a complete lattice. Define the category **Set**(L) in Struct(**Set**) of *L-fuzzy sets* ([Goguen '73], [Goguen '67]) to have as objects all (X, χ) where $\chi: X \longrightarrow L$ is any function ("degree of membership") and $f: (X, \chi) \longrightarrow (X', \chi')$ is admissible just in case $x\chi \leqslant xf\chi'$ for all x in X. (See also exercise 4.3.10.)

(a) Prove that **Set**(L) is fibre complete.

(b) Consider the algebraic theory in **Set** whose algebras are Boolean algebras. The four elements of the free Boolean algebra on one

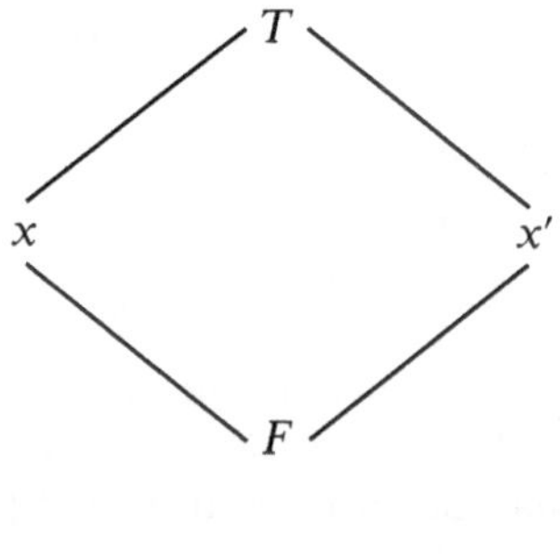

generator x has the lattice structure shown above; the meaning of these elements is: x is definitely assigned either "true" or "false," x' is then given the opposite truth value, T is "true" and F is "false." Noting that **Set** = **Set**(1) and that a single generator is the 1-fuzzy set $\mathrm{id}:1 \longrightarrow 1$, it is natural to investigate the free L-fuzzy Boolean algebra (in the sense of 5.6) over the L-fuzzy set $\mathrm{id}:L \longrightarrow L$.

(c) Let L be the two element lattice. Determine the structure of the free L-fuzzy Boolean algebra over $\mathrm{id}:L \longrightarrow L$. [Hint: it has sixteen elements, coinciding as a Boolean algebra with the free complete atomic Boolean algebra generated by a two-element set.]

8. Let $(\mathscr{A}, U)$ be fibre-complete over **Set**. Recall from exercise 3(b) of section 2 that $\mathscr{A}$ is a symmetric monoidal category.

(a) Prove that $\mathscr{A}$ is closed as defined in exercise 5 of section 2. [Hint: the exponential object B^A—which we write hence as $[A, B]$—is the set of admissible maps from A to B with the "subspace of the product" structure, i.e., the restricted projections are optimal.] Observe that $[A, -]:\mathscr{A} \longrightarrow \mathscr{A}$ and $[-, A]:\mathscr{A}^{\mathrm{op}} \longrightarrow \mathscr{A}$ are functorial. A theory **T** in $\mathscr{A}$ is *enriched* if for all A, B, C in $\mathscr{A}$ the map

$$\begin{aligned}[A, BT] \otimes [B, CT] &\xrightarrow{\ \circ\ } [A, CT] \\ \alpha, \beta &\longmapsto \alpha \circ \beta\end{aligned}$$

is admissible in $\mathscr{A}$.

(b) If **T** is enriched, show that the passages as in 1.5.6 and 1.5.7 establish a bijection from $\mathscr{A}$-morphisms $\omega:I \longrightarrow AT$ (i.e., elements of AT) to natural transformations $[A, -] \longrightarrow T$. [Hint: the only new detail is proving that $f \longmapsto \langle \omega, fT \rangle$ is admissible from $[A, X]$ to XT; to this end consider the

$$[A, X] \xrightarrow{-.X\eta} [A, XT] \xrightarrow{\mathrm{in}_{\mathrm{id}}} [AT, AT] \otimes [A, XT] \xrightarrow{\ \circ\ } [AT, XT] \xrightarrow{\mathrm{pr}_\omega} XT$$

map shown above.]

(c) For each A let U^A be the composition

$$\mathscr{A}^{\mathbf{T}} \xrightarrow{U^{\mathbf{T}}} \mathscr{A} \xrightarrow{[A,-]} \mathscr{A}$$

Show that if **T** is enriched, the passages of 1.5.8 and 1.5.9 establish a bijective correspondence between morphisms $I \longrightarrow AT$ and natural transformations from U^A to U.

(d) Using the bijections

$$\frac{\dfrac{A \longrightarrow [B, X]}{A \otimes B \longrightarrow X}}{B \longrightarrow [A, X]}$$

show that $[-, X]:\mathscr{A}^{op} \longrightarrow \mathscr{A}$ has $[-, X]:\mathscr{A} \longrightarrow \mathscr{A}^{op}$ as a left adjoint. Show that the induced algebraic theory $\mathbf{T}_{[X]}$ in $\mathscr{A}$ is enriched.

(e) Generalizing exercise 8 of section 2, show that if $\mathbf{T}$ is enriched then a $\mathbf{T}$-algebra structure on X corresponds to a theory map from $\mathbf{T}$ to $\mathbf{T}_{[X]}$.

It is clear by now that much of the theory developed in this book for the base category of sets will generalize to a fibre-complete category over **Set**. The reader may wish to develop some of this theory herself. Actually, the proper setting for a general theory of enriched algebraic theories is, at least, in a symmetric monoidal closed category. See, e.g., [Bunge '69], [Dubuc '70], [Kock '70, '71], [Linton '69-A], [Pfender '74], [Wiesler and Calugareanu '70], and the bibliographies there.

9. We continue exercise 8 by constructing a large class of enriched theories.

(a) Given a theory $\mathbf{T}$ in $\mathscr{A}$ show that $\mathbf{T}$ is enriched if and only if for all B, C the map

$$[B, CT] \xrightarrow{(-)^{\#}} [BT, CT]$$
$$\beta \longmapsto \beta^{\#}$$

is admissible in $\mathscr{A}$. [Hint: see exercise 1.3.12.]

(b) Conclude that the canonical lift $\bar{\mathbf{T}}$ of a theory $\mathbf{T}$ in **Set** is rarely enriched, but that the theory $\mathbf{S}$ in $\mathscr{A}$ corresponding to $\mathscr{C}$ as in 5.6 is always enriched. [Hint: we have

$$\begin{array}{ccc} [B, CS] & \xrightarrow{(-)^{\#}} & [BS, CS] \\ \downarrow & & \downarrow {\scriptstyle \mathrm{pr}_{\omega}} \\ CS^{|B|} & \xrightarrow[|B|\tilde{\omega}]{} & CS \end{array}$$

where $|B|$ is the underlying set of B and $CS^{|B|}$ is the cartesian power.] This result improves [Manes '67, 3.4.6].

10. (cf. [H. Neumann '67, page 9].) Let $(\mathscr{A}, U)$ be fibre complete over **Set**, let E be the class of admissible surjections, and let $\mathbf{T}$ be any enriched (see exercise 8) theory in $\mathscr{A}$ such that T preserves E. Given a $\mathbf{T}$-algebra (X, ξ), suppose given an $\mathscr{A}$-morphism $\tau: A \longrightarrow X$ such that

(i) For each $f: A \longrightarrow X$ there exists a unique $\mathbf{T}$-homomorphism $\psi:(X, \xi) \longrightarrow (X, \xi)$ such that $\tau.\psi = f$.

(ii) The family of all $\mathbf{T}$-endomorphisms of (X, ξ) is optimal in $\mathscr{A}$.

(iii) Given $x \neq y$ in X there exists a $\mathbf{T}$-endomorphism ψ of (X, ξ) with $x\psi \neq y\psi$.

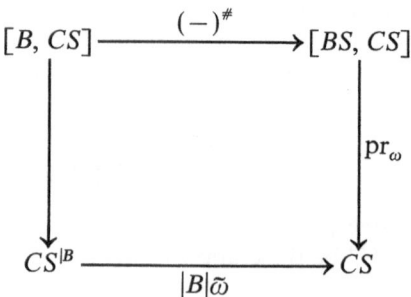

Show that (X, ξ, τ) is the free algebra over A in the E-Birkhoff subcategory of $\mathscr{A}^{\mathbf{T}}$ generated by (X, ξ). [Hint: prove the analog of exercise 3(d) of section 4 and use the diagram

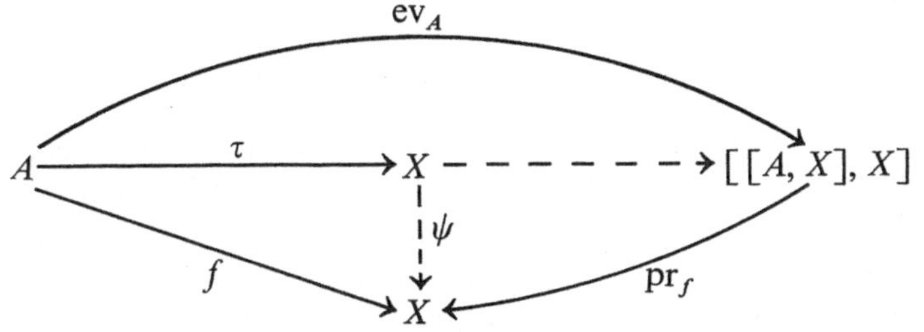

].

11. For any algebraic theory **T** in **Set**, show that compact Hausdorff **T**-algebras forms a full reflective subcategory of topological **T**-algebras. [Hint: use the general adjoint functor theorem.] Show that the forgetful functor from compact **T**-algebras to topological spaces is algebraic. [Hint: use the Beck theorem and 2.2.30.] The free compact **T**-algebra over a space generalizes the well-known *Bohr compactification* which is just the case when **T** is the theory for groups.

6. Bialgebras

We consider bialgebras which are sets that are simultaneously equipped with algebra structure from two algebraic theories in such a way that each operation of one sort commutes with those of the other sort. Unless both theories have a rank, the question of whether the bialgebra category is algebraic over sets is a delicate one. "Completely commutative" algebras are characterized.

It is clear that the definition of $\mathscr{C}$ in 5.6 works if we replace $(\mathscr{A}, U)$ with any category of sets with structure which constructs products (2.3.17). In particular, we may consider what happens when $(\mathscr{A}, U)$ is algebraic.

6.1 Bialgebras. Let **S** and **T** be algebraic theories in **Set**. An **S**-**T** *bialgebra* is a triple (X, ξ, θ) such that (X, ξ) is an **S**-algebra, (X, θ) is a **T**-algebra, and for all semantic **T**-operations (1.5.4+) $\alpha:(U^{\mathbf{T}})^n \longrightarrow U^{\mathbf{T}}$, $(X, \theta)\alpha: (X, \xi)^n \longrightarrow (X, \xi)$ is an **S**-homomorphism. A *homomorphism of* **S**-**T** bialgebras $f:(X, \xi, \theta) \longrightarrow (X', \xi', \theta')$ is a function $f:X \longrightarrow X'$ which is simultaneously an **S**-homomorphism and a **T**-homomorphism. The resulting category of sets with structure will be denoted $\mathbf{Set}^{\mathbf{S} \otimes \mathbf{T}}$ and the underlying set functor from bialgebras will be denoted $U^{\mathbf{S} \otimes \mathbf{T}}: \mathbf{Set}^{\mathbf{S} \otimes \mathbf{T}} \longrightarrow \mathbf{Set}$.

6.2 Groups with Operators. Let X be a set. A *group with operators indexed by* X is a triple (G, ξ, θ) where (G, ξ) is a group and θ assigns to each element x of X a group endomorphism $\theta_x:(G, \xi) \longrightarrow (G, \xi)$ (cf. [Van der Waerden '53, section 43]). Let **S** be the algebraic theory for groups and let **T** be the algebraic theory induced by the equational presentation with one unary operation for each element of X and no equations. Then groups with operators are just **S-T** bialgebras.

6.3 Bimodules. Let R and S be rings. A *left R-module* is an abelian group A together with a function

$$R \times A \longrightarrow A \qquad (r, a) \longmapsto ra$$

satisfying $(rr_1)a = r(r_1a)$, $r(a + b) = ra + rb$ and $(r + r')a = ra + r'a$ as well as $1a = a$. It is clear, using an equational presentation with one binary operation and a unary operation for each element of R that left R-modules constitute an algebraic category of sets with structure. Similarly a *right S-module* is an abelian group A equipped with a function $(a, s) \longrightarrow as$ satisfying $a(ss_1) = (as)s_1$, $(a + b)s = as + bs$, $a(s + s') = as + as'$, $a1 = a$. Again, right S-modules are algebraic over sets. Let the algebraic theories induced by the rings R and S in this way be denoted by **.R** and **S.** respectively. An *R-S bimodule* ([Mac Lane '63, V.3]) is a triple (A, ξ, θ) such that (A, ξ) is a left R-module, (A, θ) is a right S-module, and $r(as) = (ra)s$. Thus an R–S bimodule is the same thing as an **.R-S.** bialgebra. The proof is safely left to the reader with the hint that the next example is most of the work.

6.4 Abelian Groups. Let **T** be the theory whose algebras are abelian groups. If (X, ξ) is a **T**-algebra then (X, ξ, ξ) is a **T-T** bialgebra. To prove it, observe that "zero," "minus," and "plus" are **T**-homomorphisms. This statement would not be true for a non-abelian group. In fact let **S** be the theory whose algebras are groups and let (X, ξ, θ) be an arbitrary **S-S** bialgebra. Then $\xi = \theta$ and $(X, \xi) = (X, \theta)$ is abelian. (Write $(X, \xi) = (X, m, e), (X, \theta) = (X, m', e')$. Since the **S**-operation $e' : 1 \longrightarrow (X, \xi)$ is an **S**-homomorphism, $e' = e$. Since the **S**-operation $m' : (X, \xi)^2 \longrightarrow (X, \xi)$ is an **S**-homomorphism, we have the law $axmbymm' = abm'xym'm$. Taking $x = e = b$, we deduce $m = m'$. Taking $a = e = y$ we then deduce $xbm = bxm$.) In particular, a **T-T** bialgebra is the same thing as a **T**-algebra.

6.5 Compact Algebras. Let **T** be an algebraic theory in **Set**. Let **β** be the algebraic theory for compact Hausdorff spaces (1.5.24). A *compact* **T**-*algebra* is a **β-T** bialgebra. Because of the Tychonoff product theorem, (in the restricted form: a product of compact Hausdorff spaces is compact Hausdorff), a compact **T**-algebra is the same thing as a topological **T**-algebra (5.12) whose underlying topological space is compact Hausdorff.

6.6 Compact Compact Spaces. It is clear that the empty space and the one-element space are **β-β** bialgebras. It turns out that there are no others. For let (X, ξ, θ) be a **β-β** bialgebra and suppose that X has at least two elements. Then any two element subset of X is simultaneously a subalgebra of (X, ξ) and of (X, θ) (a finite subset of a Hausdorff space is closed) and it is a general fact that if (X, ξ, θ) is an **S-T** bialgebra and (A, ξ_0, θ_0) is such that (A, ξ_0) is an **S**-subalgebra whereas (A, θ_0) is a **T**-subalgebra of (X, ξ, θ) then (A, ξ_0, θ_0) is again an **S-T** bialgebra (see 6.9 below). Therefore the two element set $2 = \{0, 1\}$ with its unique compact Hausdorff topology α is a **β-β** bialgebra $(2, \alpha, \alpha)$. To show this is not true, contradicting the existence of (X, ξ, θ) above, we will prove that if X is any infinite set and if $\mathscr{U}$ is any non-principal ultrafilter on X then the corresponding **β**-operation $\tilde{\mathscr{U}}: 2^X \longrightarrow 2$ is not continuous. We leave the following two simple facts as exercises for the reader.

1. Capitalizing on the bijection between subsets A of X and characteristic functions $\chi_A: X \longrightarrow 2$ we regard the elements of 2^X as subsets of X. Then a typical basic open set in the Tychonoff topology for X copies of discrete 2 is $B_{(G,H)} = \{A \subset X : A \cap G = H\}$ where G is a finite subset of X and $H \subset G$.

2. $\tilde{\mathscr{U}}: 2^X \longrightarrow 2$ is the characteristic function of $\mathscr{U}$, that is $A\tilde{\mathscr{U}} = 1$ if and only if $A \in \mathscr{U}$.

Accepting these facts, it is easy to prove that $\tilde{\mathscr{U}}$ is not continuous. Consider the inclusion-ordered directed set of finite subsets of X. This may be comfortably regarded as a net in 2^X. This net converges to X since if $X \in B_{(G,H)}$ then $H = X \cap G = G$ and so $F \cap G = H$ for all finite subsets F containing G, i.e., there exists G with $F \in B_{(G,H)}$ for all $F \supset G$. On the other hand $F\tilde{\mathscr{U}} = 0$ for all finite subsets F whereas $X\tilde{\mathscr{U}} = 1$.

6.7 Symmetry Proposition. *Let* **S** *and* **T** *be algebraic theories in* **Set**. *Then* (X, ξ, θ) *is an* **S-T** *bialgebra if and only if* (X, θ, ξ) *is a* **T-S** *bialgebra.*

Proof. Earlier work in 1.4.25 and 1.5.40 introduced the principle that a function $f: X \longrightarrow Y$ is an **S**-homomorphism $(X, \xi) \longrightarrow (Y, \theta)$ if and only if f commutes with the **S**-operations. To make this perfectly clear, consider the diagram (see 1.5.5)

$$\begin{array}{ccccc} X^n & \xrightarrow{X\hat{\omega}} & XS & \xrightarrow{\xi} & X \\ \Big\downarrow{\scriptstyle f^n} & & \Big\downarrow{\scriptstyle fS} & & \Big\downarrow{\scriptstyle f} \\ Y^n & \xrightarrow[Y\hat{\omega}]{} & YS & \xrightarrow[\theta]{} & Y \end{array} \tag{6.8}$$

induced by $\omega \in nS$. The rightmost square commutes if and only if f is an S-homomorphism whereas the outer rectangle commutes if and only if f commutes with the **S**-operation ω. It is immediate that if f is an **S**-homomorphism then f commutes with all **S**-operations and the converse is true providing every element of XS is in the image of $X\hat{\omega}$ for some ω, and this is the case by 1.5.5.

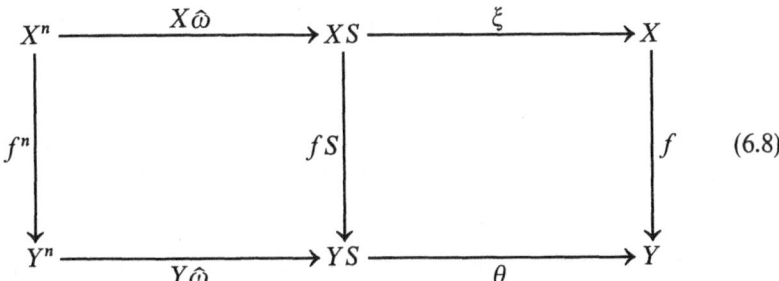

The result of 6.7 is therefore not surprising: each **T**-operation is an **S**-homomorphism if and only if each **T**-operation commutes with each **S**-operation and this surely sounds like a symmetric statement. Here is the formal proof. Let (X, ξ, θ) be an **S-T** bialgebra and let $\beta:(U^{\mathbf{S}})^m \longrightarrow U^{\mathbf{S}}$ be an **S**-operation. To prove $(X, \xi)\beta:(X, \theta)^m \longrightarrow (X, \theta)$ is a **T**-homomorphism it suffices to prove that $(X, \xi)\beta$ commutes with $(X, \theta)\alpha$ where $\alpha:(U^{\mathbf{T}})^n \longrightarrow U^{\mathbf{T}}$ is an arbitrary **T**-operation, i.e., we must show that the following diagram is commutative:

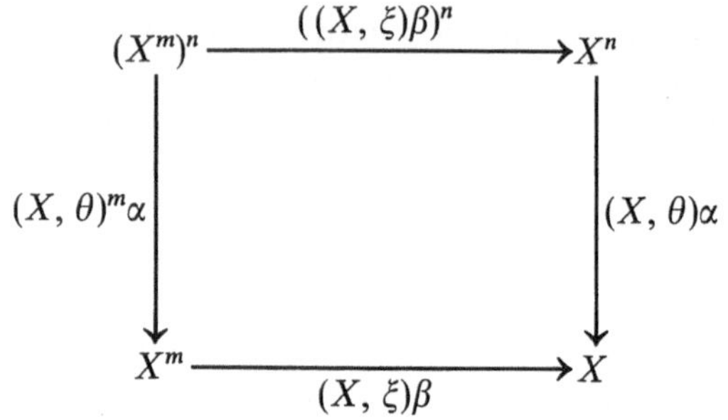

This diagram is obtained by pasting together three pieces as shown below.

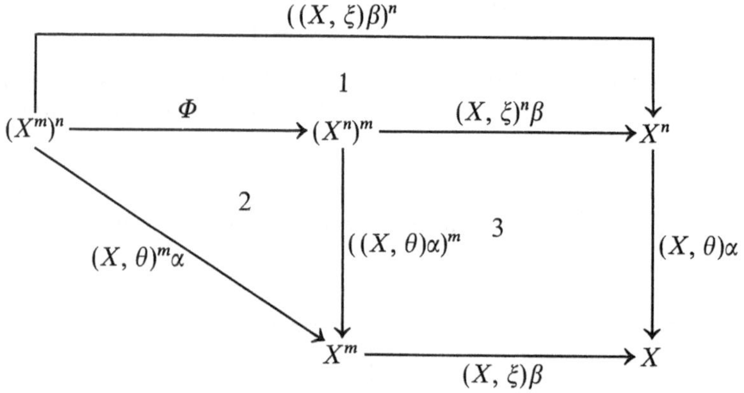

Square 3 commutes because the **T**-operation $(X, \theta)\alpha$ is an **S**-homomorphism. To explain triangles 1 and 2, Φ is the canonical isomorphism

$$(X^m)^n \cong X^{(m \times n)} \cong (X^n)^m$$

sending $f:n \longrightarrow X^m$ to $g:m \longrightarrow X^n$ where $jg_i = if_j$ for all i in m and j in n. To establish 1, consult the diagram shown below.

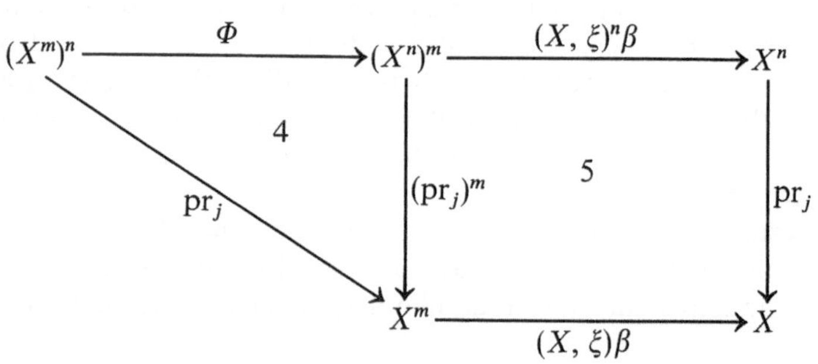

Triangle 4 is easily verified directly and 5 holds because pr_j is an **S**-homomorphism from $(X, \xi)^n$ to (X, ξ). Together, 4 and 5 verify that 1 holds followed by each product projection. Similarly, triangle 2 is verified by using the diagram

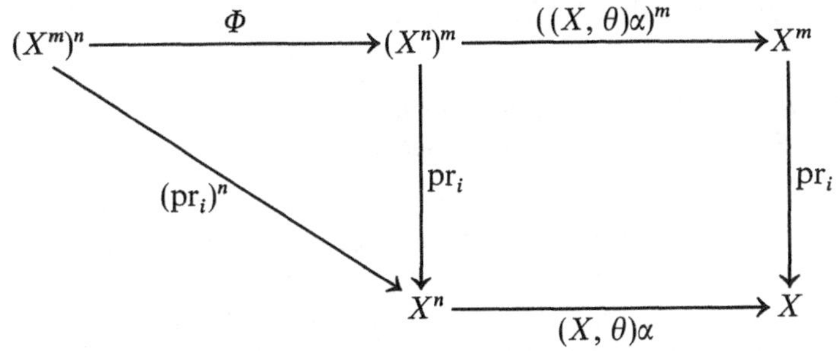

□

6.9 Lemma. *Let* **S** *and* **T** *be algebraic theories in* **Set**. *Then* $\mathbf{Set}^{\mathbf{S}\otimes\mathbf{T}}$, *which is obviously a full replete subcategory of* $\mathbf{Set}^{\mathbf{S}} \times \mathbf{Set}^{\mathbf{T}}$, *is a "Birkhoff subcategory" in the sense that (even though* $\mathbf{Set}^{\mathbf{S}} \times \mathbf{Set}^{\mathbf{T}}$ *may not be algebraic over* **Set***)* $\mathbf{Set}^{\mathbf{S}\otimes\mathbf{T}}$ *is closed under products, subalgebras, and quotients. In particular,* $U^{\mathbf{S}\otimes\mathbf{T}}:\mathbf{Set}^{\mathbf{S}\otimes\mathbf{T}} \longrightarrow \mathbf{Set}$ *is a Beck functor.*

Proof. The second statement is obvious from the first using 1.19 and 1.18. We turn to the proof of the first statement (which is hardly a surprise since the bialgebra condition is equational; cf. 1.4.22). All three closure properties can be established simultaneously by proper interpretation of the following generic cube induced by $f:(X, \xi, \theta) \longrightarrow (Y, \xi', \theta')$ in $\mathbf{Set}^{\mathbf{S}} \times \mathbf{Set}^{\mathbf{T}}$ and $\alpha:(U^{\mathbf{T}})^n \longrightarrow U^{\mathbf{T}}$ (where $(X, \xi)^n = (X^n, \xi^n)$, $(Y, \xi')^n = (Y^n, (\xi')^n)$.

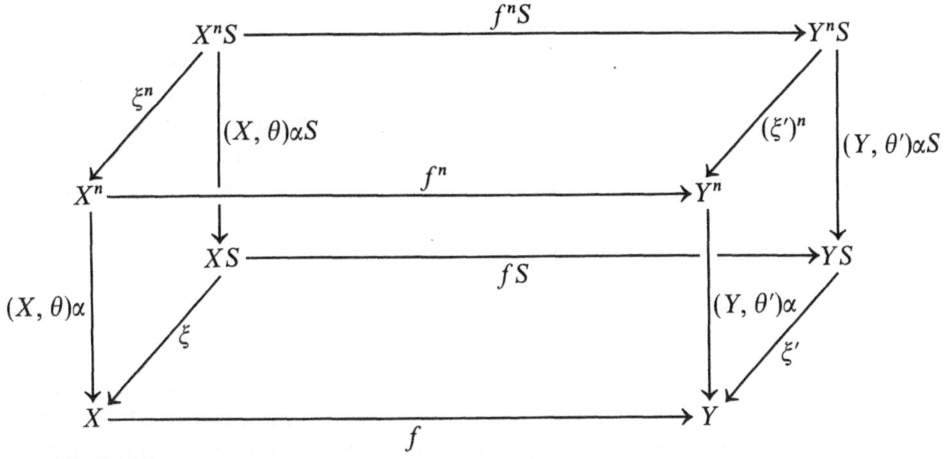

Of the six faces of the cube, all commute except possibly the left and right sides (f is both an **S**-homomorphism and a **T**-homomorphism).

If the right side commutes, that is, if (Y, ξ', θ') is an **S-T** bialgebra, then the left side at least commutes when followed by f. Therefore, if (X, ξ, θ) is the product of the bialgebras (Y_i, ξ_i, θ_i) and f runs over the product projections we deduce that (X, ξ, θ) is also a bialgebra; and, with the same reasoning, if (Y, ξ', θ') is a bialgebra and f is injective then (X, ξ, θ) is a bialgebra.

If the left side commutes then the right side commutes at least when preceded by $f^n S$. Therefore, if (X, ξ, θ) is a bialgebra and if f is surjective then $f^n S$ is also surjective (by 1.4.29 with $H = (\ )^n S$) and (Y, ξ', θ') is also a bialgebra. □

6.10 Theorem. *Let* **S** *and* **T** *be bounded (1.5.14) algebraic theories in* **Set**. *Then* $U^{\mathbf{S} \otimes \mathbf{T}}: \mathbf{Set}^{\mathbf{S} \otimes \mathbf{T}} \longrightarrow \mathbf{Set}$ *is algebraic.*

Proof. By 6.9, 1.19, 1.18, 3.6, and 3.3 it is sufficient to prove that the underlying set functor $U: \mathbf{Set}^{\mathbf{S}} \times \mathbf{Set}^{\mathbf{T}} \longrightarrow \mathbf{Set}$ is algebraic. By 1.5.40, present $\mathbf{Set}^{\mathbf{S}}$ as $(\Omega_{\mathbf{S}}, E_{\mathbf{S}})$-alg and $\mathbf{Set}^{\mathbf{T}}$ as $(\Omega_{\mathbf{T}}, E_{\mathbf{T}})$-alg. It is obvious that if $\Omega_n = (\Omega_{\mathbf{S}})_n + (\Omega_{\mathbf{T}})_n$ and $E = E_{\mathbf{S}} + E_{\mathbf{T}}$ that U is isomorphic to the underlying set functor from (Ω, E)-algebras. Since **S** and **T** are bounded, it follows from the construction in 1.5.40 that there exists a cardinal n_0 with $\Omega_n = \varnothing$ for all $n > n_0$. By 1.27, U is algebraic. □

(6.11) Let **S**, **T** be algebraic theories in **Set**. The *tensor product of* **S** *and* **T** *exists* providing $U^{\mathbf{S} \otimes \mathbf{T}}: \mathbf{Set}^{\mathbf{S} \otimes \mathbf{T}} \longrightarrow \mathbf{Set}$ is algebraic; in this case, the corresponding algebraic theory is denoted by $\mathbf{S} \otimes \mathbf{T}$. With the exception of the bounded case, as in 6.10, the question of the existence of the tensor product seems to be a subtle one. Isbell has shown ([Isbell '72, 3.11]) that $\mathbf{S} \otimes \mathbf{T}$ does not exist if **S** is the theory whose algebras are real vector spaces and **T** is suitably chosen.

(6.12) By 6.9, 1.18, 1.19, and 1.22(1) it is clear that a necessary and sufficient condition that $\mathbf{S} \otimes \mathbf{T}$ exists is that $U^{\mathbf{S} \otimes \mathbf{T}}$ satisfies the solution set condition. To verify the solution set condition at n we consider an **S**-**T** bialgebra (X, ξ, θ) and a subset A of X whose cardinal is at most that of n (i.e., we are thinking of A as the image of an n-tuple $n \longrightarrow X$ in X) and we attempt to show that there exists a cardinal α, depending only on n (and not on X) such that the **S**-**T** subalgebra of X generated by A has cardinal at most α. One possible procedure is to try to construct this subalgebra by alternately closing up under the two sorts of operations. Let $\langle B \rangle_{\mathbf{S}}$ denote the **S**-subalgebra generated by B and define $\langle B \rangle_{\mathbf{T}}$ similarly. Then we can form

$$\langle\langle A \rangle_{\mathbf{S}} \rangle_{\mathbf{T}}$$

but this may fail to be an **S**-subalgebra. Again,

$$\langle\langle\langle A \rangle_{\mathbf{S}} \rangle_{\mathbf{T}} \rangle_{\mathbf{S}}$$

may fail to be a **T**-subalgebra. This procedure may be iterated not only countably often but transfinitely often (by taking the union before continuing anew the **S**, **T** iterations) without achieving an **S**-**T** subalgebra, and indeed this is what one should expect if, in (X, ξ, θ), there were no relation between ξ and θ. One might hope, however, that the bialgebra relation is strong enough so that there exists a stage (i.e., an ordinal) before which an **S**-**T** subalgebra is found and depending only on n. In this case, the solution set condition at n is clear by 1.4.31. As a very special case we have the following

6.13 Lemma. *Let* **S**, **T** *be algebraic theories in* **Set**. *Then* $\mathbf{S} \otimes \mathbf{T}$ *exists providing whenever* (X, ξ, θ) *is an* **S-T** *bialgebra and* A *is an* **S**-*subalgebra of* (X, ξ), *the* **T**-*subalgebra of* (X, θ) *generated by* A *is still an* **S**-*subalgebra of* (X, ξ).

Proof. To fully complete the discussion preceding the lemma, we note that the **S-T** subalgebra of (X, ξ, θ) generated by the subset A_0 of X is $\langle\langle A_0\rangle_{\mathbf{S}}\rangle_{\mathbf{T}}$ which has cardinal at most that of A_0ST. □

Since the bialgebra condition of 6.1 involves operations, it is natural, before attempting to apply 6.13, that we detour to explore the relationships between the "subalgebra generated by" operator $\langle\ \rangle_{\mathbf{T}}$ and **T**-operations. The reader should recall the notations and results of 1.5.5. We will write just $\langle\ \rangle$ instead of $\langle\ \rangle_{\mathbf{T}}$ when only one algebraic theory is in the picture.

6.14 Proposition. *Let* **T** *be an algebraic theory in* **Set** *and let* $f: (X,\xi) \longrightarrow (Y,\theta)$ *be a* **T**-*homomorphism. Then for all* $A \subset X$, $\langle A\rangle f = \langle Af\rangle$.

Proof. Let $i: A \longrightarrow X$ be an inclusion map and let (p, j) be the image factorization of $i.f$. Consider the diagram

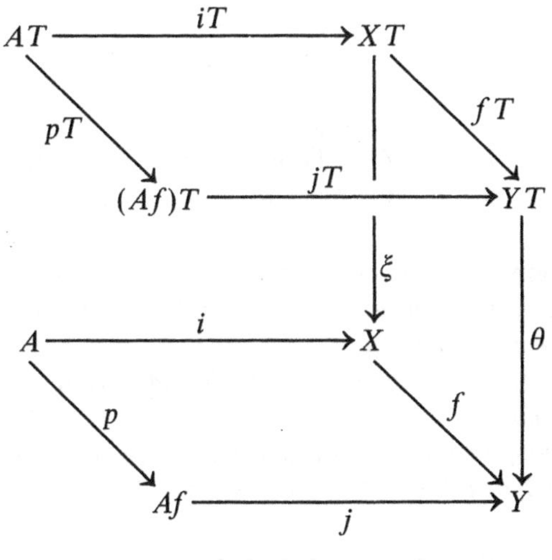

Using 1.4.31, we have $\langle Af\rangle = \mathrm{Im}(jT.\theta)$. But this is the same as $\mathrm{Im}(pT.jT.\theta)$ since pT is surjective (1.4.29). Thus $\langle Af\rangle = \mathrm{Im}(iT.\xi.f) = \langle A\rangle f$. □

(6.15) Let **T** be an algebraic theory in **Set** and let (X, ξ) be a **T**-algebra. The *n-ary operations of* (X, ξ), denoted $\mathcal{O}_n(X, \xi)$, is the subset of $X^{(X^n)}$ defined by

$$\mathcal{O}_n(X, \xi) = \{(X, \xi)\alpha : \alpha \in \mathcal{O}_n(\mathbf{T})\}$$

where $\mathcal{O}_n(\mathbf{T})$ is as in 1.5.5.

6.16 Proposition. *Let* **T** *be an algebraic theory in* **Set** *and let* (X, ξ) *be a* **T**-*algebra. Then for each set* n, $\mathcal{O}_n(X, \xi)$ *is the subalgebra of* $(X, \xi)^{(X^n)}$ *generated by the projections* $\{p_i: X^n \longrightarrow X \mid i \in n\}$.

Proof. Let (Y, θ) denote the **T**-algebra $(X, \xi)^{(X^n)}$. Let $p: n \longrightarrow Y$ be the injective passage $i \longrightarrow p_i$. Since p is isomorphic to the inclusion of $\{p_i : i \in n\}$ it follows from 1.4.31 that $\langle\{p_i : i \in n\}\rangle = \mathrm{Im}(pT.\theta)$. For arbitrary $\omega \in nT$ and $f: n \longrightarrow X$ we have the diagram

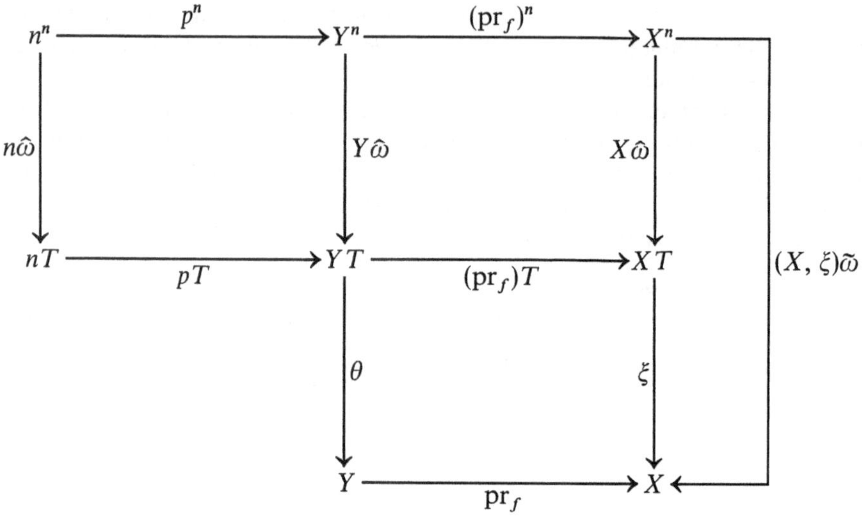

Since $p.\mathrm{pr}_f : n \longrightarrow Y \longrightarrow X = f$, $\langle \omega, pT.\theta\rangle \mathrm{pr}_f = \langle \mathrm{id}_{n^n}, n\hat{\omega}.pT.\theta\rangle \mathrm{pr}_f = \langle p.\mathrm{pr}_f, (X, \xi)\tilde{\omega}\rangle = \langle f, (X, \xi)\tilde{\omega}\rangle$. Thus, $pT.\theta$ is the map that sends ω to $(X, \xi)\tilde{\omega}$. As $pT.\theta$ is a homomorphism with image $\mathcal{O}_n(X, \xi)$, the proof is complete by 1.4.31. □

6.17 Proposition. *Let* **T** *be an algebraic theory in* **Set**, *let* (X, ξ) *be a* **T**-*algebra, and let* A *be a subset of* X *with inclusion map* $i: A \longrightarrow X$. *Then the following two statements are true:*

1. $\langle A\rangle = i\mathcal{O}_A(X, \xi)$; *that is,* $x \in \langle A\rangle$ *if and only if there exists* $\omega \in AT$ *such that* $\langle i, (X, \xi)\tilde{\omega}\rangle = x$.

2. A *is a subalgebra of* (X, ξ) *if and only if* A *is "closed under all* A-*ary operations," that is, for all* $\omega \in AT$, *there exists a factorization*

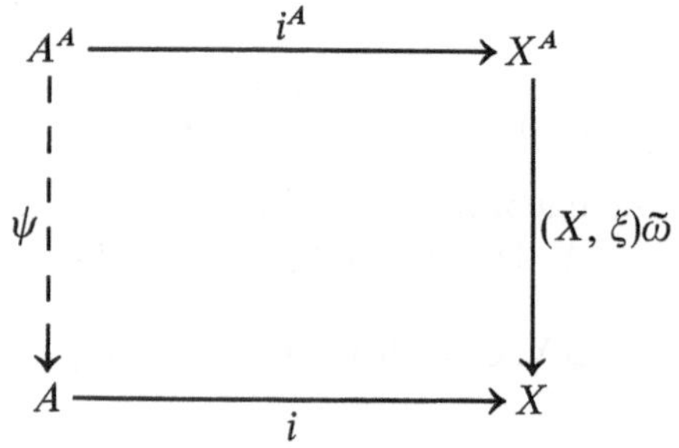

Proof. 1. For $a \in A$, $\langle p_a, \mathrm{pr}_i : X^{(X^A)} \longrightarrow X\rangle = a$. Using 6.14 and 6.16 we have $\langle A\rangle = \langle\{p_a : a \in A\}\mathrm{pr}_i\rangle\ \langle\{p_a : a \in A\}\rangle \mathrm{pr}_i = (\mathcal{O}_A(X, \xi))\mathrm{pr}_i = i\mathcal{O}_A(X, \xi)$.

2. If A is a subalgebra by virtue of $\xi_0 : AT \longrightarrow A$, the desired factorization is $\psi = (A, \xi_0)\tilde{\omega}$. Conversely, suppose ψ exists. Evaluating at id_A, $\langle i, (X, \xi)\tilde{\omega}\rangle = \langle \mathrm{id}_A, \psi\rangle \in A$ so that, by (1), $\langle A\rangle = i\mathcal{O}_A(X, \xi) \subset A$. □

(6.18) Let **T** be an algebraic theory in **Set**, let (X, ξ) be a **T**-algebra, let $A \subset X$, and let $j : \langle A\rangle \longrightarrow X$ be the inclusion map. As $j^n : \langle A\rangle^n \longrightarrow (X, \xi)^n$ is a **T**-homomorphism and an inclusion map, $\langle A\rangle^n$ is a subalgebra of $(X, \xi)^n$ (1.4.32). Since $A \subset \langle A\rangle$, $A^n \subset \langle A\rangle^n$; as the latter is a subalgebra it follows that $\langle A^n\rangle \subset \langle A\rangle^n$. Say that *subalgebras commute with powers in* **T** if for all **T**-algebras (X, ξ), subsets A of X and sets n we have $\langle A^n\rangle = \langle A\rangle^n$.

6.19 Theorem. *Let* **S** *be an algebraic theory in* **Set** *such that subalgebras commute with powers in* **S**. *Then for all algebraic theories* **T** *in* **Set**, **S** ⊗ **T** *exists.*

Proof. We use 6.13 (interchanging the roles of **S** and **T**, which is valid by 6.7). Let (X, ξ, θ) be an **S**-**T** bialgebra. Let (A, θ_0) be a **T**-subalgebra of (X, θ),

$$\begin{array}{ccccc} A^B & \longrightarrow & \langle A^B\rangle_{\mathbf{S}} = B^B & \longrightarrow & X^B \\ {\scriptstyle (A,\theta_0)\tilde{\omega}}\big\downarrow & & & & \big\downarrow{\scriptstyle (X,\theta)\tilde{\omega}} \\ A & \longrightarrow & B & \longrightarrow & X \end{array}$$

let $B = \langle A\rangle_{\mathbf{S}}$, and let $\omega \in BT$. Set $f = (X, \theta)\tilde{\omega}$. By 6.17(2) it suffices to prove that $(B^B)f \subset B$. But as f is an **S**-homomorphism, we use 6.14 and the diagram above to obtain $(B^B)f = (\langle A^B\rangle_{\mathbf{S}})f = \langle (A^B)f\rangle_{\mathbf{S}} \subset \langle A\rangle_{\mathbf{S}} = B$. □

We have already observed that **S** ⊗ **T** need not exist if **S**-algebras are real vector spaces; such S has rank $\aleph_0$. On the other hand there exist theories **S** without rank such that **S** ⊗ **T** always exists:

6.20 Theorem. *Let* **β** *be the ultrafilter theory whose algebras are compact Hausdorff spaces (1.5.24). Then for every algebraic theory* **T** *in* **Set**, **β** ⊗ **T** *exists.*

Proof. By 6.19 it suffices to prove that subalgebras commute with powers in **β**. This amounts to saying that if (X, ξ) is a compact Hausdorff space, and if A is a subset of X, then $\mathrm{Cls}(A^n) = (\mathrm{Cls}(A))^n$ for every set n, where "Cls" is the closure operator of the product space $(X, \xi)^n$. For completeness, we prove this well-known fact from topology which is true, in fact, for any topological space (X, τ) as follows. Let $(x_i : i \in n) \in (\mathrm{Cls}(A))^n$ and let U be a neighborhood of (x_i) in X^n. To show: $U \cap A^n \neq \varnothing$ (for then $(x_i) \in \mathrm{Cls}(A^n)$). By standard properties of the product topology we have $(x_i) \in \prod V_i \subset U$ where each V_i is open in $(X, \mathcal{T})$ (in fact $V_i = X$ for all but finitely-many i). Since each $x_i \in \mathrm{Cls}(A)$, there exists $y_i \in A \cap V_i$. But then $(y_i) \in U \cap A^n$. □

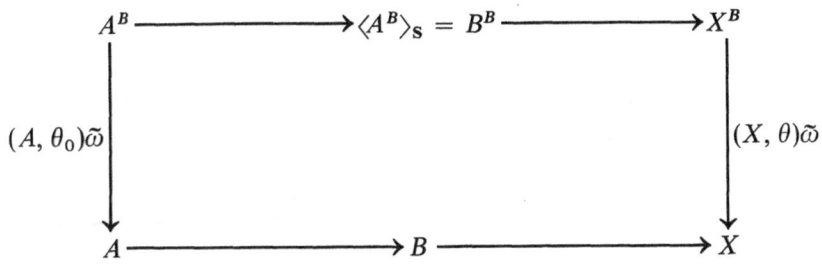

It is an immediate application of 6.20 that compact groups and compact abelian groups (cf. 1.23) are sets with algebraic structure.

6.21 Proposition. *Let* **S**, **T** *be algebraic theories in* **Set**. *Then if every synatactic* **S**-*operation has arity* 1 (*1.5.10*) *then* $\mathbf{S} \otimes \mathbf{T}$ *exists.*

Proof. By 1.5.40, $\mathbf{Set}^{\mathbf{S}}$ may be thought of as (Ω, E)-alg where Ω has only unary operation labels. Let X be a set and let $(f_u : XS \longrightarrow XS \mid u \in \Omega)$ be the free **S**-algebra on X, i.e., XS is the free algebra and (f_u) is its Ω-structure. Since every equation in E has form $u_1 \cdots u_n = v_1 \cdots v_m$ or $u_1 \cdots u_n = \mathrm{id}$ and T is a functor, $(f_u T : XST \longrightarrow XST)$ is an **S**-algebra. Since $f_u T$: $(XST, XS\mu_{\mathbf{T}}) \longrightarrow (XST, XS\mu_{\mathbf{T}})$ is a **T**-homomorphism, XST is an **S**-**T** bialgebra. Define $\eta : X \longrightarrow XST$ by $= X\eta_{\mathbf{S}}.XS\eta_{\mathbf{T}}$. Let (Y, ξ, θ) be an **S**-**T** bialgebra with **S**-structure $(\xi_u : Y \longrightarrow Y)$ and let $g : X \longrightarrow Y$ be a function. As shown below let g_1 be the **S**-homomorphic extension of

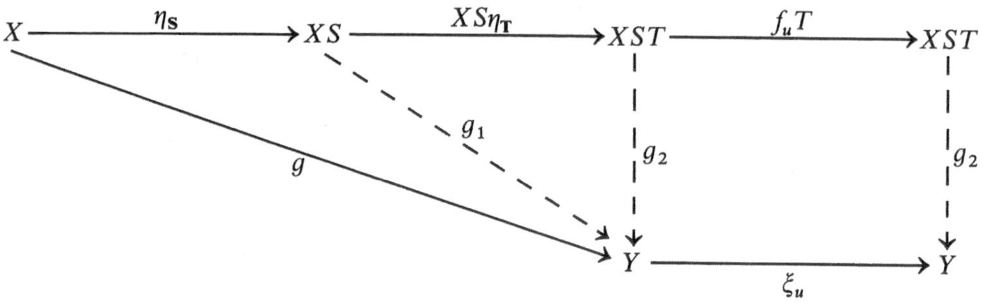

g and let g_2 be the **T**-homomorphic extension of g_1. Then $XS\eta_{\mathbf{T}}.g_2.\xi_u = g_1.\xi_u = f_u.g_1 = f_u.XS\eta_{\mathbf{T}}.g_2 = XS\eta_{\mathbf{T}}.f_u T.g_2$; as g_2, $f_u T$ and ξ_u are **T**-homomorphisms, $g_2.\xi_u = f_u T.g_2$, and g_2 is an **S**-homomorphism. We have proved that $U^{\mathbf{S} \otimes \mathbf{T}}$ satisfies the solution-set condition. □

We conclude this section with some remarks on "completely commutative" algebras.

(6.22) Let **T** be an algebraic theory in **Set** and let (X, ξ) be a **T**-algebra. (X, ξ) is *completely commutative* if (X, ξ, ξ) is a **T**-**T** bialgebra, i.e., every **T**-operation is a **T**-homomorphism. By 6.7, (X, ξ) is completely commutative if and only if

$$((x_{ij} : i \in n)\xi_\alpha : j \in m)\xi_\beta = ((x_{ij} : j \in m)\xi_\beta : i \in n)\xi_\alpha$$

holds for all n-ary $\alpha : (U^{\mathbf{T}})^n \longrightarrow U^{\mathbf{T}}$, m-ary $\beta : (U^{\mathbf{T}})^m \longrightarrow U^{\mathbf{T}}$ and $(x_{ij}) \in X^{n \times m}$ (where ξ_α abbreviates $(X, \xi)\alpha$). **T** is *commutative* if every **T**-algebra is completely commutative. Thus (6.4) the completely commutative groups are the abelian groups and "abelian groups" is a commutative theory. Commutative rings are not completely commutative in "rings" however (e.g. multiplication is not a homomorphism of the additive structure).

6.23 Proposition. *Let* **T** *be an algebraic theory in* **Set** *and let* (Y, θ) *be a* **T**-*algebra. Then* (Y, θ) *is completely commutative if and only if for every* **T**-

algebra (X, ξ) *the set* $A = \{f: X \longrightarrow Y | f$ *is a* **T**-*homomorphism* $(X, \xi) \longrightarrow (Y, \theta)\}$ *is a subalgebra of* $(Y, \theta)^X$.

Proof. Assume that (Y, θ) is completely commutative and let α be an n-ary **T**-operation. By 6.17(2) it suffices to show that $f = (f_i)\tau_\alpha$ is in A if each f_i is in A (where (Y^X, τ) denotes $(Y, \theta)^X$). Let β be an m-ary **T**-operation. Since a **T**-homomorphism is the same thing as a function which commutes with **T**-operations (6.8) it suffices to show that $(x_j)\xi_\beta f = (x_j f)\theta_\beta$ for $(x_j) \in X^m$. To this end, first observe that since $\mathrm{pr}_x: (Y, \theta)^X \longrightarrow (Y, \theta)$ is a **T**-homomorphism, operations on $(Y, \theta)^X$ are "pointwise," i.e., for $(g_i) \in (Y^X)^n$ and $x \in X$, $x(g_i)\tau_\alpha = (xg_i)\theta_\alpha$. We therefore have

$$\begin{aligned}(x_j)\xi_\beta f &= \langle (x_j)\xi_\beta, (f_i)\tau_\alpha \rangle = \langle (x_j)\xi_\beta f_i, \theta_\alpha \rangle \\ &= ((x_j f_i : j \in m)\theta_\beta : i \in n)\theta_\alpha \qquad \text{(as each } f_i \in A) \\ &= ((x_j f_i : i \in n)\theta_\alpha : j \in m)\theta_\beta \qquad \text{(as } (Y, \theta) \text{ is completely commutative)} \\ &= (\langle x_j, (f_i)\tau_\alpha \rangle : j \in m)\theta_\beta = (x_j) f \theta_\beta.\end{aligned}$$

Conversely, if A is always a sublagebra, consider the case $(X, \xi) = (Y, \theta)^n$. Since A is a subalgebra containing the n projections $p_i: Y^n \longrightarrow Y$ it follows from 6.16 that $\mathcal{O}_n(Y, \xi) \subset A$. □

The following result parallels 6.9 and is left to the reader:

6.24 Proposition. *Let* **T** *be an algebraic theory in* **Set**. *Then the completely commutative* **T**-*algebras form a Birkhoff subcategory of* $\mathbf{Set}^{\mathbf{T}}$. □

6.25 Lack of Completely Commutative Lattices. A *lattice* is a partially ordered set in which each two elements have an infimum and a supremum. Equivalently, a lattice is an (Ω, E)-algebra where Ω has two binary operation labels "Inf" and "Sup" and E consists of the following eight equations:

Inf is associative, commutative, and idempotent;
Sup is associative, commutative, and idempotent;
Inf and Sup satisfy the absorptive laws (see 1.5.46).

The verification of equivalence is left as an exercise. The empty lattice and the one-element lattice are the only completely commutative lattices. For suppose X were a completely commutative lattice with distinct elements x, y. Since at least one of x, y is strictly less than $\mathrm{Sup}(x, y)$, X possesses a sublattice isomorphic to the two-element lattice 0, 1 with $0 < 1$. By 6.24 it suffices to observe that the two-element lattice is not completely commutative. This is clear from the obervation that $1 = \mathrm{Inf}(\mathrm{Sup}(0, 1), \mathrm{Sup}(1, 0)) \neq \mathrm{Sup}(\mathrm{Inf}(0, 1), \mathrm{Inf}(1, 0) = 0$, i.e., Sup is not a lattice homomorphism.

6.26 Complete Semilattices Are Completely Commutative. Let **T** be the power-set theory whose algebras are complete semilattices (1.5.15). For $\alpha \in nT$ (i.e., α is a subset of n) the corresponding operation $(X, \xi)\alpha: X^n \longrightarrow X$ on a complete semilattice (X, ξ) sends $f: n \longrightarrow X$ to the supremum of $\alpha f \subset X$. That (X, ξ) is completely commutative is the familiar fact that suprema commute with each other.

Notes for Section 6

Tensor products of algebraic theories were considered by [Freyd '66], [Manes '67], [Lawvere '68], and [Isbell '72]. In Isbell's formulation, the tensor product of two algebraic theories always exists as some sort of theory, though is not always an algebraic theory of course. Theorems 6.19 and 6.20 appear in [Manes '69]. Proposition 6.23 was proved in [Freyd '66] and [Linton '66-A]. Commutative theories in closed categories are treated in [Kock '71].

The term "bialgebra" has an alternative meaning in the literature, namely "coalgebra in the category of algebras."

Exercises for Section 6

1. Prove that if subalgebras commute with powers in **T** then **T** is affine in the sense of exercise 5 of 1.1.3. Show that the converse fails for the stochastic matrix theory (exercise 6, 1.1.3).
2. Show that the category of theories in **Set** and theory maps does not have finite coproducts although the full subcategory of theories with rank has small coproducts. [Hint: use exercise 8 of section 2.]
3. ([Freyd '66].) Let **T**, **S** in **Set** each possess at least one constant operation (1.5.13 –). If **T** ⊗ **S** exists, prove that **T** ⊗ **S** has a unique constant.
4. Let R be a ring. Show that R-modules forms a commutative theory if and only if R is a commutative ring. Let $R[x]$ denote the ring of one-variable polynomials over R. Show that $R[x]$-modules are the algebras over the theory of R-modules tensored with the theory corresponding to a single unary operation.
5. ([Lawvere '63], [Freyd '66], [Manes '67].) A *semiadditive category* is a category $\mathscr{A}$ together with the structure of an abelian monoid on the set $\mathscr{A}(A, B)$ for each pair (A, B) in such a way that for each $f: A' \longrightarrow A$ and $g: B \longrightarrow B'$, $f.-.g: \mathscr{A}(A, B) \longrightarrow \mathscr{A}(A', B')$ is a monoid homomorphism. An *additive category* is a semiadditive category such that each monoid $\mathscr{A}(A, B)$ is even a group. Just as a ring is a one-object additive category, a *semiring* is a one-object semiadditive category (i.e., just like a ring but no additive inverses). If R is a semiring, an R-*semimodule* is an abelian monoid X on which R acts $X \otimes R \longrightarrow X$, $(x, r) \longmapsto xr$ subject to the usual laws $(x + x')r = xr + x'r$, $x(r + r') = xr + xr'$, $x(rr') = (xr)r'$; if R is a ring, then, an R-module is an R-semimodule which is a group. Of course such module categories are algebraic over **Set**. Let **AM** and **AG** denote, respectively, the theories in **Set** whose algebras are abelian monoids and abelian groups.
   (a) For any **T** in **Set** show that **T**-**AM** bialgebras is a semiadditive category and that **T**-**AG** bialgebras is an additive category. [Hint: "zero" and "plus" are **T**-homomorphisms.]
   (b) If $A \times A$ exists for an object A in the semiadditive category $\mathscr{A}$ show that

$$A \xrightarrow{\begin{pmatrix}\mathrm{id}\\ 0\end{pmatrix}} A \times A \xleftarrow{\begin{pmatrix}0\\ \mathrm{id}\end{pmatrix}} A$$

is a coproduct diagram in $\mathscr{A}$. [Hint: add the pieces.]

(c) If $\mathbf{Set}^{\mathbf{T}}$ is semiadditive (respectively, additive) show that $\mathbf{T} \otimes \mathbf{AM}$ ($\mathbf{T} \otimes \mathbf{AG}$) exists and coincides with $\mathbf{T}$. [Hint: using (b), define the necessary operations as $\mathbf{T}$-homomorphisms.]

(d) If $\mathbf{T}$ is finitary and $\mathbf{Set}^{\mathbf{T}}$ is semiadditive (respectively, additive) show that $\mathbf{Set}^{\mathbf{T}}$ coincides with the category of semimodules (modules) over the semiring (ring) $1T$. [Hint: if $(x_1, \ldots, x_n)\omega$ is a $\mathbf{T}$-operation, it decomposes as $(0, x_2, \ldots, x_n)\omega + \cdots + (x_1, \ldots, x_{n-1}, 0)\omega$.]

(e) For any theory $\mathbf{T}$ in $\mathbf{Set}$, prove that an epimorphism in the category of $\mathbf{T}$-$\mathbf{AG}$ bialgebras is a coequalizer. [Hint: given an epimorphism $f: X \longrightarrow Y$, set $q = \mathrm{coeq}(f, 0)$ in the category of abelian groups; then q is obtained by dividing out by the abelian group congruence $R = \{(y_1, y_2): y_1 - y_2 \in \mathrm{Im}(f)\}$; since "minus": $Y \times Y \longrightarrow Y$ is a $\mathbf{T}$-homomorphism, R is also a $\mathbf{T}$-congruence, and $q = \mathrm{coeq}(f, 0)$ in the category of bialgebras; as f is epi, $q = 0$.]

(f) Let $\mathscr{A}$ be the additive category of torsion-free abelian groups. Show that $\mathbf{Z} \longrightarrow \mathbf{Z}$, $n \longmapsto 2n$ is an epimorphism but not a coequalizer. Conclude that there is no algebraic functor $\mathscr{A} \longrightarrow \mathbf{Set}$.

See also [Isbell '64], [Johnson and Manes '70], and the references there.

6. A topological group is *monothetic* if one of its elements generates a dense subgroup. [Hewitt and Ross '63, Theorem 25.12] prove, using character theory, that: "There is a largest compact monothetic group G_0, in the sense that every compact monothetic group is a continuous homomorphic image of G_0." Give a proof using the theory of this section.

7. Let $\mathscr{C}$ be in Struct($\mathbf{Set}$) and let (X, s), (Y, t) be $\mathscr{C}$-structures. By a *bi-admissible map* $(X, s); (Y, t) \longrightarrow (Z, u)$ we mean a function $f: X \times Y \longrightarrow Z$ such that $(x, -)f: (Y, t) \longrightarrow (Z, u)$ and $(-, y)f: (X, s) \longrightarrow (Z, u)$ are admissible for all x in X, y in Y. A *tensor product of* (X, s) *and* (Y, t) is a pair $((X, s) \otimes (Y, t), \Gamma)$ such that $\Gamma: (X, s); (Y, t) \longrightarrow (X, s) \otimes (Y, t)$ is bi-admissible and possessing the universal property that whenever $f: (X, s); (Y, t) \longrightarrow (Z, u)$ is biadmissible, there

$$\begin{array}{ccc}
X \times Y & \xrightarrow{\ \Gamma\ } & (X, s) \otimes (Y, t) \\
 & {\scriptstyle f}\searrow \quad \swarrow{\scriptstyle \psi} & \\
 & (Z, u) &
\end{array}$$

exists a unique ψ with $\Gamma.\psi = f$.

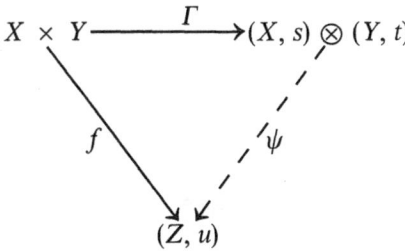

(a) Show that any fibre-complete category has tensor products by the construction of exercise 3(b) of section 2.

(b) For any algebraic theory **T** in **Set** show that every pair of **T**-algebras has a tensor product. [Hint: define $(X, \xi) \otimes (Y, \theta) = (X \times Y)T/R$ where R is the intersection of all congruences of the form $R_f = \{(x, y): xf^{\#} = yf^{\#}\}$ with $f:(X, \xi); (Y, \theta) \longrightarrow (Z, \gamma)$ bi-admissible.]

(c) ([Freyd '66], [Linton '66-A].) If **T** is a commutative theory in **Set** show that $\mathbf{Set}^{\mathbf{T}}$ is a closed category with the tensor product of (b).

(d) Show that the passage from (Z, u) to the set of bi-admissible maps $(X, s); (Y, t) \longrightarrow (Z, u)$ describes a functor $U:\mathscr{C} \longrightarrow \mathbf{Set}$ such that $(X, s) \otimes (Y, t)$ exists if and only if there exists a free $\mathscr{C}$-object over 1 with respect to U. Give an alternate proof of (b) using the adjoint functor theorem.

8. ([Kennison and Gildenhuys '71].) Let $U:\mathscr{A} \longrightarrow \mathbf{Set}$ be a functor such that AU is finite for all A in $\mathscr{A}$. Let $\mathscr{A}$ have and U preserve finite products. Suppose given, also, a finitary theory **S** in **Set** such that $\Psi:\mathscr{A} \longrightarrow \mathbf{Set}^{\mathbf{S}}$ is a full subcategory over **Set** closed under **S**-subalgebras.

(a) Show that $\mathscr{A}$ is a full subcategory of $\mathbf{S} \otimes \boldsymbol{\beta}$-algebras. [Hint: a finite product of discrete spaces is discrete.]

(b) Show that U is tractable [Hint: in the notation of exercise 13 of section 2, $D:(n, U) \longrightarrow \mathbf{Set}$ has a limit because the full subcategory $[n, U]$ of all $f:n \longrightarrow AU$ with $f^{\#}:nS \longrightarrow A\Psi$ onto is small and final; it is necessary to know that $\mathscr{A}$ is closed under **S**-subalgebras.]

We set out to identify the equational completion $\Phi:\mathscr{A} \longrightarrow \mathbf{Set}^{\mathbf{T}}$ of U as the Birkhoff subcategory $\mathscr{B}$ generated by $\mathscr{A}$ in $\mathbf{Set}^{\mathbf{S}\otimes\boldsymbol{\beta}}$.

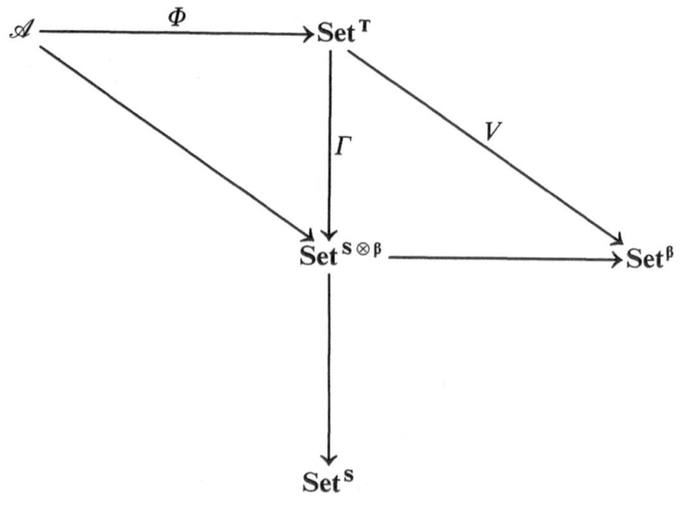

(c) Let $\lambda:\mathbf{S} \otimes \boldsymbol{\beta} \longrightarrow \mathbf{T}$ be the theory map corresponding to Γ. Prove that if each $n\lambda$ is surjective then Γ is an isomorphism onto $\mathscr{B}$. [Hint: use the universal property of Φ.]

(d) Complete the argument by proving that $n\lambda$ is surjective. [Hint: as $n\lambda$ is a $\mathbf{T} \otimes \boldsymbol{\beta}$-homomorphism, $\mathrm{Im}(n\lambda)$ is closed; if $\lambda':\mathbf{S} \longrightarrow \mathbf{T}$ corresponds to $\mathbf{Set}^{\mathbf{T}} \longrightarrow \mathbf{Set}^{\mathbf{S}}$, $\mathrm{Im}(\lambda') \subset \mathrm{Im}(\lambda)$ so it suffices to show $\mathrm{Im}(n\lambda')$ is dense; as V is limit preserving and $A\Phi V$ is finite, the topology on $n\mathrm{T}$ corresponding to $(nT, n\mu)V$ is the optimal lift of $\psi_{n,(A,f)}:nT \longrightarrow AU$ where (A, f) ranges over $[n, U]$ and AU is a finite discrete space; the subbasic open set $\{a\}\psi_{n;(A,f)}^{-1}$, if nonempty, intersects the image of $n\lambda'$ because $n\lambda'.\psi_{n;(A,f)}$ is onto; since $[n, U]$ is a poset with infima (U preserves products), subbasic open sets are basic.]

9. A *directed set* is a partially ordered set in which every two elements have a lower bound. A *profinite group* is a compact group which can be presented as the limit, in the category of compact groups, of a diagram of finite groups whose diagram scheme is a directed set. Use exercise 8 to prove that the equational completion of the category of finite groups is the category of profinite groups.
10. Show that the one-element solution set of 6.21 is in fact the free **S-T** bialgebra.
11. [Isbell '73-A.] Construct a finitary theory **T** in **Set** such that every epimorphism in $\mathbf{Set}^{\mathbf{T}}$ is onto but not every epimorphism in $\mathbf{Set}^{\mathbf{T}\otimes\mathbf{T}}$ is onto. [Hint: one binary operation will do.] In the paper cited, Isbell provides an example of a finitary **T** for which every subalgebra of a free algebra is free but for which not every **T**-epimorphism is onto.
12. Let $\mathbf{Set}^{\mathbf{T}}$ be Boolean algebras. Show that $\mathbf{Set}^{\mathbf{T}\otimes\boldsymbol{\beta}}$ = complete atomic Boolean algebras. [Hint: if **S** is the double power-set theory, define $\mathbf{Set}^{\mathbf{S}} \longrightarrow \mathbf{Set}^{\mathbf{T}\otimes\boldsymbol{\beta}}$ by recalling that $A = 2^{At(A)}$ for each complete atomic Boolean A.]
13. (We learned this from F. E. J. Linton.) Let **W** be a nontrivial algebraic theory in **Set**, let α be an infinite cardinal and let $H: \mathbf{Set}^{\mathbf{W}} \longrightarrow (\mathbf{Set})^{\mathrm{op}}$ preserve products of size $< \alpha$. (Example: the maximal ideal functor from Boolean algebras preserves finite products.) Let **T** be a theory of rank $\leqslant \alpha$. Show that if there exists a nontrivial **T-W** bialgebra then there exists a nontrivial **T-S** bialgebra for every nontrivial theory **S**. [Hint: let (X, ξ, θ) be a nontrivial **T-W** bialgebra and let (Y, γ) be a nontrivial **S**-algebra; if $Z = (X, \theta)H$, each **T**-operation $f:X^n \longrightarrow X$ induces $fH:Z \longrightarrow n \times Z$, defining

$$[(Y, \gamma)^Z]^n \cong (Y, \gamma)^{Z \times n} \longrightarrow (Y, \gamma)^Z$$

which, by checking the equations of 1.5.40, determine a **T**-structure.]
14. (M. Barr.) For **T** a nontrivial theory in **Set**, show that $1T = 1$ if and only if for every pair (X, ξ), (Y, θ) of **T**-algebras and pair of subsets $A \subset X, B \subset Y, \langle A\rangle \times \langle B\rangle = \langle A \times B\rangle$. Conclude that if $1T = 1$ then $\mathbf{T} \otimes \mathbf{S}$ exists for every finitary **S**. [Hint: given $1T = 1$ it suffices to show

$\langle A \rangle \times B \subset \langle A \times B \rangle$; if $x \in A$, $x = (a_i)\tilde{p}$ with p in AT and a_i in A; as $1T = 1$, $(a_i, b)\tilde{p} \in \langle A \times \{b\} \rangle$; conversely, consider $\{1\} \subset 1T$; the last statement is clear from the proof of 6.19.]

7. Colimits

Most categories of interest have small colimits. In this section we establish a few theorems to support this contention.

7.1 Definition. *Let Δ be a diagram scheme. The category $\mathscr{K}^\Delta$ of Δ-diagrams in $\mathscr{K}$ has diagrams (Δ, D) in $\mathscr{K}$ as objects and has as morphisms $\chi: D \longrightarrow E$, $N(\Delta)$-indexed collections $\chi_i: D_i \longrightarrow E_i$ subject to the commutativities*

$$\begin{array}{ccc} D_i & \xrightarrow{\chi_i} & E_i \\ {\scriptstyle D_\alpha}\downarrow & & \downarrow{\scriptstyle E_\alpha} \\ D_j & \xrightarrow[\chi_j]{} & E_j \end{array}$$

for all $\alpha \in \Delta(i, j)$. As diagrams are like functors, so morphisms of diagrams are like natural transformations. $\mathscr{K}^\Delta$ forms a category with pointwise composition $(\chi\chi')_i = \chi_i \cdot \chi'_i$ and pointwise identities $(\mathrm{id}_D)_i = \mathrm{id}_{D_i}$.

Each object K of $\mathscr{K}$ induces the constant diagram $\tilde{K}$ in $\mathscr{K}^\Delta$ defined by $\tilde{K}_i = K$, $\tilde{K}_\alpha = \mathrm{id}_K$. Notice that an upper bound (L, ψ) of D is the same thing as a diagram morphism $\psi: D \longrightarrow L$. The constant diagram construction defines *the embedding functor*

$$\begin{array}{ccc} \mathscr{K} & \xrightarrow{\widetilde{(\cdot)}} & \mathscr{K}^\Delta \\ & & \\ K & & \tilde{K} \\ {\scriptstyle f}\downarrow & \longmapsto & \downarrow{\scriptstyle \tilde{f}} \\ L & & \tilde{L} \end{array}$$

where $\tilde{f}: \tilde{K} \longrightarrow \tilde{L}$ is the diagram morphism $\tilde{f}_i = f$; (warning: these are not the only diagram morphisms $\tilde{K} \longrightarrow \tilde{L}$).

The existence of colimits is related to the embedding functor as follows:

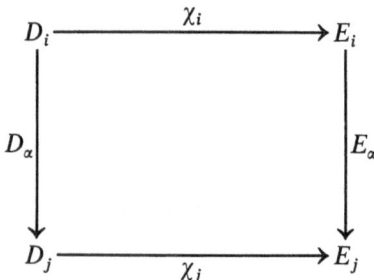

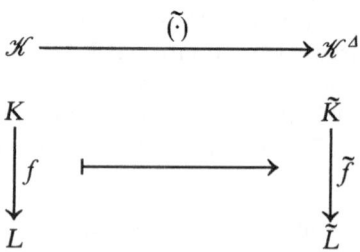

7.2 Lemma. *Let (Δ, D) be a diagram in $\mathscr{K}$, let $\widetilde{(\cdot)}: \mathscr{K} \longrightarrow \mathscr{K}^\Delta$ be the*

embedding functor of 7.1, and let (L, ψ) *be an upper bound of* D. *Then* (L, ψ) *is a colimit of* D *if and only if* (L, ψ) *is free over* D *with respect to* $(\tilde{\cdot})$.

Proof. Just compare the picture of the universal property of a free (L, ψ) over D with the definition of a colimit:

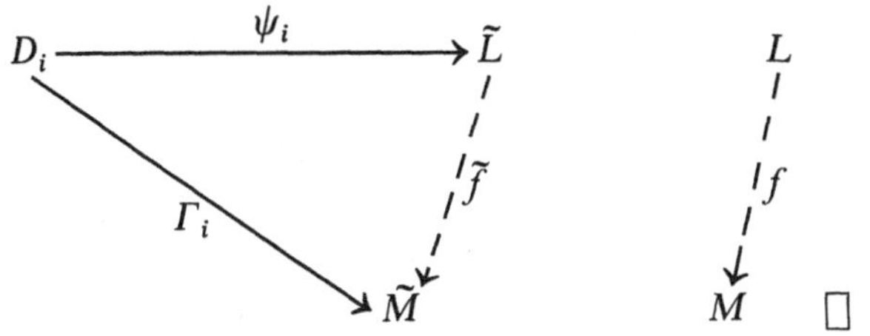

The following proposition suggests that the adjoint functor theorems can be used to prove that colimits exist:

7.3 Proposition. *Let* Δ *be a diagram scheme and let* $(\tilde{\cdot}): \mathscr{K} \longrightarrow \mathscr{K}^{\Delta}$ *be the embedding functor of 7.1. Then* $(\tilde{\cdot})$ *preserves limits.*

Proof. Let (Σ, E) be a diagram in $\mathscr{K}$ and let (L, ψ) be a limit of E. Let us write E-nodes as j and use superscripts for Σ-diagrams to avoid notational confusion with Δ-diagrams. We must show that $(\tilde{L}, (\psi^j)^{\sim} : \tilde{L} \longrightarrow (E^j)^{\sim})$ is a limit of the diagram $\tilde{E}$ in $\mathscr{K}^{\Delta}$. Let $(D, \Gamma^j : D \longrightarrow (E^j)^{\sim})$ be a lower bound of $\tilde{E}$ in $\mathscr{K}^{\Delta}$. Then for each $i \in N(\Delta)$, (D_i, Γ_i) is a lower bound of E, inducing

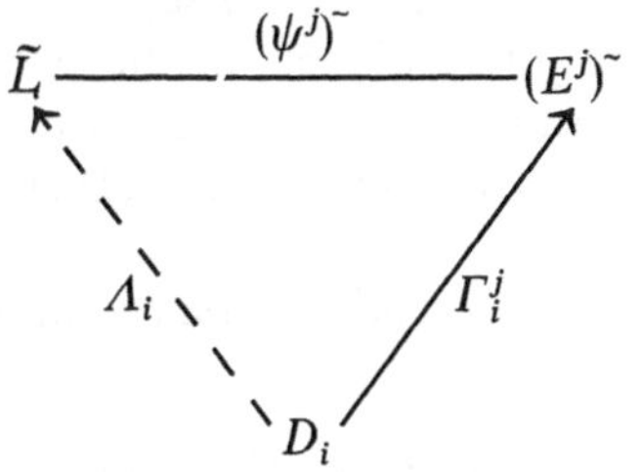

unique $\Lambda_i : D_i \longrightarrow L$ such that $\Lambda_i . \psi^j = \Gamma_i^j$. To see that $\Lambda : D \longrightarrow L$ is a diagram morphism consider

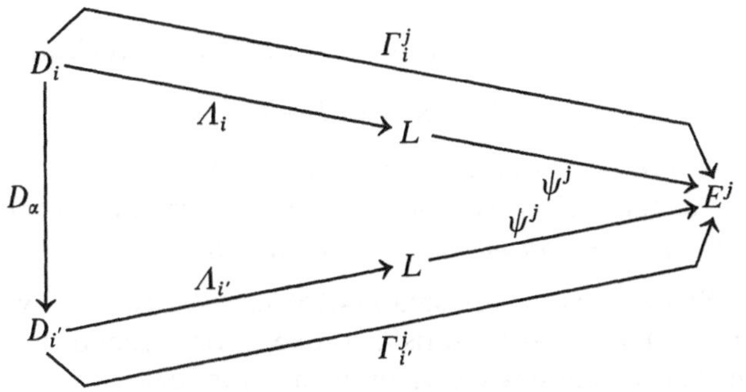

As $\Gamma^j : D \longrightarrow (E^j)$ is a diagram morphism for all $j \in N(\Sigma)$, $D_\alpha.\Lambda_{i'}.\psi^j = \Lambda_i.\psi^j$. But as $(\psi^j : j \in N(\Sigma))$ is a limit, $D_\alpha.\Lambda_{i'} = \Lambda_i$ as desired. □

The next two theorems are immediate consequences of 7.2 and 7.3:

7.4 Theorem. *Let $\mathscr{K}$ be a locally small category which has small limits and let Δ be a diagram scheme. Then necessary and sufficient for every diagram (Δ, D) in $\mathscr{K}$ to have a colimit is that the embedding functor $(\tilde{\cdot}) : \mathscr{K} \longrightarrow \mathscr{K}^\Delta$ of 7.1 satisfies the solution set condition.*

Proof. This is an immediate consequence of 2.2.24, in view of 2.1.22, 7.2, and 7.3. □

7.5 Theorem. *Let $\mathscr{K}$ be a locally small category which has small limits, is well powered, and has a cogenerator. Then $\mathscr{K}$ has small colimits.*

Proof. This is similar to the proof of 7.4 with the special adjoint functor theorem of 2.2.29 replacing 2.2.24. We remind the reader that here it is important to know that $\mathscr{K}^\Delta$ is locally small and this is why we must require that Δ be a small diagram scheme. □

A familiar fact about partially ordered sets is that if all infima exist (small limits!) then so do all suprema. This is easily seen to be a corollary of 7.5.

As a prelude to the next theorem we need a definition:

7.6 Definition. *Let (Δ, D) be a diagram in $\mathscr{K}$. A quasi-colimit of D is an $N(\Delta)$-indexed family $(\tau_i : D_i \longrightarrow C)$ of $\mathscr{K}$-morphisms with the property that for each upper bound (A, Γ) of D there exists (not necessarily unique) $f : C \longrightarrow A$ such that $\tau_i.f = \Gamma_i$ for all i. In case (τ_i) is itself a lower bound of D, it is a weak colimit of D* (cf. the weakly free objects in the proof of 2.2.24). Thus every colimit of D is a weak colimit of D and every weak colimit of D is a quasi-colimit of D. Notice that a weak coproduct of $(D_i : i \in N(\Delta))$ is a quasi-colimit of D.

We now present what we consider to be the most useful colimit theorem in practice:

7.7 Colimit Theorem. *Let $(\mathscr{K}, \mathsf{E}, \mathsf{M})$ be a regular category (4.15) and let $\mathscr{C}$ be a category of $\mathscr{K}$-objects with structure (2.3.1) with forgetful functor $U : \mathscr{C} \longrightarrow \mathscr{K}$ satisfying the following three conditions:*

1. *$\mathscr{C}$ has small limits.*
2. *U has a left adjoint.*
3. *Given $f : (K, t) \longrightarrow (L, s)$ in $\mathscr{C}$ with* E-M *factorization*

$$f = K \xrightarrow{e} I \xrightarrow{m} L$$

in $\mathscr{K}$, $m : I \longrightarrow (L, s)$ has an optimal lift (2.3.14).

Let Δ be a small diagram scheme. Then if each Δ-diagram in $\mathscr{K}$ has a quasi-colimit, each Δ-diagram in $\mathscr{C}$ has a colimit.

Proof. Since $\mathscr{K}$ is locally small (4.15) so is $\mathscr{C}$. Hence, by 7.4, we need only show that $(\tilde{\cdot}) : \mathscr{C} \longrightarrow \mathscr{C}^\Delta$ satisfies the solution set condition at the object D of $\mathscr{C}^\Delta$. The construction is shown in the diagram below:

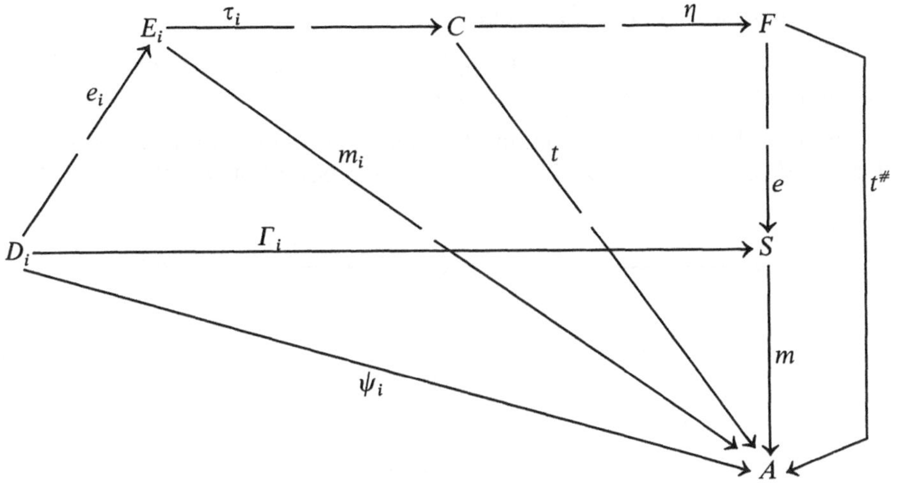

We are using broken arrows to denote $\mathscr{K}$-morphisms and solid arrows to denote morphisms admissible in $\mathscr{C}$. Here $\psi: D \longrightarrow \tilde{A}$ in $\mathscr{C}^\Delta$ is arbitrary (i.e., (A, ψ) is a lower bound of D) and (Γ, m) is the factorization to be constructed with (S, Γ) ranging over a small set depending only on D. It suffices to show that such S ranges over a small set $\mathscr{S}$, for then Γ ranges over the small set

$$\bigcup_{S \in \mathscr{S}} \bigcup_{i \in N(\Delta)} \mathscr{C}(D_i, S)$$

Here we have used the fact that Δ is small. Let (e_i, m_i) be an E-M factorization of ψ_i. E becomes a Δ-diagram in $\mathscr{K}$ by diagonal fill-in:

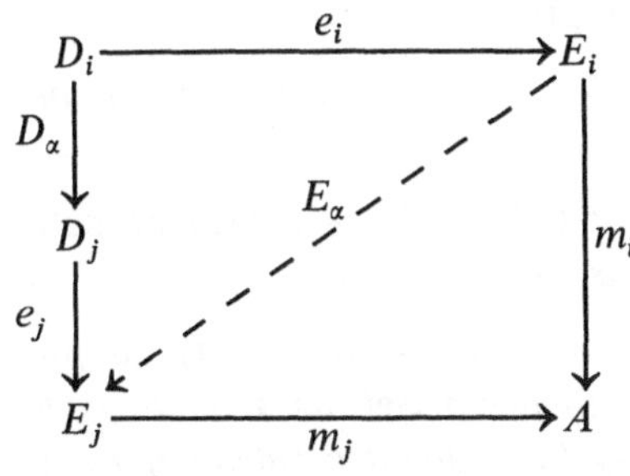

and it is clear that (A, m_i) is a lower bound of E. Hence, if (C, τ) is a quasi-colimit of E there exists $t: C \longrightarrow A$ with $\tau_i.t = m_i$. Let (F, η) be free over C with respect to U and let $t^\#: F \longrightarrow A$ be the unique $\mathscr{C}$-admissible extension of t. Let (e, m) be an E-M factorization of $t^\#$ in $\mathscr{K}$. By hypothesis, m admits an optimal lift S. Define $\Gamma_i = e_i.\tau_i.\eta.e$. As $\Gamma_i.m = \psi_i$ is admissible and m is optimal, Γ_i is admissible. Moreover, (S, Γ) is a lower bound of D because (A, ψ) is and m is a monomorphism. Finally, we must be sure that S ranges over a small set. But S is determined by the family $(E_i: i \in N(\Delta))$ and the latter ranges over a small set because $\mathscr{K}$ is E co-well-powered. □

Another useful colimit theorem is:

7.8 Theorem. *Let $\mathscr{L}$ be a full reflective subcategory of $\mathscr{K}$ and let (Δ, D) be a diagram in $\mathscr{L}$. Then if D has a colimit in $\mathscr{K}$, D has a colimit in $\mathscr{L}$.*

Proof. Let (K, ψ) be a colimit of D qua diagram in $\mathscr{K}$ and let $\eta: K \longrightarrow L$ be a reflection of K in $\mathscr{L}$. If (L', Γ) is a lower bound of D with $L' \in \mathscr{L}$

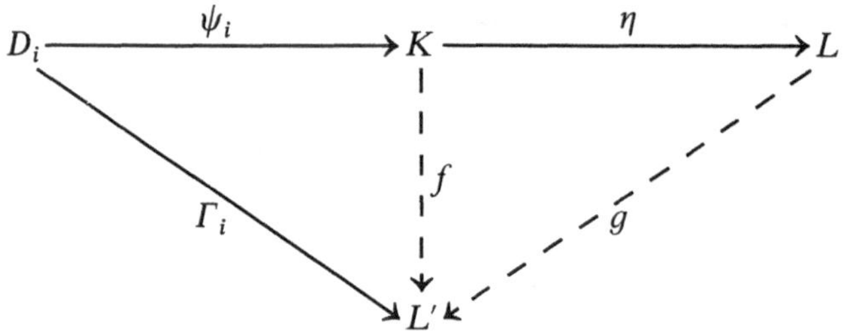

there exists unique $f: K \longrightarrow L'$ with $\psi_i.f = \Gamma_i$ and, as $L' \in \mathscr{L}$, there exists unique $g: L \longrightarrow L'$ with $\eta.g = f$. □

The following corollaries of 7.7 and 7.8 show that most of the categories mentioned in this book have small colimits.

7.9 Corollary. *Let $(\mathscr{K}, \mathsf{E}, \mathsf{M})$ be a regular category and let $\mathbf{T} = (T, \eta, \mu)$ be an algebraic theory in $\mathscr{K}$ such that T preserves E. Then if $\mathscr{K}$ has small weak coproducts (7.6), $\mathscr{K}^{\mathbf{T}}$ has small colimits.*

Proof. It is straightforward to apply 7.7. $\mathscr{K}^{\mathbf{T}}$ has small limits by 2.1.11, 2.1.14, and 2.1.22. Of course, $U^{\mathbf{T}}: \mathscr{K}^{\mathbf{T}} \longrightarrow \mathscr{K}$ has a left adjoint. Condition 3 of 7.7 is clear from the proof of 4.17. □

7.10 Corollary. *If* $\mathbf{T}$ *is an algebraic theory in* **Set**, $\mathbf{Set}^{\mathbf{T}}$ *has small colimits.* □

7.11 Corollary. *Let $(\mathscr{K}, \mathsf{E}, \mathsf{M})$ be a regular category with small weak coproducts. Then $\mathscr{K}$ has small colimits.* □

7.12 Corollary. *Let* $(\mathscr{A}, U) \in \mathrm{Struct}(\mathbf{Set})$ *be fibre complete, let* $\mathbf{T} = (T, \eta, \mu)$ *be an algebraic theory in* **Set**, *let* $\bar{\mathbf{T}}$ *be the canonical lift of* $\mathbf{T}$ *to* $\mathscr{A}$ *as in 5.3, and let* $\mathscr{B}$ *be any full reflective subcategory of* $\mathscr{A}^{\mathbf{T}}$ *(e.g., the various Birkhoff subcategories considered in section 5). Then* $\mathscr{B}$ *has small colimits.*

Proof. As noted in the proof of 5.6, if E = admissible surjections and if M = admissible injections then $(\mathscr{A}, \mathsf{E}, \mathsf{M})$ is a regular category and $\bar{T}$ preserves E. By 7.9, $\mathscr{A}^{\bar{\mathbf{T}}}$ has small colimits. By 7.8, $\mathscr{B}$ has small colimits. □

7.13 Example. Let $\mathscr{K}$ be the category of abelian groups with no element of order 4 as in 2.1.58. Thought of as a full subcategory of the category of abelian groups, $\mathscr{K}$ is clearly closed under products and subgroups, so is a quasivariety by 4.22. It follows from 7.8 that $\mathscr{K}$ has small colimits. □

Notes for Section 7

Theorem 7.7 is new although similar to many results in the folklore; see [Tholen '74]; the astute reader will observe that part of the hypothesis there is unnecessary. Concerning co-completeness in $\mathscr{K}^{\mathbf{T}}$, results have been obtained by [Barr '70-A], [Linton '69], [Schubert '72, section 21.3], and [Ulmer '69]. It is still an open problem whether or not $\mathscr{K}^{\mathbf{T}}$ must have all small colimits if $\mathscr{K}$ has all small limits and colimits.

Exercises for Section 7

1. Prove that the category of complete Boolean algebras does not have countable copowers but does have coequalizers.
2. Prove that if $U^{\mathbf{T}}:\mathbf{Set}^{\mathbf{T}} \longrightarrow \mathbf{Set}$ preserves coproducts, then $\mathbf{T}$ is the algebraic theory corresponding to the monoid $1T$. [Hint: since T preserves coproducts and every set is the coproduct of its elements, T is naturally equivalent to $1T \times -$.]
3. Show that the category of small categories and functors has small colimits. [Hint: use 7.11 with E the class of all functors $H:\mathscr{A} \longrightarrow \mathscr{B}$ such that $\mathscr{B}$ is the subcategory generated by the set of morphisms of form fH.]
4. Show that the category of Banach spaces and norm-decreasing linear maps has small colimits.
5. Let $\mathbf{T}$ be an algebraic theory in a category $\mathscr{K}$ and let Δ be a diagram scheme. Prove that if $\mathbf{T}$ preserves colimits of type Δ then $\mathscr{K}^{\mathbf{T}}$ has colimits of type Δ.
6. (a) Show that the forgetful functor from the category of small categories and functors to the category of small diagram schemes and morphisms of diagram schemes has a left adjoint. [Hint: the general adjoint functor theorem works; better, show that the free category over a diagram retains the same nodes as objects and generates compositions freely.]
   (b) If (Δ, D) is a small diagram in an arbitrary category $\mathscr{K}$ and if $\tilde{\Delta}$ is the free category over Δ, show that there exists a unique functor $\tilde{D}$

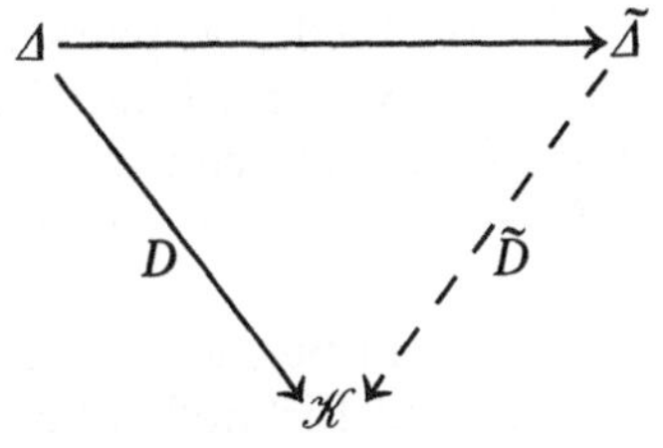

   and that the colimit of D exists if and only if the colimit of $\tilde{D}$ exists.
   (c) Prove that a category has small colimits if and only if it has small coproducts and every reflexive (see exercise 8 of section 1) pair has a coequalizer. [Hint: in the dual of the proof of 2.1.22, f, g is reflexive if the diagram is a functor.]

7. (Linton.) Let $\mathbf{T}$ be an algebraic theory in a category $\mathscr{K}$ with small coproducts and assume that, in $\mathscr{K}^{\mathbf{T}}$, every reflexive pair of homomorphisms between free algebras has a coequalizer. Prove that $\mathscr{K}^{\mathbf{T}}$ has small colimits. [Hint: use exercise 6(c); given reflexive $f, g:(K, \xi) \longrightarrow (L, \theta)$, $(K + LT, \mu) = (KT, \mu) + (LTT, \mu)$ in $\mathscr{K}^{\mathbf{T}}$ and the pair

$$(K + LT, \mu) \underset{\begin{pmatrix} gT \\ \theta T \end{pmatrix}}{\overset{\begin{pmatrix} fT \\ L\mu \end{pmatrix}}{\rightrightarrows}} (LT, \mu)$$

is reflexive so has coequalizer (Q, γ); $(L, \theta) \longrightarrow (Q, \gamma)$, the desired coequalizer of f and g, is the factorization resulting from the fact that $(L, \theta) = \text{coeq}(L\mu, \theta T)$ in $\mathscr{K}^{\mathbf{T}}$; the free algebra over the initial object is initial in $\mathscr{K}^{\mathbf{T}}$; to construct the coproduct of the nonempty family (A_i, ξ_i), define a $\mathscr{K}$-morphism u and $\mathbf{T}$-homomorphisms f, g: $(\coprod(A_i T))T \longrightarrow (\coprod A_i)T$ by

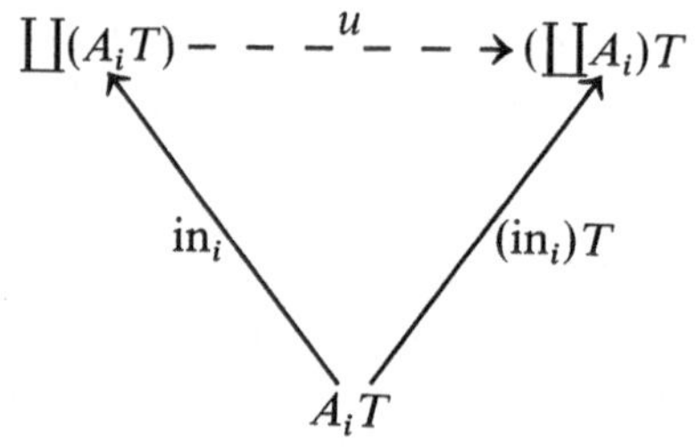

$$f = u^{\#}$$
$$g = (\coprod \xi_i)T$$

f, g is reflexive and $q = \text{coeq}(f, g)$ exists in $\mathscr{K}^{\mathbf{T}}$; the desired coproduct has injections $\text{in}_i.(\coprod A_i)\eta.q.$]

8. Let $\mathscr{K}$ have small colimits and let $\mathbf{T}$ be an algebraic theory in $\mathscr{K}$ such that T preserves coequalizers of reflexive pairs. Prove that $\mathscr{K}^{\mathbf{T}}$ has small colimits. [Hint: this combines exercises 5 and 7 but there is a subtle point; observe that in the hint of 7 for the construction of the coequalizer in $\mathscr{K}^{\mathbf{T}}$ of f and g the proof goes through if f, g are known only to be reflexive in $\mathscr{K}$.]

9. Let $\mathbf{T}$ be an algebraic theory in $\mathbf{Set}$, let $(\mathscr{A}, U)$ in Struct($\mathbf{Set}$) be fibre complete, and let $\mathscr{C}$ be any taut Birkhoff subcategory of $\mathscr{A}^{\mathbf{T}}$ as in 5.10. Show that the colimit of any small diagram in $\mathscr{C}$ is obtained by providing the colimit in $\mathbf{Set}^{\mathbf{T}}$ with the appropriate $\mathscr{A}$-structure. [Hint: in $\mathscr{A}^{\mathbf{T}}$ the co-optimal lift works; if (X, s, ξ) is in $\mathscr{A}^{\mathbf{T}}$, set t to be the optimal lift of the family $f:X \longrightarrow (Y, u)$ indexed by all (f, Y, θ, u) such that f: $(X, s, \xi) \longrightarrow (Y, u, \theta)$ is in $\mathscr{A}^{\mathbf{T}}$ with (Y, u, θ) in $\mathscr{C}$ and show that $\text{id}_X:(X, s, \xi) \longrightarrow (X, t, \xi)$ is the reflection in $\mathscr{C}$].

10. [Banaschewski and Nelson '75.] Give an example of a co-complete intractable equational class for which the free algebra on n generators fails to exist unless n is empty. [Hint: consider small modules over a large ring.]
11. Let $i:\mathscr{A} \longrightarrow \mathscr{B}$ be a subcategory for which there exists $U:\mathscr{B} \longrightarrow \mathscr{A}$ with $i.U = \mathrm{id}_{\mathscr{A}}$. Let Δ be a diagram scheme. Prove that if $\mathscr{B}$ has weak colimits of type Δ then so does $\mathscr{A}$.

Chapter 4

Some Applications and Interactions

Diverse applications of algebraic theories have already been offered in the text and exercises. In this chapter we present a detailed account of some interactions between algebraic theories and problems originating in topological dynamics and in automata theory. The latter is at the forefront of the research frontier.

1. Minimal Algebras: Interactions with Topological Dynamics

A central problem in abstract topological dynamics is the classification of compact minimal orbit closures. We show that similar questions apply to "dynamic" algebraic theories of sets—the motivating example being a special case—and observe that the problems center upon the nature of the monoid of unary operations.

Fix an algebraic theory $\mathbf{T}$ in **Set**.

1.1 Definition. *A* $\mathbf{T}$*-algebra* (X, ξ) *is minimal if* X *is nonempty and if* (X, ξ) *contains no proper, nonempty* $\mathbf{T}$*-subalgebra.*

If $\varnothing T \neq \varnothing$ then every $\mathbf{T}$-algebra is nonempty and possesses a unique minimal subalgebra, namely the subalgebra generated by the empty set. For example, if $\mathbf{T}$ corresponds to "rings with unit," $\{0, 1\}$ is the unique minimal subring of every ring. This situation is very uninteresting. If $\mathbf{T}$ is the theory corresponding to "semigroups" then $\varnothing T = \varnothing$. Here it is not the case that every semigroup possesses a minimal semigroup; consider the semigroup of natural numbers greater than 0 under addition.

1.2 Lemma. *Let* $f:(X, \xi) \longrightarrow (Y, \theta)$ *be a* $\mathbf{T}$*-homomorphism. If* f *is onto and* (X, ξ) *is minimal then* (Y, θ) *is minimal. Conversely, if* (Y, θ) *is minimal and* X *is nonempty then* f *is onto.*

Proof. If B is a nonempty subalgebra of (Y, θ), $A = Bf^{-1}$ is a nonempty subalgebra of (X, ξ) since $U^{\mathbf{T}}$ creates pullbacks. Similarly, if X is nonempty, $\mathrm{Im}(f)$ is a nonempty subalgebra of (Y, θ). □

We now introduce the motivating example.

1.3 Compact Transformation Groups. Each monoid M gives rise to an algebraic theory $\mathbf{T}_M$ whose algebras are M-sets (cf. exercise 2.1.19d, 1.3.7, 1.4.16, and exercise 3.2.4). Specifically, $X\mathbf{T}_M = X \times M$, $X\eta : X \longrightarrow X \times M$ sends x to (x, e), and $X\mu : X \times M \times M \longrightarrow X \times M$ sends (x, g, h) to (x, gh); a $\mathbf{T}_M$-algebra $=$ M-set $(X, \xi : X \times M \longrightarrow X)$ satisfies $xe = x$, $x(gh) = (xg)h$, where $(x, g)\xi$ is denoted xg.

Let G be a group. A *compact transformation group with phase group* G is simply a $\boldsymbol{\beta} \otimes \mathbf{T}_G$-algebra, that is, a compact Hausdorff G-set X such that $x \longmapsto xg$ is continuous for all g in G. Note that $\xi : X \times G \longrightarrow X$ is continuous if G is considered as a discrete space and $X \times G$ has the product topology. The above definitions may be repeated where now G is a topological group and ξ is required to be jointly continuous. The resulting transformation groups then form a Birkhoff subcategory of $\boldsymbol{\beta} \otimes \mathbf{T}_G$-algebras (see exercise 1). Since all concepts we study in this section (e.g. "minimal subalgebra") have the same meaning relative to a Birkhoff subcategory as opposed to the ambient category of algebras, the topology on the phase group is unimportant to us. We note, however, that many problems of abstract topological dynamics were motivated by qualitative problems arising from the local solutions to differential equations in Euclidean space where G is the topological group $\mathbf{R}$.

If X is a compact transformation group, the *orbit* of x in X is the subset $xG = \{xg : g \in G\}$ of X; thus xG is the $\mathbf{T}_G$-subalgebra generated by x. X is a *minimal orbit closure* if every orbit is dense; since the proofs of 3.6.19, 3.6.20 show that the $\boldsymbol{\beta} \otimes \mathbf{T}_G$-subalgebra generated by x is the closure $(xG)^-$ of xG, this coincides with the definition in 1.1.

We now give an example of a minimal orbit closure:

1.4 The Spinning Circle. Let $G = \mathbf{Z}$, the discrete group of integers under addition. Let the space X be a circle of circumference 1 and let $0 < \theta < 1$ be an irrational number. Set $f : X \longrightarrow X$ to be the homeomorphism "rotate θ units of circumference counterclockwise." Define $xn = xf^n$. Then X is a compact transformation group. Let x, y be arbitrary elements of X. Roll X along the real axis starting with x at the origin:

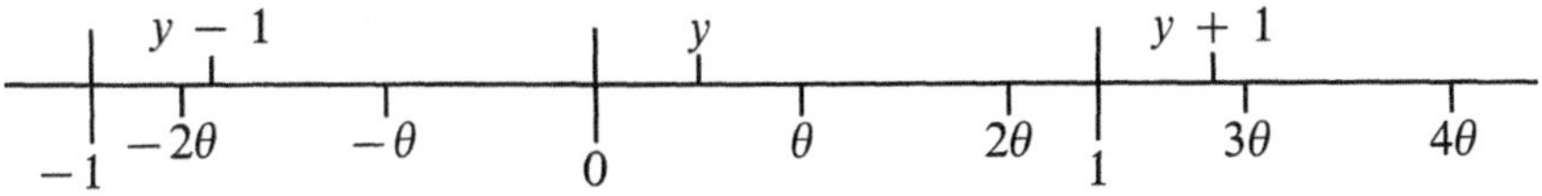

Then the orbit xG of x touches the axis at the points $n\theta$ whereas y touches the axis at the points $y + m$. To prove that X is minimal, it suffices to find, given small $\varepsilon > 0$, integers n and m such that $|(y + m) - n\theta| < \varepsilon$.

Let H be the topological subgroup $\{m + n\theta : m, n \in \mathbf{Z}\}$ of $\mathbf{R}$. Suppose 0 were an isolated point of H, i.e., $\{0\}$ is open. Since for all h in H, $h' \mapsto hh'$ is a homeomorphism, $\{h\}$ is open, and H is discrete. The intersection of H with the unit interval $[0, 1]$, being compact and discrete, is finite. Let $[\lambda]$ denote the largest integer $\leqslant \lambda$. Since $n\theta - [n\theta]$ is in $H \cap [0, 1]$ for all n in $\mathbf{Z}$ there exists $n \neq n'$ with $(n - n')\theta = [n\theta] - [n'\theta]$, contradicting the irrationality of θ. Thus 0 is in fact not an isolated point of H and there exists h in H with $0 < h < \varepsilon$. Let M be the integer such that $Mh \leqslant y \leqslant (M + 1)h$. Then $|y - Mh| < \varepsilon$. Write $Mh = -m + n\theta$. Then $|(y + m) - n\theta| = |y - (-m + n\theta)| < \varepsilon$ as desired. □

For x in X let $\langle x\rangle$ denote the $\mathbf{T}$-subalgebra generated by x. Then (X, ξ) is minimal if and only if X is nonempty and $X = \langle x\rangle$ for all x in X. Intuitively, a typical element of $\langle x\rangle$ is obtained as the value of an n-ary operation δ_ω on the n-tuple constantly x, that is the value of x under the *unary* operation $x\delta_{\omega'} = (x)\delta_\omega$. We now set forth to formalize this observation. (See 1.8 below.)

1.5 Definition. *Given the* $\mathbf{T}$*-algebra* (X, ξ), *the enveloping monoid* $E(X, \xi)$ *of* (X, ξ) *is the* $\mathbf{T}$*-subalgebra of* $(X, \xi)^X$ *generated by* id_X.

1.6 Theorem. *Let* $\lambda:(1T, 1\mu) \longrightarrow E(X, \xi)$ *be the unique* $\mathbf{T}$*-homomorphism such that* $\langle 1\eta, \lambda\rangle = \mathrm{id}_X$. *Then* λ *is the reflection of* $(1T, 1\mu)$ *in the Birkhoff subcategory* $\mathsf{SMP}(X, \xi)$ *generated by* (X, ξ) (see 3.4.25; we use S for $\mathsf{S}^{\mathbf{T}}$ and M for $\mathsf{M}^{\mathbf{T}}$; here S, M mean "onto," "injective").

Proof. Clearly $E(X, \xi)$ is in $\mathsf{MP}(X, \xi)$. Let $f:(1T, 1\mu) \longrightarrow (Z, \gamma)$ be a $\mathbf{T}$-homomorphism with $q:(Y, \theta) \longrightarrow (Z, \gamma)$ a surjective $\mathbf{T}$-homomorphism from the subalgebra (Y, θ) of the product algebra $(X, \xi)^I$ and consider the diagram below:

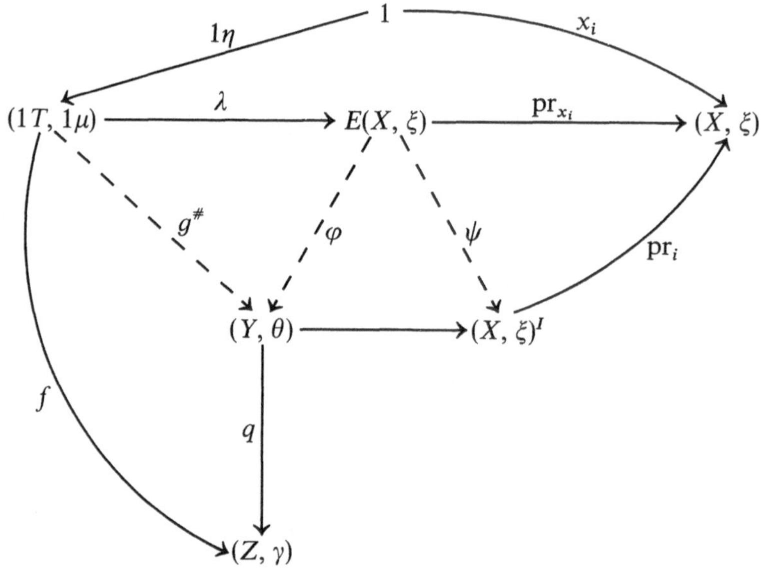

There exists $g:1 \longrightarrow Y$ with $\langle g, q\rangle = \langle 1\eta, f\rangle$ so that $g^{\#}.q = f$. For each i in I define $x_i = \langle 1\eta, g^{\#}.\mathrm{pr}_i\rangle$. There exists unique $\mathbf{T}$-homomorphism ψ: $E(X, \xi) \longrightarrow (X, \xi)^I$ such that $\psi.\mathrm{pr}_i$ is the restricted projection pr_{x_i}. Since $\langle \mathrm{id}_X, \psi\rangle \mathrm{pr}_i = \langle \mathrm{id}_X, \mathrm{pr}_{x_i}\rangle = x_i$, $\langle \mathrm{id}_X, \psi\rangle = \langle 1\eta, \lambda\psi\rangle = \langle 1\eta, g^{\#}\rangle \in Y$ so that $\mathrm{Im}(\psi) \subset Y$ giving rise to φ as shown. Thus $\varphi.q$ is a $\mathbf{T}$-homomorphism such that $\lambda.\varphi.q = f$. Uniqueness is clear because λ is onto. $\square$

As an immediate corollary of 1.6 we see that $(E(X, \xi), \mathrm{id}_X)$ is the free algebra on one generator in the Birkhoff subcategory $\mathsf{SMP}(X, \xi)$ of $\mathbf{Set}^{\mathbf{T}}$.

For each g in $1T$, let us write ξ^g for the more cumbersome $(X, \xi)\tilde{g}$: $X \longrightarrow X$ induced by the passage of 1.5.8. We have:

1.7 Proposition. *The map* $\lambda:(1T, 1\mu) \longrightarrow E(X, \xi)$ *of* 1.6 *is in fact the function* $g \longmapsto \xi^g$.

Proof. Let $g:1 \longrightarrow 1T$ and $x:1 \longrightarrow X$ be arbitrary elements. Using 1.5.8, 1.5.6, and 1.4.13, $\langle x, \xi^g\rangle = \langle x, 1\hat{g}.\xi\rangle = \langle g, xT.\xi\rangle = \langle g, x^{\#}\rangle$. As $\langle 1\eta, x^{\#}\rangle = x = \langle \mathrm{id}_X, \mathrm{pr}_x\rangle$ (where $\mathrm{pr}_x:E(X, \xi) \longrightarrow (X, \xi)$ is the restricted projection) $= \langle 1\eta, \lambda.\mathrm{pr}_x\rangle$, $x^{\#} = \lambda.\mathrm{pr}_x$. Thus $x\langle g, \lambda\rangle = \langle g, \lambda.\mathrm{pr}_x\rangle = \langle g, x^{\#}\rangle = x\xi^g$. □

It follows that $E(X, \xi)$ is not only a **T**-algebra, but a submonoid of X^X—i.e., $\xi^g\xi^h = \xi^{gh}$ where $(gh)\tilde{}$ is the horizontal composition $\tilde{g}.\tilde{h}$—thereby accounting for the term "enveloping *monoid*." $E(X, \xi)$ is, in short, the monoid $1S$ of unary operations of the theory **S** corresponding to $\mathsf{SMP}(X, \xi)$. The map $\lambda:(1T, 1\mu) \longrightarrow E(X, \xi)$ is a **T**-homomorphism and a monoid homomorphism. $\lambda:(1T, 1\mu) \longrightarrow E(1T, 1\mu)$ is an isomorphism (since both are free on one generator in $\mathsf{SMP}(1T, 1\mu)$). If (Y, θ) is in $\mathsf{SMP}(X, \xi)$ there is a canonical **T**- and monoid surjective homomorphism $E(X, \xi) \longrightarrow E(Y, \theta)$; it is just a λ, replacing **T** by **S**, and is described as the passage $\xi^g \longmapsto \theta^g$.

Monoid multiplication $E(X, \xi) \times E(X, \xi) \longrightarrow E(X, \xi)$ is *not* a **T**-homomorphism in general. For p in $E = E(X, \xi)$, let $L_p:E \longrightarrow E$ be the left multiplication $q \longmapsto pq$ and, similarly, let $R_p:E \longrightarrow E$ be the right multiplication $q \longmapsto qp$. The diagrams

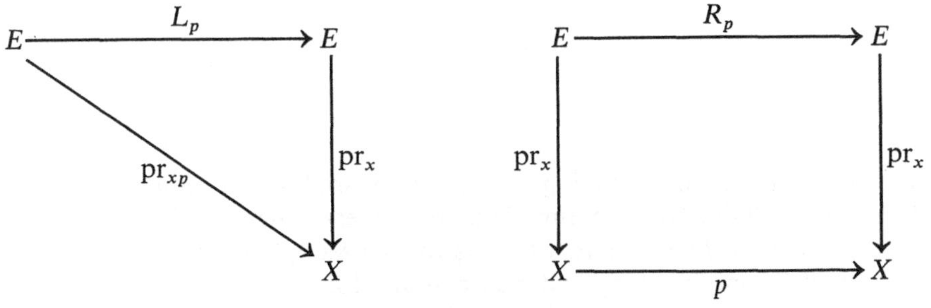

show that L_p is always a **T**-homomorphism and that R_p is a **T**-homomorphism whenever $p:(X, \xi) \longrightarrow (X, \xi)$ is a **T**-homomorphism.

1.8 Proposition. *Let* (X, ξ) *be a* **T**-*algebra, set* $E = E(X, \xi)$, *and let* x *be an element of* X. *Then* $\langle x\rangle = xE$ *where* $xE = \{xp:p \in E\}$.

Proof. $\mathrm{pr}_x:(X, \xi)^X \longrightarrow (X, \xi)$ maps E onto xE so $\langle x\rangle \subset xE$. By 1.7, $xE = \{x\xi^g:g \in 1T\}$ so that $xE \subset \langle x\rangle$ by 3.6.17. □

The reduction of the study of minimal algebras to theories with only unary operations can be formalized. Let $E_{\mathbf{T}}$ be the monoid $E(1T, 1\mu)$ inducing (as in 1.3) the algebraic theory $\mathbf{E_T}$. The passage $Q \times E_{\mathbf{T}} \longrightarrow QT$, $(q, g) \longmapsto \langle g, qT\rangle$ is a theory map whose induced functor from **T**-algebras to $E_{\mathbf{T}}$-sets interprets the **T**-algebra (X, ξ) as the $E_{\mathbf{T}}$-set $(X, (\xi^g))$. The important observation is that (X, ξ) is a minimal **T**-algebra if and only if $(X, (\xi^g))$ is a minimal $E_{\mathbf{T}}$-set since, by 1.8, the singly-generated subalgebras are the same. In some sense, then, the study of minimal algebras could be carried out entirely in the context of M-sets. We remark here that if $\mathbf{T} = \mathbf{T}_M$ corresponds to M-sets then, as is easily verified, $E_{\mathbf{T}} = M$, so that $E_{\mathbf{T}}$ is an arbitrary monoid.

Our approach will be to use the monoid structure of $E(X, \xi)$ to study minimal subalgebras of (X, ξ). Recall that a *right ideal* of a monoid M is a subset I of M such that $IM \subset I$. A *minimal right ideal* is a nonempty right ideal which properly contains no other nonempty right ideals. We have:

1.9 Lemma. *Let (X, ξ) be a $\mathbf{T}$-algebra and set $E = E(X, \xi)$. Let I be a subset of E. The following three statements are true:*

1. *If I is a $\mathbf{T}$-subalgebra of E then I is a right ideal of E.*
2. *For p in E, the $\mathbf{T}$-subalgebra $\langle p \rangle$ generated by p is the set $pE = \{pq : q \in E\}$.*
3. *I is a minimal subalgebra of E if and only if I is a minimal right ideal of E.*

Proof. 1. The diagram

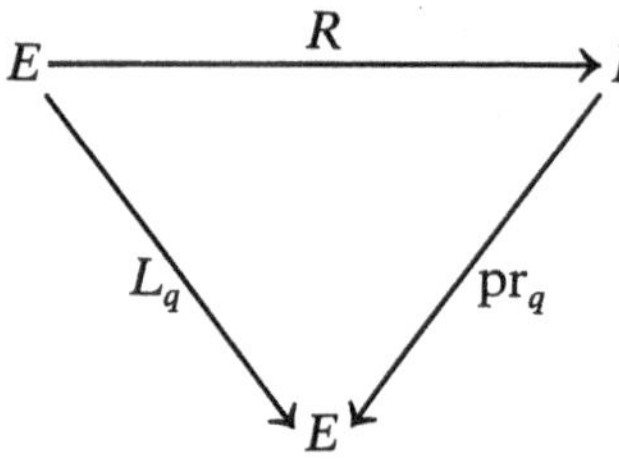

shows that the map R sending p to the right multiplication $q \longmapsto qp$ from I into E is a $\mathbf{T}$-homomorphism. Thus the inverse image (pullback!) of the subalgebra I^I of E^I under R, which is the subset $\{p \in E : Ip \subset I\}$ is a subalgebra of E containing id_X and, hence, is all of E.

2. Thinking of (E, id_X) as the free algebra on one generator in $\mathsf{SMP}(X, \xi)$, $L_p : E \longrightarrow E$ is the unique homomorphism sending id_X to p and hence its image pE is $\langle p \rangle$. Note that "$\langle - \rangle$" has the same effect in different Birkhoff subcategories.

3. I is a minimal right ideal if and only if $I = pE$ for all p in I whereas I is a minimal subalgebra if and only if $I = \langle p \rangle$ for all p in I. □

In the case of compact transformation groups, two special properties hold. Every nonempty compact transformation group contains a minimal orbit closure; this follows easily from Zorn's lemma since in a compact space the intersection of a nest of nonempty closed sets is nonempty. Also, there exists a "universal" minimal orbit closure in the sense of the following definition:

1.10 Definition. *The $\mathbf{T}$-algebra (X, ξ) is a universal minimal $\mathbf{T}$-algebra if the following three conditions hold:*

1. *(X, ξ) is a minimal $\mathbf{T}$-algebra.*

2. *If* (Y, θ) *is a minimal* **T**-*algebra there exists a* **T**-*homomorphism from* (X, ξ) *to* (Y, θ); (*such homomorphisms must be onto by 1.2*).

3. *Every* **T**-*endomorphism* $(X, \xi) \longrightarrow (X, \xi)$ *is an isomorphism.*

It is clear from the definition that any algebra satisfying the first two properties is isomorphic to the universal minimal one if it exists.

The corresponding monoid-theoretic notions are:

1.11 Definition. *Let* M *be a monoid. A coalescent ideal of* M *is a minimal right ideal* I *of* M *with the property that whenever* p, q, r *are in* I *and satisfy* $pq = pr$ *then* $q = r$; *i.e., the left multiplications of the monoid* I *are injective.* M *is dynamic if* M *has a coalescent ideal.* **T** *is dynamic if* $E(1T, 1\mu)$ *is dynamic.* We have:

1.12 Proposition. *In order that every nonempty* **T**-*algebra possess a minimal subalgebra and that there exist a universal minimal* **T**-*algebra it is necessary and sufficient that* **T** *be dynamic.*

Proof. We first prove necessity. Let $E = E(1T, 1\mu)$ and let I be a minimal subalgebra of E. Since I satisfies the first two properties of 1.10, I is the universal minimal **T**-algebra. By 1.9, I is a minimal right ideal. For all p in M, $L_p\colon E \longrightarrow E$ is a **T**-endomorphism and hence its restriction $L_p\colon I \longrightarrow I$ is an isomorphism. In particular, $L_p\colon I \longrightarrow I$ is injective for p in I, and I is a coalescent ideal of E.

Conversely, let I be a coalescent ideal of $E = E(1T, 1\mu)$. I is a minimal algebra by 1.9. Since every nonempty **T**-algebra admits a homomorphism from $(1T, 1\mu)$ and $(1T, 1\mu)$ is isomorphic to E, every nonempty **T**-algebra contains a minimal subalgebra. It is similarly clear that I admits a **T**-homomorphism to every minimal **T**-algebra. Let $f\colon I \longrightarrow I$ be a **T**-homomorphism. Then f is onto. To prove that f is injective we need the following lemma:

1.13 Lemma. *Let* M *be a dynamic monoid with coalescent ideal* J, *and let* p *be in* J. *Then the unique* u *in* J *with* $pu = p$ *is a left unit of* J.

Proof. The existence of u with $pu = p$ follows from $pJ = J$ by minimality and such u is unique because $L_p\colon J \longrightarrow J$ is injective. We must show that $uq = q$ for arbitrary q in J. As $uJ = J$ there exists r in J with $ur = q$; but $pq = pur = pr$ so that $q = r$. □

Returning to the proof of 1.12, let u in I be a left unit of I. As in the proof of 1.9, $R\colon E \longrightarrow I^I$ sending p to the restricted right multiplication $R_p\colon I \longrightarrow I$ is a **T**-homomorphism whose image, by 3.6.14, is the enveloping monoid of I. Since these R_p are then the unary operations of I (by 1.7), we have $f.R_p = R_p.f$ for all p in E. In particular, for p in I, $pf = (up)f = uR_pf = ufR_p = pL_{uf}$ so that $f = L_{uf}$ is injective. □

We now turn to a property which provides many examples of dynamic theories.

1.14 Definition. *A monoid* M *is compactible if there exists a compact Hausdorff topology on* M *such that each left multiplication* $L_p\colon M \longrightarrow M$ *is continuous.* **T** *is compactible if* $E(1T, 1\mu)$ *is compactible.*

1.15 Proposition. *If there exists a theory map* $\boldsymbol{\beta} \longrightarrow \mathbf{T}$, $\mathbf{T}$ *is compactible.*

Proof. For any $\mathbf{T}$-algebra (X, ξ) and p in $E = E(X, \xi)$, $L_p : E \longrightarrow E$ is a $\mathbf{T}$-homomorphism. Now use the corresponding $V : \mathbf{Set}^{\mathbf{T}} \longrightarrow \mathbf{Set}^{\boldsymbol{\beta}}$.

1.16 Theorem *Every compactible monoid is dynamic. Hence every compactible theory is dynamic.*

Proof. Let M be compactible. Provide M with a compact Hausdorff topology rendering the left multiplications continuous (and hence closed). M is a nonempty closed right ideal and, by compactness, the intersection of a chain of nonempty closed right ideals is a nonempty closed right ideal. By Zorn's lemma there exists a minimal closed right ideal I. We will show that I is a coalescent ideal. Since pE is closed for any p in E, $pE = I$ for all p in I and I is a minimal right ideal. Let p be in I. We must show that the restricted left multiplication $L_p : I \longrightarrow I$ is injective. We first need the lemma:

1.17 Lemma. *Every nonempty closed subsemigroup of* M *possesses an idempotent.*

Proof. S is a *subsemigroup* if $SS \subset S$. Let F be a nonempty closed subsemigroup. By a similar proof to the one above, F possesses a minimal closed subsemigroup S. For p in S, $\varnothing \neq pS \subset S$, $(pS)(pS) \subset pSSS \subset pS$ and pS is closed, so $pS = S$ and there exists u in S with $pu = p$. Set $R = \{s \in S : us = u\}$. R is clearly a subsemigroup. As the equalizer of the continuous functions L_u and "constantly u," R is closed. Since $uS = S$ (as argued above), R is nonempty. Thus $R = S$ and $uu = u$ as desired. □

Returning to the proof of 1.16, let q, r be in I and such that $pq = pr$. Define $F = \{s \in I : ps = p\}$. Then F is a closed subsemigroup of M which is nonempty since $pI = I$. By the lemma, F contains an idempotent u. Thus

1. $uu = u$, $pu = p$.

Next we observe

2. $ut = t$ for all t in I.

To prove it, let $ut' = t$ (as $uI = I$) and observe $ut = uut' = ut' = t$. Moreover,

3. There exists a in I with $ap = u$.

To prove this, let $pa = u$ with a in I and let $ab = u$ similarly; then $p = pu = pab = ub = b$, and $ap = u$. Finally, we have $q = uq = apq = apr = ur = r$. □

Summing up some of the consequences of 1.10–1.16, for any algebraic theory $\mathbf{T}$, $\boldsymbol{\beta} \otimes \mathbf{T}$ (which exists by 3.6.20) is dynamic (there is always a theory map from $\mathbf{S}$ to $\mathbf{S} \otimes \mathbf{T}$ because $\mathbf{S}$-$\mathbf{T}$ bialgebras are $\mathbf{S}$-algebras) and we have that every $\boldsymbol{\beta} \otimes \mathbf{T}$-algebra possesses a minimal subalgebra and there exists a universal minimal $\boldsymbol{\beta} \otimes \mathbf{T}$-algebra. The same comments apply to any Birkhoff subcategory of $\mathbf{Set}^{\boldsymbol{\beta} \otimes \mathbf{T}}$. This fully includes the case of compact transformation groups.

1.18 A Noncompactible Dynamic Monoid. Let G be any countably infinite group. Then G is a coalescent ideal of G so G is dynamic, but G is not compactible. For suppose G were provided with a compact Hausdorff topology with all L_p continuous. Since $(L_p)^{-1} = L_q$ with $q = p^{-1}$, L_p is a homeomorphism. Since $\{p\}L_r = \{q\}$ if $r = p^{-1}q$, if any point of G is isolated then all are and G is discrete, a contradiction. It suffices, then, to make the following observation: *Every countably infinite compact Hausdorff space has at least one isolated point.* For let X be such a space. By the Baire theorem [Kelley '55, Theorem 6.34] every countable family of dense open sets has dense intersection. Since the family of all sets with a finite complement has empty intersection, there exists a finite set F whose complement is not dense. Consequently, there exists a nonempty open subset G of F. If x is in G, x is an isolated point since $\{x\}$ is the intersection of the open sets $G, X - \{x_1\}, \ldots, X - \{x_n\}$ where $G - \{x\} = \{x_1, \ldots, x_n\}$.

1.19 Distal Transformation Groups. A compact transformation group X with phase group G is *distal* if whenever $x \neq y$ in X there exists a neighborhood α of the diagonal in $X \times X$ such that $(xg, yg) \notin \alpha$ for all g in G. The reader should recall that a compact Hausdorff space is uniquely uniformizeable with the neighborhoods of the diagonal as the entourages. In the case when X is metric, the condition is "there exists $\varepsilon > 0$ such that $d(xg, yg) > \varepsilon$ for all g." In other words, "distinct points were far apart in the past and will remain far apart in the future." For example, let X be the *annulus*, that is the subset of the plane of all points whose polar coordinates (r, θ) satisfy $1/2 \leqslant r \leqslant 1$. Let G be the group of real numbers under addition, representing "time." A particularly simple dynamical system is "X spins at a constant rate," i.e., $(r_0, \theta_0)t = (r_0, \theta_0 + \gamma t)$ where γ is constant. This system is the solution of the differential equation

$$x' = -\gamma y$$
$$y' = \gamma x$$

with initial value conditions $x(0) = x_0$, $y(0) = y_0$, as can be checked by differentiating

$$x(t) = r_0 \cos(\theta_0 + \gamma t)$$
$$y(t) = r_0 \sin(\theta_0 + \gamma t)$$

This system is distal because $d(x, y) = d(xt, yt)$ holds for all x, y, t. On the other hand, consider the system $(r_0, \theta_0)t = (1 - e^{-t^2}(1 - r_0), \theta_0)$ which solves

$$x' = 2t(\cos\theta_0 - x)$$
$$y' = 2t(\sin\theta_0 - y)$$

This system is not distal because $(r_0, \theta_0)t$ approaches $(1, \theta_0)t$ as the absolute value of t gets large.

The content of the next proposition is that pointwise limits of the homeomorphisms $\xi^g : X \longrightarrow X$ for g in G determine asymptotic behavior.

1.20 Proposition. *Let (X, ξ) be a $\mathbf{\beta} \otimes \mathbf{T}_G$-algebra, that is a compact transformation group with phase group G, and set $E = E(X, \xi)$. Then (X, ξ) is distal if and only if each $p : X \longrightarrow X$ in E is injective.*

Proof. First suppose (X, ξ) is distal and consider p in E and $x \neq y$ in X. There exists an open neighborhood of the diagonal α such that $(xg, yg) \notin \alpha$ for all g in G. p is the pointwise limit of a net g_i with each g_i in G. As α is open, $(xp, yp) \notin \alpha$ and in particular $xp \neq yp$. Conversely, suppose each p in E is injective and consider $x \neq y$ in X. Suppose that for all neighborhoods of the diagonal α, there were some g_α in G with $(xg_\alpha, yg_\alpha) \in \alpha$. By compactness, there exists a subnet g_β which converges to some p in E. Since the intersection of all closed β is the diagonal and since (xp, yp) is in every closed β, $x = y$. □

We have motivated:

1.21 Definition and Theorem. *Let* **T** *be a dynamic theory (1.11). Then the following five conditions on a* **T**-*algebra* (X, ξ) *are equivalent and define a distal* **T**-*algebra*:

1. $E = E(X, \xi)$ *is a subgroup of bijections of* X.
2. *Every element of* E *is injective.*
3. E *is a minimal* **T**-*algebra.*
4. *If* (Y, θ) *is in the Birkhoff subcategory,* SMP(X, ξ), *generated by* (X, ξ) *then the subalgebra generated by any element of* Y *is minimal.*
5. *The subalgebra generated by any element of* $(X, \xi) \times (X, \xi)$ *is minimal.*

Proof. We begin with a useful lemma:

1.22 Lemma. *Let* M *be a monoid and let* I *be a minimal right ideal of* M. *Then* I *is coalescent if and only if for every* p *in* I *there exists* a *in* I *such that* ap *is a left unit of* I.

Proof. If I is coalescent then, given p, we may define u as in 1.13 and then a as in (3) of the proof of 1.16. Conversely, let $pq = pr$ with p, q, r in I and let ap be a left unit. Then $q = apq = apr = r$. □

As an immediate consequence of 1.22 we have that any monoid quotient of a dynamic monoid is dynamic; for any surjective monoid homomorphism preserves minimal right ideals (cf. 1.2) and the condition "for all p there exists a such that ap is a left unit." It follows also that the theory corresponding to any Birkhoff subcategory of a dynamic theory is dynamic.

We return to the proof of 1.21:

1 implies 2: obvious

2 implies 3: Since E is a quotient of $E(1T, 1\mu)$, E is dynamic and possesses a coalescent ideal I. It suffices to show that $E = I$ in view of 1.9. Let u in I be idempotent. As $I \supset uE$ it suffices to show $u = \mathrm{id}_X$; but this is clear since u is injective.

3 implies 4: Since $E(Y, \theta)$ is a **T**-quotient of E, $E(Y, \theta)$ is a minimal **T**-algebra. Now use 1.2 and 1.8. Note: Y might have been empty.

4 implies 5: obvious

5 implies 1: Suppose $x \neq y$ in X. The subalgebra $\langle x, y\rangle$ of $(X, \xi) \times (X, \xi)$ is just $\{(xp, yp): p \in E\}$ as is clear from consideration of the **T**-homomorphism $p \longmapsto (xp, yp)$; since this subalgebra is minimal by hypothesis and is not contained in the diagonal subalgebra, it cannot intersect the diagonal

so that, for all p in E, $xp \neq yp$. Thus 2 holds so, as proved above, 3 holds, and using 1.9, $pE = E$ for all p in E. Given p, then, there is a q in E with $pq = \mathrm{id}_X$. As q is injective, p and q are mutually inverse bijections so that $q = p^{-1}$ and q is in E. □

1.23 Corollary. *Let* **T** *be a dynamic theory. Then the class of all distal* **T**-*algebras is a Birkhoff subcategory of* $\mathbf{Set}^{\mathbf{T}}$.

Proof. That "distal" is closed under subalgebras and quotient algebras is immediate from 1.21 (4) and (5). Now let (X, ξ) be the product of the family (X_i, ξ_i) of distal algebras. The diagram

$$\begin{array}{ccc} X & \xrightarrow{\mathrm{pr}_i} & X_i \\ {\scriptstyle \xi^g}\downarrow & & \downarrow{\scriptstyle \xi_i^g} \\ X & \xrightarrow[\mathrm{pr}_i]{} & X_i \end{array}$$

shows that $\xi^g = \prod \xi_i^g$ so that ξ^g is injective. □

Notes for Section 1

The theory of this section originates with the work of Robert Ellis. See [Ellis '58, '60, '60-A, '69] and [Ellis and Gottschalk '60]. Minimal **T**-algebras were studied in [Manes '67, '69-A]. For more on this sort of topological dynamics we refer the reader to [Ellis '69]; for relationships with the qualitative theory of differential equations see [Nemytskii and Stepanov '60].

Exercises for Section 1

1. (a) Let X, H, Y be topological spaces with X compact and Y Hausdorff and consider a commutative square of functions

$$\begin{array}{ccc} X \times H & \xrightarrow{a} & X \\ {\scriptstyle f \times \mathrm{id}}\downarrow & & \downarrow{\scriptstyle f} \\ Y \times H & \xrightarrow[b]{} & Y \end{array}$$

where f is continuous and onto and a is continuous. Prove that b

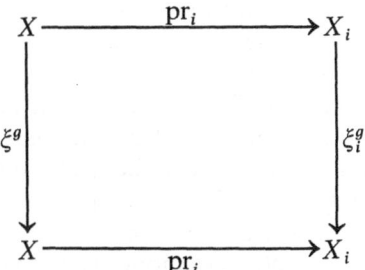

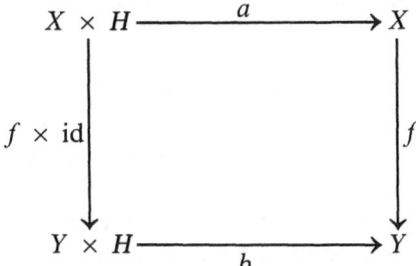

is continuous. [Hint: let $\mathscr{U}$ be an ultrafilter on $Y \times H$ converging to (y, h); the inverse image of $\mathscr{U}$ in $X \times H$ contains an ultrafilter converging to a point of form (x, h), $a.f$ of which is the desired convergent point in Y.]

(b) Let $\mathscr{T}$ be any topology on the monoid M and let $\mathscr{B}$ be the class of all $\boldsymbol{\beta} \otimes \mathbf{T}_M$-algebras (X, ξ, θ) such that $\theta:(X, \xi) \times M \longrightarrow (X, \xi)$ is jointly continuous. Prove that $\mathscr{B}$ is a Birkhoff subcategory. [Hint: for quotients, use (a).]

2. Show that $E(X, \xi)$ is a submonoid of X^X by proving that $\{p \in X^X: Fp \subset E(X, \xi)\}$ is a subalgebra of X^X for any subset F of X^X.
3. For any $\mathbf{T}$-algebra (X, ξ), show that $E(X, \xi) \longrightarrow E(E(X, \xi))$, $\xi^g \longmapsto -.\xi^g$ is a $\mathbf{T}$-monoid isomorphism.
4. Show that every minimal semigroup has one element. [Hint: the semigroup generated by an element of a minimal one cannot be infinite; a finite semigroup is compactible, hence has an idempotent.] Show that the multiplicative semigroup of a Boolean ring is a distal semigroup.
5. ([Ellis '60], [Manes '69-A].) Let M be a monoid and consider the set $M\beta$ of ultrafilters on the set M. For $\mathscr{U}$, $\mathscr{V}$ in $M\beta$ define $\mathscr{U} \cdot \mathscr{V} = \{A \subset M: \exists V \text{ in } \mathscr{V} \forall v \text{ in } V \exists U \text{ in } \mathscr{U} \text{ with } Uv \subset A\}$.

(a) Prove that $(M\beta, \cdot, \text{prin}(e))$ is a monoid and that $\text{prin}: M \longrightarrow M\beta$ is a monoid homomorphism.

(b) Show that $\mathscr{U} \cdot \text{prin}(p) = \langle \mathscr{U}, R_p\beta \rangle$ and $\text{prin}(p) \cdot \mathscr{U} = \langle \mathscr{U}, L_p\beta \rangle$.

(c) Prove that $M\beta$ is the free compact M-set on one generator. [Hint: the M-action is induced by $\mathscr{U} \longmapsto \mathscr{U} \cdot \text{prin}(p)$ for p in M; given $f: M \longrightarrow X$ equivariant with X a compact M-set, with continuous extension $f^{\#}: M\beta \longrightarrow X$, $f^{\#}$ is equivariant because for all p in M the set $\{\mathscr{U} \in M\beta: (\mathscr{U} \cdot \text{prin}(p))f^{\#} = \langle \mathscr{U}, f^{\#}.L_{pf} \rangle\}$ is dense and closed.]

Let $\mathscr{C}$ in Struct(**Set**) be the category whose objects are monoids equipped with compact Hausdorff topology such that each left multiplication is continuous and whose morphisms are continuous monoid homomorphisms.

(d) Prove that $M\beta$ is the free $\mathscr{C}$-object over M with respect to the forgetful functor from $\mathscr{C}$ to monoids. [Hint: if $f: M \longrightarrow N$ is a monoid homomorphism with N in $\mathscr{C}$, N is a compact M-set via $nm = \langle n, mf \rangle$ so that the continuous extension $f^{\#}: M\beta \longrightarrow N$ is equivariant by (c); for fixed $\mathscr{U}$, the set $\{\mathscr{V} \in M\beta: (\mathscr{U} \cdot \mathscr{V})f^{\#} = \mathscr{U}f^{\#} \cdot \mathscr{V}f^{\#}\}$ is dense and closed.] Conclude that $\mathscr{C}$ is algebraic over **Set** via a compactible theory.

6. (a) Prove the *Cayley theorem*: for any monoid M, $p \longmapsto R_p$ embeds M as a submonoid of M^M.

(b) Prove that, for any set X, the monoid X^X is compactible. [Hint: consider the one-point compactification of $X - \{x\}$.] Thus every monoid is a submonoid of a compactible one.

(c) For X an infinite set, set $M = \{f \in X^X: f = \text{id}_X \text{ or } f \text{ is injective and the complement of the image of } f \text{ is countably infinite}\}$. Show that M is a submonoid of X^X and that M has a unique minimal ideal which, however, has no idempotents (so that M is not dynamic).

7. Let M be a compactible monoid. Show that all minimal right ideals of M are closed (in any compact Hausdorff topology rendering the left multiplications continuous) and that if I, J are two minimal right ideals of M then there exists p in J such that L_p restricts to an M-set isomorphism of I onto J.
8. Let G be a discrete group and let $\mathcal{S}$ be a family of normal subgroups of G of pairwise relatively prime finite indices.
   (a) Prove the *generalized Chinese remainder theorem*: given $S_1, \ldots, S_n$ in $\mathcal{S}$ and $g_1, \ldots, g_n$ in G then there exists x in G such that xg_i^{-1} is in S_i for all $i = 1, \ldots, n$. [Hint: it is an easy consequence of [Hall '59, 1.5.3 and 1.5.6] that whenever $S \neq S'$ in $\mathcal{S}$ then $\#(S \cap S') = (\# S)(\# S')$, where $\#$ denotes index in G; if $\bar{S}_i$ is defined to be $\cap(S_j: j \neq i)$ then $\#\bar{S}_i$ and $\#S_i$ are relatively prime and there exists, by Euler's theorem, an integer n_i such that $(\#\bar{S}_i)n_i - 1$ is a multiple of $\#S_i$; set $x_i = g_i^{N_i}$ where $N_i = (\#\bar{S}_i)n_i$; by Lagrange's theorem, x_i is in g_iS_i whereas for $j \neq i$, x_i is in S_j; set $x = x_1 \cdots x_n$.]
   (b) Look up and recover the Chinese remainder theorem as a corollary of the result of (a).
   (c) (W. H. Gottschalk.) For each S in $\mathcal{S}$, G/S is a compact transformation group with G-action $[s]g = [sg]$. Show that the product transformation group
   $$\prod(G/S : S \in \mathcal{S})$$
   is a minimal orbit closure. [Hint: this statement is just a rewording of the result of (a).]
9. ([Manes 69-A].) For each monoid M and set A, M^A is an M-set with action $fm = f.R_m$. M is *quasicompactible* if M has a minimal right ideal and if for every set A and function $f: A \longrightarrow M$, $fM = (fM)^*$, where $*$ denotes the closure operator of the product topology on M^A induced by the discrete topology on M. A subset Δ of a monoid M is a *division set* if for all p, q in M there exists x in M such that $\delta px = \delta q$ holds for all $\delta \in \Delta$.
   (a) Prove that every compactible monoid is quasicompactible. [Hint: fM is even closed in the compact product topology.]
   (b) Prove that every quasicompactible monoid has a minimal right ideal and a maximal division set. [Hint: if I is a minimal right ideal, each singleton subset of I is a division set; if Δ_α is a chain of division sets and f is the inclusion function of $\cup\Delta_\alpha$ then for all p, q in M, fq is in $(fpM)^*$.]
   (c) Prove that every monoid which has a minimal right ideal and a maximal division set is dynamic. [Hint: given p in I there exists x in M with $\delta px = \delta$ for all δ in Δ; set $u = px$ in I; as $\Delta \cup \{u\}$ is a division set, u is idempotent, hence a left unit of I; ap is a left unit if $pa = u$ with a in I.]

It is shown in [Manes '69-A] that none of the three implications can be reversed.

10. Let $\mathbf{T}$ be a dynamic theory, let (X, ξ) be a $\mathbf{T}$-algebra, let $E = E(X, \xi)$ and let I be a minimal right ideal of E. x, y in X are *distal* if $xp \neq yp$ for all p in E and x, y are *proximal* if x, y are not distal.
    (a) ([Auslander '63, Theorem 2].) Show that if (X, ξ) is minimal and if $f:(X, \xi) \longrightarrow (X, \xi)$ is a $\mathbf{T}$-endomorphism with $f \neq \mathrm{id}_X$, then x, xf are distal for all x in X. [Hint: if there exists x with x, xf proximal then $xp = xfp$ for some p in E; as $xpE = X$, $xpq = x$ for some q in E; as x is in $\mathrm{eq}(f, \mathrm{id}_X)$, $f = \mathrm{id}_X$, a contradiction.]
    (b) ([Auslander '63, Theorem 4].) Let (X, ξ) be minimal. Prove that (X, ξ) is a universal minimal set in $\mathsf{SMP}(X, \xi)$ if and only if for every x, y in X there exists a $\mathbf{T}$-endomorphism $f:(X, \xi) \longrightarrow (X, \xi)$ such that xf, y are proximal. [Hint: if p, q are in I let ap be a left unit of I; then $(pL_{qa})ap = (q)ap$, L_{qa} is a $\mathbf{T}$-endomorphism of I and $-.ap$ is in $E(I)$ using exercise 3; conversely, it suffices to show $I \longrightarrow X$, $p \longmapsto xp$ is injective for any fixed x in X; if $xp = xq$ and y is in X let xf, y be proximal; as $xfr' = yr'$ for some r' in E, $xfr = yr$ for r in $r'I \subset I$; if $rs = q$, $yq = xfrs = xpr$; as f is independent of q, $yp = xpf$ also.]
    (c) Let (X, ξ) be minimal. Prove that the following three conditions are equivalent: (i) (X, ξ) is *potentially free on one generator* in the sense that there exists x in X such that for all y in X there exists a $\mathbf{T}$-endomorphism $f:(X, \xi) \longrightarrow (X, \xi)$ with $xf = y$; (ii) (X, ξ) is *homogeneous* in the sense that given y, z in X there exists a $\mathbf{T}$-automorphism $f.(X, \xi) \longrightarrow (X, \xi)$ with $yf = z$; (iii) (X, ξ) is distal and free on one generator in $\mathsf{SMP}(X, \xi)$. [Hint: for '(i) implies (ii)', let $xg = y$, $xg' = z$; if $wg = x$ and $xg'' = w$ then $gg'' = \mathrm{id}_X$; set $f = g^{-1}g'$; for '(ii) implies (iii)' use (a) and (b).]

2. Free Algebraic Theories: The Minimal Realization of Systems

A problem common to systems engineering and finite-state automata theory is to build an optimal system realizing a prescribed input/output response function. In this section we describe systems as algebras over a free algebraic theory and present a minimal realization theorem.

2.1 Minimal Realization of Automata. For a fixed set X of *input symbols* and a fixed set Y of *outputs*, an *automaton is* (Q, δ, τ, β) where Q is a set of (*internal*) *states*, $\delta: Q \times X \longrightarrow Q$ is the *dynamics*, or *state-transition function*, $\tau \in Q$ is the *initial state*, and $\beta: Q \longrightarrow Y$ is the *output function.* For example, a coin-operated vending machine is an automaton where a possible X is {quarter, dime, nickel, slug; choice 1, choice 2; coin return} and a possible Y is {candy bar, peanuts, contents of coin hopper, nothing}. To run the vending machine, one puts in an appropriate sequence of input symbols such as "dime, nickel, dime, choice 1" obtaining, hopefully, the sequence of outputs "nothing, nothing, nothing, candy bar" of which the last output is regarded as the response of the automaton to the sequence of inputs. In general, let X^* be the free monoid generated by X of all strings $x_1 \cdots x_n$ including the empty string Λ. The *reachability map* of the automaton is the map $r: X^* \longrightarrow Q$ with the interpretation that the internal state resulting from the input string w is wr, i.e., (using algebraic general recursion (1.1.14))

$$\Lambda r = \tau$$
$$(wx)r = (wr, x)\delta \qquad (\text{for } w \text{ in } X^*, x \text{ in } X)$$

The input/output response of the automaton is the map $f: X^* \longrightarrow Y$ defined by $f = r.\beta$. The *observability map* of the automaton is the function $\sigma: Q \longrightarrow Y^{X^*}$ such that $q\sigma$ is the input/output response which would result if q (rather than τ) were the initial state. The automaton is *reachable* if r is onto, i.e., "every state is used." The automation is *observable* if σ is injective; since $q\sigma$ represents the total effect of the state q on the input/output response, observability guarantees that "different states have different effect."

The vending machine example illustrates the principle that we may know the response of a system without knowing the details of its dynamics. An arbitrary function $f: X^* \longrightarrow Y$ is called a *response*. An automaton $M = (Q, \delta, \tau, \beta)$ *realizes* f, or is a *realization of* f, if its input/output response coincides with f. It is intuitively clear that any optimal realization of f must be at least reachable and observable. The general theory of this section will establish that all reachable and observable realizations of f are isomorphic and that one always exists (although it need not have only finitely many states even if X and Y are finite).

2.2 Example. Let $X = \{a, b\}$, $Y = \{t, u, v\}$, $Q = \{q_0, q_1, q_2, q_2', q_3, \hat{q}\}$, $\tau = q_0$ and consider the automaton M whose *state graph* is shown below:

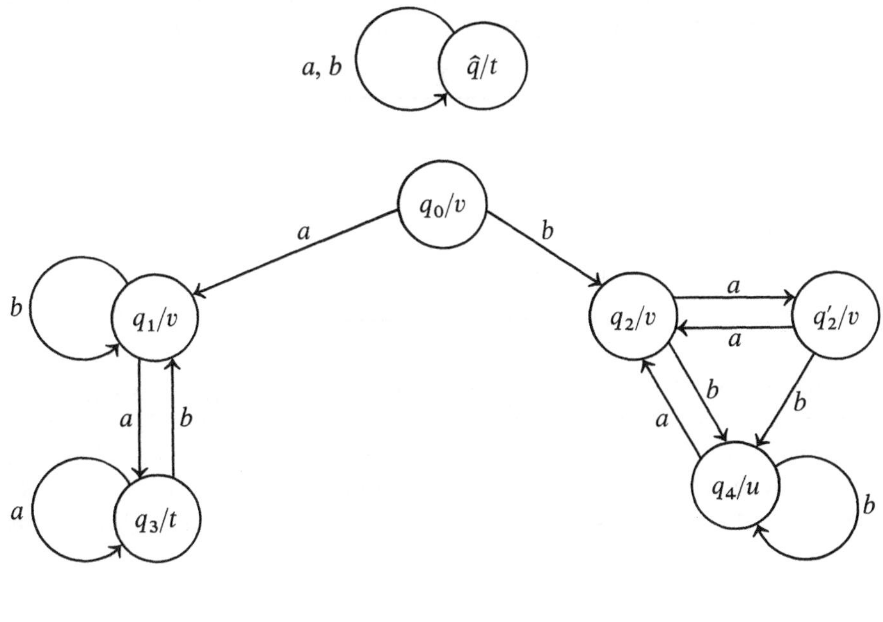

Here the label q/y denotes that the state labelled q has output $q\beta = y$, whereas $(q, x)\delta$ is the unique state at the terminus of the arrow labelled x emanating from the state labelled by q. This automaton is not reachable since there is no way to get to the state $\hat{q}$. The response $f = q_0\sigma$ is the function $X^* \longrightarrow Y$ which sends elements of $aX^*a = \{awa : w \in X^*\}$ to t, elements of bX^*b to u and everything else to v. M is not observable since $q_2\sigma = q'_2\sigma$, both being the function sending X^*b to u and everything else to v. The following state graph depicts the (unique) reachable and observable realization of f:

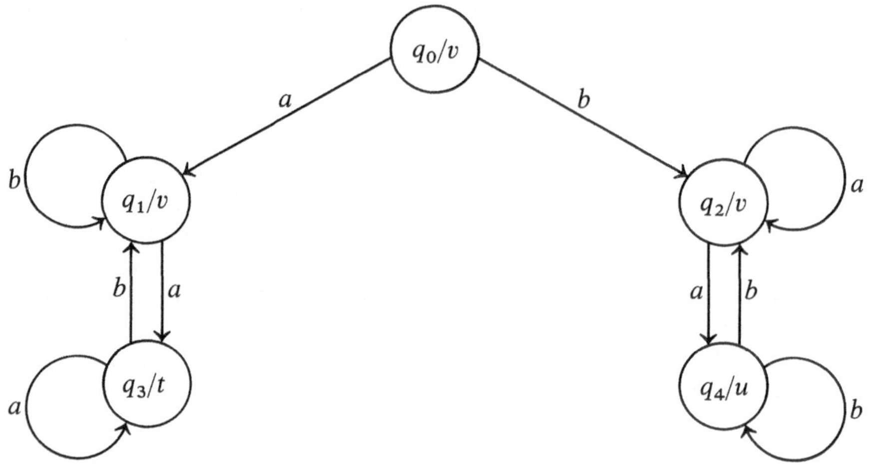

We now provide the definitions to categorize the constructions of 2.1:

2.3 Input Processes and Output Processes. Let $\mathscr{K}$ be an arbitrary category. A *process in* $\mathscr{K}$ is an endofunctor $X:\mathscr{K} \longrightarrow \mathscr{K}$. For any process X, the category Dyn(X) *of* X*-dynamics* has as objects all pairs (Q, δ) where $\delta: QX \longrightarrow Q$ and as morphisms all $f:(Q, \delta) \longrightarrow (Q', \delta')$ such that

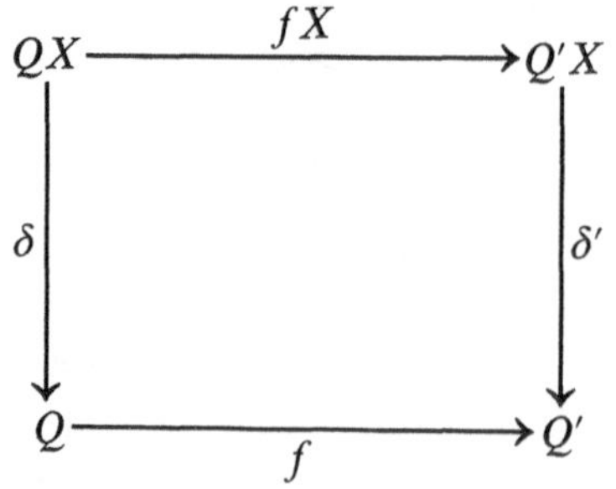

(2.4)

Morphisms in Dyn(X) are called X*-dynamorphisms.* It is evident that $U:\text{Dyn}(X) \longrightarrow \mathscr{K}$ is in Struct($\mathscr{K}$). An X*-automaton* is $(Q, \delta, I, \tau, Y, \beta)$ where (Q, δ) is an X-dynamics and $\tau: I \longrightarrow Q$, $\beta: Q \longrightarrow Y$ are morphisms in $\mathscr{K}$. The automata of 2.1 are recaptured by setting $\mathscr{K} = \mathbf{Set}$, using $- \times X$: $\mathbf{Set} \longrightarrow \mathbf{Set}$ for the process and setting $I = 1$. In general, Q, I, Y are, respectively, the *state object*, *initial state object*, and *output object*, τ is the *initial state*, and β is the *output morphism*.

X is an *input process* if $U:\text{Dyn}(X) \longrightarrow \mathscr{K}$ has a left adjoint. The free dynamics over A with respect to U will be denoted $(AX^{@}, A\mu_0; A\eta)$. The unique dynamorphic extension of the initial state, as shown below, is the

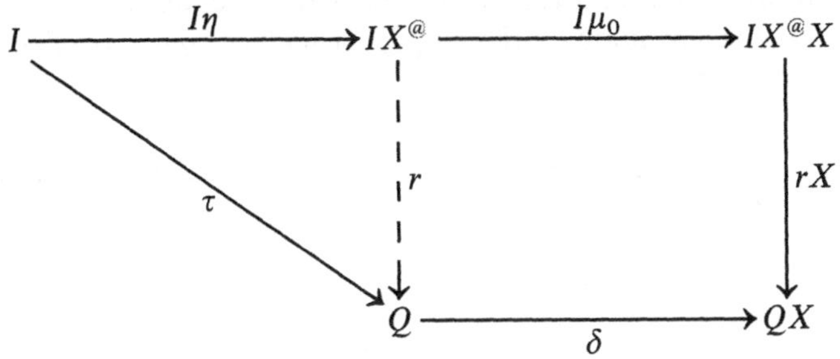

(2.5)

reachability map $r:(IX^{@}, I\mu_0) \longrightarrow (Q, \delta)$. The *response map* is then defined to be the $\mathscr{K}$-morphism $r.\beta:IX^{@} \longrightarrow Y$. Accordingly, $IX^{@}$ is called the *object of inputs*.

X is an *output process* if $U:\text{Dyn}(X) \longrightarrow \mathscr{K}$ has a right adjoint. In this case, the cofree dynamics over A with respect to U will be denoted $(AX_{@}, AL; A\Lambda)$. The unique dynamorphic coextension of the output morphism, as shown below, is the *observability map* $\sigma:(Q, \delta) \longrightarrow (YX_{@}, YL)$.

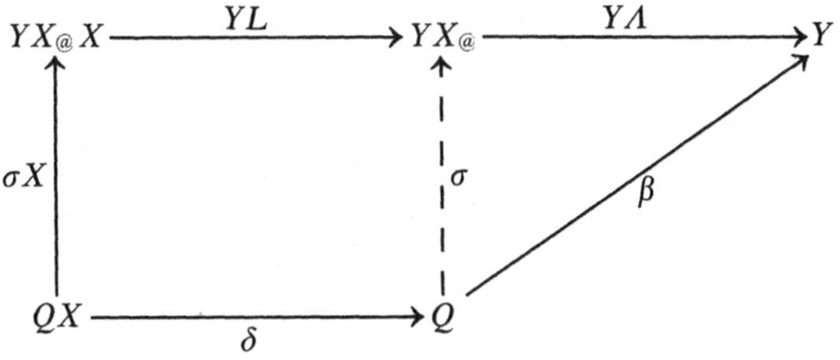

(2.6)

$X = - \times X_0$ is both an input- and an output process in **Set**. $AX^{@} = A \times X_0^*$, $A\mu_0:A \times X_0^* \times X_0 \longrightarrow A \times X_0^*$ sends (a, w, x) to (a, wx) and $A\eta:A \longrightarrow A \times X_0^*$ sends a to (a, Λ). We observe that the two diagrams

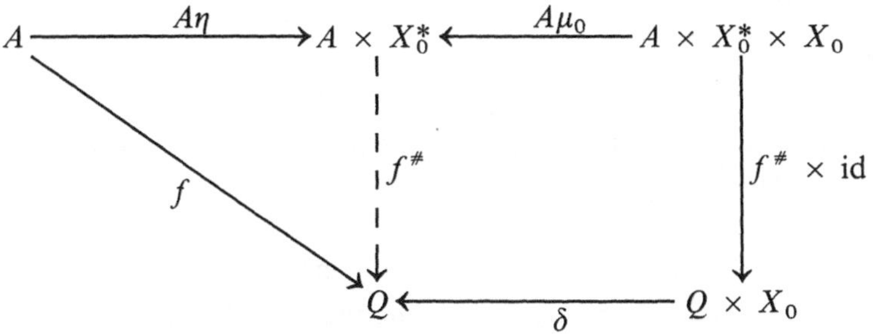

amount to the algebraic recursion

$$(a, \Lambda)f^{\#} = af$$
$$(a, wx)f^{\#} = ((a, wf^{\#}), x)\delta$$

$AX_@ = A^{(X_0^*)}$, $AL\colon AX_@ \times X_0 \longrightarrow AX_@$ sends (g, x) to $L_x.g$ where L_x is the left translation endomorphism of X_0^* which sends w to xw, and $A\Lambda$: $AX_@ \longrightarrow A$ is "evaluate on the empty string," i.e., $g \mapsto \Lambda g$. The diagrams expressing the couniversal property:

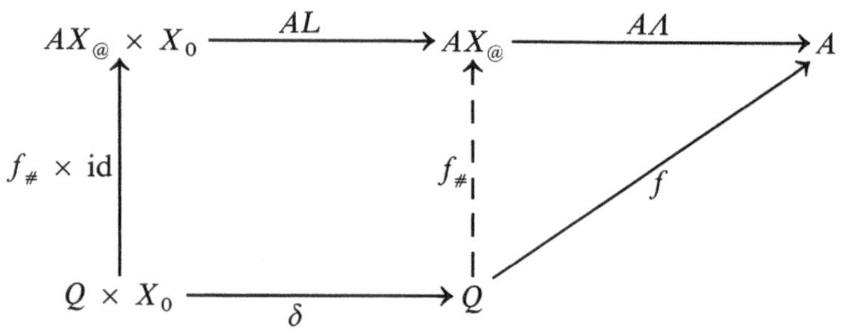

amount to the algebraic recursion

$$\langle \Lambda, qf_{\#}\rangle = \langle q, f\rangle$$
$$\langle xw, qf_{\#}\rangle = \langle w, \langle (q, x)\delta, f_{\#}\rangle\rangle$$

which proceeds simultaneously on all elements of Q. It is easy to check that the reachability and observability maps of 2.1 are given by 2.5 and 2.6.

2.7 Decomposable Systems. Let $\mathscr{K}$ be any category. If every object A has a countably infinite copower $A^{\S}$ and countably infinite power $A_{\S}$ then $X = \mathrm{id}_{\mathscr{K}}$ is both an input- and an output process in $\mathscr{K}$, with $AX^@ = A^{\S}$ and $AX_@ = A_{\S}$. Here $A\eta = \mathrm{in}_0$, $A\Lambda = \mathrm{pr}_0$, and μ_0 and L are defined by

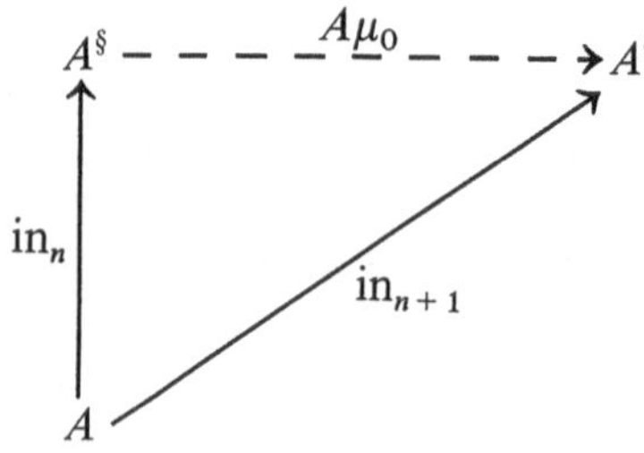

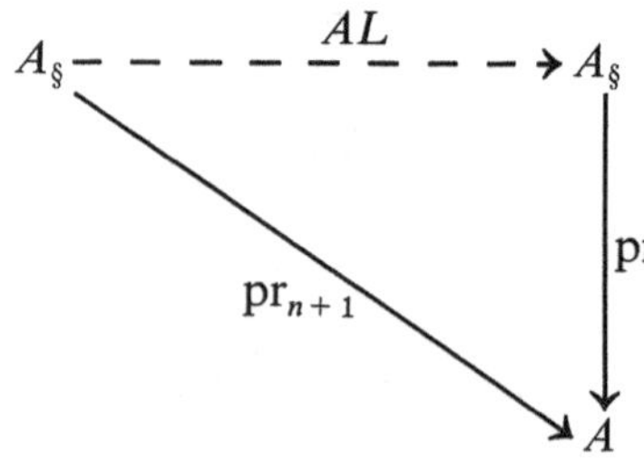

The universal and couniversal properties may be read from the following diagram in which $f^{\#}$ and $g_{\#}$ are defined using ordinary simple recursion (see 1.1.20+):

$$\mathrm{in}_0.f^{\#} = f \qquad g_{\#}.\mathrm{pr}_0 = g$$
$$\mathrm{in}_{n+1}.f^{\#} = \mathrm{in}_n.f^{\#}.\delta \qquad g_{\#}.\mathrm{pr}_{n+1} = \delta.g_{\#}.\mathrm{pr}_n$$

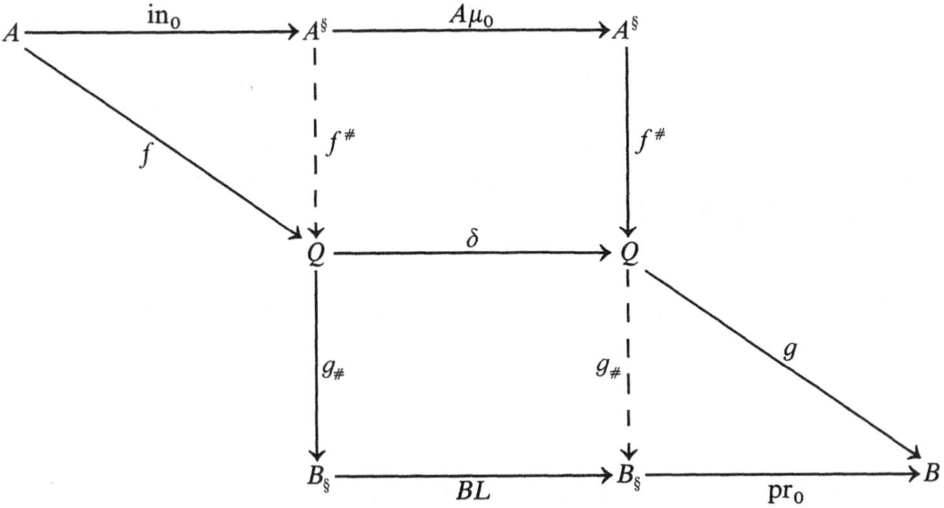

As explained in the notes, the term "decomposable" is for historical reasons.

2.8 Controlling a Physical System. Due largely to the efforts of *R.* Kalman (see [Kalman, Falb and Arbib '69]) a major portion of the qualitative study of controlled dynamical systems is subsumed in the study of decomposable systems in the category of real vector spaces and linear maps. For an excellent elementary introduction see [Padulo and Arbib '74]. Some of these ideas are illustrated in the following example. Consider a spring with constant k and mass m, influenced by gravity and the motions of the piston to which it is attached. An input to this system is a function v defined

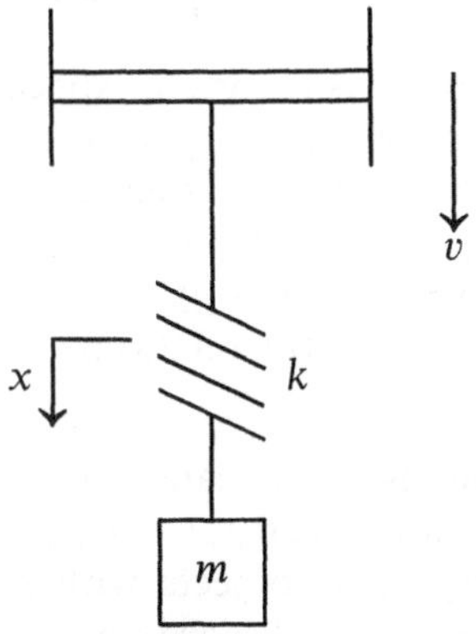

on the closed interval $[0, t^*]$ (subject to natural limitations such as "piecewise continuous" and "$|v(t)| < A$ for some maximum amplitude A") whose interpretation is that $v(t)$ is the force acting on the piston at time t. The output of the system is the displacement, x, of the spring from its equilibrium position at time t^*. A possible control problem associated with this system is to bring the system to rest in minimum time given the initial state of the system.

The differential equation expressing Newton's second law for this system is

$$m\ddot{x} = mg - kx + v$$

where $(\dot{\ })$ denotes "derivative with respect to time" and g is the acceleration of the Earth's gravitational field. Thus

$$\ddot{x} = g - \omega^2 x + (1/m)v$$

where $\omega^2 = k/m$. Set $q = \begin{pmatrix} x \\ \dot{x} \end{pmatrix}$, the *state-vector* of the system. Then

$$\dot{q} = \begin{pmatrix} \dot{x} \\ \ddot{x} \end{pmatrix} = Aq + Bu, \quad A = \begin{pmatrix} 0 & 1 \\ -\omega^2 & 0 \end{pmatrix}, \quad B = \begin{pmatrix} 0 \\ 1 \end{pmatrix}, \quad u = g + (1/m)v$$

We next assume that the system can be *discretized* in the sense that we assume there exists a "quantum" of time, Δt, such that

$$q(t + \Delta t) = q(t) + \dot{q}(t)\,\Delta t$$

is approximately true at every time t. Such Δt is often called the *cycle time* since, in practice, it represents the internal clock time of computers monitoring the system. We have

$$q(t + \Delta t) = \begin{pmatrix} 1 & 1 \\ -\omega^2 & 1 \end{pmatrix} q(t) + \begin{pmatrix} 0 \\ 1 \end{pmatrix} u(t)$$

The system has been reduced to the decomposable system $(Q, \delta, I, \tau, Y, \beta)$ in the category of vector spaces and linear maps where $Q = \mathbf{R}^2$, $I = \mathbf{R} = Y$ and

$$\delta = \begin{pmatrix} 1 & 1 \\ -\omega^2 & 1 \end{pmatrix}, \qquad \tau = \begin{pmatrix} 0 \\ 1 \end{pmatrix} \qquad \text{and} \qquad \beta = (1, 0)$$

The object of inputs, $IX^{@} = \mathbf{R}^{\S}$ is the vector space of all sequences $(u_n : n = 0, 1, 2, \ldots)$ of real numbers for which $u_n = 0$ for all but finitely many n. Each such input (u_n) is regarded as a step approximation of a function $u(t)$—with u_n corresponding to the input $u(t^* - n\,\Delta t)$—which in turn determines the input $v(t)$ by $v(t) = m(u(t) - g)$.

We now turn our attention to the relationship between input processes and free algebraic theories.

2.9 Free Algebraic Theories. The category $\mathscr{M}$ of monoids and monoid homomorphisms in the monoidal category $(\mathscr{K}, \otimes, I, a, b, c)$, as in 3.2.3 and 3.2.4, is (clearly) a category of $\mathscr{K}$-objects with structure with underlying $\mathscr{K}$-object functor $U : \mathscr{M} \longrightarrow \mathscr{K}$. Accordingly, a *free monoid over* K is, simply, a free $\mathscr{M}$-object over K with respect to U. As discussed in 3.2.6, free algebraic theories are defined as a special case. Hence, if $X : \mathscr{K} \longrightarrow \mathscr{K}$ is an endofunctor in the arbitrary category $\mathscr{K}$, a free algebraic theory over X is an algebraic theory $\mathbf{X}^{@} = (X^{@}, \eta, \mu)$ in $\mathscr{K}$ together with a natural transformation $\rho : X \longrightarrow X^{@}$ possessing the universal property that for each algebraic theory $\mathbf{T} = (T, \eta_{\mathbf{T}}, \mu_{\mathbf{T}})$ in $\mathscr{K}$ and each natural transformation

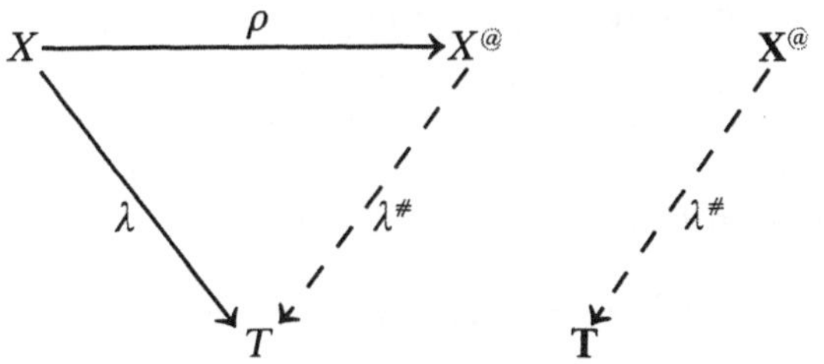

(2.10)

$\lambda: X \longrightarrow T$ there exists a unique morphism of theories $\lambda^{\#}: \mathbf{X}^{@} \longrightarrow \mathbf{T}$ such that the horizontal composition $\rho.\lambda^{\#}$ is just λ.

Before stating the next theorem it is helpful to establish some notations. Let X be an input process in $\mathscr{K}$. For each (Q, δ) in $\mathrm{Dyn}(X)$, the *run map of* (Q, δ) is defined to be the unique dynamorphic extension $\delta^{@}$ of id_Q:

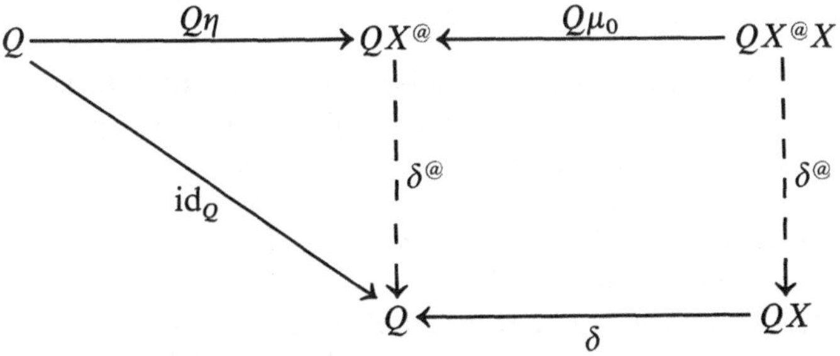

(2.11)

This includes the $\delta^{@}$ of 1.1.18 by 2.16 below. In the context of 2.1 for example, $\delta^{@}$ is conventionally written δ^*, and is the function $\delta^*: Q \times X^* \longrightarrow Q$ which sends $(q, x_1 \cdots x_n)$ to the state that would be reached if the automaton were started in state q and inputs $x_1, \ldots, x_n$ were applied in that order.

From the definition of ε in the proof of 2.2.18, we have at once that $(Q, \delta)\varepsilon = \delta^{@}: (QX^{@}, Q\mu_0) \longrightarrow (Q, \delta)$. It then follows at once from the constructions in 2.2.20 that

2.12 Lemma. *If X is an input process in $\mathscr{K}$ then the algebraic theory in $\mathscr{K}$ induced by the adjointness*

$$\mathrm{Dyn}(X) \rightleftarrows \mathscr{K}$$

is $\mathbf{X}^{@} = (X^{@}, \eta, \mu)$ where $\mu = \mu_0^{@}$. The semantics comparison functor

$$\Phi: \mathrm{Dyn}(X) \longrightarrow \mathscr{K}^{\mathbf{X}^{@}}$$

is given by $(Q, \delta)\Phi = (Q, \delta^{@})$. □

We then have:

2.13 Theorem. *Let X be an input process in $\mathscr{K}$ with induced algebraic theory $\mathbf{X}^{@}$ as in 2.12. Define $A\rho: AX \longrightarrow AX^{@}$ by*

$$A\rho = AX \xrightarrow{A\eta X} AX^{@}X \xrightarrow{A\mu_0} AX^{@} \tag{2.14}$$

Then $U: \mathrm{Dyn}(X) \longrightarrow \mathscr{K}$ is algebraic, the inverse $\Phi^{-1}: \mathscr{K}^{\mathbf{X}^{@}} \longrightarrow \mathrm{Dyn}(X)$ to the semantics comparison functor being given by $(Q, \xi)\Phi^{-1} = (Q, Q\rho.\xi)$. Further, $(\mathbf{X}^{@}, \rho)$ is the free algebraic theory over X.

Proof. By the Beck theorem of 3.1.9, to prove that Dyn(X) is algebraic over $\mathscr{K}$ we need only establish that U creates coequalizers of U-contractible pairs; here, the details are so similar to the proof of 3.1.9 "1 implies 2" that we will omit them. By the proof of 3.1.9 "3 implies 1," the semantics comparison functor is an isomorphism. The diagram

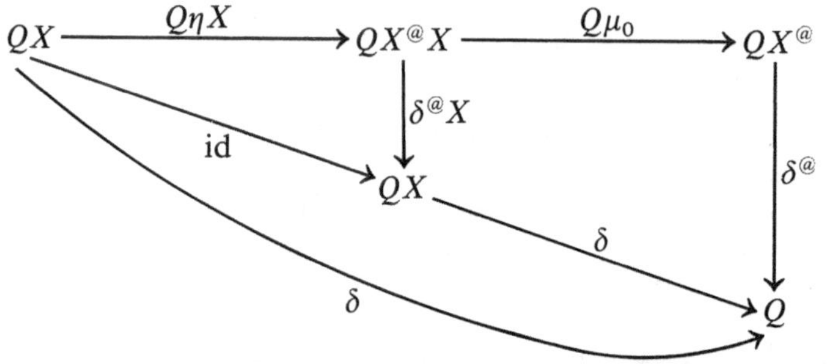

proves that $Q\rho.\delta^{@} = \delta$ so that the inverse to Φ is as advertised (functoriality is clear from the discussion below). To show that $\mathbf{X}^{@}$ is the free algebraic theory over X we must first observe that ρ is natural; but ηX is natural and μ_0 is natural since $fX^{@}$ is a dynamorphism by definition (see 2.2.18 "(i) implies (ii)"). Now consider 2.10 with $\mathbf{T}$ and λ arbitrary. Define V: $\mathscr{K}^{\mathbf{T}} \longrightarrow \text{Dyn}(X)$ by $(Q, \xi)V = (Q, Q\lambda.\xi)$. V is a well-defined homomorphism in Struct($\mathscr{K}$) because λ is natural. Let $\lambda^{\#}: \mathbf{X}^{@} \longrightarrow \mathbf{T}$ be the theory map corresponding to $V.\Phi: \mathscr{K}^{\mathbf{T}} \longrightarrow \mathscr{K}^{\mathbf{X}^{@}}$ as in 3.2.9; then we have

$$A\lambda^{\#} = AX^{@} \xrightarrow{A\eta_{\mathbf{T}}X^{@}} ATX^{@} \xrightarrow{(AT\lambda.A\mu_{\mathbf{T}})^{@}} AT$$

The diagram

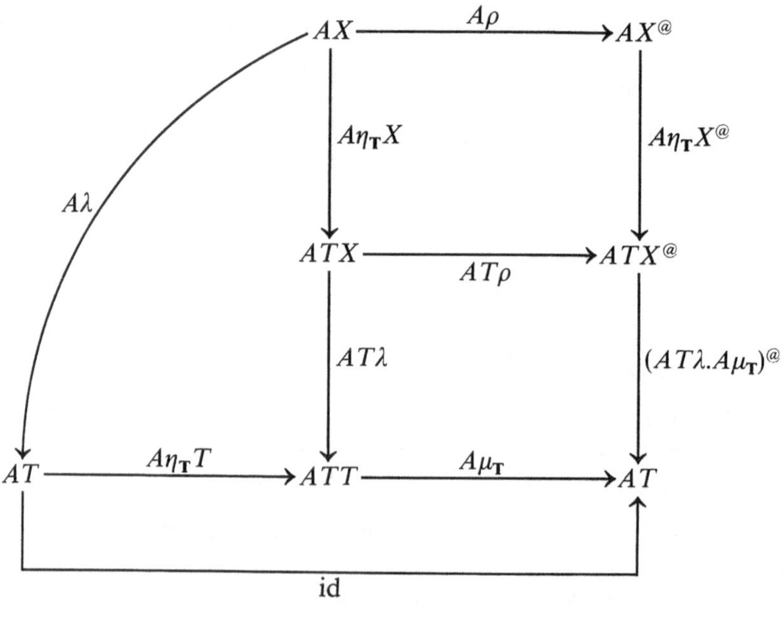

proves that $\rho.\lambda^{\#} = \lambda$. If also $\rho.\psi = \lambda$, the corresponding homomorphism $V':\mathscr{K}^{\mathbf{T}} \longrightarrow \mathscr{K}^{\mathbf{X}^{@}}$ is such that $V.\Phi^{-1}$ is the passage

$$(Q, \xi) \longmapsto (Q, QX \xrightarrow{Q\rho} QX^{@} \xrightarrow{Q\psi} QT \xrightarrow{\xi} Q)$$

which proves, since $Q\rho.Q\psi = Q\lambda$, that $V = V'$ and hence $\lambda^{\#} = \psi$. □

The converse is often true:

2.15 Theorem. *Let $\mathscr{K}$ be a locally small well-powered category which has small limits. Then if $X:\mathscr{K} \longrightarrow \mathscr{K}$ has a free algebraic theory, X is an input process.*

Proof. Let $(X^{@}, \eta, \mu; \rho)$ be the free algebraic theory over X and let A be an object in $\mathscr{K}$. Since $U:\mathrm{Dyn}(X) \longrightarrow \mathscr{K}$ clearly creates limits it suffices to show—by the general adjoint functor theorem—that $(AX^{@}, AX^{@}\rho.A\mu; A\eta)$ is a one-element solution set for A (and is then in fact the free dynamics over A by the proof of 2.13). Let (Q, δ) be an X-dynamics and let $f:A \longrightarrow Q$. Let $\mathscr{P}$ be the full subcategory of $\mathrm{Dyn}(X)$ of all X-dynamics admitting a dynamorphism, U of which is mono, into a power of copies of (Q, δ). Essentially the same arguments as in 3.4.24 make it clear that $\mathscr{P}$ is closed under limits. By construction, (Q, δ) is a cogenerator in $\mathscr{P}$. By the special adjoint functor theorem (it is not necessary to first prove that $\mathscr{P}$-monomorphisms are monomorphisms in $\mathscr{K}$ since the crucial monomorphism $i:S \longrightarrow P$ of the proof of 2.2.29 is created as a monomorphism in $\mathscr{K}$) $\mathscr{P} \rightarrow \mathscr{K}$ has a left adjoint. Let $\mathbf{T}$ be the corresponding theory in $\mathscr{K}$ and let $\xi:QT \longrightarrow Q$ be the corresponding structure map, $\xi = (\mathrm{id}_Q)^{\#}$, so that

$$\begin{array}{ccccc} & & QTX & \xrightarrow{\quad \xi X \quad} & QX \\ & & \Big\downarrow{\scriptstyle \delta_Q} & & \Big\downarrow{\scriptstyle \delta} \\ Q & \xrightarrow{\quad Q\eta_{\mathbf{T}} \quad} & QT & \xrightarrow{\quad \xi \quad} & Q \\ \end{array} \qquad Q \xrightarrow{\quad \mathrm{id} \quad} Q$$

where δ_Q is the X-dynamical structure of the free $\mathscr{P}$-object over Q. Since δ_Q describes a natural transformation $TX \longrightarrow T$, so too does $\lambda:X \longrightarrow T$ defined by

$$Q\lambda = QX \xrightarrow{A\eta_{\mathbf{T}}X} QTX \xrightarrow{\delta_Q} QT$$

There exists a theory map $\bar{\lambda}:\mathbf{X}^{@} \longrightarrow \mathbf{T}$ such that $\rho.\bar{\lambda} = \lambda$. Thus, by 3.2.9, $(Q, Q\bar{\lambda}.\xi)$ is an $\mathbf{X}^{@}$-algebra and there exists a unique $\mathbf{X}^{@}$-homomorphism g extending f as shown below. But then $Q\rho.Q\lambda.\xi = Q\lambda.\xi = Q\eta_{\mathbf{T}}X.\delta_Q.\xi = (Q\eta_{\mathbf{T}}.\xi)X.\delta = \delta$ so that g is an X-dynamorphic extension of f as desired. □

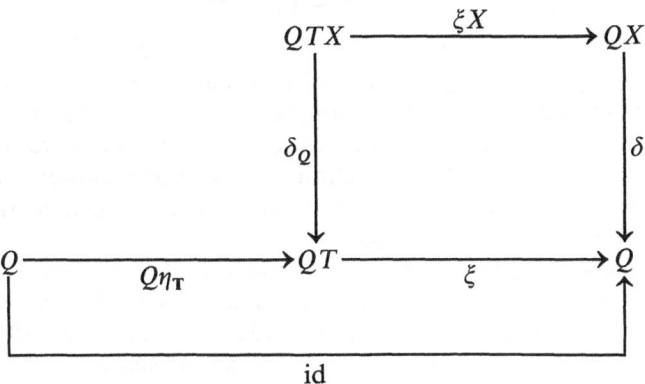

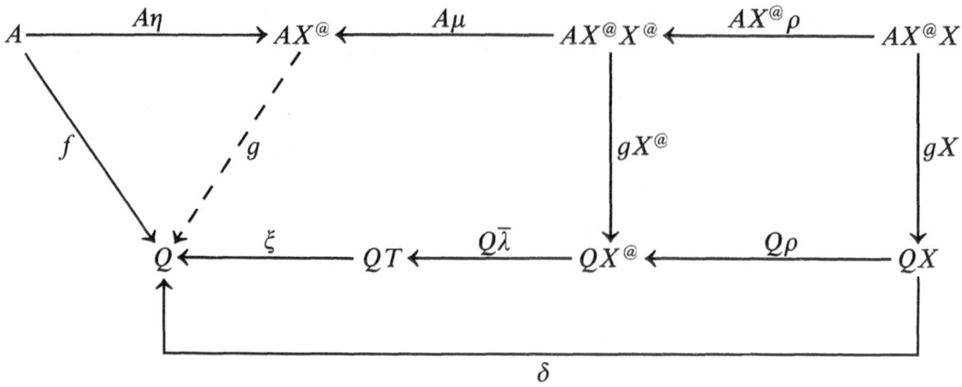

2.16 Algebra Automata. Let Ω be an operator domain as in 1.5.34 and define X_Ω:**Set** $\longrightarrow$ **Set** by

$$QX_\Omega = \coprod_{\omega \in \Omega_n} Q^n$$

Then Dyn(X_Ω) is, essentially, the category of Ω-algebras and Ω-homomorphisms. By 3.1.27, if Ω is bounded then X_Ω is an input process. Unless Ω has only unary operations—in which case $X_\Omega = - \times \Omega_1$ recaptures 2.1—X_Ω is not an output process (see exercise 1).

For finitary Ω, X_Ω-automata are interpreted as *tree processors* in computer science (see [Bobrow and Arbib '74, section 3–4]). We hint at the reason. Consider the arithmetic expression

$$p = \sqrt{x^2 + (x - 5)^2}$$

Let $\Omega_0 = \{5\}$, $\Omega_1 = \{(\;)^2, \sqrt{\;}\}$ and $\Omega_2 = \{+, -\}$. Set $Q = \mathbf{R}$, $I = \{x\}$. For each real number $\tau: I \longrightarrow Q$, the reachability map $r: I\mathbf{X}_\Omega^@ \longrightarrow Q$ evaluates expressions in one variable. For example, $p(\tau) = \langle p, r \rangle$. More realistically, Q should be the finite set of all internal computer bit configurations used to code real numbers and $\beta: Q \longrightarrow Y$ should reflect internal-to-external coding.

We conclude this section with a minimal realization theorem, pausing to define morphisms of automata:

2.17 Definition. *Let $X:\mathcal{K} \longrightarrow \mathcal{K}$ be a process and let $M = (Q, \delta, I, \tau, Y, \beta)$, $M' = (Q', \delta', I, \tau', Y, \beta')$ be two X-automata with the same initial state and output objects. A simulation $\psi: M \longrightarrow M'$ from M to M' is a dynamorphism $\psi:(Q, \delta) \longrightarrow (Q', \delta')$ which commutes with input and output*:

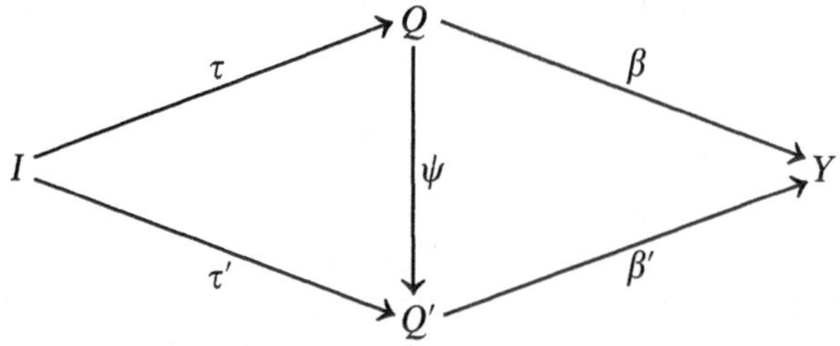

This defines a category of $\mathscr{K}$-objects with structure.

If (E, M) *is an image factorization system in* $\mathscr{K}$ *then, if* X *is an input process,* M *is reachable if* $r: IX^{@} \longrightarrow Q$ *is in* E; *if* X *is an output process,* M *is observable if* $\sigma: Q \longrightarrow YX_{@}$ *is in* M.

2.18 Minimal Realization Theorem. *Let* (E, M) *be an image factorization system in* $\mathscr{K}$, *let* X *be an input process and an output process in* $\mathscr{K}$, *and assume that at least one of the following conditions holds:*

1. X *preserves* E;
2. $X^{@}$ *preserves* E;
3. E *is the class of all coequalizers; or*
4. E *is the class of all epimorphisms.*

Then for every response $f: IX^{@} \longrightarrow Y$ *there exists a reachable and observable realization* $M_f = (Q_f, \delta_f, I, \tau_f, Y, \beta_f)$ *of* f. *Any such* M_f *is a terminal object in the category of reachable realizations of* f *and simulations and any such* M_f *is an initial object in the category of observable realizations of* f *and simulations; thus* M_f *is unique up to isomorphism.*

Proof. If $\psi: (Q, \delta) \longrightarrow (Q', \delta')$ is a dynamorphism and if $\psi = e.m$ is an E-M factorization of ψ, then there exists a unique X-dynamical structure making e and m dynamorphisms. This is clear from the diagram

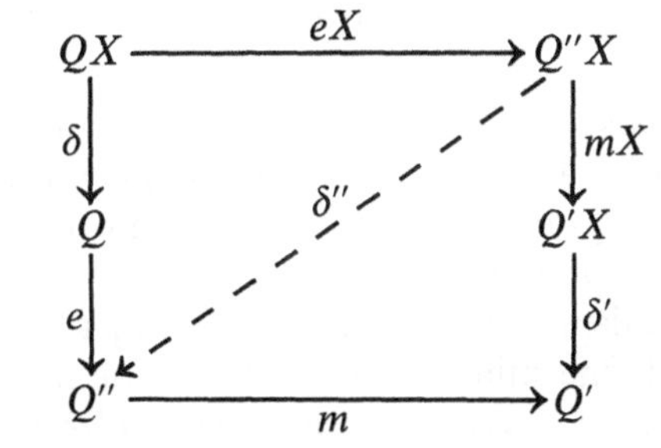

if X preserves E and, similarly, follows at once from 2.13 and the proof of 3.4.17 if $X^{@}$ preserves E (i.e., the structure map $\xi'': Q''X^{@} \longrightarrow Q''$ induced by diagonal fill-in satisfies the algebra axioms because $\xi''X^{@}$ and $\xi''X^{@}X^{@}$ are epi). The reader may easily formalize the argument

$$\begin{array}{cr} \underline{AX^{@} \longrightarrow B} & \qquad \mathscr{K}\text{-morphism} \\ \underline{AX^{@} \longrightarrow BX_{@}} & \qquad \text{dynamorphism} \\ A \longrightarrow BX_{@} & \qquad \mathscr{K}\text{-morphism} \end{array}$$

to prove that $X^{@}$ has $X_{@}$ as a right adjoint; thus $X^{@}$ preserves E if E is as in (3) or in (4).

Given $f: IX^{@} \longrightarrow Y$, let $f_{\#}: (IX^{@}, I\mu_0) \longrightarrow (YX_{@}, YL)$ be the unique dynamorphic coextension of f. Define Q_f, r_f, σ_f as the E-M factorization of $f_{\#}$ as shown below. Then, as we have just observed, there exists unique $\delta_f: Q_f X \longrightarrow Q_f$ admitting r_f and σ_f as dynamorphisms.

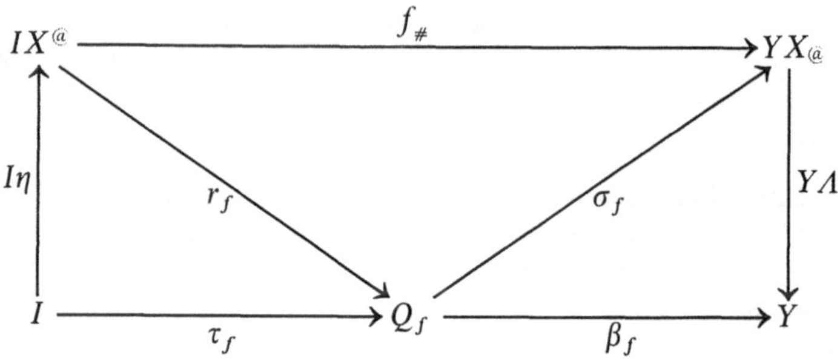

Then $r_f = (\tau_f)^\#$ if $\tau_f = I\eta.r_f$, $\sigma_f = (\beta_f)_\#$ if $\beta_f = \sigma_f.Y\Lambda$ and $M_f = (Q_f, \delta_f, I, \tau_f, Y, \beta_f)$ is a reachable and observable realization of f. Now suppose that $M = (Q, \delta, I, \tau, Y, \beta)$ is another reachable realization of f. By

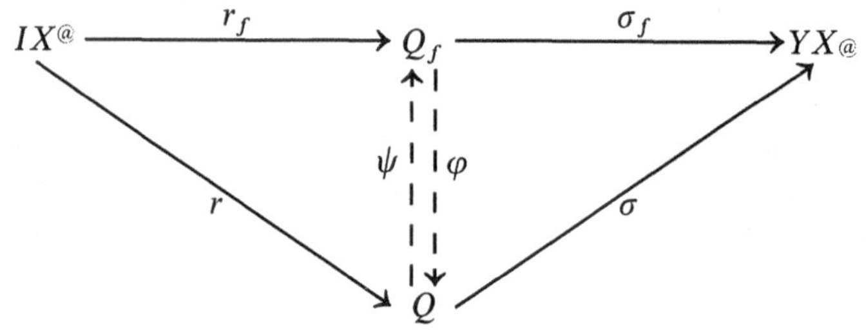

diagonal fill-in there exists unique $\psi: Q \longrightarrow Q_f$ with $r.\psi = r_f$ and $\psi.\sigma_f = \sigma$. ψ is a dynamorphism because rX or $rX^@$ is epi and r and $r.\psi$ are dynamorphisms, and ψ is a simulation. Similarly, if M is an observable realization of f then there exists unique $\varphi: Q_f \longrightarrow Q$ with $r_f.\varphi = r$ and $\varphi.\sigma = \sigma_f$ and such φ is a dynamorphism because either $r_f X$ or $r_f X^@$ is epi. □

It is possible to define minimal realizations for input processes which are not output, namely as terminal reachable realizations. Theorem 2.18 shows that these coincide with reachable and observable realizations for input processes which are output as well. It is possible to show that, for $X = X_\Omega$ as in 2.16, every response $f: IX^@ \longrightarrow Y$ has a minimal realization if Ω is finitary, but not so if Ω has infinitary operation labels (see exercises 10, 11). This shows that the existence of minimal realizations is not an automatic consequence of small limits and colimits and suggests it is a "finitary" condition.

Notes for Section 2

Categorical automata theory is a new field of research interest (see [Manes '74-A] and the bibliography there). The approach to realization theory discussed in this section is due to M. A. Arbib and the author (see [Arbib and Manes, '74-B, '74-C] and their sequels).

In the literature of algebraic linear system theory it is conventional to view the linear system $\delta: Q \longrightarrow Q$, $\tau: I \longrightarrow Q$ as a special case of a $(- \times I)$-

automaton in **Set** via the injective passage

$$(\delta, \tau) \longmapsto \begin{pmatrix} \delta \\ \tau \end{pmatrix} : Q \times I \longrightarrow Q$$

which capitalizes on the fact that binary products are also coproducts in the category of vector spaces and linear maps. In this sense, a linear system is an automaton $\bar{\delta}: Q \times I \longrightarrow Q$ which is "decomposable": $(q, x)\bar{\delta} = q\delta + x\tau$.

Theorem 2.15 is adapted from [Barr 70-A, 5.10].

Exercises for Section 2

1. (a) Prove that if X_Ω is an output process in **Set** then Ω_n is empty whenever $n \neq 1$. [Hint: by the Beck theorem, $\text{Dyn}(X_\Omega) \longrightarrow$ **Set** is co-algebraic, so creates colimits.]
   (b) ([Arbib and Manes '75, 5.8].) Generalizing (a), prove that if X is an input- and an output process in **Set** then X is naturally equivalent to $- \times 1X$. [Hint: the desired natural equivalence $\Gamma: - \times 1X \longrightarrow X$ is defined by $(q, x)Q\Gamma = \langle x, qX \rangle$; since $X^@$ preserves coproducts, we may assume $X^@ = - \times 1X^@$; $\Gamma.\rho: - \times 1X \longrightarrow - \times 1X^@$ is natural, so corresponds to a function $1X \longrightarrow 1X^@$; prove directly that $Q\Gamma$ is injective and surjective.]
2. ([Barr '70-A, 5.11, 5.12].) Let $X: \mathscr{K} \longrightarrow \mathscr{K}$ be an arbitrary endo-functor.
   (a) Prove that if (Q, δ) is an initial object in $\text{Dyn}(X)$ then $\delta: QX \longrightarrow Q$ is an isomorphism. [Hint: there exists unique $\psi: (Q, \delta) \longrightarrow (QX, \delta X)$ and $\delta.\psi = (\psi.\delta)X = \text{id}$.]
   (b) Let Q be such that $(QX^@, Q\mu_0; Q\eta)$ exists and $Q + R$ exists for all R. Prove that

   $$Q \xrightarrow{Q\eta} QX^@ \xleftarrow{Q\mu_0} QX^@X$$

   is a coproduct diagram in $\mathscr{K}$. [Hint: it suffices to show that $\bar{\delta} = (Q\eta, \mu_0): Q + QX^@X \longrightarrow QX^@$ is an isomorphism; define a new process $\bar{X}: \mathscr{K} \longrightarrow \mathscr{K}$ by $R\bar{X} = Q + RX$; then $(Q, \bar{\delta})$ is an initial object in $\text{Dyn}(\bar{X})$.]
3. This result has been long in the folklore. It was first published in an automata-theoretic context by [Goguen '72]. Let $\mathscr{K}$ have and let $X: \mathscr{K} \longrightarrow \mathscr{K}$ preserve countable coproducts. Prove that X is an input process. [Hint: $QX^@ = \coprod QX^n$ is the "classical" free monoid construction.]
4. [Adámek '74], [Trnková, Adámek, Koubek and Reiterman '75]; cf. [Barr 70-A], [Dubuc '73]. Let $\mathscr{K}$ be a small co-complete category and let M be a subclass of monomorphisms satisfying the following four conditions:

   Coproduct injections are in M.

   If f and g are in M *so is their coproduct $f + g$.*

Given an ordinal-indexed ascending chain in M, *the injections to the colimit are in* M *and the colimit-induced map owing to an upper bound whose components are in* M *is again in* M.

$\mathscr{K}$ *is* M *well-powered.*

For $Y:\mathscr{K} \longrightarrow \mathscr{K}$ any functor, any $f:I \longrightarrow IY$ may be transfinitely iterated through the ordinals:

$$A \xrightarrow{f_0=f} AY \xrightarrow{f_1=f_0Y} AY^2 \longrightarrow \cdots AY^\omega = \text{colim } AY^n$$

$$AY^\omega \xrightarrow{f_\omega} AY^{\omega+1} \text{ is colimit-induced} \cdots$$

Notice that if Y preserves M then each f is in M.

(a) Let $X:\mathscr{K} \longrightarrow \mathscr{K}$ preserve M and fix I in $\mathscr{K}$. Prove that the following three conditions are equivalent:
  (i) The free dynamics $(IX^@, I\mu_0)$ over I exists.
  (ii) If $Y:\mathscr{K} \longrightarrow \mathscr{K}$ is defined by $AY = I + AX$ and if $f:I \longrightarrow IY$ is the first injection then the iteration of f (as above) *stops*, i.e., f_α is an isomorphism for some α.
  (iii) There exists A in $\mathscr{K}$ with $A \cong I + AX$.
  [Hint: for (ii) implies (i) set $IX = IY^\alpha$, $I\mu_0 = in_2.f_\alpha^{-1}$; for (i) implies (iii) use exercise 2 (b); for (iii) implies (i) construct canonical morphisms $IY^\alpha \longrightarrow A$ in M and use well-poweredness.]

(b) As a corollary to (a), show that X: **Set** $\longrightarrow$ **Set** is an input process if and only if $\text{card}(\alpha X) \leqslant \alpha$ for arbitrarily large cardinals α.

(c) Show that the power-set functor is not an input process in **Set**.

(d) In **Set**, use (b) to show that a quotient functor of an input process is an input process. Generalize this result by using a Birkhoff subcategory argument.

(e) Construct an example of an input process in **Set** such that $X^@$ is unbounded. [Hint: define $AX = \{\psi : \psi$ is an injective function from a singular cardinal into $A\} \cup \{\infty\}$, $\langle \psi, fX \rangle = \psi.f$ if $\psi.f$ is injective, all other values of $fX = \infty$; $\text{card}(AX) = \text{card}(A)$ if $\text{card}(A)$ is an infinite regular cardinal.] This example justifies the need to have Y depend on I in part (a).

5. For arbitrary $X:\mathscr{K} \longrightarrow \mathscr{K}$, a *Peano algebra* is a diagram

$$Q \xrightarrow{\eta} R \xleftarrow{\delta} RX$$

satisfying (i) the diagram is a coproduct; (ii) whenever (Q', δ') is in $\text{Dyn}(X)$ and $\mathscr{K}(Q, Q')$ is nonempty then $\mathscr{K}(R, Q')$ is nonempty (this condition always holds if $\mathscr{K}$ = **Set**); (iii) Q generates (R, δ) in the sense that any two dynamorphisms from (R, δ) which agree preceded by η are equal. Let M be the class of all $\mathscr{K}$-morphisms which are the collective equalizer of a countable family of maps. Assume that $\mathscr{K}$ has and X preserves colimits of ascending chains $f_n : A_n \longrightarrow A_{n+1}$ with each f_n in M. Prove that each Peano algebra is free, i.e., prove that $R = QX^@$. [Hint: consider

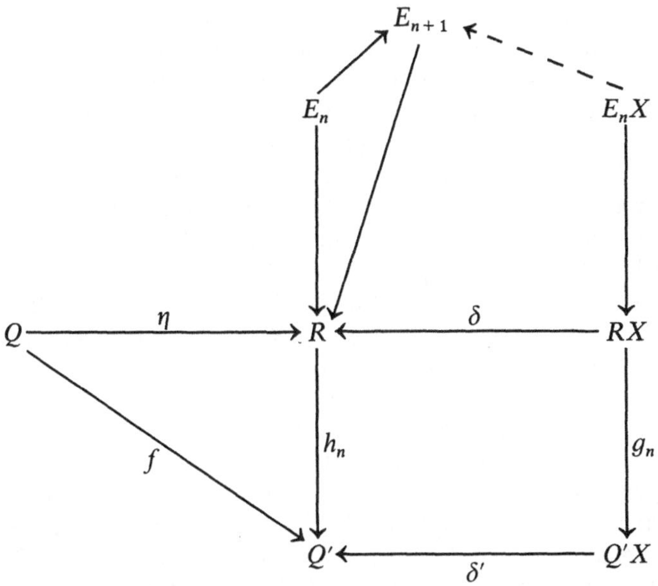

here, g_0 is arbitrary, h_n is defined by "$\eta.h_n = f$, $\delta.h_n = g_{n-1}.\delta'$" and $g_{n+1} = h_nX$; define $E_n = \text{eq}(h_m : m \geqslant n)$; then $E = \text{colim}\, E_n$ is a dynamics through which f factors because X preserves the colimit; $E = R$ because Q generates R.]

It is standard to say that the Ω-algebra (R, δ) is a *Peano algebra* if (i) whenever $\omega \in \Omega_n$, $\omega' \in \Omega_{n'}$, $f \in Q^n$, $f' \in Q^{n'}$ are such that $f\delta_\omega = f'\delta_{\omega'}$, then $n = n'$, $\omega = \omega'$, and $f = f'$ and (ii) the set Q of elements not of form $f\delta_\omega$ generates (R, δ). It is clear that this concept coincides with the Peano algebras relative to X_Ω in the sense of exercise 5. It has been noted by [Lowig '52], [Słomiński '55], [Diener '66], and [Felscher '72] that Peano Ω-algebras are free; this can be proved for finitary Ω as in exercise 5. It is interesting to ask which input processes not of the form X_Ω are such that "Peano implies free" is true, and to speculate upon the possibility that such new Peano algebras will play a role in the syntax of universal algebra.

6. X: **Set** $\longrightarrow$ **Set** is *finitary* if, for all A, every element of AX has finite support, i.e., is in the image of fX for some $f: n \longrightarrow A$ with n finite.
   (a) Show that X is finitary if and only if X is a quotient functor of X_Ω for some finitary operator domain Ω. [Hint: use $\Omega_n = nX$.] Conclude from exercise 4 (d) that a finitary functor is an input process.
   (b) For $n > 1$, show that $AX_n = \{S \subset A : S$ is nonempty and has at most n elements$\}$ is finitary and that Dyn(X) is the category of sets equipped with a symmetric n-ary operation.

   Note that $AX_2^@$ can be defined "syntactically" by the scheme

$$A \subset AX_2^@$$
$$\text{If } p, q \in AX_2^@ \text{ then } \{p, q\}\omega \in AX_2^@$$

where $\{p, q\} = \{q, p\}$ is literally a doubleton. It is interesting to ask for which finitary X such a syntactic construction of $X^{@}$ is possible.

(c) If X is finitary, show that the Adámek algorithm for $AX^{@}$ [exercise 4(a) (ii)] stops at $\aleph_0$ for all A. Study this algorithm in detail for $- \times X_0, (-)^2$ and X_2 as in (b).

7. Let $f:X^* \longrightarrow Y$ be a function. Using the proof of 2.18, prove that the minimal realization of f (with respect to $- \times X:\mathbf{Set} \longrightarrow \mathbf{Set}, I = 1$) is constructed as the subdynamics $Q_f = \{L_w f : w \in X^*\} \subset Y^{X^*}$ with initial state f and output pr_Λ, where $L_w:X^* \longrightarrow X^*$ is the map $w' \longmapsto ww'$. Derive M_f for f as in 2.2 using this approach.

8. Let $X = \{a\}$, $Y = \{0, 1\}$, $f = \chi_L:X^* \longrightarrow Y$ where $L = \{a^m : m$ is prime$\} \subset X^*$. Prove that Q_f is infinite (and hence that all realizations of f have infinitely many states).

9. Let $X = \{$dime, choice 1, choice 2$\} = \{d, 1, 2\}$, $Y = \{$candy bar, peanuts, nothing$\} = \{c, p, n\}$, and let $f:X^* \longrightarrow Y$ be the vending machine response $(d1X^*)f = c$, $(d2X^*)f = p$, otherwise $f = n$. Use exercise 7 to construct the state graph of M_f. [Hint: there are five states.]

10. (See [Anderson, Arbib and Manes '74, 3.12] for a more abstract treatment.) Let Ω be finitary and let I, Y be arbitrary sets. Prove that any $f:IX_\Omega^@ \longrightarrow Y$ has a minimal realization, that is a terminal reachable realization as discussed in 2.18+. [Hint: define $E_f = \{(p, q): p\tau f = q\tau f$ for all translations $\tau\} \subset IX_\Omega^@ \times IX_\Omega^@$; here, an *elementary translation* is an endomorphism $u:IX_\Omega^@ \longrightarrow IX_\Omega^@$ for which there exists $n > 0$ and $\omega \in \Omega_n$ such that u is obtained from $(\mu_0)_\omega:(IX_\Omega^@)^n \longrightarrow IX_\Omega^@$ by fixing $n - 1$ of the arguments, and a *translation* is the identity or is a composition of elementary translations; use the fact that Ω is finitary to prove that E_f is a congruence; set $Q_f = IX_\Omega^@/E_f$; if M is another reachable realization with reachability map r, then if $pr = qr$ and if τ is an elementary translation, $p\tau f = pr\tau\beta = q\tau f$, so that $(p, q) \in E_f$.] For the case of $- \times X:\mathbf{Set} \longrightarrow \mathbf{Set}$, $I = 1$, show that $u:X^* \longrightarrow X^*$ is a translation if and only if $w'u = w'w$ for some w in X^*. Conclude that, for any $f:X^* \longrightarrow Y$, $E_f = \{(v, w): L_v f = L_w f\}$. This is known in automata theory as the *Nerode/Raney/Myhill equivalence of f*.

11. ([Anderson, Arbib and Manes '74, 3.15].)

(a) Let $X:\mathbf{Set} \longrightarrow \mathbf{Set}$ be an input process. Suppose given an ascending chain (R_n) of congruences on $IX^@$ such that the equivalence relation $R = \cup R_n$ is not a congruence. Prove that the canonical projection $f:IX^@ \longrightarrow Y = IX^@/R$ has no minimal realization. [Hint: if M_f exists, its universal property guarantees that each R_n is contained in the kernel pair of r_f; this implies that $R = \ker \text{pair } r_f$ is a congruence.]

(b) Let Ω have a single operation ω of countably-infinite arity and let $I = \{0, 1, 2 \ldots\}$. Prove that there exists $f:IX_\Omega^@ \longrightarrow Y$ with no minimal realization. [Hint: define $\psi:IX^@ \longrightarrow IX^@ = hX^@$ where $nh = \mathrm{Max}(n - 1, 0)$; define $R_n = \ker \text{pair } \psi^n$ and apply (a); to show that R is not a subalgebra, define $p_n = (012\ldots n0000\ldots)\omega$ and observe that $(p_0, p_n) \in R_n$ but that $((p_0p_0p_0\ldots)\omega, (p_0p_1p_2\ldots)\omega)$ is not in R.] Our construction requires the sort of infinitary syntax we avoided in section 1.5.

The question of when $f: IX^@ \longrightarrow Y$ has a minimal realization has been considered by [Adámek '74] and [Trnková '74, 75]. [Trnková '75, Proposition 3] improves (b) above:

> Suppose there exists an infinite cardinal $\alpha \leqslant \text{card}(IX^@)$ for which there exists p in αX such that p is not in the image of fX for any $f: A \longrightarrow \alpha$ with $\text{card}(A) < \alpha$. Then there exists a subset of $IX^@$ whose characteristic function has no minimal realization.

The hypotheses of Trnková's theorem hold in (b) with $\alpha = \aleph_0$; but her construction, even for this X, is difficult.

3. Nondeterminism

Algebraic theories are used to model "fuzzy theories." We use this approach to extend the ideas of the previous section to "nondeterministic automata theory," discovering enroute a rapprochement with Beck distributive laws.

Given a set Q and two elements q, q_0 of Q, there are many ways to measure the "degree of certainty" that $q = q_0$. We may know that q is in A for some subset A of Q; that is, "subset" is such a measure. Then again, for each r in Q we may consider the probability λ_r that $q = r$ with respect to which λ_{q_0} is the "certainty" that $q = q_0$. Without further ado, we have:

3.1 Definition. *Let $\mathscr{K}$ be a category. A fuzzy theory in $\mathscr{K}$ is an algebraic theory* $\mathbf{T} = (T, e, \circ, m)$ *in* $\mathscr{K}$. (Since other algebraic theories enter the picture later, we write e for η and m for μ.)

The heuristics motivating 3.1 are as follows. For each object Q, QT is the "object of degrees of $\mathbf{T}$-uncertainty over Q." We may view QT as a "cloud of fuzzy states, the values of $Qe: Q \longrightarrow QT$ representing the 'crisp' or 'pure' states." A morphism $\alpha: A \longrightarrow B = \alpha: A \longrightarrow BT$ is a "$\mathbf{T}$-fuzzy morphism from A to B." The fundamental assumption is that the structure of "uncertainty" lies in knowing how fuzzy morphisms compose which is, of course, the role of $\circ$. The interpretation of a $\mathbf{T}$-algebra (Q, ξ) is as a "decider"; that is, ξ is "a way of assigning a pure state to each fuzzy one in such a way that pure states are assigned themselves and consistent with the causality implied by the operation of fuzzy morphism composition."

The "subset" uncertainty measure mentioned above corresponds to $\mathbf{T}$ as in 1.3.5, where fuzzy morphisms are relations. Interesting subtheories are "finite subsets," "nonempty subsets," "$QT = Q + \{\varnothing\}$;" in the first three cases the algebras are suitably complete semilattices and in the last case the algebras are sets with base point. The "probability" uncertainty corresponds to the subtheory of the matrices theory of 1.3.6 given by QT = all $(\lambda_b : b \in B)$ with each $\lambda_b \geqslant 0$ and $\sum \lambda_b = 1$, that is, QT is the set of probability distributions on Q. Fuzzy morphisms are then row-stochastic matrices and the proof that 'probability distributions' forms a subtheory reduces to the well-known result that a composition of row-stochastic matrices is again row-stochastic. While every convex subset of a real vector space is an algebra over this theory there are other examples (see exercise 1).

3.2 Credibility Values. Let M be a monoid "of credibility values" and let $\mathbf{T_M}$ be the associated algebraic theory in **Set** as in 1.3. Given the fuzzy morphism $\alpha : A \longrightarrow B \times M$, if $a\alpha = (b, m)$ then "$a\alpha = b$ with credibility m." The monoid multiplication determines how credibilities combine. This example seems apt for M the unit interval under multiplication.

3.3 L-Relations ([Goguen '67, Section 6]). Let L be a *complete distributive lattice*, that is, a complete lattice in which for each x in L the map $x \wedge (-) : L \longrightarrow L$ preserves suprema. L induces a fuzzy theory $(T, e, \circ)$ as follows. $AT = L^A$. $Ae : A \longrightarrow AT$ sends a to its "characteristic function" $\chi_a : A \longrightarrow L$ defined by $a'\chi_a = 1$ (the greatest element of L) if $a' = a$, $a'\chi_a = 0$ (the least element of L) otherwise. Given $\alpha : A \longrightarrow L^B$, $\beta : B \longrightarrow L^C$ then $\alpha \circ \beta : A \longrightarrow L^C$ is given by

$$c(\alpha \circ \beta)_a = \mathrm{Sup}((c\beta_b) \wedge (b\alpha_a) : b \in B)$$

Complete distributivity is used to prove that the theory axioms are satisfied. Notice that sets and relations is recovered by taking L to be the two element lattice.

The remainder of this section is concerned with the extension of the realization theory of the previous section to "fuzzy systems." We begin with a motivating example.

3.4 Example. Let X have one element, let Y be the set of subsets of $\{a, b, c, d\}$, and let $f : X^* \longrightarrow Y$ be the response

$$f : a/b/c/d//ab/ac/ad$$

(i.e., f is a sequence, $f_0 = \{a\}$, $f_4 = f_7 = \{a, b\}$; // denotes a return point for cycling). Using 2.18 it is clear that the state graph of the minimal realization of f is

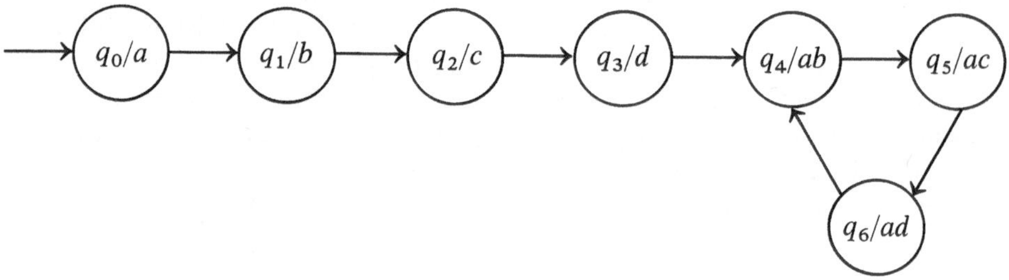

It is possible to build the following "multiple branching" realization of f which has fewer states:

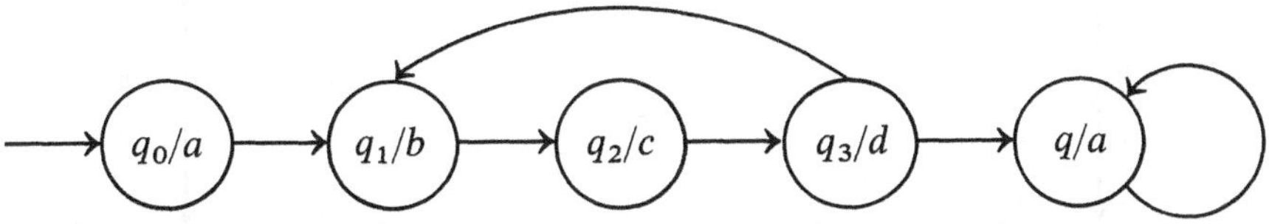

For the first four cycle times this system runs as before; but then the state q_3 splits into the "fuzzy state" $\{q_1, q\}$ emitting output $\{a, b\}$ thence transiting to the fuzzy state $\{q_2, q\}$ emitting output $\{a, c\}$ and so forth, so that the response is again f.

The dynamics of this multiple branching system takes the form $\delta: Q \times X \longrightarrow QT$ where QT is the set of nonempty subsets of Q. This motivates the study of dynamics of form $\delta: QX \longrightarrow QT$ where X is an input process in a category $\mathscr{K}$ and $\mathbf{T}$ is a fuzzy theory in $\mathscr{K}$. As pointed out in the notes, the concepts introduced in this abstract setting have elucidated certain aspects of realization theory even for the much-studied situation where $\delta: Q \times X \longrightarrow Q$ is a relation.

Let us begin the general discussion in the specific context of the multiple branching system above by pointing out a specific structural fact: if X is a set and $\mathbf{T}$ is the "nonempty subset" theory, there is a canonical map

$$\begin{aligned} QT \times X &\xrightarrow{Q\lambda} (Q \times X)T \\ (A, x) &\longmapsto A \times \{x\} \end{aligned} \tag{3.5}$$

Each multiple branching dynamics $\delta: Q \times X \longrightarrow QT$ induces the ordinary dynamics $\delta^{\bullet}: QT \times X \longrightarrow QT$ by

$$\delta^{\bullet} = QT \times X \xrightarrow{Q\lambda} (Q \times X)T \xrightarrow{\delta^{\#}} QT$$

Here, $\delta^{\#}$ is the unique union-preserving extension of δ. The reader should check that, for the multiple branching machine of 3.4, $\delta^{\bullet}$ is as expected, e.g. $(\{q_2, q_3\})\delta^{\bullet} = \{q_3, q_1, q\}$.

For the case of general X and $\mathbf{T}$ we have:

3.6 Definition. Let $X: \mathscr{K} \longrightarrow \mathscr{K}$ be a functor and let $\mathbf{T} = (T, e, m)$ be an algebraic theory in $\mathscr{K}$. *A distributive law of X over $\mathbf{T}$ is a natural transformation $\lambda: TX \longrightarrow XT$ satisfying*

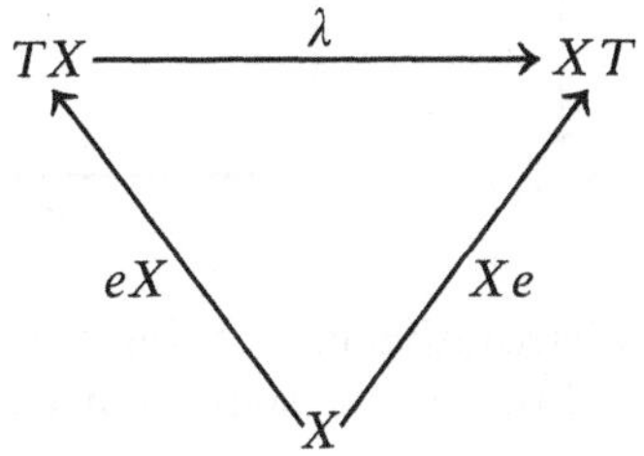

$$\begin{array}{ccccc}
TTX & \xrightarrow{T\lambda} & TXT & \xrightarrow{\lambda T} & XTT \\
{\scriptstyle mX}\downarrow & & & & \downarrow{\scriptstyle Xm} \\
TX & & \xrightarrow[\lambda]{\qquad\qquad\qquad} & & XT
\end{array}$$

The term "distributive law" comes from the distributive law of ring theory as will gradually be explained below.

The following proposition generalizes 3.5:

3.7 Proposition. *Let* X *be a set and let* $\mathbf{T} = (T, e, m)$ *be an algebraic theory in* **Set**. *Then*

$$\begin{array}{ccc}
QT \times X & \xrightarrow{Q\lambda} & (Q \times X)T \\
(p, x) & \longmapsto & \langle p, \mathrm{in}_x T\rangle
\end{array}$$

(*where* $\mathrm{in}_x : Q \longrightarrow Q \times X$ *sends* q *to* (q, x)) *is a distributive law of* $- \times X$ *over* **T**.

Proof. This is an easy consequence of the diagrams

$$\begin{array}{ccc}
Q & \xrightarrow{\mathrm{in}_x} & Q \times X \\
{\scriptstyle f}\downarrow & & \downarrow{\scriptstyle f \times \mathrm{id}} \\
Q' & \xrightarrow[\mathrm{in}_x]{} & Q' \times X
\end{array}
\qquad
\begin{array}{ccc}
Q & \xrightarrow{Qe} & QT \\
{\scriptstyle \mathrm{in}_x}\downarrow & & \downarrow{\scriptstyle \mathrm{in}_x T} \\
Q \times X & \xrightarrow[(Q \times X)e]{} & (Q \times X)T
\end{array}$$

$$\begin{array}{ccc}
QT & \xrightarrow{\mathrm{in}_x} & QT \times X \\
 & {\scriptstyle \mathrm{in}_x T}\searrow & \downarrow{\scriptstyle Q\lambda} \\
 & & (Q \times X)T
\end{array}
\qquad
\begin{array}{ccc}
QTT & \xrightarrow{\mathrm{in}_x TT} & (Q \times X)TT \\
{\scriptstyle Qm}\downarrow & & \downarrow{\scriptstyle (Q \times X)m} \\
QT & \xrightarrow[\mathrm{in}_x T]{} & (Q \times X)T
\end{array} \quad \square$$

3.8 Example. The distributive law of ring theory asserts that multiplication distributes over addition in the sense that we have

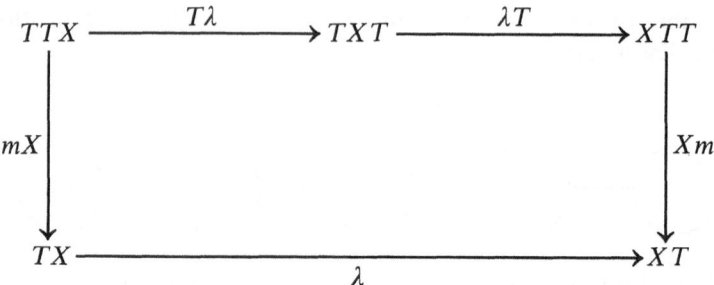

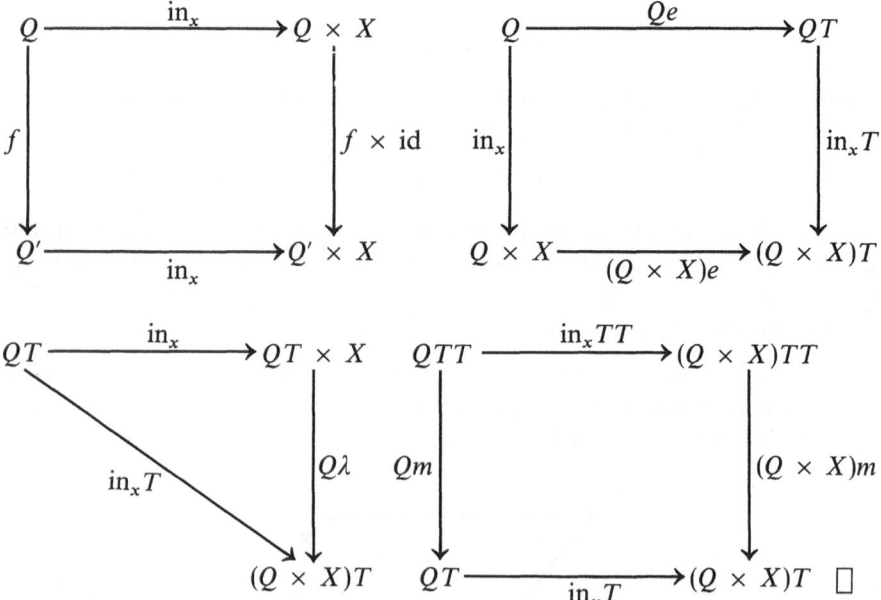

$$(x_1 + \cdots + x_n) \cdot (y_1, + \cdots + y_n) = x_1 y_1 + x_1 y_2 + \cdots + x_n y_m$$

The corresponding distributive law in the sense of 3.6 is constructed as follows. Let **T** be the theory for abelian groups and write a typical element of the free abelian group QT as $\sum n_q q$ where $n_q \in \mathbf{Z}$ (e.g. the formal sum $2q - r$ has $n_q = 2$, $n_r = -1$, all other $n_t = 0$). Let $QX = Q \times Q$ so that $X = X_\Omega$ where $\Omega_2 = \{\cdot\}$, $\Omega_n = \varnothing$ if $n \neq 2$; accordingly, we write a typical element of QX as $q \cdot r$ rather than (q, r). Define $\lambda: TX \longrightarrow XT$ by

$$\begin{array}{ccc} QT \times QT & \xrightarrow{Q\lambda} & (Q \times Q)T \\ \sum_q n_q q \cdot \sum_r n_r r & \longmapsto & \sum_{q,r} n_q n_r (q \cdot r) \end{array}$$

We leave it as an exercise to check that λ is a distributive law. The precise sense in which a ring is a **T**-algebra and an X-dynamics which satisfies the above distributive law is discussed in exercise 6(e) below.

As motivated by our earlier discussion, distributive laws provide the key to "running" fuzzy systems. We have:

3.9 Fuzzy Systems. A *λ-automaton with respect to* a fuzzy theory (i.e., an algebraic theory) **T** in $\mathscr{K}$, functor $X: \mathscr{K} \longrightarrow \mathscr{K}$ and distributive law $\lambda: TX \longrightarrow XT$, is a 7-tuple $M = (Q, \delta, I, \tau, Y, \theta, \beta)$ where (Y, θ) is a **T**-algebra and

$$I \xrightarrow{\tau} QT \qquad QX \xrightarrow{\delta} QT \qquad Q \xrightarrow{\beta} Y$$

Notice that β generalizes a map of form $Q \longrightarrow Y$.

Associated with the λ-automaton M as above is the X-automaton $M^\bullet = (QT, \delta^\bullet, I, \tau, Y, \beta^\#)$ where

$$\delta^\bullet = QTX \xrightarrow{Q\lambda} QXT \xrightarrow{\delta^\#} QT$$

In case $IX^@$ (the free X-dynamics over I) exists, the *response of M* is defined to be the response of $M^\bullet$ (as defined in 2.3). Thus, while the definition of a λ-automaton is independent of λ, its response is not.

The reader should check that the deterministic and multiple branching machines of 3.4 have the same response (using the distributive law of 3.5 to run the latter).

3.10 A Probabilistic System. Consider the problem of designing a vending machine subsystem to accept fifteen cents. This means that the machine should reach a "successful" state s from its initial state q_0 just in case the input sequence is one of *dn*, *nd*, *nnn* where d represents "dime" (ten cents) and n represents "nickel" (five cents). The following state graph describes the deterministic minimal realization of this subsystem; (it is minimal regardless of whether we regard the output to be the state, $\beta = \mathrm{id}_Q$ for $Q = \{q_0, q_1, q_2, s, v\}$ (as is natural for a subsystem) or choose β to be the characteristic function of $\{s\}$). Suppose now that there is a probability d' for a dime and n' for a nickel of premature transition to the "overflow" state v. The deterministic diagram below must be modified to take account of these

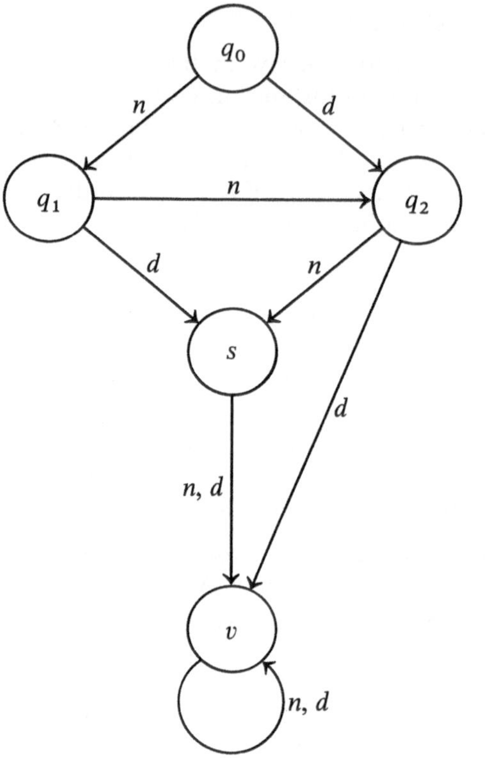

probabilities; thus, for example, the n-transition from q_1 to q_2 now occurs only with probability $1 - n'$ and there is an n-transition with probability n' from q_1 to v, and so forth. If $\mathbf{T}$ is the stochastic matrix theory discussed in 3.1 and λ is the canonical distributive law of $X = - \times \{d, n\}$ over $\mathbf{T}$ of 3.7, then our probabilistic system is sensibly modelled as the $(\mathbf{T}, X, \lambda)$-automaton $M = (Q, \delta, 1, \tau, [0, 1], \beta)$ described as follows. $\delta: Q \times \{d, n\} \longrightarrow QT$ assigns to (q, x) the probability distribution on Q of transition from q with input x so that, for example,

$$(q_1, n)\delta = (0 \quad 0 \quad 1 - n' \quad 0 \quad n')$$

where the ordering on Q is q_0, q_1, q_2, s, v. As is easily checked, $\delta^\bullet$ assigns the convex transformations corresponding the row-stochastic matrices given by δ, so that

$$(-, d)\delta^\bullet = \begin{bmatrix} 0 & 0 & 1-d' & 0 & d' \\ 0 & 0 & 0 & 1-d' & d' \\ 0 & 0 & 0 & 0 & 1 \\ 0 & 0 & 0 & 0 & 1 \\ 0 & 0 & 0 & 0 & 1 \end{bmatrix} \quad (-, n)\delta^\bullet = \begin{bmatrix} 0 & 1-n' & 1 & 0 & n' \\ 0 & 0 & 1-n' & 0 & n' \\ 0 & 0 & 0 & 1-n' & n' \\ 0 & 0 & 0 & 0 & 1 \\ 0 & 0 & 0 & 0 & 1 \end{bmatrix}$$

A natural initial state is $\tau = (1\ 0\ 0\ 0\ 0)$ (i.e. we start in state q_0 with probability 1). If $r:\{d, n\}^* \longrightarrow Q$ denotes the reachability map of $M^\bullet$, it is easily checked that $(dn)r = (nd)r = (0\ 0\ 0\ a\ 1-a)$ whereas $(nnn)r = (0\ 0\ 0\ b\ 1-b)$, where $a = (1-d')(1-n')$ and $b = (1-n')^3$. The unit interval $[0, 1]$ is a convex set in the usual way, and hence a **T**-algebra. Let $\beta:Q \longrightarrow [0, 1]$ send s to 1 and everything else to 0. Then if $f = r.\beta^\#$ is the response of M, we have, e.g., $(nnn)f = (1-n')^3$. In general, wf is the probability that w will lead to state s.

In the previous section we saw that the reachability and observability morphisms, fundamental in the discussion of system response, were X-dynamorphisms. For λ a distributive law of X over **T**, we shall see that the appropriate generalization for the needs of fuzzy system response is from "X-dynamorphism" to "homomorphism of λ-algebras." We begin with a definition.

3.11 Definition. Let $\lambda:TX \longrightarrow XT$ be a distributive law of X over **T** as in 3.6. *A λ-algebra is a triple (Q, δ, ξ) where (Q, δ) is an X-dynamics and (Q, ξ) is a* **T**-*algebra subject to the λ-law that $\xi:(QT, Q\lambda.\delta T) \longrightarrow (Q, \xi)$ is an X-dynamorphism*:

$$\begin{array}{ccc} QTX & \xrightarrow{\ \xi X\ } & QX \\ {\scriptstyle Q\lambda}\downarrow & & \\ QXT & & \downarrow{\scriptstyle \delta} \\ {\scriptstyle \delta T}\downarrow & & \\ QT & \xrightarrow[\ \xi\ ]{} & Q \end{array} \qquad (3.12)$$

The category $\mathscr{K}^\lambda$ of λ-algebras may be regarded as a category of $\mathscr{K}$-objects with structure—indeed a full subcategory of $\mathrm{Dyn}(X) \times \mathscr{K}^{\mathbf{T}}$ in $\mathrm{Struct}(\mathscr{K})$—by defining a *$\lambda$-homomorphism* $f:(Q, \delta, \xi) \longrightarrow (Q, \delta', \xi')$ to be a simultaneous X-dynamorphism and **T**-homomorphism.

3.13 Example. Let λ be the distributive law of 3.7. Writing $q\delta_x = (q, x)\delta$, the diagram

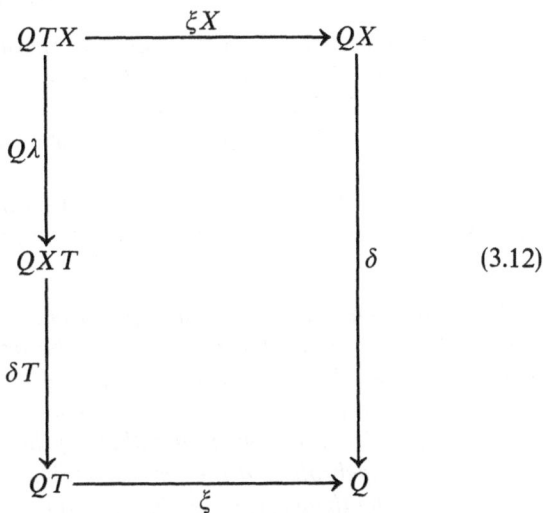

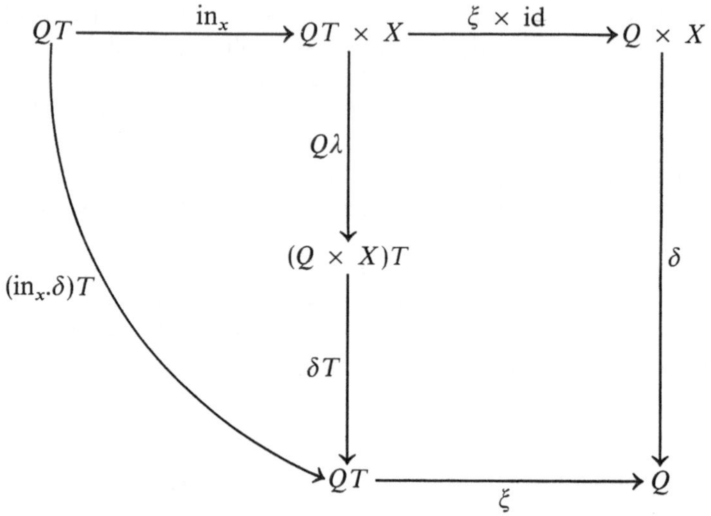

shows that the λ-law holds if and only if δ_x is a $\mathbf{T}$-homomorphism for all x in X. Identifying Dyn(X) with $\mathbf{Set}^{\mathbf{X}^{@}}$ as in 2.13, we see that $\mathbf{Set}^{\lambda}$ is isomorphic in Struct(**Set**) to $\mathbf{Set}^{\mathbf{X}^{@} \otimes \mathbf{T}}$ which *is* algebraic over **Set** (see 3.6.21). We shall prove below in 3.15 that $\mathscr{K}^{\lambda}$ is always algebraic over $\mathscr{K}$.

3.14 Example. Let $\mathbf{T}$ be the identity theory in $\mathscr{K}$, $QT = Q$, $Qe = \text{id}_Q$, $\alpha \circ \beta = \alpha.\beta$. For any $X: \mathscr{K} \longrightarrow \mathscr{K}$, $\lambda = \text{id}_X: TX \longrightarrow XT$ is a distributive law and, since (Q, ξ) is a $\mathbf{T}$-algebra if and only if $\xi = \text{id}_Q$, $\mathscr{K}^{\lambda}$ is isomorphic to Dyn(X) in Struct($\mathscr{K}$). Thus the theory of λ-algebras generalizes Dyn(X).

The next two theorems present the fundamental facts about λ-algebras. The proofs of these theorems make crucial use of the axioms assumed about λ in 3.6.

3.15 Theorem. *Let X be an input process in $\mathscr{K}$ with induced algebraic theory $\mathbf{X}^{@} = (X^{@}, \eta, \mu)$, let $\mathbf{T} = (T, e, m)$ be an algebraic theory in $\mathscr{K}$, and let $\lambda: TX \longrightarrow XT$ be a distributive law of X over $\mathbf{T}$. Let $U_1: \mathscr{K}^{\lambda} \longrightarrow$ Dyn(X), U_2:Dyn(X) $\longrightarrow \mathscr{K}$, and $U = U_1 U_2: \mathscr{K}^{\lambda} \longrightarrow \mathscr{K}$ be the forgetful functors. Then the following three statements are valid:*

1. *U_1 is algebraic and the corresponding algebraic theory in* Dyn(X) *is isomorphic to the theory $\hat{\mathbf{T}} = (\hat{T}, \hat{e}, \hat{m})$ defined as follows:*

$$(Q, \delta)\hat{T} = (QT, QTX \xrightarrow{Q\lambda} QXT \xrightarrow{\delta T} QT)$$
$$(Q, \delta)\hat{e} = Qe:(Q, \delta) \longrightarrow (Q, \delta)\hat{T}$$
$$(Q, \delta)\hat{m} = Qm:(Q, \delta)\hat{T}\hat{T} \longrightarrow (Q, \delta)\hat{T}$$

2. *The algebraic theory in $\mathscr{K}$ induced by the fact that U has a left adjoint is isomorphic to the theory $\mathbf{T}^{\star} = (T^{\star}, e^{\star}, m^{\star})$ defined by*

$$QT^{\star} = QX^{@}T$$
$$Qe^{\star} = Q\eta.QX^{@}e = Q\eta e = Qe.Q\eta T$$
$$Qm^{\star} = QX^{@}TX^{@}T \xrightarrow{(QX^{@}\lambda.Q\mu_0 T)^{@}T} QX^{@}TT \xrightarrow{QX^{@}m} QX^{@}T$$

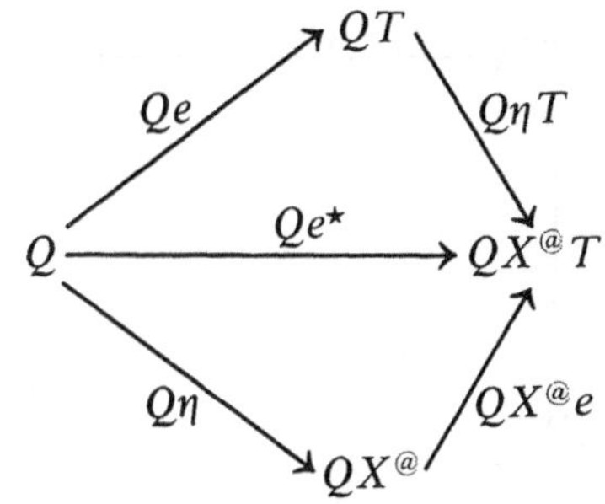

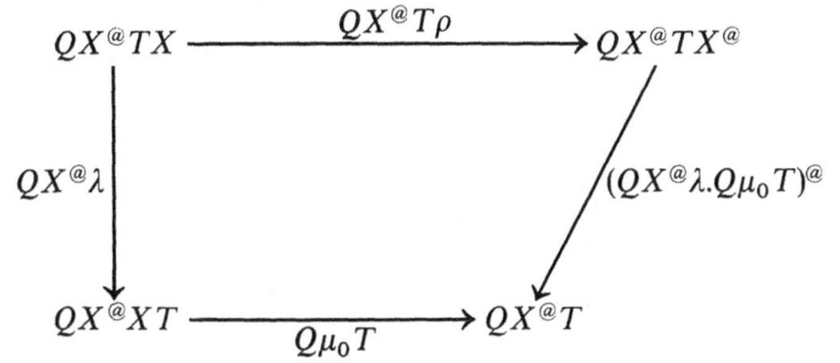

(*see 2.13 for the definition of* ρ *and see 2.11*).

3. *U is in fact algebraic, the semantic comparison functor* $\Phi\colon \mathscr{K}^{\lambda} \longrightarrow \mathscr{K}^{\mathbf{T}^{\star}}$ *being given by*

$$(Q, \delta, \xi)\Phi = (Q, QX^{@}T \xrightarrow{\delta^{@}T} QT \xrightarrow{\xi} Q)$$

and the inverse passage Φ^{-1} *being given by*

$$(Q, \psi)\Phi^{-1}$$
$$= (Q, QX \xrightarrow{Q\rho} QX^{@} \xrightarrow{QX^{@}e} QX^{@}T \xrightarrow{\psi} Q, QT \xrightarrow{Q\eta T} QX^{@}T \xrightarrow{\psi} Q)$$

Proof. Since $\mathbf{T}$ is an algebraic theory in $\mathscr{K}$, the proof that $\hat{\mathbf{T}}$ is an algebraic theory in Dyn(X) requires only that we check that $\hat{\mathbf{T}}$ is functorial and that $\hat{e}$ and $\hat{m}$ are dynamorphisms. This is clear from the three diagrams:

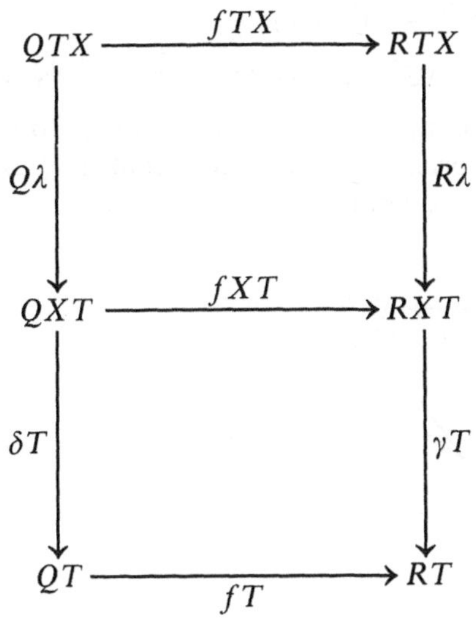

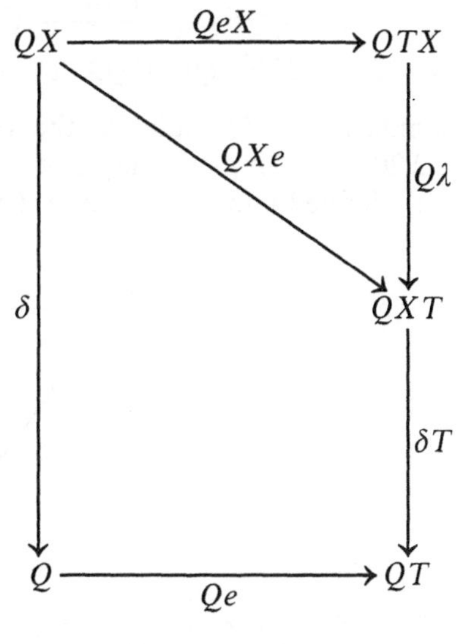

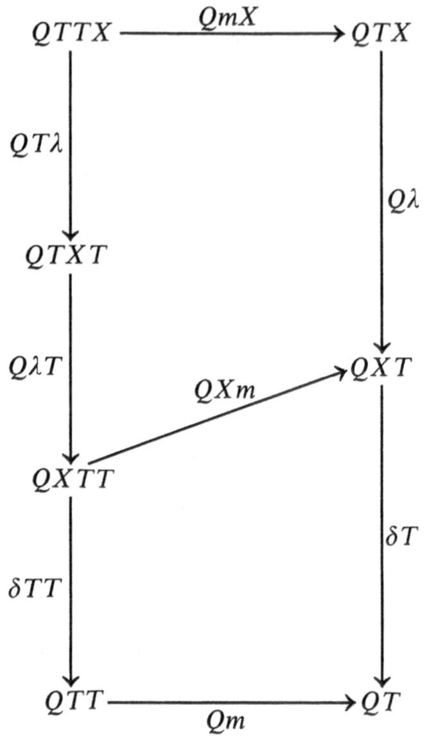

in which all three axioms for a distributive law have been used. It is clear that a $\hat{\mathbf{T}}$-algebra is the same thing as (Q, δ, ξ) where (Q, δ) is an X-dynamics, (Q, ξ) is a $\mathbf{T}$-algebra, and $\xi:(Q, \delta)\hat{T} \longrightarrow (Q, \delta)$ is a dynamorphism; but this last condition is just the λ-law 3.12. This completes the proof of (1).

The formulas for $\mathbf{T}^\star$ are direct consequences of 2.2.30 and 2.2.20. Thus $m^\star = F_2F_1\varepsilon^\star U_1U_2$ where

$$(Q, \delta, \xi)\varepsilon^\star = QF_2F_1 \xrightarrow{U_1\varepsilon_2F_1} (Q, \delta)F_1 \xrightarrow{\varepsilon_1} (Q, \delta, \xi)$$
$$= (QX^@T, QX^@\lambda.Q\mu_0T, QX^@e) \xrightarrow{\delta^@T} (QT, Q\lambda.\delta T, Qe) \xrightarrow{\xi} (Q, \delta, \xi)$$

Similarly, the formula for the semantics comparison functor follows from 2.2.20. To complete the proof of the theorem we must show that Φ^{-1} is well defined and inverse to Φ. To begin, the diagrams

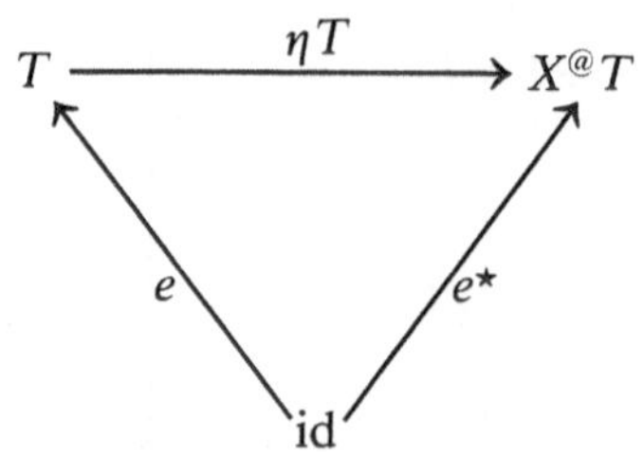

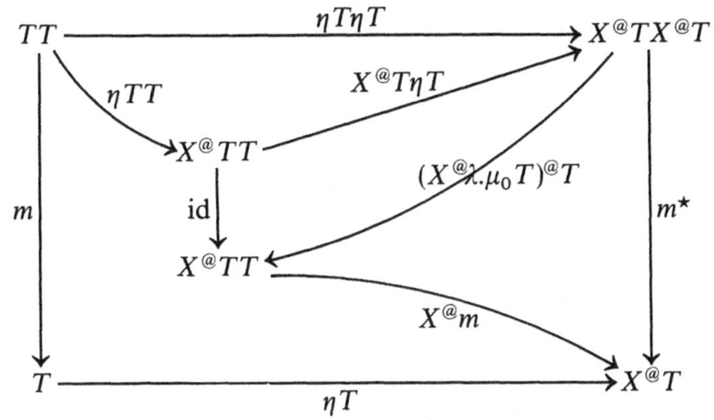

prove that $\eta T:\mathbf{T} \longrightarrow \mathbf{T}^\star$ is a theory map so that, by 3.2.9, $(Q, \psi)\Phi^{-1}$ is at least an X-dynamics and a $\mathbf{T}$-algebra. Now consider the diagram

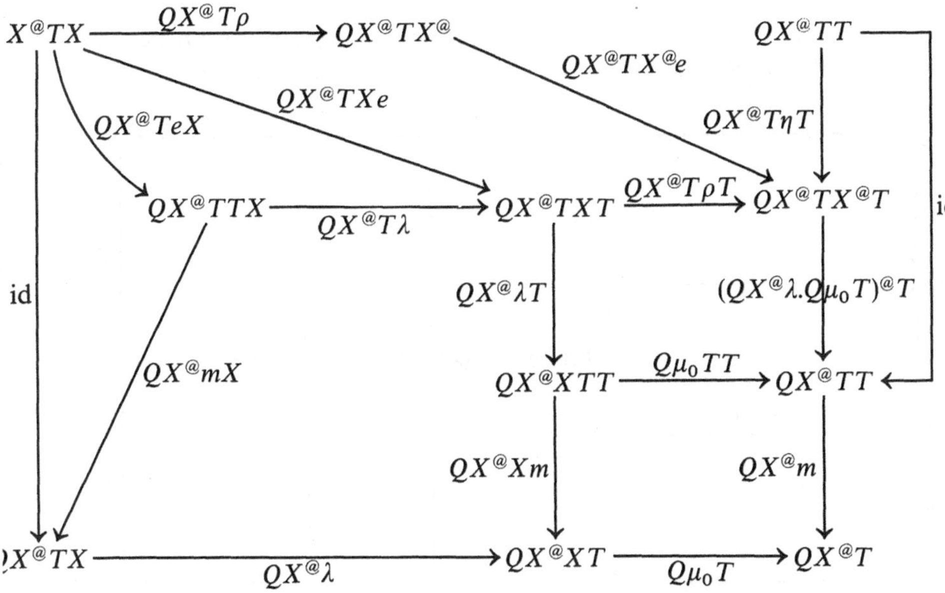

(3.16)

As a first use of diagram 3.16, we have $QX^@T\lambda.(QX^@\lambda.Q\mu_0 T)T.QX^@m = QX^@mX.(QX^@\lambda.Q\mu_0 T)$ which establishes the λ-law at least for $(QX^@T, Qm^\star)\Phi^{-1}$. Now, one expects the full subcategory of (Q, δ, ξ)'s satisfying the λ-law to be closed under products, subobjects, and epimorphisms split at the level of $\mathscr{K}$ because the λ-law is "an equation." That this is so follows from diagram 3.17. (The only case important to this proof is that in which (Q, δ, ξ) satisfies the λ-law and the X-dynamorphism and $\mathbf{T}$-homomorphism

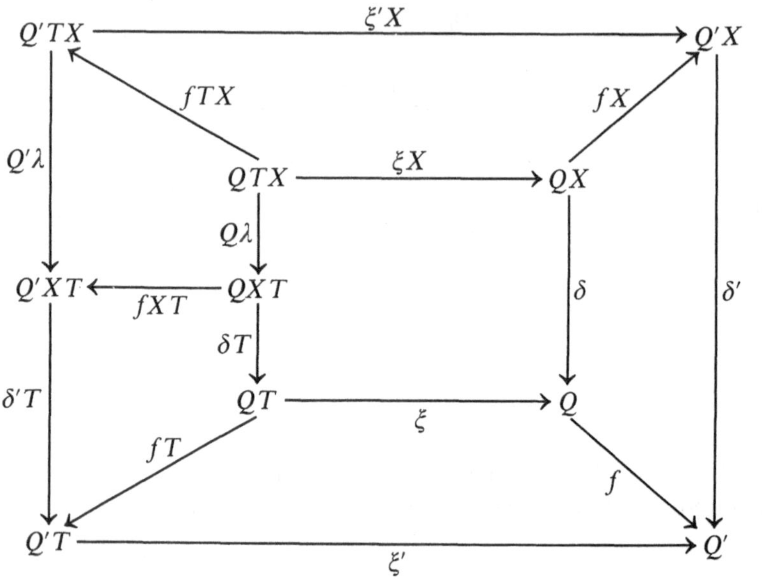

(3.17)

f is split epi in $\mathscr{K}$; since fTX is epi, we deduce that (Q', δ', ξ') satisfies the λ-law as well.) But Φ^{-1} is functorial, as is clear from the diagrams

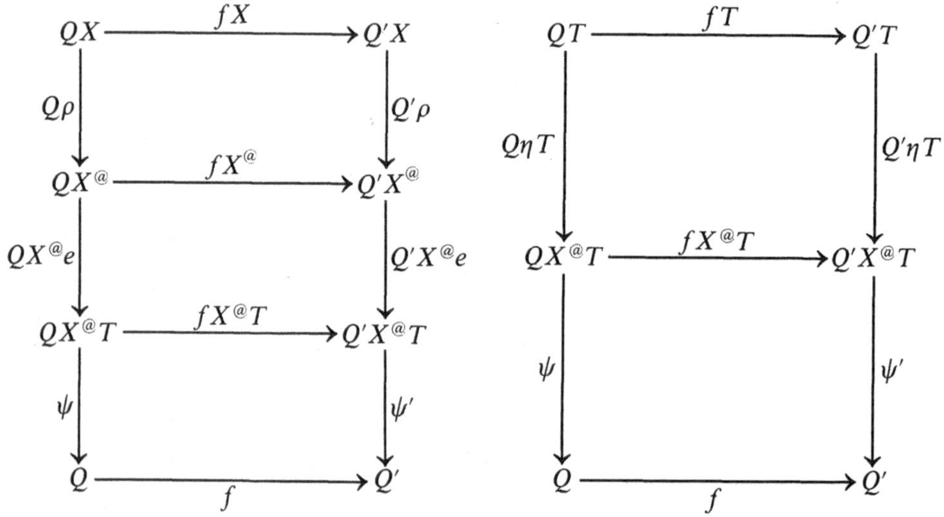

Since $\psi:(QX^{@}T, Qm^{\star})\Phi^{-1} \longrightarrow (Q, \psi)\Phi^{-1}$ is an X-dynamorphism $\mathbf{T}$-homomorphism split epi in $\mathscr{K}$, it is now clear that Φ^{-1} is a well-defined functor. Very similar arguments are used to show that $(Q, \psi)\Phi^{-1}\Phi = (Q, \psi)$. For we read from 3.16 that $(QX^{@}T, Qm^{\star})\Phi^{-1} = (QX^{@}T, QX^{@}\lambda.Q\mu_0 T, QX^{@}m)$ so that $(QX^{@}T, Qm^{\star})\Phi^{-1}\Phi = (QX^{@}T, Qm^{\star})$; then, because $\psi:(QX^{@}T, Qm^{\star}) \longrightarrow (Q, \psi)\Phi^{-1}\Phi$ is a $\mathbf{T}^{\star}$-homomorphism, it follows immediately from 3.1.9 that $(Q, \psi)\Phi^{-1}\Phi = (Q, \psi)$.

The last detail to check is that if $(Q, \delta, \xi)\Phi = (Q, \psi)$, so that $\psi = \delta^{@}T.\xi$, that $(Q, \psi)\Phi^{-1} = (Q, \delta, \xi)$. This follows from the diagrams

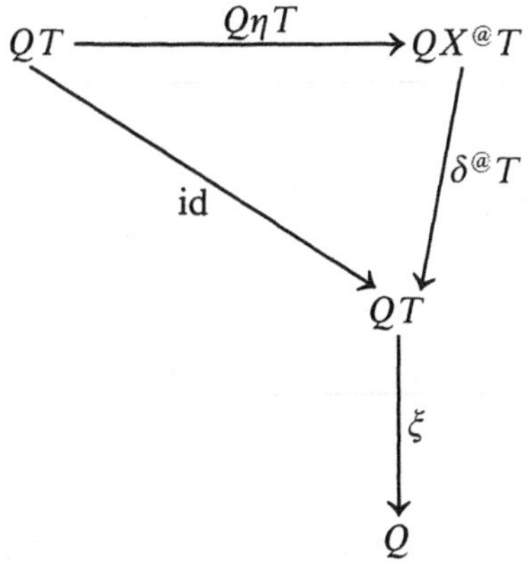

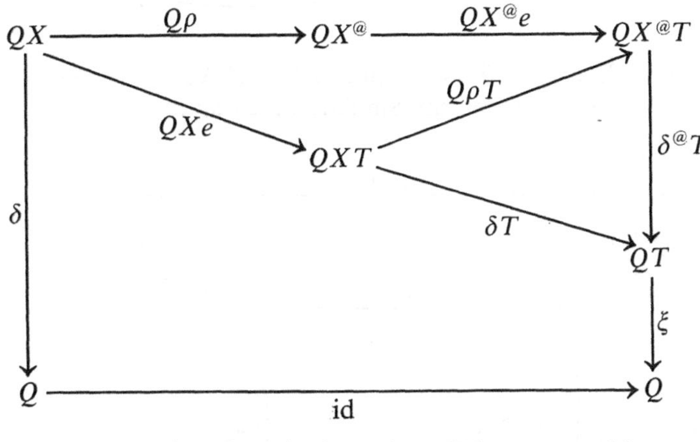

recalling from 1.4.13 that $\delta T.\xi$ is the unique **T**-homomorphic extension of δ. □

3.18 Theorem. *Let X be an output process process in $\mathscr{K}$, let $\mathbf{T} = (T, e, m)$ be an algebraic theory in $\mathscr{K}$, and let $\lambda: TX \longrightarrow XT$ be a distributive law of X over **T**. Then the forgetful functor $\mathscr{K}^{\lambda} \longrightarrow \mathscr{K}^{\mathbf{T}}$ has a right adjoint. Using notations as in 3.15 and 2.3, the cofree λ-algebra over the **T**-algebra (Y, θ) is $(YX_{@}, YL, \theta_{\#})$ with coinclusion of the generators $Y\Lambda: (YX_{@}, \theta_{\#}) \longrightarrow (Y, \theta)$, where $\theta_{\#}$ is obtained from θ as the unique dynamorphic coextension of the unique **T**-homomorphic extension $(Y\Lambda)^{\#}$ as shown in the diagram below:*

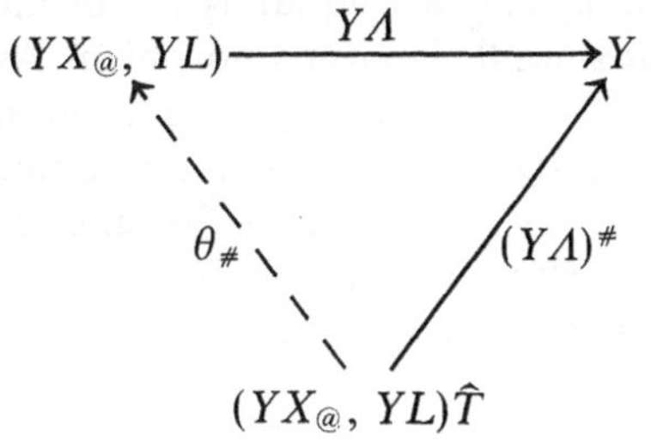

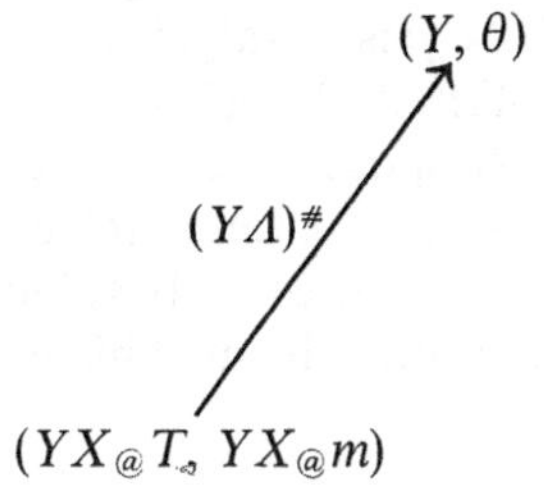

Proof. We first show that $(YX_{@}, \theta_{\#})$ is a **T**-algebra. The diagram

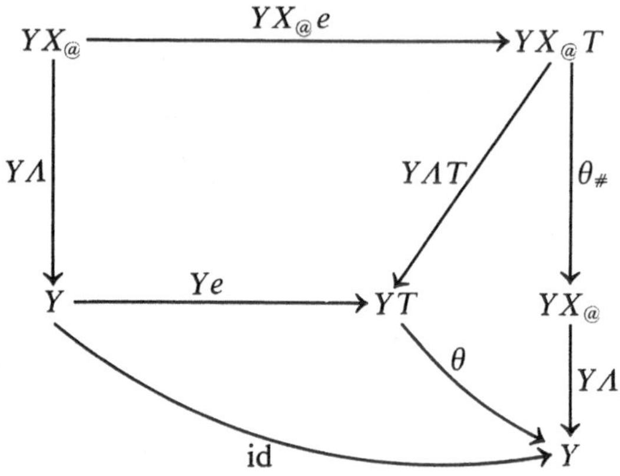

shows that $YX_{@}e.\theta_{\#} = \mathrm{id}_{YX_{@}}$, since $YX_{@}e:(YX_{@}, YL) \longrightarrow (YX_{@}, YL)\hat{T}$ is a dynamorphism. Similarly, the diagram

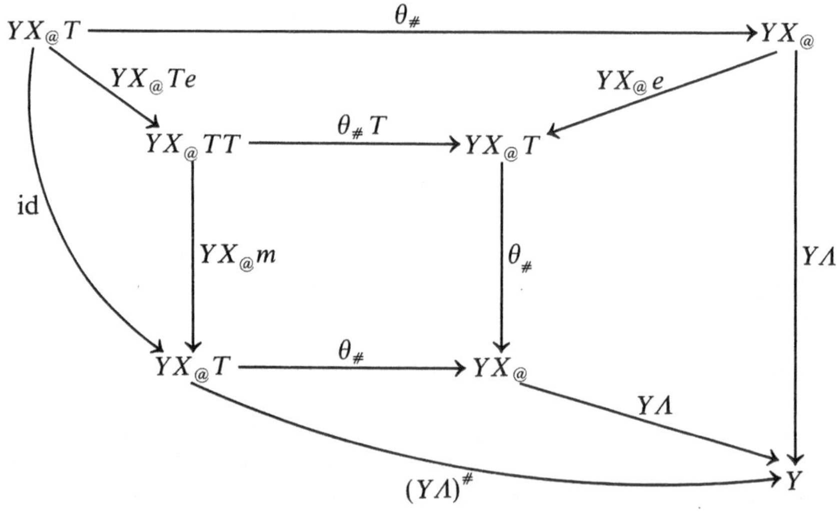

establishes first that $YX_{@}m.\theta_{\#}.Y\Lambda$ and $\theta_{\#}T.\theta_{\#}.Y\Lambda$ are equal **T**-homomorphisms and thus that $YX_{@}m.\theta_{\#}$ and $\theta_{\#}T.\theta_{\#}$ are equal dynamorphisms. That $(YX_{@}, YL, \theta_{\#})$ is a λ-algebra is clear since $\theta_{\#}$ is a dynamorphism. As $\theta_{\#}.Y\Lambda = (Y\Lambda)^{\#} = Y\Lambda T.\theta$, $Y\Lambda:(YX_{@}, \theta_{\#}) \longrightarrow (Y, \theta)$ is a **T**-homomorphism.

Now suppose given a λ-algebra (Q, δ, ξ) and a **T**-homomorphism f: $(Q, \delta) \longrightarrow (Y, \theta)$. Let $\Gamma:(Q, \delta) \longrightarrow (YX_{@}, YL)$ be the unique dynamorphic coextension of f. It suffices to show that $\Gamma:(Q, \xi) \longrightarrow (YX^{@}, \theta_{\#})$ is a **T**-homomorphism. This is clear from the diagram

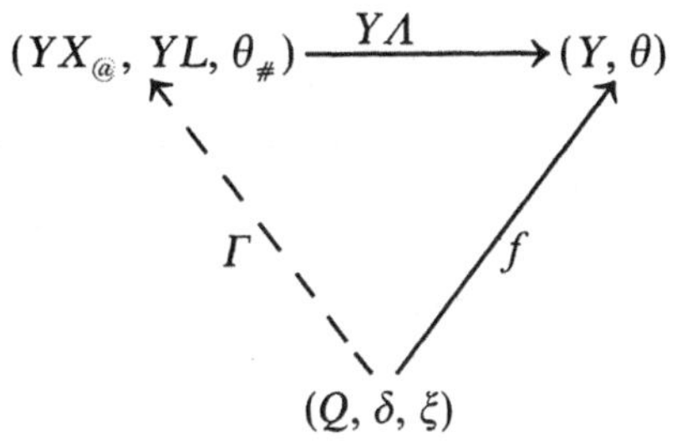

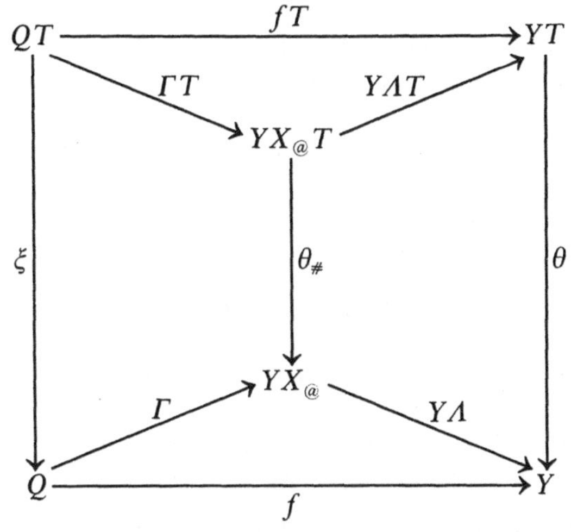

since $\xi.\Gamma$ and $\Gamma\hat{T}.\theta_\#$ are dynamorphisms. □

For the balance of this section we fix an input and output process X: $\mathcal{K} \longrightarrow \mathcal{K}$, an algebraic theory $\mathbf{T} = (T, e, m)$ in $\mathcal{K}$ and a distributive law $\lambda: TX \longrightarrow XT$ of X over $\mathbf{T}$. We use the notations introduced after 3.15 without special mention.

Generalizing the $M^\bullet$ of 3.9 (see 3.20), we now present a "state-free" algebraic version of a fuzzy system:

3.19 Implicit λ-Automata. An *implicit λ-automaton* is an 8-tuple $\bar{M} = (\bar{Q}, \bar{\delta}, \xi, I, \bar{\tau}, Y, \theta, \bar{\beta})$ where $(\bar{Q}, \bar{\delta}, \xi)$ is a λ-algebra, $\bar{\tau}: I \longrightarrow \bar{Q}$ is a $\mathcal{K}$-morphism and, $\bar{\beta}: (\bar{Q}, \xi) \longrightarrow (Y, \theta)$ is a $\mathbf{T}$-homomorphism.

Observe that when $\mathbf{T}$ = (id, id, id) and λ = id, "implicit λ-automaton" and "λ-automaton" coincide with "X-automaton."

The *X-reachability map* of the implicit λ-automaton $\bar{M}$ is the unique dynamorphic extension $r: (IX^@, I\mu_0) \longrightarrow (Q, \delta)$ of τ. The *reachability map of $\bar{M}$* is the $\mathbf{T}$-homomorphic extension $r^\#: (IX^@T, I\mu_0, IX^@m) \longrightarrow (\bar{Q}, \bar{\delta}, \xi)$ of r. As is clear from the proof of 3.15, $r^\#$ may be viewed as the unique

λ-homomorphic extension of $\bar{\tau}$:

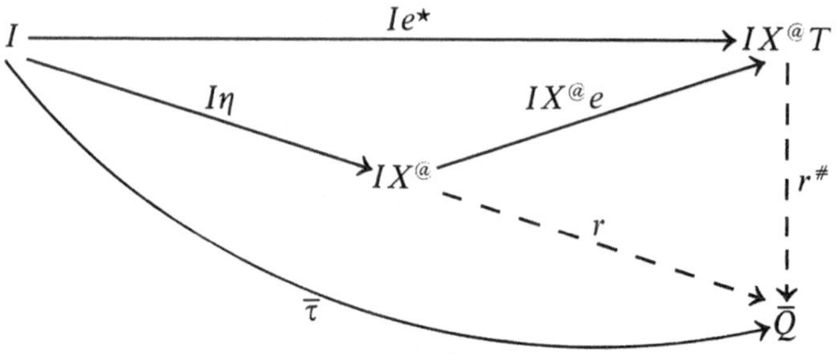

The *response* of $\bar{M}$ is then the composition $f = r.\bar{\beta}: IX^{@} \longrightarrow Y$.

The *observability map* of $\bar{M}$ is the unique λ-homomorphic coextension of $\bar{\beta}$:

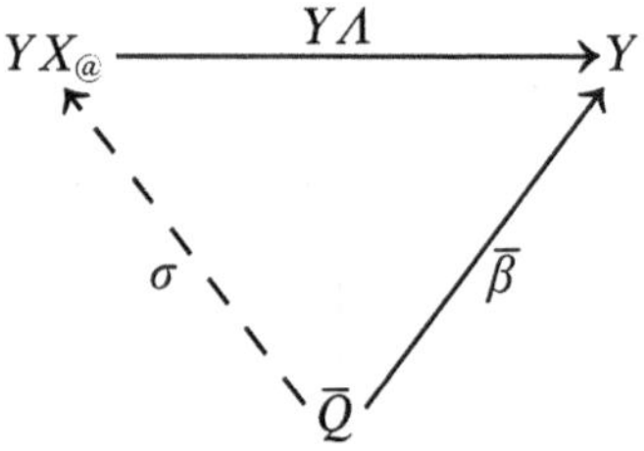

In the next two propositions we show that λ-automata and implicit λ-automata compute the same responses.

3.20 Proposition. *Let* $M = (Q, \delta: QX \longrightarrow QT, I, \tau, Y, \theta, \beta)$ *be a λ-automaton. Then* $M^{\bullet} = (QT, \delta^{\bullet}, Qm, I, \tau, Y, \beta^{\#})$ *as defined in 3.9 is an implicit λ-automaton and the response of M (as defined in 3.9, i.e., as in 2.3) is the response of $M^{\bullet}$ (as defined in 3.19).*

Proof. That $(QT, \delta^{\bullet}, Qm)$, with $\delta^{\bullet} = Q\lambda.\delta^{\#}$, is a λ-algebra follows from

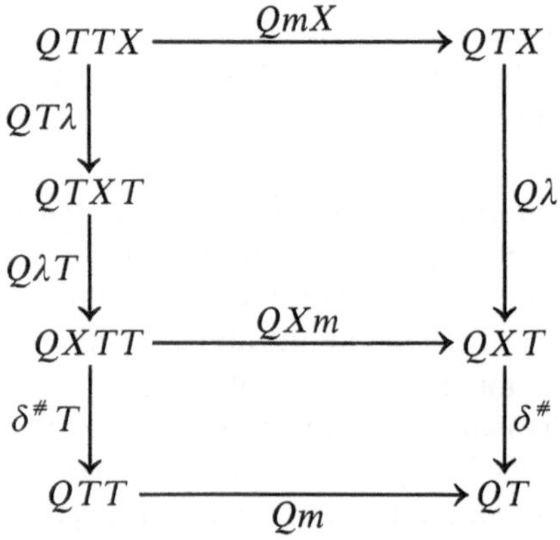

The assertion about the responses is clear from the definitions. □

To prove the converse of 3.20 we need to have a way of associating a λ-automaton M to an implicit λ-automaton $\bar{M}$. This can be done in such a way that the state-object can be taken to be "any subobject generating $\bar{Q}$ as a **T**-algebra" in the following precise sense:

3.21 Definition. Let $(\bar{Q}, \xi)$ be a **T**-algebra. *A scoop of* $(\bar{Q}, \xi)$ *is a triple* (Q, i, c) *where* $i:Q \longrightarrow \bar{Q}$, $c:\bar{Q} \longrightarrow QT$ *are such that* $c.i.^{\#} = \mathrm{id}$. By 1.4.31, if $\mathscr{K} = \mathbf{Set}$ then for each (Q, i) there exists a scoop of form (Q, i, c) if and only

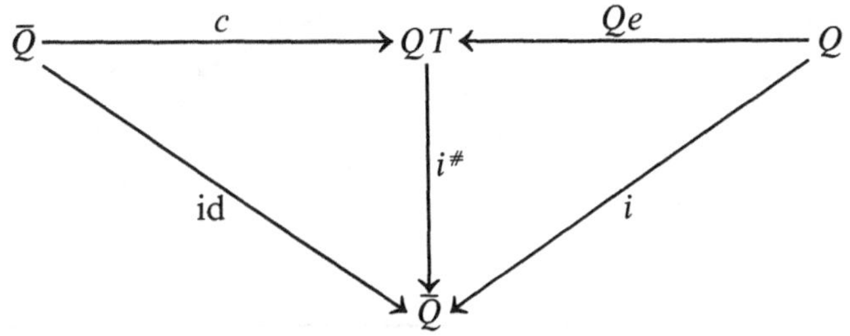

if the image of Q in $\bar{Q}$ generates $\bar{Q}$ as a **T**-algebra. Of course there always exists at least one scoop, namely $(\bar{Q}, \mathrm{id}, Qe)$.

3.22 Proposition. *Let* $\bar{M} = (\bar{Q}, \bar{\delta}, \xi, I, \bar{\tau}, Y, \theta, \bar{\beta})$ *be an implicit* λ-*automaton and let* (Q, i, c) *be any scoop of* $\bar{Q}$. *Then the* λ-*automaton* $M = (Q, \delta, I, \tau, Y, \theta, \beta)$ *where*

$$\delta = QX \xrightarrow{iX} \bar{Q}X \xrightarrow{\bar{\delta}} \bar{Q} \xrightarrow{c} QT$$
$$\tau = I \xrightarrow{\bar{\tau}} \bar{Q} \xrightarrow{c} QT$$
$$\beta = Q \xrightarrow{i} \bar{Q} \xrightarrow{\bar{\beta}} Y$$

has the same response as $\bar{M}$.

Proof. Since $c.i^{\#} = \mathrm{id}_{\bar{Q}}$ we have

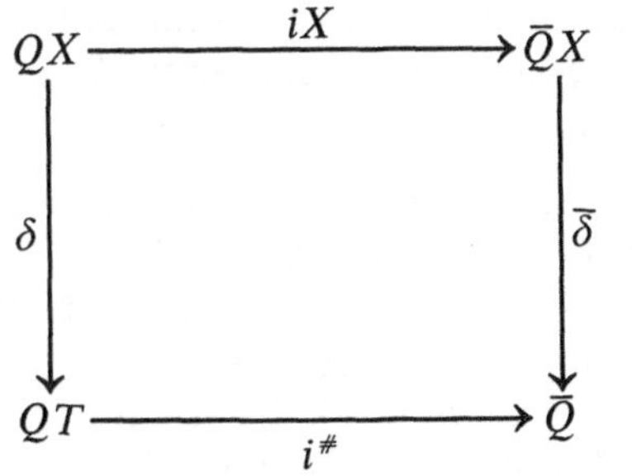

It follows that $iXT.\bar{\delta}T.\xi = (iX.\bar{\delta})^{\#} = (\delta.i^{\#})^{\#} = \delta^{\#}.i^{\#}$. We therefore have the

diagram

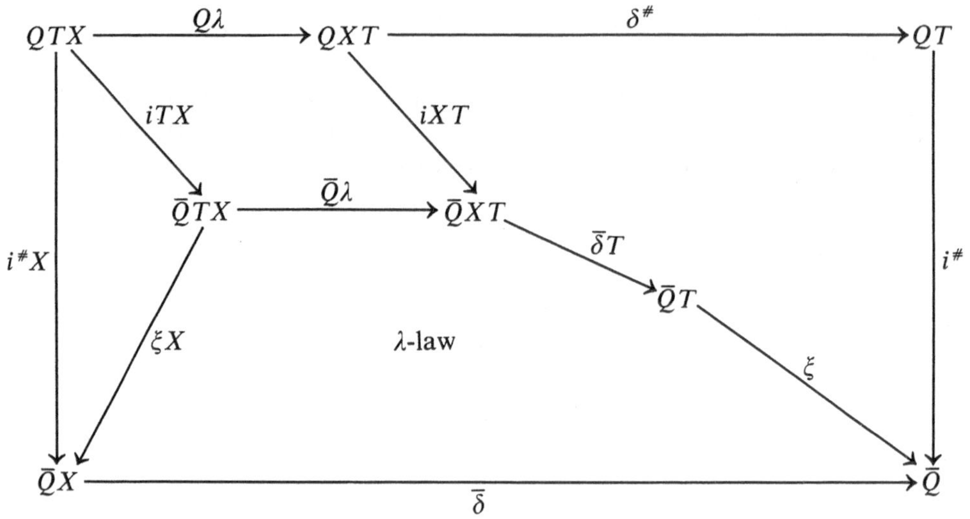

We also clearly have (since $\overline{\beta}$ is a **T**-homomorphism)

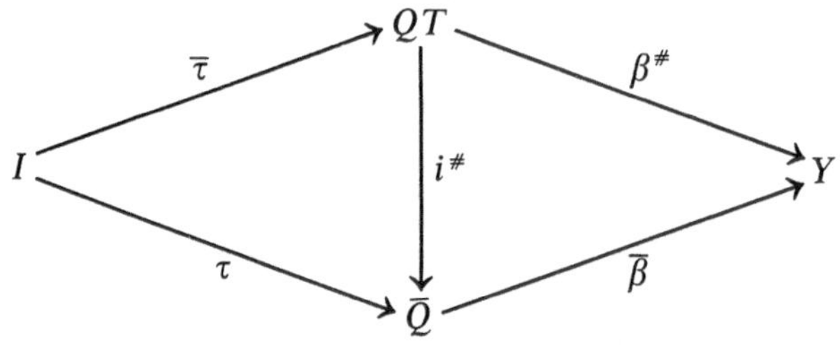

Thus $i^{\#}:(QT, \delta^{\bullet}, I, \tau, Y, \beta^{\#}) \longrightarrow (\overline{Q}, \overline{\delta}, I, \overline{\tau}, Y, \overline{\beta})$ is a simulation of X-automata and, in particular, M and $\overline{M}$ have the same response. □

Although its significance is smaller than we might have hoped, the minimal realization theorem for X-automata (2.18) has a straightforward generalization to implicit λ-automata:

3.23 Proposition. *Let* (E, M) *be an image factorization system in* $\mathcal{K}$ *such that* $X^{@}T$ *preserves* E. *Let* I, (Y, θ) *be fixed. Then for every* $f:IX^{@} \longrightarrow Y$ *there exists an implicit* λ*-automaton* $\overline{M} = (\overline{Q}, \overline{\delta}, \xi, \overline{\tau}, \overline{\beta})$ *such that the response of* $\overline{M}$ *is* f, *the reachability map* $r^{\#}:IX^{@}T \longrightarrow \overline{Q}$ *is in* E, *and the observability map* $\sigma:Q \longrightarrow YX_{@}$ *is in* M. *If* $\overline{M}'$ *also satisfies these three conditions then* $\overline{M}$ *and* $\overline{M}'$ *are isomorphic (i.e., there exists an isomorphism* ψ: $(\overline{Q}, \overline{\delta}, \xi) \longrightarrow (\overline{Q}', \overline{\delta}', \xi')$ *of* λ*-algebras such that* $\overline{\tau}.\psi = \overline{\tau}'$, $\psi.\overline{\beta}' = \overline{\beta}$).

Proof. Using the notations of 3.18, f extends to a λ-homomorphism

$\bar{f}:(IX^{@}T, I\mu_0, Im) \longrightarrow (YX_{@}, YL, \theta_{\#})$ as shown below:

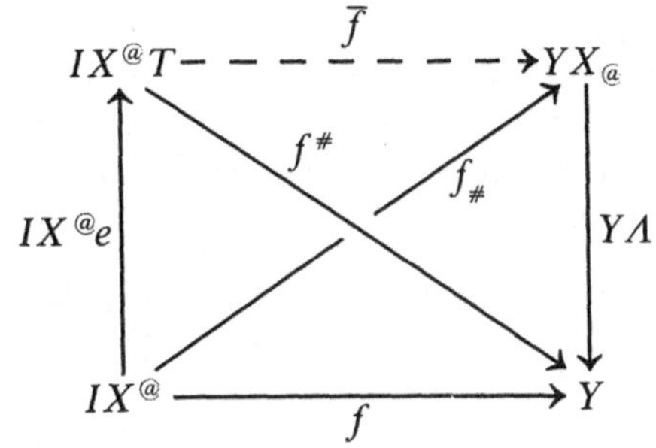

(where it is easily checked that $(f^{\#})_{\#} = \bar{f} = (f_{\#})^{\#}$). Using the same reasoning as in 2.18 (2), as $X^{@}T$ preserves E, the E-M image

$$IX^{@}T \xrightarrow{r^{\#}} \bar{Q} \xrightarrow{\sigma} YX_{@}$$

of $\bar{f}$ lifts uniquely to λ-algebra homomorphisms. The remaining details follow the proof of 2.18. □

For the remainder of the section we fix $\mathscr{K}$ = **Set**. Example 3.31 below will clarify why the previous proposition is at best a first step towards a minimal realization theory for fuzzy systems.

3.24 Isolated Elements. Let $(\bar{Q}, \xi)$ be a **T**-algebra. Recall from 1.4.31 that, if A is a subset of Q, then $\langle A\rangle = \text{Im}(iT.\xi)$ is the subalgebra of $(\bar{Q}, \xi)$ generated by A. Say that q in $\bar{Q}$ is an *isolated element of* $(\bar{Q}, \xi)$ if q cannot be generated by applying the **T**-operations to the other elements, that is, if $q \notin \langle Q - \{q\}\rangle$. When **T** is the theory whose algebras are semilattices (1.5.15), i.e., QT is the set of finite subsets of Q qua subtheory of the theory of 1.3.5, isolated elements are more commonly called *join-irreducibles*. In the semilattice shown below

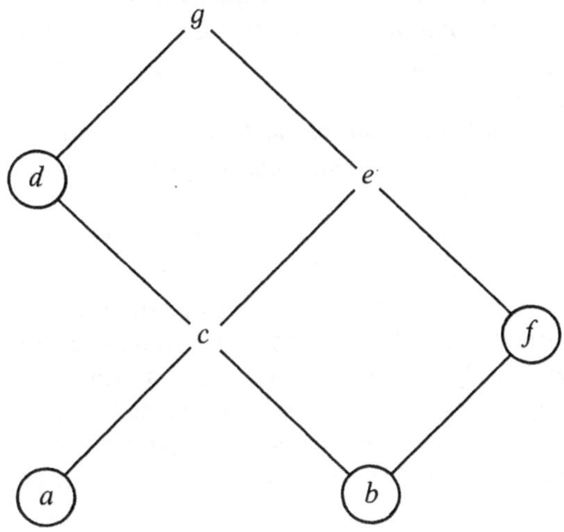

(i.e., $a \vee b = c$, $e \leqslant g$, etc.), the circled elements are the join-irreducibles. Now let **T** be the stochastic matrix theory of 3.1+. Each convex subset Q of a real linear space may be viewed as a **T**-algebra. In this context, isolated elements are called *extreme points*. For example, the extreme points of a plane convex polygon are its vertices. In both of these examples we may speculate upon the observation that the isolated elements form a minimal set of generators. We always have:

3.25 Proposition. *Let Q be the subset of isolated elements of the* **T**-*algebra* $(\bar{Q}, \xi)$ *and let* $R \subset \bar{Q}$ *generate* $(\bar{Q}, \xi)$, *i.e.*, $\langle R \rangle = \bar{Q}$. *Then* $Q \subset R$.

Proof. For $q \in Q$, since $q \in \langle R \rangle$ it is false that R is a subset of $Q - \{q\}$. □

3.26 Extremal-State Algebra. A large part of the theory of finite-dimensional vector spaces concerns "change of coordinates." At the other extreme we have: The **T**-algebra $(\bar{Q}, \xi)$ is *extremal state* if the subset Q of isolated elements of $(\bar{Q}, \xi)$ generates $(\bar{Q}, \xi)$. In this case, elements of Q are called the *states* of $(\bar{Q}, \xi)$. By 3.25, such Q is the unique minimal set of generators of $(\bar{Q}, \xi)$. A nonzero vector space is not extremal state. Every free monoid is extremal state, the states being the words of length one. Since the three-element group is the submonoid generated by either of its nonunit elements, it is false that a quotient of an extremal-state algebra need be extremal state. A subalgebra of an extremal-state algebra may fail to be extremal state (e.g. the interior of a convex polygon has no extreme points). Moreover, an extremal-state algebra can sometimes be embedded in an extremal state algebra with fewer states (e.g. any plane convex polygon can be embedded as a convex subset of the interior of a triangle).

The theory **T** is *extremal state* if every finitely-generated **T**-algebra is extremal state.

We are now ready for:

3.27 Definition. *Let* (Y, θ) *be a* **T**-*algebra and let* $f: IX^{@} \longrightarrow Y$ *be a response. A finite-state minimal realization of* f *is an implicit* λ-*automaton* $\bar{M} = (\bar{Q}, \bar{\delta}, \xi, \bar{\tau}, \bar{\beta})$ *satisfying*

1. *f is the response of $\bar{M}$;*
2. *$(\bar{Q}, \xi)$ is extremal state and the set Q of states is finite; and*
3. *No implicit λ-automaton satisfying (1), (2) has fewer states than the number of elements of Q.*

Some observations about finite-state minimal $\bar{M}$:

(3.28) As Q generates $\bar{Q}$, $i^{\#}: (QT, Qm) \longrightarrow (\bar{Q}, \xi)$ is onto (where $i: Q \longrightarrow \bar{Q}$ is inclusion) and so there exist scoops (Q, i, c) giving rise to a class of λ-automata with state set Q and response f. If a λ-automata realization of f had state set R strictly smaller than Q then—assuming the free **T**-algebra on a finite set of generators is extremal state (which is true for the classical examples of fuzzy automata theory, as discussed below)—we would contradict the minimality of Q. Thus 3.27 captures the flavor of "realizing f with fewest states."

(3.29) If $\mathbf{T}$ is the identity theory then every $\mathbf{T}$-algebra has form (Q, id_Q) and is extremal state with state set Q and the concept of a λ-algebra is just the concept of an X-dynamics. Given a finite-state minimal realization, $\bar{M}$, of f as in 3.27, $\bar{M}$ is reachable (else the image of r becomes the state-set of a smaller realization), and $\bar{M}$ is observable (else the image of σ becomes the state-set of a smaller realization). It follows from the proof of 2.18 that $\bar{M}$ is isomorphic to M_f. This shows that 3.27 generalizes the deterministic minimal realization.

We will see shortly that even if the realization of 3.23 is extremal state with finitely many states, it need *not* be minimal; and that, moreover, the minimal realization need not have surjective reachability map. The explanation is not difficult: as pointed out in 3.26, an extremal state λ-algebra may well be a λ-subalgebra of an extremal state λ-algebra with even fewer states. We first pause to prove:

3.30 Proposition. *Let* $\mathbf{T}$ *be the subset theory of 1.3.5 (the algebras are complete semilattices, 1.5.15) or either of the subtheories "finite subsets," "finite nonempty subsets" (whose algebras are, respectively, finitely complete semilattices and semilattices). Then* $\mathbf{T}$ *is extremal state.*

Proof. The proof is essentially the same in all cases. A finitely-generated $\mathbf{T}$-algebra $(\bar{Q}, \xi)$ is necessarily finite and we may define subsets S_n, Q_n according to the following inductive scheme:

$$S_0 = \langle \varnothing \rangle$$

Q_n is the set of minimal elements of $\bar{Q} - S_n$ if $S_n \neq \bar{Q}$, and need not be defined if $S_n = \bar{Q}$

$$S_{n+1} = \langle Q_0 \cup \cdots \cup Q_n \rangle$$

By finiteness we reach $S_N = \bar{Q}$. Clearly $\bar{Q}$ is generated by $Q_0 \cup \cdots \cup Q_{N-1}$ so that it suffices to show that if q is in Q_n with $0 < n < N$, then q is join irreducible. Suppose that $q = a_1 \vee \cdots \vee a_m$ with no $a_i = q$. If $m = 0$, then (depending on whether or not the empty supremum is a $\mathbf{T}$-operation) either $q \in \langle \varnothing \rangle$ contradicts that q is in Q_n or else $Q_0 = \{q\}$ and q is not in $\langle Q - \{q\} \rangle$. Otherwise $m > 0$ and each $a_i < q$. It follows that no a_i is in $\bar{Q} - S_n$ so that q is in $\langle S_n \rangle = S_n$, the desired contradiction. □

Notice that the above proof gives a simple algorithm to find join-irreducibles in a finite semilattice! Notice, too, that the semilattice of infinite subsets of an infinite set has no join-irreducibles.

3.31 Example. Let us return to the response $f: X^* \longrightarrow Y$ of 3.4. We will show that the realization of 3.23 has six states, whereas the five-state realization given in 3.4 turns out to be minimal. Here, we are dealing with "multiple-branching" realizations and we take $\mathbf{T}$ to be the "nonempty finite subsets" theory and λ as in 3.5; (Y, θ) is the inclusion semilattice of subsets of $\{a, b, c, d\}$. By 3.13, a λ-algebra is a semilattice (Q, ξ) with dynamical structure $\delta: Q \times X \longrightarrow Q$ in such a way that $\delta: Q \longrightarrow Q$ preserves nonempty finite suprema (recall that X has only one element here). It is easy to check that the **T**-algebra $(Y^{X^*}, \theta_{\#})$ of 3.18 is just the product **T**-algebra $(Y, \theta)^{X^*}$ (see exercise 3).

Let us compute the realization $\bar{M}$ of 3.23. Here, $\bar{Q}$ is the image of $\bar{f}$: $X^*T \longrightarrow Y^{X^*}$ which, thinking of $\bar{f}$ as $(f^{\#})_{\#}$, is just the closure under nonempty unions of the X^*-closure of $f \in Y^{X^*}$. Recalling from 2.3 that the X-dynamical structure on Y^{X^*} is $g, x \longmapsto L_{x}.g$ (where $X = \{x\}$), the X^*-closure of f is routinely computed:

$$
\begin{aligned}
f &= a/b/c/d//ab/ac/ad \\
fx &= b/c/d//ab/ac/ad \\
fx^2 &= c/d//ab/ac/ad \\
fx^3 &= d//ab/ac/ad \\
fx^{4+3k} &= //ab/ac/ad \\
fx^{5+3k} &= //ac/ad/ab \\
fx^{6+3k} &= //ad/ab/ac
\end{aligned}
$$

The inclusion relationships among $f, \ldots, fx^6$ are as shown below:

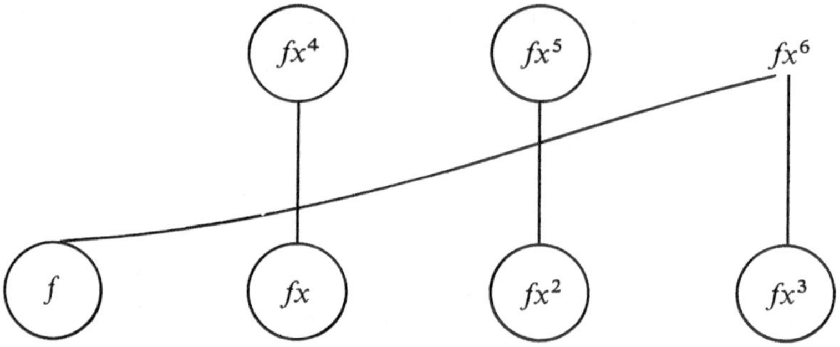

It is clear from 3.25 that the join-irreducibles of $\bar{Q}$ are those elements of $f, \ldots, fx^6$ which cannot be written as the supremum of a subset of the remaining six. Checking that $fx^6 = f \cup fx^3$ (as remarked above, the semilattice structure in Y^{X^*} is elementwise union), the join-irreducibles are then $Q = \{f, \ldots, fx^5\}$ as circled above.

A scoop $c: \bar{Q} \longrightarrow QT$, is just a function which assigns to each element

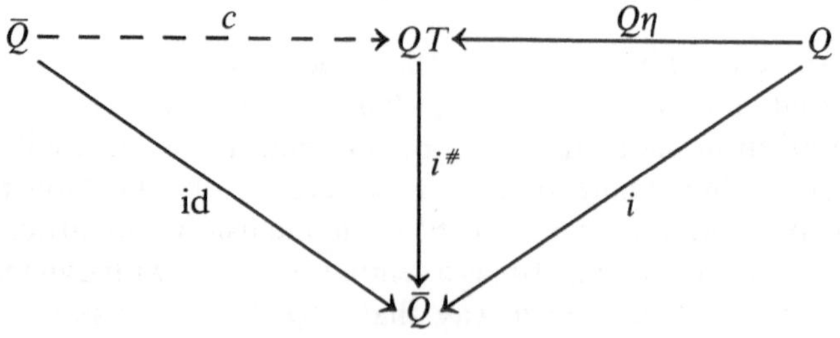

of $\bar{Q}$ a choice of representation as a supremum of elements of Q. To compute the corresponding λ-automaton with state set Q as in 3.22 it is not necessary to know the value of c on all elements of $\bar{Q}$ (which is fortunate since it requires considerable combinatorics to compute $\bar{Q}$). Thus, to define $\delta: Q \times X \longrightarrow QT$ we need only know the values of c on elements of form $L_{x} \cdot q$, that is, on the set $\{fx, \ldots, fx^6\}$. Recalling the proof of 3.23 (specifically, we refer to the formulas "$\tau_f = I\eta . r_f$, $\beta_f = \sigma_f . Y\Lambda$" of 2.18), $\bar{M}$ has $f \in \bar{Q}$ as initial state and has output map $g \longmapsto (\Lambda)g$; thus to define the initial state of the corresponding λ-automaton we need only the value of c on f, and c is not needed at all to define the output map. Of course, $qc = \{q\}$ (for q in Q) is the only reasonable choice in view of the definition of "isolated element." Moreover, $(fx^6)c = \{f, fx^3\}$ is forced since, as it happens, there is no other way to write fx^6 as a supremum of elements of Q. Hence there is only one λ-automaton which can be obtained as in 3.22 by scooping $\bar{M}$ (even though c possibly extends in more than one way to $\bar{Q}$); it is given by:

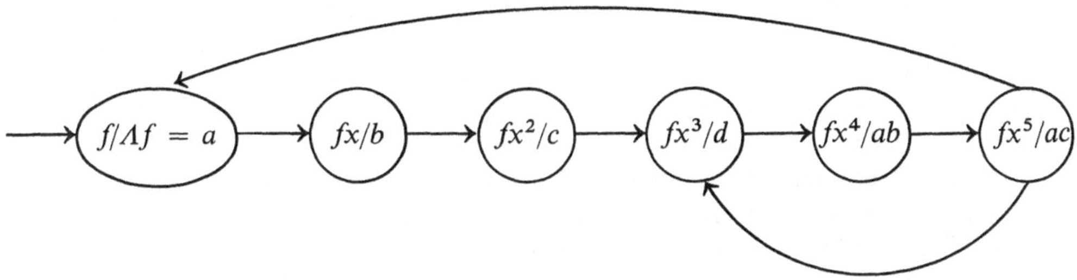

It is already clear from 3.4 that $\bar{M}$ is not minimal in the sense of 3.27. Indeed, we can discover this five-state realization as follows. Let $\tilde{a}$ in Y^{X^*} be constantly $\{a\}$. Then $\{f, \ldots, fx^6, \tilde{a}\}$ is X^*-closed. Since $L_{x}.(-): Y^{X^*} \longrightarrow Y^{X^*}$ preserves nonempty suprema, the union-closure $\bar{R}$ of $\{f, \ldots, fx^6, \tilde{a}\}$ is a λ-subalgebra of $(Y^{X^*}, YL, \theta_{\#})$ (see exercise 4). The inclusion relations on $\{f, \ldots, fx^6, \tilde{a}\}$ are:

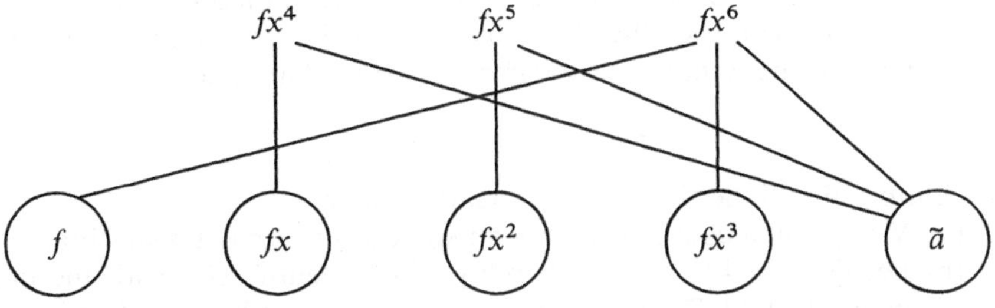

where indeed, $fx^4 = fx \cup \tilde{a}$ and $fx^5 = fx^6 \cup \tilde{a}$. Thus $R = \{f, fx, fx^2, fx^3, \tilde{a}\}$ are the states of $\bar{R}$. The values of a scoop $c: \bar{R} \longrightarrow RT$ restricted to $\{f\} \cup L_x R = \{f, fx, fx^2, fx^3, fx^4, \tilde{a}\}$ are clearly forced to be $rc = \{r\}$ (for r in R), $(fx^4)c = \{fx, \tilde{a}\}$. (Notice that c can take either of two values on fx^6.) The corresponding λ-automaton is then

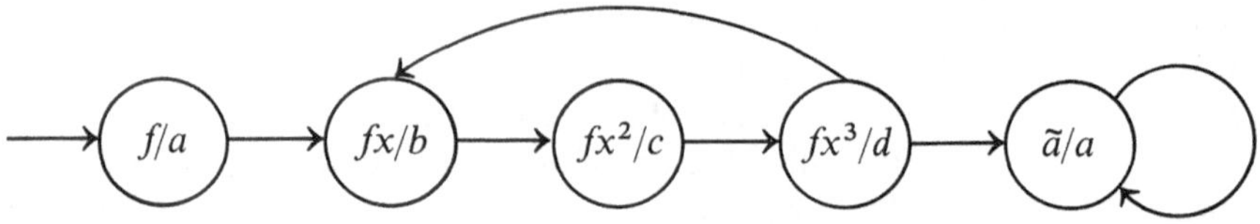

which is precisely the system of 3.4. The implicit λ-automaton $\bar{R}$ is not reachable as $\tilde{a}$ is not in $\bar{Q}$. The implicit λ-automaton $R^\bullet$ which runs R (3.20) is not reachable either since, e.g., $\{f, \tilde{a}\}$ cannot be reached. Nonetheless, R is minimal (we leave to the reader the ad hoc argument that f cannot be realized with only four states).

Notes for Section 3

"Classical fuzzy automata theory" is studied in [Starke '72] and [Paz '71]; for further references, consult the bibliographies there. Fuzzy set theory (a *fuzzy set* is a pair (Q, p) where Q is a set and $p:Q \longrightarrow [0, 1]$ is a unit interval valued "degree of membership" function) was invented by Zadeh ([Zadeh '65]) and has found many applications (see [Goguen '67, '74]). A general theory of fuzzy systems, based on "pseudoclosed categories" rather than algebraic theories, is found in [Ehrig *et al* '74] and Example 3.4–3.31 is their 10.6; the concept of "scoop" was modified from their 10.3 (see exercise 12). The idea of running a fuzzy automaton using a "fuzzy theory" interpretation of **T** and a Beck distributive law was posed by Elisabeth Burroni ([Burroni '73]), but this was not pursued to a minimal realization theory. The approach of this section is due to M. A. Arbib and the author ([Arbib and Manes '75-A]).

Almost all problems concerning the minimal realization of fuzzy systems are uninvestigated at this writing. For example, the ideas of this section effect a rapproachement between multiple-branching realizations and combinatorial problems of finite semilattices. We conjecture that the stochastic matrix theory is extremal-state. Perhaps a modification along the lines of exercise 1 below can establish this; observe that the free stochastic algebra on a finite set is the standard affine simplex in Euclidian space.

Exercises for Section 3

1. (Linton.) Let **T** be the stochastic matrix theory.
   (a) Verify that a convex subset of a real vector space is a **T**-algebra.
   (b) Let $Q = \{a, b\}$. Via "probability of a," identify QT with the unit interval $[0, 1]$. Define $\xi:QT \longrightarrow Q$ by $\lambda\xi = 1$ if and only if $\lambda = 1$. Show that (Q, ξ) is a **T**-algebra. [Hint: if (p_λ) is in $[0, 1]T$, $\Sigma p_\lambda \lambda = 1$ if and only if $p_1 = 1$.] Conclude that not every **T**-algebra may be embedded in a real linear space.
   (c) Doubly-generated **T**-algebras can be classified by observing that a

congruence on $2T$ is an equivalence relation on $2T$ which is convex (in the usual sense) as a subset of the unit square $2T \times 2T$. Observing that whenever R is a congruence and (λ, μ) is in R then $[\lambda, \mu] \times [\lambda, \mu] \subset R$, each such R is seen to be the union of the diagonal and $A \times A$ where A is a (closed, open, or half-open) interval in $[0, 1]$. Verify these facts and show that every two-element **T**-algebra is isomorphic to the one in (b). Show that, similarly, there is only one three-element doubly-generated **T**-algebra and that there are no other finite doubly-generated **T**-algebras.

One interpretation of these doubly-generated stochastic algebras is as an immediate generalization of Zadeh's unit interval.

2. Let $Y = \{0, 1\}$. Each response $f: X^* \longrightarrow Y$ is then the characteristic function χ_A of a subset ("language") A of X^*. A system realizing such a response is called an *acceptor* (i.e., M "accepts" w in X^* if w is in A). Let $X = \{x, y\}$ and let $A = y^*\{x, y\}$.
   (a) Show that the deterministic minimal realization of χ_A has four states. [Hint: for z in X, $B \subset X^*$, $L_z B = \{w : zw \in B\}$; the four states are A, $y^*x \cup y^*$, $\{\Lambda\}$, $\varnothing$.]
   (b) Show that the canonical "multiple-branching" realization of 3.23 and 3.22 (where **T** is "nonempty finite subsets") has unique scoop, has three states, and is minimal in the sense of 3.27.
   (c) Let **T** be "all finite subsets". Show that the canonical realization of 3.23 and 3.22 has only two states and is minimal; indeed, is given by

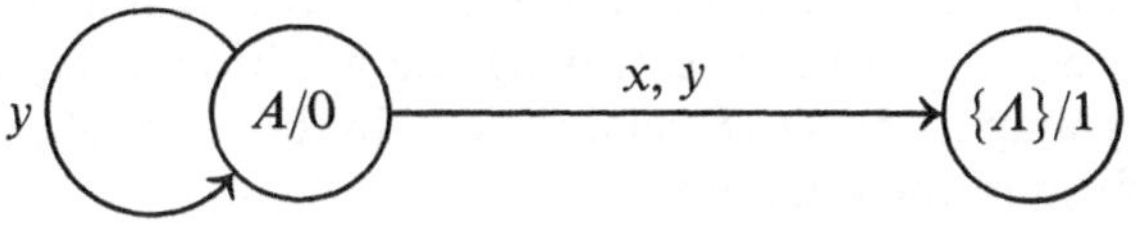

(formally, writing $q_0 = A$ and $q_1 = \{\Lambda\}$, $(q_0, x)\delta = \{q_1\}$, $(q_0, y)\delta = \{q_0, q_1\}$, $(q_1, x)\delta = (q_1, y)\delta = \varnothing$). According to 3.9, the interpretation of "$\delta = \varnothing$" is that "the computation halts."

3. When $\mathscr{K} = \mathbf{Set}$, the only input- and output process is of the form $- \times X$ (see exercise 2.1(b)). For arbitrary **T**, show that $(Y^{X^*}, YL)\hat{T}$ has dynamics $Y^{X^*}T \times X \longrightarrow Y^{X^*}T$, $(p, x) \longmapsto \langle p, (L_{x^\cdot} -)T\rangle$. Establish the diagram

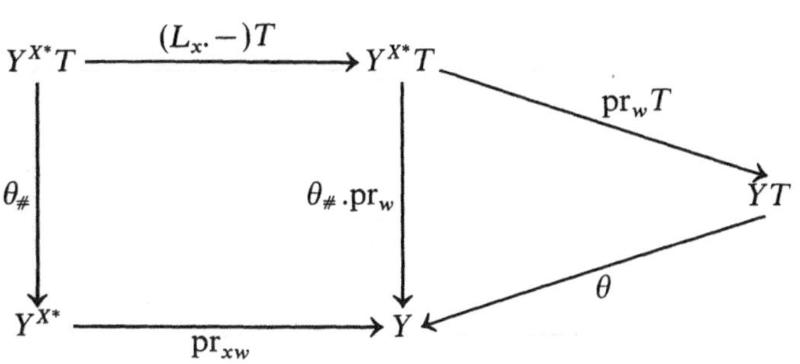

[Hint: the square is the definition of $\theta_{\#}$ of 2.3 and the triangle is the inductive hypothesis.] Conclude that $(Y^{X^*}, \theta_{\#})$ is the product **T**-algebra $(Y, \theta)^{X^*}$.

4. Let λ be a distributive law of X over **T**. Show that the unique λ-homomorphic extension of $f: A \longrightarrow (Q, \delta, \xi)$ can be constructed as the unique **T**-homomorphic extension of the unique dynamorphic extension as indi-

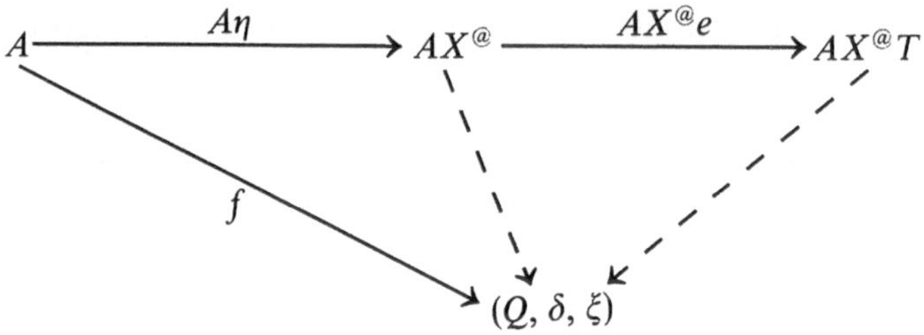

cated above. Conclude that, when $\mathscr{K} = $ **Set**, if A is a subset of a λ-algebra, then the λ-subalgebra generated by A is just the **T**-subalgebra of the $X^{@}$-subalgebra generated by A. Can you generalize this statement to regular categories?

5. Show that the semilattice $\bar{R}$ of 3.31 can be embedded as a subsemilattice of a finite semilattice with only four join-irreducibles.

6. In this exercise we relate the distributive laws of 3.6 to the distributive laws studied by Beck ([Beck '69]). Let $\mathbf{S} = (S, \eta, \mu)$ and $\mathbf{T} = (T, e, m)$ be algebraic theories in $\mathscr{K}$. A *distributive law of* **S** *over* **T** is a distributive law $\lambda: TS \longrightarrow ST$ of S over **T** satisfying the additional requirement that λ is a theory map from **S** to **S** relative to T in the sense of exercise 3.2.10, that is we require the diagrams:

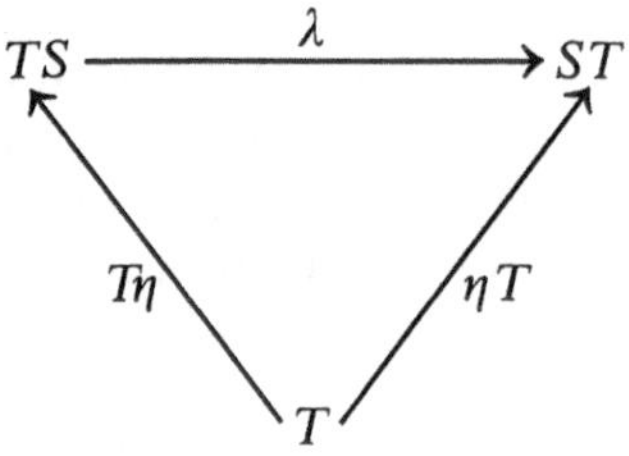

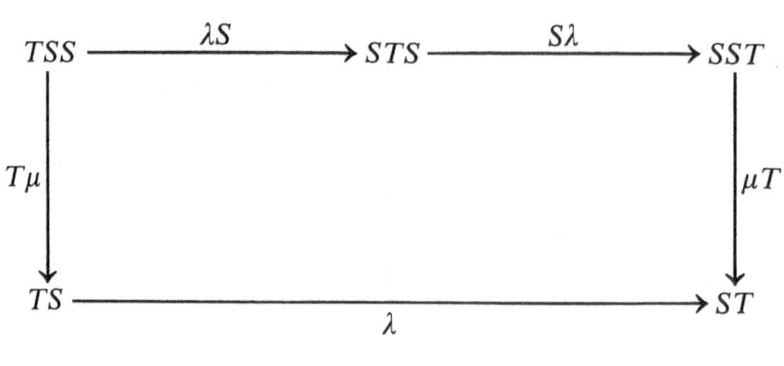

(a) Let X be an input process in $\mathscr{K}$ and let $\lambda: TX \longrightarrow XT$ be a distributive law of X over $\mathbf{T}$. Recalling that $AX^{@}T$ has X-dynamical structure $AX^{@}\lambda . A\mu_0 T : AX^{@}TX \longrightarrow AX^{@}T$, define $\bar{\lambda}$: $TX^{@} \longrightarrow X^{@}T$ as the unique dynamorphic extension

$$\begin{array}{ccc} TX^{@} & \overset{\bar{\lambda}}{\dashrightarrow} & X^{@}T \\ {\scriptstyle T\eta}\uparrow & & \uparrow{\scriptstyle \eta T} \\ TX & \underset{\lambda}{\longrightarrow} & XT \end{array}$$

Prove that $\lambda \longmapsto \bar{\lambda}$ establishes a bijective correspondence between distributive laws of X over $\mathbf{T}$ and distributive laws of $\mathbf{X}^{@}$ over $\mathbf{T}$.

If $\lambda: TS \longrightarrow ST$ is a distributive law of $\mathbf{S}$ over $\mathbf{T}$, a *λ-algebra* is (Q, γ, ξ) where (Q, γ) is an $\mathbf{S}$-algebra and (Q, ξ) is a $\mathbf{T}$-algebra subject to the *λ-law*

$$\begin{array}{ccc} QTS & \overset{\xi S}{\longrightarrow} & QS \\ {\scriptstyle Q\lambda}\downarrow & & \\ QST & & \downarrow{\scriptstyle \gamma} \\ {\scriptstyle \gamma T}\downarrow & & \\ QT & \underset{\xi}{\longrightarrow} & Q \end{array}$$

(b) In the context of (a), prove that the category of λ-algebras is isomorphic to the category of $\bar{\lambda}$-algebras.

(c) If λ is a distributive law of $\mathbf{S}$ over $\mathbf{T}$ then, mimicking the proof of 3.15, prove that $\mathscr{K}^{\lambda} \longrightarrow \mathscr{K}^{\mathbf{S}}$ is algebraic and that $\mathscr{K}^{\lambda} \longrightarrow \mathscr{K}$ is algebraic with algebraic theory

$$\mathbf{ST} = (ST, \mathrm{id} \xrightarrow{\eta e} ST, STST \xrightarrow{S\lambda T} SSTT \xrightarrow{\mu m} ST)$$

(d) If λ is a distributive law of $\mathbf{S}$ over $\mathbf{T}$, prove that $\mathscr{K}^{\lambda} \longrightarrow \mathscr{K}^{\mathbf{T}}$ is algebraic.

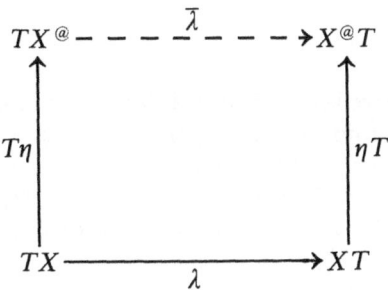

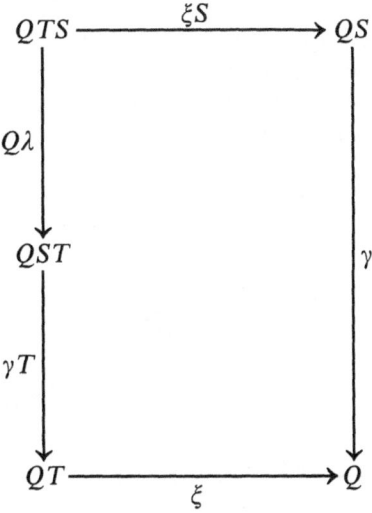

(e) This is the proper formulation of 3.8. Let $\mathscr{K} = \mathbf{Set}$, let $\mathbf{S}$ be the theory for monoids, and let $\mathbf{T}$ be the theory for abelian groups. Prove that $\lambda: TS \longrightarrow ST$ defined by

$$QTS \xrightarrow{Q\lambda} QST$$

$$\prod_{i=1}^{m} \sum_{Q} n_{q,i}q \longmapsto \sum_{Q^m} n_{q_1} \cdots n_{q_m}(q_1 \cdots q_m)$$

is a distributive law of $\mathbf{S}$ over $\mathbf{T}$. Verify that λ-algebras are rings.

7. Let M be a monoid and let $\mathbf{T}_M$ be the corresponding theory in $\mathbf{Set}$ of 1.3. For any theory $\mathbf{S}$ in $\mathbf{Set}$, $\mathbf{S} \otimes \mathbf{T}_M$ exists by 3.6.21.
   (a) Show that $A\lambda: AS \times M \longrightarrow (A \times M)S$ as in 3.7 is a distributive law of $\mathbf{T}_M$ over $\mathbf{S}$ (as in exercise 6) and that $\mathbf{Set}^{\lambda} = \mathbf{Set}^{\mathbf{S} \otimes \mathbf{T}_M}$.
   (b) Let $\mathbf{T} = (2^{(- \times M)}, \eta, \circ)$ be the "relations with credibility" theory obtained by setting $\mathbf{S}$ to be the "subsets" theory. Verify that

$$Q \xrightarrow{Q\eta} 2^{Q \times M}$$

$$q \longmapsto \{(q, e)\}$$

and that, given $\alpha: A \longrightarrow 2^{B \times M}$ and $\beta: B \longrightarrow 2^{C \times M}$,

$$(\alpha \circ \beta)_a = \{(c, mm'): \exists\, b, m, m' \text{ with } (b, m) \in \alpha_a, (c, m') \in \beta_b\}$$

8. ([Alagić '74]; cf. [Applegate '65], [Manes '67, 1.7], [Meyer '72].) Let $X: \mathscr{K} \longrightarrow \mathscr{K}$ be an endofunctor and let $\mathbf{T} = (T, e, m)$ and $\mathbf{S} = (S, e', m')$ be theories in $\mathscr{K}$. A natural transformation $\lambda: SX \longrightarrow XT$ is an *inverse-state transformation* providing the following diagrams are commutative:

$$\begin{array}{ccc} X & \xrightarrow{e'X} & SX \\ & \searrow^{Xe} & \downarrow \lambda \\ & & XT \end{array}$$

$$\begin{array}{ccccc} SSX & \xrightarrow{S\lambda} & SXT & \xrightarrow{\lambda T} & XTT \\ \downarrow m'X & & & & \downarrow Xm \\ SX & & \xrightarrow{\lambda} & & XT \end{array}$$

   (a) Let $(-)^{\Delta}: \mathscr{K} \longrightarrow \mathscr{K}_{\mathbf{T}}$ be the functor of 1.3.9. Prove that the passage $\bar{X} \longrightarrow \lambda$, $Q\lambda: QSX \longrightarrow QXT = (\mathrm{id}_{QS}: QS \longrightarrow Q)\bar{X}$

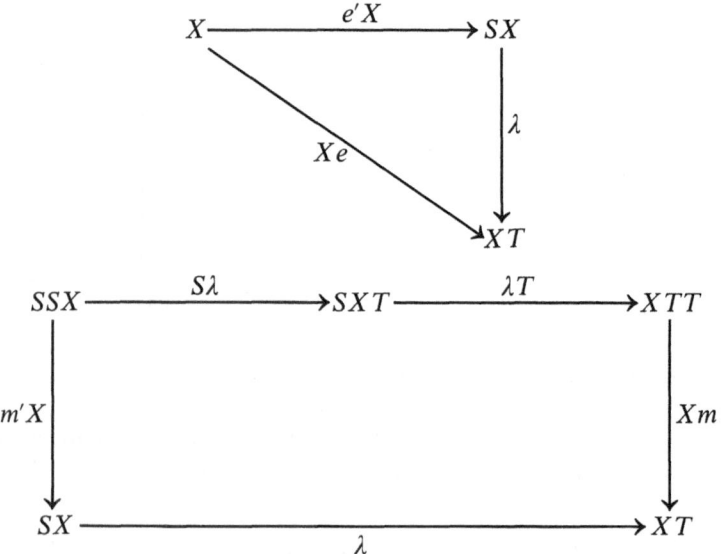

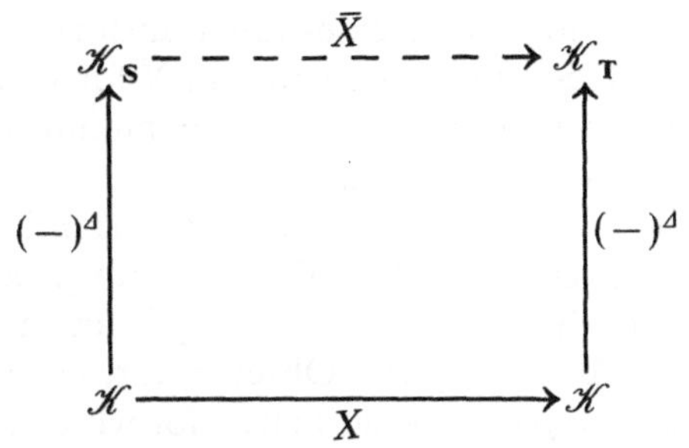

establishes a bijective correspondence from liftings $\bar{X}$ as above to inverse-state transformations, with inverse passage

$$\lambda \longrightarrow \bar{X}, (A \xrightarrow{\alpha} B)\bar{X} = AX \xrightarrow{\alpha X} BSX \xrightarrow{B\lambda} BXT$$

(b) Let $Z:\mathscr{K} \longrightarrow \mathscr{K}$ be an input process, let X have a right adjoint $X^{\bullet}$, and let $\lambda_{\bullet}:ZX \longrightarrow XT$ be a natural transformation. Prove that there exists a unique inverse-state transformation $\lambda:Z^{@}X \longrightarrow XT$ such that

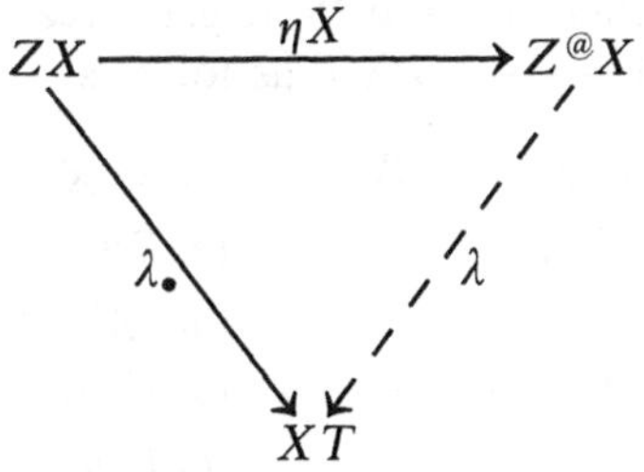

[Hint:

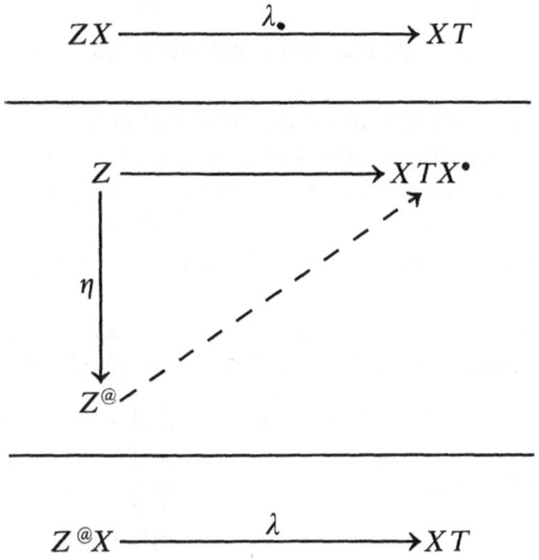

where $Z^{@} \longrightarrow XTX^{\bullet}$ is a theory map.]

(c) When $X = \mathrm{id}_{\mathscr{K}}$, an inverse-state transformation is just a theory map $\lambda:\mathbf{S} \longrightarrow \mathbf{T}$. Study the structure of $\bar{X}$ for various examples of theory maps. Observe that $\bar{X}$ does *not* commute with the underlying $\mathscr{K}$-object functors in general.

(d) When $\mathbf{S} = \mathbf{T}$, an inverse-state transformation λ is just a distributive law of X over T. If, also, X is an input process, prove that the corresponding lift $\bar{X}:\mathscr{K}_{\mathbf{T}} \longrightarrow \mathscr{K}_{\mathbf{T}}$ is an input process in $\mathscr{K}_{\mathbf{T}}$ with $A\bar{X}^{@} = AX^{@}$ on objects. Observe, further, that an $\bar{X}$-automaton is the same thing as a λ-automaton for which the output algebra (Y, θ) is a free $\mathbf{T}$-algebra, and that the responses coincide.

(e) The following example illustrates the computer science origins of inverse-state transformations. See the bibliography of [Alagić '74] for more about applications of inverse-state (and also direct-state) natural transformations in computer science. We thank Jim Thatcher for calling our attention to this example. In the context of (d), let $X = - \times X_0$ with $X_0 = \{D, I\}$ (for "derivative" and "identity"), let $Z = X_\Omega$ where $\Omega_1 = \{S\}$ (for "sine") and $\Omega_2 = \{+, m\}$ and let $\mathbf{T}$ be the theory whose algebras are rings equipped with two unary functions labelled "C" and "S" (e.g. the reals with "cosine" and "sine"). Define $\lambda_0:ZX \longrightarrow XT$ as follows:

$$
\begin{array}{lcl}
AZ \times X_0 & \xrightarrow{A\lambda_0} & (A \times X_0)T \\
(ab+, D) & & (a, D)(b, D)+ \\
(abm, D) & & (a, I)(b, D)m(a, D)(b, I)m+ \\
(aS, D) & \longmapsto & (a, I)C(a, D)m \\
(ab+, I) & & (a, I)(b, I)+ \\
(abm, I) & & (a, I)(b, I)m \\
(aS, I) & & (a, I)S
\end{array}
$$

Verify that λ_0 is natural and construct its extension $\lambda:Z^{@}X \longrightarrow XT$.

9. In this exercise we speculate on "internal equality" for fuzzy states. Let $\mathbf{T}$ be a fuzzy theory in **Set**. Let $2 = \{0, 1\}$. $2T$ is the *object of* $\mathbf{T}$-*truth values*. For any set A, define the *coefficient map* $\mathrm{coeff}:AT \longrightarrow 2T^A$ and the *internal equality map* $\mathrm{eq}:AT \times AT \longrightarrow 2T$ as follows. Set $\mathrm{coeff} = \alpha^{\#}$ where $\alpha:A \longrightarrow 2T^A$, $a \longmapsto \chi_{\{a\}}.2e$. Define $\mathrm{eq}(p, q) = \langle q, (\mathrm{coeff}(p))^{\#}\rangle$.

(a) Let $\mathbf{T}$ be the subsets theory. Interpret $2T = \{\phi, \{0\}, \{1\}, 2\}$ as "no information," "false," "true," and "maybe." Show that

$$
\mathrm{eq}(A, B) = \begin{cases} \text{no information} & \text{if } A = \phi \text{ or } B = \phi \\ \text{false} & \text{if } A \neq \phi \neq B \text{ and } A \cap B = \phi \\ \text{true} & \text{if } A = B = \{x\} \text{ for some } x \\ \text{maybe} & \text{otherwise} \end{cases}
$$

(b) Compute coeff and eq for $\mathbf{T} = L$-fuzzy relations. [Hint: an element of $2T$ is a "degree of membership" together with a "degree of non-membership."]

(c) For $\mathbf{T}$ = stochastic matrices, show that coeff(λ_a) sends a to λ_a and that eq(p, q) is "the probability that independent observations of p, q are equal." Compare this interpretation with (a).

10. Zadeh's fuzzy sets and Goguen's generalizations (see the notes) support the philosophy that the set QT of fuzzy states should be ordered by "degree of membership." In this exercise we provide some axioms and give a common framework for universal relational algebra and Goguen's category of L-fuzzy sets.

Let $\mathbf{T}$ be an algebraic theory in **Set**. A *degree relation on* $\mathbf{T}$ is an assignment to each set A, a reflexive and transitive relation $\leqslant$ on AT subject to:

(Uniformity axiom) For all $\alpha: A \longrightarrow BT$, $\alpha^{\#}: AT \longrightarrow BT$ preserves $\leqslant$.

(Causality axiom) If $\alpha: A \longrightarrow AT$ and p in AT are such that $\alpha_a \leqslant p$ for all a in A, then $p\alpha^{\#} \leqslant p$.

The uniformity axiom ensures, for example, that isomorphic sets induce isomorphic orders. The causality axiom states that the causal operations on fuzzy states (which "build trees of trees") cannot increase the degree.

For any functor $X: \mathbf{Set} \longrightarrow \mathbf{Set}$, let $[X, \mathbf{T}, \leqslant]$ be the category of sets with structure with objects all pairs (A, δ) with $\delta: AX \longrightarrow AT$, and with morphisms $f:(A, \delta) \longrightarrow (A', \delta')$ such that

$$\begin{array}{ccc} AX & \xrightarrow{fX} & BX \\ {\scriptstyle\delta}\big\downarrow & \leqslant & \big\downarrow{\scriptstyle\delta'} \\ AT & \xrightarrow[fT]{} & BT \end{array} \qquad \begin{array}{l} \langle t, \delta.fT\rangle \leqslant \langle t, fX.\delta'\rangle \\ \text{for all } t \text{ in } AX \end{array}$$

In the examples below, verify that $\leqslant$ is a degree relation on $\mathbf{T}$. In the first three, Ω refers to an arbitrary operator domain.

(a) Let $X = X_\Omega$, let $\mathbf{T}$ be the subsets theory, and let $S_1 \leqslant S_2$ be inclusion. Show that $[X, \mathbf{T}, \leqslant]$ may be identified with the category of *relational Ω-algebras* whose objects are pairs (A, δ) such that δ assigns to each $\omega \in \Omega_n$ an n-ary relation $\delta_\omega: A^n \longrightarrow A$ and whose admissible maps $f:(A, \delta) \longrightarrow (A', \delta')$ are defined by the requirement that whenever $((a_i), a) \in \delta_\omega$, $((a_i f), af) \in \delta'_\omega$.

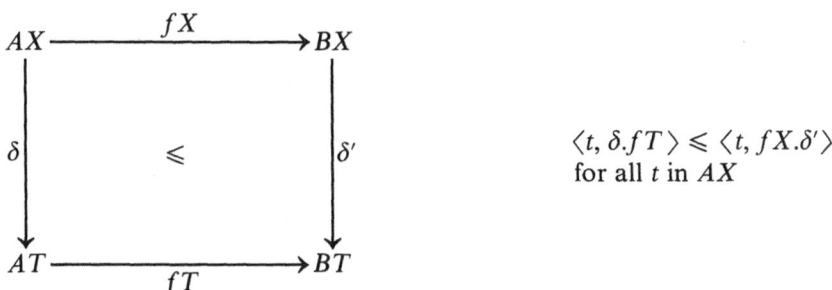

(b) Let $X = X_\Omega$, let $BT = B + 1$ as the subtheory "at most one element" of the subsets theory, and let $\leqslant$ be inclusion. Show that $[X, \mathbf{T}, \leqslant]$ may be identified with the category of *partial Ω-algebras* whose objects are pairs (A, δ) where δ assigns to each $\omega \in \Omega_n$ a partial function $\delta_\omega : A^n \longrightarrow A$ and whose admissible maps $f:(A, \delta) \longrightarrow (A', \delta')$ are defined by the requirement that whenever $(a_i)\delta_\omega$ is defined then so, too, is $(a_i f)\delta'_\omega$, and, moreover, $(a_i)\delta_\omega f = (a_i f)\delta'_\omega$.

(c) Let $X = X_\Omega$, let $\mathbf{T}$ be the stochastic matrix theory, and define $p \leqslant q$ to mean the support of p (in the sense of 1.5.10) is contained in the support of q; specifically, $(\lambda_a) \leqslant (\lambda'_a)$ if whenever $\lambda_a \neq 0$ then also $\lambda'_a \neq 0$. Show that the morphisms in $[X, \mathbf{T}, \leqslant]$ are characterized by the requirement that "*if* $(a_i)\delta_\omega = a$ with nonzero probability then also $(a_i f)\delta'_\omega = af$ with nonzero probability."

(d) Let $AX = 1$ for all A, let $\mathbf{T}$ be the theory induced by a complete distributive lattice L as in 3.3, and let $\leqslant$ be pointwise. Show that $[X, \mathbf{T}, \leqslant] =$ Goguen's $\mathbf{Set}(L)$ (as in exercise 3.5.7).

(e) Generalizing exercise 3.5.7, show that $[X, \mathbf{T}, \leqslant]$ is fibre complete whenever $(AT, \leqslant)$ is a complete lattice for every A.

11. Let $\mathbf{T}$ be the stochastic matrix theory. Show that an element in a convex subset of a real vector space is an extreme point if and only if it is an isolated point in the topology of exercise 2.3.8(c). To what extent can this be generalized to arbitrary $\mathbf{T}$-algebras?

12. This exercise explicates how scoops in the sense of [Ehrig *et. al.* '74, 10.3] give rise to scoops as in 3.21. Let X be an output process, let M be a λ-automaton as in 3.9 and let $f:A \longrightarrow Q$, $g:Q \longrightarrow AT$ satisfy

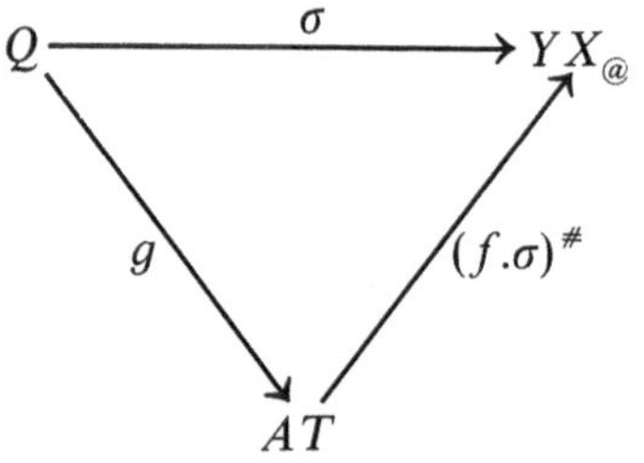

Assume that $\sigma^\#:QT \longrightarrow YX_@$ has a split epi-mono factorization

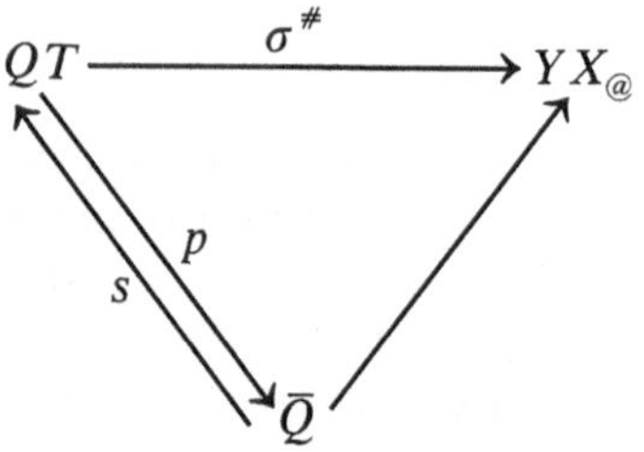

Prove that (A, i, c) is a scoop of $\bar{Q}$ (which is a $\mathbf{T}$-subalgebra of $(YX_@, \theta_\#)$ where

$$i = A \xrightarrow{\ f\ } Q \xrightarrow{\ Qe\ } QT \xrightarrow{\ p\ } \bar{Q}$$

$$c = \bar{Q} \xrightarrow{\ s\ } QT \xrightarrow{\ g^\#\ } AT$$

Bibliography

J. Adámek [1974]: Free algebras and automata realizations in the language of categories, *Comm. Math. Univ. Carolinae* **15**, 589–602.

J. Adámek [1975]: see Trnková, Adámek, Koubek and Reiterman.

S. Alagić [1975]: Natural state transformations, *J. Comp. Sys. Sci.* **10**, 266–307.

B. D. O. Anderson, M. A. Arbib, and E. G. Manes [1974]: Foundations of system theory: finitary and infinitary conditions, Lecture Notes in Economics and Mathematical Systems **115**, Springer-Verlag; to appear.

H. Appelgate [1965]: Acyclic models and resolvent functors, dissertation, Columbia University.

H. Appelgate and M. Tierney [1969]: Categories with models, *Lecture Notes in Mathematics* **80**, Springer-Verlag, 156–243.

M. A. Arbib [1969]: see Kalman, Falb, and Arbib.

—[1974]: see Bobrow and Arbib.

—[1974]: see Padulo and Arbib.

M. A. Arbib and E. G. Manes [1974–A]: Machines in a category: an expository introduction, *SIAM Rev.* **16**, 163–192.

—[1974–B]: Foundations of system theory: decomposable systems, *Automatica* **10**, 285–302.

—[1975–A]: *Arrows, Structures, and Functors: The Categorical Imperative*, Academic Press.

—[1975–B]: Adjoint machines, state behavior machines and duality, *J. Pure Appl. Alg.*, to appear.

—[1975–C]: Fuzzy machines in a category, Bull. Austral. Math. Soc. **13**, 169–210.

R. F. Arens and J. Eells Jr. [1956]: On embedding uniform and topological spaces, *Pacific J. Math.* **6**, 397–403.

J. Auslander [1963]: Endomorphisms of minimal sets, *Duke Math. J.* **30**, 605–614.

B. Banaschewski [1953]: Uber die Konstruktion wohlgeordeter Mengen, *Math. Nach.* **10**, 239–245.

B. Banaschewski and E. Nelson [1975]: Tensor products and bimorphisms, preprint.

Y. Bar-Hillel [1973]: see Fraenkel, Bar-Hillel, and Levy.

M. Barr [1969]: Coalgebras in a category of algebras, *Lecture Notes in Mathematics* **86**, Springer-Verlag, 1–12.

—[1970–A]: Relational algebras, *Lecture Notes in Mathematics* **137**, Springer-Verlag, 39–55.

—[1970–B]: Coequalizers and free triples, *Math. Zeit.* **116**, 307–322.

—[1971]: Factorizations, generators and rank, unpublished manuscript.

M. Barr and J. Beck [1966]: Acyclic models and triples, *Proceedings of the Conference on Categorical Algebra at La Jolla*, Springer-Verlag, 336–344.

—[1969]: Homology and standard constructions, *Lecture Notes in Mathematics* **80**, Springer- Verlag, 245–336.

J. Beck [1966, 1969]: see Barr and Beck

—[1967]: Triples, algebras and cohomology, dissertation, Columbia University.

—[1969]: Distributive laws, *Lecture Notes in Mathematics* **80**, Springer-Verlag, 119–140.

J. C. Bell and A. B. Slomson [1971]: *Models and Ultraproducts*, North Holland/American Elsevier.

J. Bénabou [1963]: Catégories avec multiplication, *C. R. Acad. Sci. Paris* **256**, 1887–1890.

—[1966]: Structures algebriques dans les catégories, dissertation, Université de Paris, 1966.

D. B. Benson [1974]: An abstract machine theory for formal language parsers, *Acta Informatica* **3**, 187–202.

G. Bergman [1975]: Some category-theoretic ideas in algebra, preprint.
G. Birkhoff [1935]: On the structure of abstract algebras, *Proc. Camb. Phil. Soc.* **31**, 433–454.
—[1946]: Universal algebra, *Proc. Can. Math. Cong. Montreal*, 310–326.
—[1967]: see Mac Lane and Birkhoff.
—[1967]: Lattice Theory, *AMS Colloquium Publications* **25**, Providence, American Mathematical Society.
G. Birkhoff and J. D. Lipson [1970]: Heterogeneous algebras, *J. Comb. Thy.* **8**, 115–133.
A. Blanchard [1971]: Structure species and forgetful functors, unpublished manuscript.
L. S. Bobrow and M. A. Arbib [1974]: *Discrete Mathematics*, Saunders.
N. Bourbaki [1957]: Théorie des Ensembles, Livre I, Chapitre 4, Hermann.
H.-B. Brinkmann and D. Puppe [1966]: Kategorien und funktoren, *Lecture Notes in Mathematics* **18**, Springer-Verlag.
I. Bucur and A. Deleanu [1968]: *Introduction to the Theory of Categories and Functors*, Wiley Interscience.
M. Bunge [1969]: Relative functor categories and categories of algebra, *J. Alg.* **2**, 64–101.
E. Burroni [1973]: Algèbres relatives à une loi distributive, *C. R. Acad. Sc. Paris* **276** (5 fév.) Ser. A, 443–446.
G. Calugareanu Jr. [1970]: see Wiesler and Calugareanu.
H. Cartan and S. Eilenberg [1956]: *Homological Algebra*, Princeton.
G. Choquet [1948]: Convergences, *Ann. Univ. Grenoble Sect. Sci. Math. Phy.* **23**, 57–112.
P. M. Cohn [1965]: *Universal Algebra*, Harper and Row.
R. Davis [1967]: Equational systems of functors, *Lecture Notes in Mathematics* **47**, Springer-Verlag, 1967, 92–109.
A. Day [1975]: Filter monads, continuous lattices and closure systems, *Canad. J. Math.* **XXVII**, 50–59.
A. Deleanu [1968]: see Bucur and Deleanu.
K. H. Diener [1966]: Order in absolutely free and related algebras, *Coll. Math.* **14**, 62–72.
J. A. Dieudonné [1970]: The work of Nicholas Bourbaki, *Amer. Math. Monthly* **77**, 134–145.
E. Dubuc [1970]: Kan extensions in enriched category theory, *Lecture Notes in Mathematics* **145**, Springer-Verlag.
—[1973]: Free monoids, unpublished manuscript.
E. J. Dubuc and H. Porta [1971]: Convenient categories of topological algebras and their duality theory, *J. Pure Appl. Alg.* **1**, 275–279.
J. Duskin [1969]: Variations on Beck's tripleability criterion, *Lecture Notes in Mathematics* **106**, Springer-Verlag, 74–129.
—[1974]: $K(\pi, n)$-torsors and the interpretation of "triple" cohomology, *Proc. Nat. Acad. Sci. USA* **71**, 2554–2557.
—[1975]: On the interpretation of cohomology theories definable through standard constructions ("triple cohomology"), manuscript, 135 pp.; to appear as an *Amer. Math. Soc. Memoir*.
B. Eckmann and P. J. Hilton [1962]: Group-like structures in general categories, I, *Math. Ann.* **145**, 227–255.
J. Eells Jr. [1956]: see Arens and Eells.
C. Ehresmann [1965]: *Catégories et Structures*, Dunod, Paris.
H. Ehrig (with K.-D. Kiermeier, H.-J. Kreowski, and W. Kühnel) [1974]: *Universal Theory of Automata: A Categorical Approach*, Teubner.
S. Eilenberg [1956]: see Cartan and Eilenberg.
—[1974]: *Automata, Languages and Machines, Vol. A*, Academic Press.
S. Eilenberg and C. C. Elgot [1970]: *Recursiveness*, Academic Press.
S. Eilenberg and G. M. Kelly [1966]: Closed categories, *Proceedings of the Conference on Categorical Algebra at La Jolla*, Springer-Verlag, 421–562.

—[1966–A]: A generalization of the functorial calculus, *J. Algebra* **3**, 366–375.
S. Eilenberg and S. Mac Lane [1945]: General theory of natural equivalences, *Trans. Amer. Math. Soc.* **58**, 231–294.
S. Eilenberg and J. C. Moore [1965]: Adjoint functors and triples, *Ill. J. Math.* **9**, 381–398.
S. Eilenberg and N. Steenrod [1952]: Foundations of algebraic topology, Princeton.
S. Eilenberg and J. Wright [1967]: Automata in general algebras, *Inf. Cont.* **11**, 52–70.
C. C. Elgot [1970]: see Eilenberg and Elgot
—[1975]: Monadic computation and iterative algebraic theories, in Proceedings of the Logic Colloquium, Bristol, July 1973, H. E. Rose and J. C. Shepherdson (ed.), North-Holland, 1975.
R. Ellis [1958]: Distal transformation groups, *Pacific J. Math.* **8**, 401–405.
—[1960–A]: Universal minimal sets, *Proc. Amer. Math. Soc.* **11**, 540–543.
—[1960–B]: A semigroup associated with a transformation group, *Trans. Amer. Math. Soc.* **94**, 272–281.
—[1969]: *Lectures on Topological Dynamics*, W. A. Benjamin.
R. Ellis and W. H. Gottschalk [1960]: Homomorphisms of transformation groups, *Trans. Amer. Math. Soc.* **94**, 258–271.
H. B. Enderton [1972]: *A Mathematical Introduction to Logic*, Academic Press.
P. Falb [1969]: see Kalman, Falb, and Arbib.
W. Felscher [1965–A]: Zur algebra unendlich langer Zeichenreihen, *Zeit. Math. Logik. Grundlagen Math.* **11**.
—[1965–B]: *Adjungierte Funktoren und Primitive Klassen*, Heidelberg, Springer, 1965.
—[1968–A]: Equational maps, in *Contributions to Mathematical Logic* (K. Schütte, ed.) North Holland, 121–161.
—[1968–B]: Kennzeichnung von primativen und quasiprimitiven kategorien von algebren, *Arch. d. Math.* **19**, 390–397.
—[1969]: Birkhoffsche und kategorische algebra, *Math. Ann.* **180**, 1–25.
—[1972]: Equational classes, clones, theories, and triples, unpublished manuscript.
H. R. Fischer [1974]: Convergence structures, preprint.
A. A. Fraenkel, Y. Bar-Hillel, and A. Levy [1973]: *Foundations of Set Theory*, North Holland.
P. Freyd [1964]: *Abelian Categories*, Harper and Row.
—[1966]: Algebra-valued functors in general and tensor products in particular, *Coll. Math.* **14**, 89–106.
—[1973]: Concreteness, *J. Pure Appl. Alg.* **3**, 171–191.
P. Gabriel and F. Ulmer [1971]: Lokal präsentierbare kategorien, *Lecture Notes in Mathematics* **221**, Springer-Verlag.
H. Gaifmann [1964]: Infinite Boolean polynomials I, *Fund. Math.* **54**, 229–250.
D. C. Gerneth [1948]: Generalization of Menger's result on the structure of logical formulas, *Bull. Amer. Math. Soc.* **54**, 803–804.
D. Gildenhuys [1971]: see Kennison and Gildenhuys.
R. Godement [1958]: *Théorie des Faisceaux*, Hermann.
J. A. Goguen [1967]: L-fuzzy sets, *J. Math. Anal. Appl.* **18**, 145–174.
— [1969]: The logic of inexact concepts, *Synthese* **19**, 325–373.
—[1972]: Minimal realization of machines in closed categories, *Bull. Amer. Math. Soc.* **78**, 777–783.
—[1974]: Concept representation in natural and artificial languages: axioms, extensions and applications for fuzzy sets, *Int. J. Man-Machine Studies* **6**, 513–561.
J. A. Goguen, J. W. Thatcher, E. G. Wagner and J. B. Wright [1975]: An introduction to categories, algebraic theories and algebras, *IBM Technical Report* RC 5369, T. J. Watson Research Center, Yorktown Heights, New York, 85 pp.
H. Gonshor [1974]: An application of nonstandard analysis to category theory, preprint.

W. H. Gottschalk [1960]: see Ellis and Gottschalk.
W. H. Gottschalk and G. A. Hedlund [1955]: Topological dynamics, *Amer. Math. Soc. Coll.* **XXXVI**.
G. Grätzer [1967]: *Universal Algebra*, Van Nostrand.
—[1969]: Free $\sum$- structures, *Trans. Amer. Math. Soc.* **135**, 517–542.
R. Guitart [1974]: Remarques sur les machines et les structures, *Cahiers de Top. et Géom. Diff.* **XV**, 1974, 113–144.
—[1975]: Monades involutives complementees; to appear in *Cahiers de Top. et Géom. Diff.* **XVI-1**.
A. W. Hales [1964]: On the non-existence of free complete Boolean algebras, *Fund. Math.* **54**, 45–66.
M. Hall Jr. [1959]: *The Theory of Groups*, MacMillan.
P. Halmos [1963]: *Lectures on Boolean Algebras*, Van Nostrand.
E. Harzheim [1966]: Über die Grundlagen der universellen algebra, *Math. Nach.* **31**, 39–52.
M. Hasse and L. Michler [1966]: Theorie der kategorien, *Mathematische Monographien* **7**, VEB Deutscher Verlag der Wissenschaften, Berlin.
W. S. Hatcher [1970]: Quasiprimitive categories, *Math. Ann.* **190**, 93–96.
G. A. Hedlund [1955]: see Gottschalk and Hedlund.
L. Henkin, J. D. Monk, and A. Tarski [1971]: *Cylindric Algebras I*, North Holland.
H. Hermes [1965]: *Enumerability, Decideability, Computability*, Springer-Verlag.
H. Herrlich and C. M. Ringel [1972]: Identities in categories, *Can. Math. Bull.* **15**, 297–99.
H. Herrlich and G. E. Strecker [1974]: *Category Theory*, Allyn and Bacon.
E. Hewitt and K. A. Ross [1963]: *Abstract Harmonic Analysis I*, Springer-Verlag.
P. J. Higgins [1963]: Algebras with a scheme of operators, *Math. Nachr.* **27**, 115–32.
G. Higman and B. H. Neumann [1952]: Groups as groupoids with one law, *Publ. Math. Debrecen* **2**, 215–221.
P. J. Hilton [1962]: see Eckmann and Hilton.
H.-J. Hoehnke [1966]: Zur strukturgleicheit axiomatischer klassen, *Z. Math. Logik Grundlagen Math.* **12**, 69–83.
—[1974]: Struktursätze der Algebra und Kompliziertheit logischer Schemata II, universale Algebren, *Math. Nach.* **63**, 337–351.
P. J. Huber [1961]: Homotopy theory in general categories, *Math. Ann.* **144**, 361–385.
— [1962]: Standard constructions in abelian categories, *Math. Ann.* **146**, 321–325.
S. A. Huq [1970]: An interpolation theorem for adjoint functors, *Proc. Amer. Math. Soc.* **25**, 880–883.
J. R. Isbell [1960]: Adequate subcategories, *Ill. J. Math.* **4**, 541–552.
— [1963]: Two set-theoretical theorems in categories, *Fund. Math.* **53**, 43–49.
—[1964]: Natural sums and abelianizing, *Pacific. J. Math.* **14**, 1265–1281.
—[1964–A]: Subobjects, adequacy, completeness and categories of algebras, *Rozprawy Math.* **36**, 3–33.
—[1972]: General functorial semantics I, *Amer. J. Math.* **XCIV**, 535–596.
—[1973–A]: Functorial implicit operations, *Israel J. Math* **15**, 185–188.
—[1973–B]: Epimorphisms and dominions, V, *Algebra Universalis* **3**, 318–320.
—[1974]: The unit ball of $C(X)$ as an abstract algebra, Notes from lectures delivered at the Banach Center in Warsaw, 1974.
N. Jacobson [1962]: *Lie Algebras*, John Wiley.
T. J. Jech [1973]: The axiom of choice, *Studies in Logic* **75**, American Elsevier.
J. Ježek [1970]: On categories of structures and classes of algebras, *Dissertationes Math.* **LXXV**, Warsaw.
J. S. Johnson and E. G. Manes [1970]: On modules over a semiring, *J. Alg.* **15**, 57–67.
B. Jónsson and A. Tarski [1961]: On two properties of free algebras, *Math. Scand.* **9**, 95–101.
R. Kalman, P. Falb, and M. A. Arbib [1969]: *Topics in Mathematical System Theory*, McGraw-Hill.

D. M. Kan [1958]: Adjoint functors, *Trans. Amer. Math. Soc.* **87**, 294–329.
J. L. Kelley [1955]: *General Topology*, Van Nostrand.
G. M. Kelly [1966, 1966–A]: see Eilenberg and Kelly
—[1969]: Monomorphisms, epimorphisms and pull-backs, *Austral. Math. J.* **9**, 124–142.
G. M. Kelly, M. Laplaza, G. Lewis, and S. Mac Lane [1972]: Coherence in categories, *Lecture Notes in Mathematics* **281**, Springer-Verlag.
J. F. Kennison [1975]: Triples, separated sheaf representation and comparison algebras, preprint.
J. F. Kennison and D. Gildenhuys [1971]: Equational completion, model induced triples and pro-objects, *J. Pure Appl. Alg.* **1**, 317–346.
R. Kerkhoff [1965]: Eine Konstruktion absolut frier algebren, *Math. Ann.* **158**, 109–112.
F. Klein [1934]: Beiträge zur theorie der verbände, *Math. Zeit.* **39**, 227–239.
H. Kleisli [1965]: Every standard construction is induced by a pair of adjoint functors, *Proc. Amer. Math. Soc.* **16**, 544–546.
K. Kneser [1950]: Eine direkte ableitung des zornschen lemmas aus dem auswahlaxiom, *Math. Z.* **53**, 110–113.
A. Kock [1968]: Monader og universel algebra, Mathematisk Institut, Aarhus Universitet, *Lecture Notes* **17**.
—[1969]: On double dualization monads, Mathematisk Institut, Aarhus Universitet, *Lecture Notes* **38**.
—[1970]: Monads on symmetric monoidal closed categories, *Arch. d. Math.* **XXI**, 1–10.
—[1971]: Closed categories generated by commutative monads, *J. Austral. Math. Soc.* **XII**, 405–424.
V. Koubek [1975]: see Trnková, Adámek, Koubek and Reiterman.
V. S. Krishnan [1951]: l'Équivalence de quelque reprèsentations d'une structure abstraite, *Bull. Soc. Math. France* **79**, 106–120.
S. Lang [1972]: *Differential Manifolds*, Addison-Wesley.
M. Laplaza [1972]: see Kelly, Laplaza, Lewis, and Mac Lane.
F. W. Lawvere [1963]: Functorial semantics of algebraic theories, dissertation, Columbia University, 1963.
—[1964]: An elementary theory of the category of sets, *Proc. Nat. Acad. Sci. USA* **52**, 1506–1511.
—[1966]: The category of categories as a foundation for mathematics, *Proceedings of the Conference on Categorical Algebra at La Jolla*, Springer-Verlag, 1–21.
—[1968]: Some algebraic problems in the context of functorial semantics of algebraic theories, *Lecture Notes in Mathematics* **61**, Springer-Verlag, 41–61.
A. Levy [1973]: see Fraenkel, Bar-Hillel, and Levy.
G. Lewis [1972]: see Kelly, Laplaza, Lewis, and Mac Lane.
F. E. J. Linton [1966-A]: Some aspects of equational categories, *Proceedings of the Conference on Categorical Algebra*, Springer-Verlag, 84–94.
—[1966–B]: Autonomous equational categories, *J. Math. Mech.* **15**, 637–642.
—[1969–A]: An outline of functorial semantics; Applied functorial semantics; Coequalizers in categories of algebras (three papers), *Lecture Notes in Mathematics* **80**, Springer-Verlag, 7–90.
—[1969–B]: Relative functorial semantics: adjointness results, *Lecture Notes in Mathematics* **99**, Springer-Verlag, 384–418.
—[1970]: Applied functorial semantics, 1, *Annali di Mat. Pura ed Appl.* **LXXXVI**, 1–14; also in Edgar Raymond Lorch: Sixtieth Anniversary Volume, N. Zanichelli Ed., Bologna, 1970, 1–13.
J. D. Lipson [1970]: See Birkhoff and Lipson.
H. F. J. Lowig [1952]: On properties of freely-generated algebras, *J. Reine. Angew. Math.* **190**, 65–74.
—[1957]: On the existence of freely-generated algebras, *Proc. Camb. Phil. Soc.* **53**, 790–795.

S. Mac Lane [1945]: see Eilenberg and Mac Lane.
—[1972]: see Kelly, Laplaza, Lewis, and Mac Lane.
—[1948]: Groups, categories and duality, *Proc. Nat. Acad. Sci. USA* **34**, 263–267.
—[1963]: *Homology*, Springer-Verlag.
—[1965]: Categorical algebra, *Bull. Amer. Math. Soc.* 71, **40**–106.
—[1969]: One universe as a foundation for category theory, *Lecture Notes in Mathematics* **106**, Springer-Verlag, 192–201.
—[1971]: *Categories for the Working Mathematician*, Springer-Verlag.
S. Mac Lane and G. Birkhoff [1967]: *Algebra*, MacMillan.
H. MacNeille [1937]: Partially ordered sets, *Trans. Amer. Math. Soc.* **42**, 416–460.
E. J. McShane [1934]: Extension of range of functions, *Bull. Amer. Math. Soc.* **40**, 837–842.
A. I. Mal'cev [1958]: Structural characteristics of certain classes of algebras (Russian), *Dokl. Akad. Nauk. SSSR* **120**, 29–32.
—[1971]: *The Metamathematics of Algebraic Systems* (collected papers; 1936–67), North Holland.
E. G. Manes [1970]: see Johnson and Manes.
—[1974]: see Arbib and Manes.
—[1975]: See Arbib and Manes.
E. G. Manes [1967]: A triple miscellany: some aspects of the theory of algebras over a triple, dissertation, Wesleyan University, 1967.
—[1969–A]: A triple-theoretic construction of compact algebras, *Lecture Notes in Mathematics* **80**, Springer-Verlag, 91–118.
—[1969–B]: Minimal subalgebras for dynamic triples, *Lecture Notes in Mathematics* **99**, Springer-Verlag, 1969.
—[1972]: A pullback theorem for triples in a lattice fibering with applications to algebra and analysis, *Algebra Universalis* **2**, 7–17.
—[1975]: (editor) Proceedings of the First International Symposium: Category Theory Applied to Computation and Control, *Lecture Notes in Computer Science* **25**, Springer-Verlag.
J. M. Maranda [1966]: On fundamental constructions and adjoint functors, *Canad. Math. Bull.* **9**, 581–591.
A. A. Markov [1945]: On free topological groups, *Amer. Math. Soc. Translations, Series* 1, **8**, 1962, 195–272. Originally published in Russian in *Izvestiya Akademii Nauk SSSR, Seriya Mat.* **9**, 3–64, 1945.
K. Menger [1931]: Eine elementare bermerkung über die struktur logischer formeln, *Ergebnisse Mathematischen Kolloquiums* **3**, 22–23.
—[1946]: General algebra of analysis, *Reports Math. Coll., Notre Dame Ind.* **7**, 46–60.
—[1959]: Axiomatic theory of functions and fluents, *Symposium on the Axiomatic Method* (L. Henkin, ed.), North Holland, 454–473.
J.-P. Meyer [1972]: Induced functors on categories of algebras, John Hopkins University, unpublished manuscript.
E. Michael [1951]: Topologies on spaces of subsets, *Trans. Amer. Math. Soc.* **71**, 152–182.
L. Michler [1966]: see Hasse and Michler.
B. Mitchell [1965]: *Theory of Categories*, Academic Press.
J. D. Monk [1971]: see Henkin, Monk, and Tarski.
—[1969]: *Introduction to Set Theory*, McGraw-Hill.
J. C. Moore [1965]: see Eilenberg and Moore.
S. A. Morris [1970]: Varieties of topological groups II, *Bull. Austral. Math. Soc.* **2**, 1–13.
—[1973]: Varieties of topological groups and left adjoint functors, *J. Austral. Math. Soc.* **XVI**, 220–227.
J. Negrepontis [1971]: Duality in analysis from the point of view of triples, *J. Alg.* **19**, 228–253.

E. Nelson [1974]: Not every equational class of infinitary algebras contains a simple algebra, *Coll. Math.* **XXX**, 27–30.
E. Nelson [1975]: see Banaschewski and Nelson.
V. V. Nemytskii and V. V. Stepanov [1960]: *Qualitative Theory of Differential Equations*, Princeton.
H. Neumann [1967]: *Varieties of Groups*, Springer-Verlag.
L. Padulo and M. A. Arbib [1974]: *System Theory*, Saunders.
R. Paré [1971]: On absolute colimits, *J. Alg.* **19**, 80–95.
—[1974]: Colimits in topoi, *Bull. Amer. Math. Soc.* **80**, 556–561.
B. Pareigis [1970]: *Categories and Functors*, Academic Press.
A. Paz [1971]: *Introduction to Probabilistic Automata*, Academic Press.
M. Pfender [1974]: Universal algebra in *S*-monoidal categories, preprint.
R. S. Pierce [1968]: *Introduction to the Theory of Abstract Algebras*, Holt Rinehart and Winston.
L. Pontrjagin [1966]: *Topological Groups*, Gordon and Breach.
H. Porta [1971]: see Dubuc and Porta
D. Puppe [1966]: see Brinkmann and Puppe.
J. Reiterman [1975]: see Trnková, Adámek, Koubek and Reiterman.
G. D. Reynolds [1974]: Adequacy in topology and uniform spaces, *Lecture Notes in Mathematics* **378**, Springer-Verlag.
F. Riesz [1908]: Stetigkeitsbegriff und abstrakte mengenlehre, *Proceedings of the International Congress of Mathematicians, Series I, Fourth Congress, Rome*, **2**, 1908, 18–24; Kraus Reprint Limited, Nendeln/Liechtenstein, 1967.
C. M. Ringel [1972]: see Herrlich and Ringel.
K. A. Ross [1963]: see Hewitt and Ross.
J. D. Rutledge [1964]: On Ianov's program schemata, *J. Assoc. Comp. Mach.* **11**, 1–19.
P. Samuel [1948]: On universal mappings and free topological groups, *Bull. Amer. Math. Soc.* **54**, 591–598.
J. Schmidt [1962]: On the definition of algebraic operations in finitary algebras, *Coll. Math.* **9**, 189–197.
—[1966]: A general existence theorem on partial algebras and its special cases, *Coll. Math.* **14**, 73–87.
K. Schröter [1943]: Axiomatisierung der fregeschen aussagenkalküle, *Forschungen Logik Grundleg. Exakt. Wiss., N.S.* **8**, Leipzig.
H. Schubert [1972]: *Categories*, Springer-Verlag.
B. Schweizer and A. Sklar [1969]: A grammar of functions II, *Aequationes Math.* **3**, 15–43.
D. Scott [1956]: Equationally complete extensions of finite algebras, *Indag. Math.* **18**, 35–38.
Z. Semadeni [1974–A]: Some categorical characterizations of algebras of continuous functions, to appear.
—[1974–B]: A simple topological proof that the underlying set functor for compact spaces is monadic, *Lecture Notes in Mathematics* **378**, Springer-Verlag, 429–435.
A. Sklar [1969]: see Schweizer and Sklar.
J. Słomiński [1959]: The theory of abstract algebras with infinitary operations, *Rozprawy Mat.* (*Dissertationes Math.*) **18**.
A. B. Slomson [1971]: see Bell and Slomson.
R. M. Solovay [1966]: New proof of a theorem of Gaifmann and Hales, *Bull. Amer. Math. Soc.* **72**, 282–284.
P. H. Starke [1972]: *Abstract Automata*, Elseview/North Holland.
N. Steenrod [1952]: see Eilenberg and Steenrod.
—[1967]: A convenient category of topological spaces, *Michigan Math. J.* **14**, 133–152.
G. E. Strecker [1974]: see Herrlich and Strecker.

M. H. Stone [1936]: The theory of representations of Boolean algebras, *Trans. Amer. Math. Soc.* **40**, 37–111.

S. Świerczkowski [1964]: Topologies in free algebras, *Proc. London Math. Soc.* **14**, 566–76.

A. Tarski [1961]: see Jónsson and Tarski.

A. Tarski [1971]: see Henkin, Monk, and Tarski.

J. W. Thatcher [1975]: see Goguen, Thatcher, Wagner and Wright.

T. Thode [1970]: Kategorielle Form des Satzes von G. Birkhoff über die Charakterisierung von Varietäten, dissertation, Universität Düsseldorf.

W. Tholen [1974]: Adjungierte Dreiecke, Colimites und Kan-Erweiterungen; to appear in *Math. Ann.*

B. V. S. Thomas [1974]: Free topological groups, *General Topology and Its Applications* **4**, 51–72.

M. Tierney [1969]: see Appelgate and Tierney.

V. Trnková [1974]: On minimal realizations of behavior maps in categorial automata theory, *Comm. Math. Univ. Carolinae* **15**, 555–566.

—[1975]: Minimal realizations for finite sets in categorial automata theory, *Comm. Math. Univ. Carolinae* **16**, 21–35.

V. Trnková, J. Adámek, V. Koubek and J. Reiterman [1975]: Free algebras, input processes and free monads, *Comm. Math. Univ. Carolinae* **16**, 339–351.

F. Ulmer [1971]: see Gabriel and Ulmer.

F. Ulmer [1969]: Triples in algebraic categories, unpublished manuscript.

B. L. Van der Waerden [1953]: *Modern Algebra, Volume I*, Frederick Ungar.

D. H. Van Osdol [1973]: Bicohomology theory, *Trans. Amer. Math. Soc.* **183**, 449–476.

—[1975]: Principal homogeneous objects as representable functors, preprint.

R. F. C. Walters [1969]: An alternative approach to universal algebra, *Lecture Notes in Mathematics* **106**, 64–73.

E. G. Wagner [1975]: see Goguen, Thatcher, Wagner and Wright.

A. N. Whitehead [1897]: *A Treatise on Universal Algebra with Applications*, New York; Hafner, 1960; first published 1897.

H. I. Whitlock [1964]: A composition algebra for multiplace functions, *Math. Ann.* **157**, 167–178.

H. Wiesler and G. Calugareanu Jr. [1970]: Remarks on triples in enriched categories, *Bull. Austral. Math. Soc.* **3**, 375–383.

M. Wischnewski [1973]: Generalized universal algebra in initialstructure categories, Bericht 10, Seminar F. Kasch, B. Pareigis, Mathematisches Institut der Universitat München, 1973.

G. C. Wraith [1970]: Algebraic theories, *Lecture Notes* **22**, Aarhus Universitet.

J. B. Wright [1975]: see Goguen, Thatcher, Wagner and Wright.

O. Wyler [1971–A]: On the categories of general topology and topological algebra, *Arch. Math.* **22**, 7–17.

—[1971–B]: Top categories and categorical topology, *General Topology and Its Applications* **1**, 17–28.

—[1973]: Filter space monads, regularity, completions, *Report* 73–1, Department of Mathematics, Carnegie-Mellon University; to appear in the *Proceedings of the Second Pittsburgh International Conference on General Topology.*

N. Yoneda [1954]: On the homology theory of modules, *J. Fac. Sci. Tokyo, Sec. I.* **7**, 193–227.

L. A. Zadeh [1965]: Fuzzy sets, *Inf. Cont.* **8**, 338–353.

M. Zorn [1935]: A remark on method in transfinite algebra, *Bull. Amer. Math. Soc.* **41**, 667–670.

Index

Symbol Index

Author/Subject Index